Media Piracy in Emerging Economies can be found online at
http://piracy.ssrc.org.

Published by the Social Science Research Council
Printed in the United States of America

References to Internet websites (URLs) were accurate at the time of writing. Neither the author
nor the Social Science Research Council is responsible for URLs that may have expired or changed
since the manuscript was prepared.

Designed by Rosten Woo
Maps and layout by Mark Swindle
Cover photo: AFP/Getty Images

1. Electronic industries--Developing countries. 2. Product counterfeiting--Developing countries.
3. Piracy (Copyright)--Developing countries. 4. Intellectual property infringement--Economic
aspects--Cross-cultural studies. 5. Copyright--Electronic information resources--Developing
countries. 6. Mass media--Moral and ethical aspects--Developing countries. I. Karaganis, Joe.

HD9696.A3D4466 2011
364.16´62091724--dc22

Partnering Organizations

SSRC

The Social Science Research Council
New York, NY, USA

OVERMUNDO

The Overmundo Institute
Rio de Janeiro, Brazil

The Center for Technology and Society
Getulio Vargas Foundation
Rio de Janeiro, Brazil

Sarai
The Centre for the Study of Developing Societies
Delhi, India

The Alternative Law Forum
Bangalore, India

The Association for Progressive Communications
Johannesburg, South Africa

The Centre for Independent Social Research
St. Petersburg, Russia

The Moscow Institute of Physics and Technology
Moscow State University
Moscow, Russia

**The Program on Information Justice
and Intellectual Property**
Washington College of Law, American University
Washington, DC, USA

Funders

IDRC ✳ CRDI   **The International Development Research Centre**
Ottawa, Canada

FORDFOUNDATION   **The Ford Foundation**
New York, NY, USA

Table of Contents

i Introduction

1 Chapter 1: Rethinking Piracy
Joe Karaganis

75 Chapter 2: Networked Governance and the USTR
Joe Karaganis and Sean Flynn

99 Chapter 3: South Africa
Natasha Primo and Libby Lloyd

149 Chapter 4: Russia
Olga Sezneva and Joe Karaganis

219 Chapter 5: Brazil
Pedro N. Mizukami, Oona Castro, Luiz Fernando Moncau,
and Ronaldo Lemos

305 Chapter 6: Mexico
John C. Cross

327 Chapter 7: Bolivia
Henry Stobart

339 Chapter 8: India
Lawrence Liang and Ravi Sundaram

399 Coda: A Short History of Book Piracy
Bodó Balázs

Introduction: Piracy and Enforcement in Global Perspective

Media piracy has been called "a global scourge," "an international plague," and "nirvana for criminals,"[1] but it is probably better described as a global pricing problem. High prices for media goods, low incomes, and cheap digital technologies are the main ingredients of global media piracy. If piracy is ubiquitous in most parts of the world, it is because these conditions are ubiquitous. Relative to local incomes in Brazil, Russia, or South Africa, the price of a CD, DVD, or copy of Microsoft Office is five to ten times higher than in the United States or Europe. Licit media goods are luxury items in most parts of the world, and licit media markets are correspondingly tiny. Industry estimates of high rates of piracy in emerging markets—68% for software in Russia, 82% for music in Mexico, 90% for movies in India—reflect this disparity and may even understate the prevalence of pirated goods.

Acknowledging these price effects is to view piracy from the consumption side rather than the production side of the global media economy. Piracy imposes an array of costs on producers and distributors—both domestic and international—but it also provides the main form of access in developing countries to a wide range of media goods, from recorded music, to film, to software. This last point is critical to understanding the tradeoffs that define piracy and enforcement in emerging markets. The enormously successful globalization of media culture has not been accompanied by a comparable democratization of media access—at least in its legal forms. The flood of legal media goods available in high-income countries over the past two decades has been a trickle in most parts of the world.

The growth of digital piracy since the mid-1990s has undermined a wide range of media business models, but it has also disrupted this bad market equilibrium and created opportunities in emerging economies for price and service innovations that leverage the new technologies. In our view, the most important question is not whether stronger enforcement can reduce piracy and preserve the existing market structure—our research offers no reassurance on this front—but whether stable cultural and business models can emerge at the low end of these media markets that are capable of addressing the next several billion media consumers. Our country studies provide glimpses of this reinvention as costs of production and distribution decline and as producers and distributors compete and innovate.

---

1   Respectively, by the USTR (2003), Dan Glickman of the MPAA (Boliek 2004), and Jack Valenti (2004) of the MPAA.

Invariably, industry groups invoke similar arguments on behalf of stronger enforcement: lower piracy will lead to greater investment in legal markets, and greater investment will lead to economic growth, jobs, innovation, and expanded access. This is the logic that has made intellectual property a central subject of trade negotiations since the 1980s. But while we see this mechanism operating in some contexts in emerging markets, we think that other forces play a far larger role.

The factor common to successful low-cost models, our work suggests, is neither strong enforcement against pirates nor the creative use of digital distribution, but rather the presence of firms that actively compete on price and services for local customers. Such competition is endemic in some media sectors in the United States and Europe, where digital distribution is reshaping media access around lower price points. It is widespread in India, where large domestic film and music industries dominate the national market, set prices to attract mass audiences, and in some cases compete directly with pirate distribution. And it is a small but persistent factor in the business software sector, where open-source software alternatives (and increasingly, Google and other free online services) limit the market power of commercial vendors.[2]

But with a handful of exceptions, it is marginal everywhere else in the developing world, where multinational firms dominate domestic markets. Here, our work suggests that local ownership matters. Domestic firms are more likely to leverage the fall in production and distribution costs to expand markets beyond high-income segments of the population. The domestic market is their primary market, and they will compete for it. Multinational pricing in emerging economies, in contrast, signals two rather different goals: (1) to protect the pricing structure in the high-income countries that generate most of their profits and (2) to maintain dominant positions in developing markets as local incomes slowly rise. Such strategies are profit maximizing across a global market rather than a domestic one, and this difference has precluded real price competition in middle- and low-income countries. Outside some very narrow contexts, multinationals have not challenged the high-price/small-market dynamic common to emerging markets. They haven't had to.

The chief defect of this approach in the past decade is that technology prices have fallen much faster than incomes have risen, creating a broad-based infrastructure for digital media consumption that the dominant companies have made little effort to serve. Fast technological diffusion rather than slowly rising incomes will, in our view, remain the relevant framework for thinking about the relationship between global media markets and global media piracy. Media businesses, in our view, will either learn to compete downmarket or continue to settle for the very unequal splits between low-priced pirated goods and high-priced legal sales. This status quo, it is worth noting, appears viable for most sectors of the multinational-driven media business.

---

2   Industry and trade representatives would characterize many of the same forces as "market access barriers" to foreign firms—generally ignoring the monopolization of markets that rapidly follows such access. Such issues have been central to international debates over cultural policy for some time, with much of the attention currently focused on China, where strict controls on cultural imports ensure that domestic, state-controlled companies control the market.

Software, DVD, and box office revenues in most middle-income countries have risen in the past decade—in some cases dramatically. Sales of CDs have fallen, but the overall music business, including performance, has grown.

The centrality of pricing problems to this dynamic is obvious, yet strikingly absent from policy discussions. When it comes to piracy, the boundaries of domestic and international policy conversation are exceedingly narrow. The structure of the licit media economy is almost never discussed. Instead, policy conversations focus on enforcement—on strengthening police powers, streamlining judicial procedures, increasing criminal penalties, and extending surveillance and punitive measures to the Internet. Although new thinking is visible in many corners of the media sector, as companies adapt to the realities of the digital media environment, it is hard to see much impact of these developments on IP policy—and most particularly on US trade policy, which has been the main channel for the international dialogue on enforcement.

In our view, this narrowness is increasingly counterproductive for all parties, from developing-country governments, to consumers, to the copyright interests that drive the global enforcement debate. The failure to ask broader questions about the structural determinants of piracy and the larger purposes of enforcement imposes intellectual, policy, and ultimately social costs. These are particularly high, we would argue, in the context of ambitious new proposals for national and international enforcement—notably ACTA, the Anti-Counterfeiting Trade Agreement recently finalized by the United States, the European Commission, and a handful of other countries.

To be more concrete about these limitations, we have seen little evidence—and indeed few claims—that enforcement efforts to date have had any impact whatsoever on the overall supply of pirated goods. Our work suggests, rather, that piracy has grown dramatically by most measures in the past decade, driven by the exogenous factors described above—high media prices, low local incomes, technological diffusion, and fast-changing consumer and cultural practices.

The debate is also notable for its lack of discussion of the endgame: of how expanded enforcement, whether directed against Internet-based piracy in the form of proposed "three-strikes" laws or physical piracy in the form of stronger policing, will significantly change this underlying dynamic. Much of what counts as long-term thinking in this debate involves hopes that education will build a stronger "culture of intellectual property" over time. We see no evidence of the emergence of this culture in our work or in the numerous consumer-opinion surveys conducted on the subject. Nor have we seen any attempts by industry actors to articulate credible benchmarks for success or desirable limits on expanded criminal liability, enforcement powers, and public investment. The strong moralization of the debate makes such compromises difficult.

Perhaps most important, we see little connection between these enforcement discussions and the larger problem of how to foster rich, accessible, legal cultural markets in developing countries—the problem that motivates much of our work. The key question for media access and the legalization of media markets, as we see it, has less to do with enforcement than with fostering competition at the low end of media markets—in the mass market that has been largely ceded to piracy. We take it as self-evident, at this point, that $15 DVDs, $12 CDs, and $150 copies of Microsoft Office are not going to be part of broad-based legal solutions—and in fact, we find this view commonplace in the industry itself. The choice we face is not, in the end, between high rates of piracy and low rates of piracy for media goods. It is between high-piracy/high-price markets and high-piracy/low-price markets. The public-policy question, in our view, is how to move efficiently from one to the other. The enforcement question, then, is how to support legal markets for media goods without impeding that transition.

The Media Piracy Project was created in 2007 to open up the conversation about these issues. Fundamentally, the project is an investigation of music, film, and software piracy in emerging economies and of the multinational and local enforcement efforts to combat it.[3] The primary contributions to this report are country studies of Brazil, India, Russia, and South Africa—key battlegrounds in the anti-piracy wars and frequent counterweights to US and EU dominance of international policymaking. The report also includes shorter studies of Mexico and Bolivia, drawing on the work of individual scholars whose interests aligned with the larger project.

At its broadest level, this report provides a window on digital convergence in emerging economies—a process for which piracy has been, with cell-phone use, arguably the lead application. It explores the fifteen-year arc of optical disc piracy, as discs replaced cassettes and, later, as small-scale cottage industries replaced large-scale industrial production. It traces the first real challenge to that distribution channel in the form of Internet-based services and other forms of large-scale personal sharing. It looks at the organization and practice of enforcement—from street raids, to partnerships between industry and government, to industry reporting and policy lobbying. And it explores consumer demand and changing consumer practices, including the consistent indifference or hostility to enforcement efforts of large majorities of developing-country populations.

The report consists of nine chapters: a broad introduction to piracy and enforcement; an introduction to the international politics of IP governance; country studies of South Africa, Russia, Brazil, Mexico, Bolivia, and India; and a concluding chapter that looks back to the history of the international book market for lessons about the relationship between present-day pirates and incumbent cultural producers.

Some thirty-five researchers and nine institutions were centrally involved in this project over its three-year arc, though a full accounting would include dozens of sources, readers, and reviewers who contributed generously, and sometimes anonymously. A lengthy, but inevitably partial, list of credits appears in the back matter of this report.

---

3   Because of our primary interest in the digital transition in the markets we studied, this report offers only brief treatments of contemporary book piracy. Modern book piracy in most countries is concentrated in the textbook market and is not yet an area in which digital distribution and digital consumer practices have played a large role. As global markets for laptops and e-readers grow, that will change quickly, and book piracy will assume a very important place in these discussions.

Media Piracy in Emerging Economies was made possible by support from the Ford Foundation and the Canadian International Development Research Centre.

References

Boliek, Brooks. 2004. "Dialogue: Dan Glickman." Hollywood Reporter, January 9.

USTR (Office of the US Trade Representative). 2003. 2003 Special 301 Report. Washington, DC: USTR.

Valenti, Jack. 2004. Testimony of the MPAA president to the US Senate Committee on Foreign Relations, Hearing on Evaluating International Intellectual Property, 108th Cong., June 9. http://foreign.senate.gov/hearings/hearing/?id=f3231d45-0eea-030c-e369-aa64c16e8b87.

Chapter 1: Rethinking Piracy

Joe Karaganis

Introduction

What we know about media piracy usually begins, and often ends, with industry-sponsored research. There is good reason for this. US software, film, and music industry associations have funded extensive research efforts on global piracy over the past two decades and, for the most part, have had the topic to themselves. Despite its ubiquity, piracy has been fallow terrain for independent research. With the partial exception of file sharing studies in the last ten years, empirical work has been infrequent and narrow in scope. The community of interest has been small—so much so that, when we began planning this project in 2006, a substantial part of it was enlisted in our work.

That community has grown, but there is still nothing on a scale comparable to the global, comparative, persistent attention of the industry groups. And perhaps more important, there is nothing comparable to the tight integration of industry research with lobbying and media campaigns, which amplify its presence in public and policy discussions.

Industry research consequently casts a long shadow on the piracy conversation—as it was intended to do. Our study is not envisioned as an alternative to that work but as an effort to articulate a wider framework for understanding piracy in relation to economic development and changing media economies. This perspective implies looking beyond the calculation of rights-holder losses toward the evaluation of the broader social roles and impacts of piracy. In so doing, it provides a basis for rethinking key questions raised—and often left hanging—by the industry studies: What role does piracy play in cultural markets and in larger media ecosystems? What consumer demand does it serve? How much piracy is there? What are losses? How effective is enforcement? How do software, music, and film industries differ in their exposure to piracy and in their strategies to combat it? Is education a meaningful strategy in anti-piracy efforts? What role does organized crime (or terrorism) play in pirate networks? Because such questions provide the foundation for the larger piracy debate and for the specific case studies that follow, they are the subject of the balance of this chapter.

Global factors shape many of our answers, from multinational pricing strategies, to international trade agreements, to the waves of technological diffusion that are transforming cultural economies. But the organization of piracy and the politics of enforcement are also strongly marked by local factors, from the power of domestic copyright industries, to the structure and role

Chapter Contents

1   Introduction

2   What Is Piracy?

4   How Good (or Bad) Is Industry Research?

7   What Drives the Numbers Game?

8   How Much Piracy Is There?

11   What Is a Loss?

14   Substitution Effects

16   Countervailing Benefits

18   How Is Enforcement Organized?

21   The Confiscation Regime

25   Selective Enforcement

29   How Effective Is Enforcement?

32   Does Education Work?

35   What Is Consumption?

37   Does Crime Pay?

40   Disaggregating Industry Exposure

41   Music

45   Movies and TV Shows

48   Entertainment Software

51   Why Is Business Software Piracy Different?

56   Pricing

63   Distribution

65   Looking Forward

67   About the Chapter

67   References

of the informal economy, to differing traditions of jurisprudence and policing. This report's most original contributions, in our view, are its explorations of these differences and their impact on the cultural life of their respective countries and regions.

What Is Piracy?

We use the word "piracy" to describe the ubiquitous, increasingly digital practices of copying that fall outside the boundaries of copyright law—up to 95% of it, if industry estimates of online music piracy are taken as an indicator (IFPI 2006). We do so advisedly. Piracy has never had a stable legal definition and is almost certainly better understood as a product of enforcement debates than as a description of specific behavior.[1] The term blurs, and is often used intentionally to blur, important distinctions between types of uncompensated use. These range from the clearly illegal, such as commercial-scale, unauthorized copying for resale, to disputes over the boundaries of fair use and first sale as applied to digital goods, to the wide range of practices of personal copying that have traditionally fallen below the practical threshold of enforcement. Despite fifteen years of harmonization of IP (intellectual property) laws in the wake of the Agreement on Trade-Related Aspects of Intellectual Property Rights (TRIPS), there is still a great deal of variation and uncertainty in national law regarding many of these practices, including the legality of making backups and breaking encryption; the extent of third-party liability for ISPs (Internet service providers) or search engines linking to infringing material; the evidentiary requirements for prosecution; and the meaning of "commercial scale," which under TRIPS has marked the boundary between civil and criminal liability.

The massive growth of personal copying and Internet distribution has thrown many of these categories into disarray and prompted industry efforts to bring stronger criminal and

---

1   The most thorough excavation of the term, going back to the seventeenth century, is certainly Johns (2010). We take up this history as well in the Coda to this report. The term does have recent definitions in international IP law—most notably in TRIPS, where it refers to the infringement of copyright and related rights (for example, the rights of performers, producers of phonograms, and broadcasting organizations).

civil penalties to bear on end-user infringement. The context in which most people use—and hear—the word piracy is the context created by these enforcement campaigns. We have continued to use the term because it is the inevitable locus communis of this conversation and because such discursive spaces are subject to drift and reinvention. One need look no further than the emergence of "Pirate" political parties in Europe organized around broad digital-rights agendas. As the Recording Industry Association of America recently suggested, piracy is now "too benign" a term to encompass its full range of harms (RIAA 2010).

We have wanted, consequently, to avoid moral judgments in exploring the "culture of the copy," to borrow Sundaram's (2007) more nuanced and inclusive terminology. One person's piracy has always been someone else's market opportunity, and the boundary between the two has always been a matter of social and political negotiation. The history of copyright—so extensively excavated in the past two decades[2]—is largely a history of struggles against (and later incorporation of) disruptive market innovations, often linked to the emergence of new technologies. Although there is much that is novel in the present circumstances, it is hard not to see the recurring dynamic among incumbents, pirate markets, and the new legal players who have begun to operate in the gap between them. Its current form is familiar, at this point, to anyone with an iPod.

Some further parsing of terms is necessary. Since the Berne and Paris accords of the late nineteenth century, national and international law have distinguished between piracy and counterfeiting, drawing—sometimes loosely—on the distinction between copyright infringement and trademark infringement. Traditionally, books were pirated and other branded manufactured goods were counterfeited. The value of the pirated good consisted of its reproduction of the expressive content of a work—the text rather than the pages and cover. The value of the counterfeited good lay, in contrast, in its resemblance to more expensive branded goods. The two forms of copying shared, broadly speaking, modes of production and distribution. Both required industrial-scale manufacturing.

---

2   A wave of historical inquiry on copyright emerged in the 1990s, with Goldstein (1994), Woodmansee and Jaszi (1993), and Rose (1993) among the most prominent.

Acronyms and Abbreviations

ACTA	Anti-Counterfeiting Trade Agreement
APCM	(Brazil) Associação Anti-Pirataria de Cinema e Música (Association for the Protection of Movies and Music)
BASCAP	Business Coalition to Stop Counterfeiting and Piracy
BSA	Business Software Alliance
ESA	Entertainment Software Association
GAO	(US) Government Accountability Office
GDP	gross domestic product
ICC	International Chamber of Commerce
IDC	International Data Corporation
IFPI	International Federation of the Phonographic Industry
IIPA	International Intellectual Property Alliance
IP	intellectual property
ISP	Internet service provider
IT	information technology
MPAA	Motion Picture Association of America
OECD	Organisation for Economic Co-operation and Development
P2P	peer-to-peer
PRO-IP Act	Prioritizing Resources and Organization for Intellectual Property Act
RIAA	Recording Industry Association of America
TRIPS	Agreement on Trade-Related Aspects of Intellectual Property Rights
USTR	Office of the United States Trade Representative
VCD	video compact disc
WTO	World Trade Organization
WIPO	World Intellectual Property Organization

Both relied on clandestine distribution networks and often transborder smuggling. Both were most easily subject to interdiction at the border, and consequently to the efforts of customs services.

These common roots continue to shape the law and enforcement landscape to the extent that piracy and counterfeiting are often treated as a single phenomenon. But the practices that define them have increasingly diverged. Industrial-scale manufacture and transborder smuggling represent a rapidly diminishing share of the digital culture of the copy. Border enforcement is increasingly irrelevant to this culture, as—we will argue—is organized crime. Today, the conflation of piracy and counterfeiting has little to do with shared contexts or policy solutions and more to do, in our view, with the effort to "level up" the harms attributed to copyright infringement—most notably in relation to the health and safety hazards associated with substandard products and the social costs of "harder" forms of trafficking in drugs, arms, and people. The reflexive linking of the two in research and policymaking has become an impediment to understanding either phenomenon, and it is time to pry them apart.

How Good (or Bad) Is Industry Research?

At the risk of over generalizing, we see a serious and increasingly sophisticated industry research enterprise embedded in a lobbying effort with a historically very loose relationship to evidence. Criticizing RIAA, MPAA (Motion Picture Association of America), and BSA (Business Software Alliance) claims about piracy has become a cottage industry in the past few years, driven by the relative ease with which headline piracy numbers have been shown to be wrong or impossible to source. The BSA's annual estimate of losses to software piracy—US$51 billion in 2009—dwarfs other industry estimates and has been an example of the commitment to big numbers in the face of obvious methodological problems regarding how losses are estimated.[3] Widely circulating estimates of 750,000 US jobs lost and $200 billion in annual economic losses to piracy have proved similarly ungrounded, with origins in decades-old guesses about the total impact of piracy and counterfeiting (Sanchez 2008; GAO 2010).[4]

The preference for attention-grabbing numbers is inevitable when lobbying efforts drive the use of evidence. In the piracy field, this headline approach also drowns out a more circumspect body of industry findings and the considerable diversity of methods and core assumptions in the work of industry researchers. Several major industry groups—notably the IFPI (International Federation of the Phonographic Industry) and the ESA (Entertainment Software Alliance)—do not estimate monetary losses to industry in their regular reporting but only characterize the street value of pirate sales. A pirated CD purchased on the street for $2 is valued at $2 in this model, not $12. Consumer surveys, moreover, have largely supplanted earlier "supply-side" efforts to estimate the quantity of pirated goods in circulation—a practice that relied heavily on the

---

3   See later in this chapter for a more detailed discussion. The BSA stopped calling these numbers "losses" in 2010 and now refers only to the "commercial value" of pirated software.

4   Circulated, an industry source noted, by the US Chamber of Commerce and government officials, not by the copyright industry groups.

observation of points of sale. These earlier methods drew together the opinions of local industry representatives and enforcement officers, producing interesting qualitative reporting that added significantly to our understanding of optical disc piracy. But as estimates of rates and losses, these methods represented a "best guess" rather than a serious quantitative method, and they rapidly became obsolete as the channels of media piracy expanded beyond the retail sale of pirated discs.[5] This era of subjective estimates closed in 2004, when the MPAA rolled out an elaborate, multi-country consumer-survey methodology that mapped different types of piracy against the various release windows in the life of a film. In the process, the MPAA dropped its assumption of a one-to-one equivalence between pirated discs and lost sales in favor of a more complex estimation of "displacement effects" across the different types and periods of movie exhibition.

Several of the industry groups have pulled back from reporting altogether as they explore how to analyze the shift from optical disc to online piracy. The ESA conducted its last consumer survey in 2007 and is just beginning to release results from new online monitoring efforts. The MPAA demoed its consumer-survey methodology in a massive 2005 study of twenty-two countries, but the high cost of the effort (involving some 25,000 people surveyed) has thus far precluded a follow-up. The BSA's method for measuring rates of software piracy, for its part, was developed in the late 1990s and is uniquely robust in the industry—in sharp contrast to its long-standing approach to losses. The IIPA (International Intellectual Property Alliance) consistently produces rich qualitative reporting and legal analysis on the countries it surveys as part of its Special 301 submissions to the Office of the United States Trade Representative (USTR). Overall, the industry record is both interesting and, arguably, improving.

Although all these efforts have their origins in industry lobbying, they are not simply subordinate to it. Industry research is shaped by a variety of pressures—including demands from sponsoring companies seeking to better understand the changing media markets in which they work. In this context, we see pressure for greater autonomy in the research efforts of these organizations, driven by a number of factors:

- Overlap with the market research needs of corporate sponsors, who in many instances are more interested in the analysis of consumer behavior than in reinforcing moral imperatives against piracy. Despite the RIAA's very high profile in suing file sharers, for example, its domestic US research is focused primarily on understanding behavioral changes around music consumption. None of its domestic research, according to RIAA research staff, focuses on measuring monetary losses.

- Pressure from within research units to improve methods and the quality of findings. The

---

5   A related supply-side approach, used by the MPAA in "high-piracy" countries such as Russia and Brazil, offered a still more ambiguous basis for quantitative estimates. The number of pirated discs in circulation, the MPAA argued, equaled the total productive capacity of optical disc factories in a given country minus the number of known licensed copies. According to an IFPI representative, a more reasonable estimate of total production is 60%–70% of capacity.

professionalization of research staff over, in some cases, twenty years of piracy research and the challenge of analyzing the digital transition in media piracy, in particular, have placed a premium on methodological innovation and prompted a reconstruction of industry research strategies in the past half decade.[6]

- The diminishing returns of outsized piracy claims. The rise of an Internet-based public sphere has eroded the industry's ability to shape the representation and reception of its research. Industry research is now part of a larger—and in many contexts, highly skeptical—debate about the scope and impact of piracy and, more generally, the future of media business models. In our view, the lack of industry transparency and the advocacy-driven representation of findings have significantly devalued the industry research brand, to a point where greater independence, transparency, and dialogue are strongly in the industry's interest.

The basis of credibility in this context is transparency. The main industry associations publish general descriptions of their methods but little about the assumptions, practices, or data underlying their work. It is impossible to evaluate BSA findings on rates of piracy, for example, without understanding the key inputs into the model, such as their estimates of the number of computers in a country, average software prices, or the "average software load" on machines in different contexts. It is impossible to evaluate the MPAA's claims without knowing what questions the surveys ask and how they calculate key variables, such as the substitution effects between pirate and licit sales—a critical variable at the center of debates about the net impact of piracy. The IFPI aggregates consumer surveys from its local affiliates but indicates that each affiliate makes its own choices about how to conduct its research. There is no general template for the surveys—nor, for outsiders, any clarity about how the IFPI manages the obvious challenges of aggregating the studies.

Every report has its own secret sauce, including the underlying data and often the assumptions that anchor the methodology and inform the results. The typical rationale for withholding such information is its commercial sensitivity. This is certainly possible in some cases—notably around sales figures, which in some sectors are treated as trade secrets. But it can hardly explain the across-the-board reluctance of industry groups to show their work.[7] This is a key difference

---

6   In 2008, for example, the MPAA disclosed a three-fold overestimate in its claims about the incidence of piracy on college campuses. The initial report, based on survey results, attributed 44% of domestic piracy to college students. The revised statement listed it at 15%. Critics noted that 80% of college students live off campus, making campus networks arguably responsible for something closer to 3%. The initial figure was nonetheless used to justify anti-piracy provisions in the College Opportunity and Affordability Act of 2008, which have resulted, inter alia, in the introduction of spyware by campus ISPs and the termination of service on receipt of infringement notifications from rights holders.

7   It is worth mentioning the handful of studies we encountered that take both data disclosure and methodological description seriously—even if they rely in part on data or methods from other studies that cannot be adequately sourced. Work by Ernst & Young (USIBC/Ernst & Young 2008), StrategyOne (BASCAP/StrategyOne 2009), and TERA Consultants (BASCAP/TERA Consultants 2010)—all funded

between an advocacy research culture, built on private consulting, and an academic or scientific research culture whose credibility depends on transparency and reproducibility. It also departs—we note—from what governments increasingly require in the evidentiary standards that support policymaking. We explore this question in the next chapter in relation to the evidentiary requirements of the USTR and its Special 301 process, which for over twenty years has been the primary audience for industry research.

In our view, this secrecy has become counterproductive in an environment in which hyperbolic claims have undermined confidence in the industry research enterprise. The copyright industries no longer enjoy the benefit of the doubt. Openness and disclosure of the research underlying industry claims is an obvious response, and one that was supported by every industry researcher we spoke with. All were prepared to stand by their work. All were frank about the difficulty of studying piracy, the limitations of their methods, and the desirability of improving them. It is time, in our view, to let that impulse shape the industry research culture and the policymaking process.

What Drives the Numbers Game?

Industry investments in piracy research emerged in the context of growing corporate activism on IP issues in the late 1980s and 1990s—a period marked by the establishment of the USTR's Special 301 process in 1988 and the WTO (World Trade Organization) in 1994. Special 301 created a means for industry groups to formally complain about perceived deficiencies in the IP law and enforcement practices of other countries. The IIPA, a copyright industry association founded in 1984 to advocate for stronger global IP policies, became the main intermediary between industry research and the Special 301 process. By the early 1990s, the annual Special 301 report had become, at least with respect to copyright, a vessel for IIPA-compiled findings and policy recommendations and the primary means of translating industry views into official US trade positions. For nearly two decades, the IIPA and the USTR have been, in key respects, symbiotic organizations—the research and policy wings of a larger enterprise.

Industry research went global in the wake of Special 301. The Special 301 process created demand for studies that could ground USTR recommendations, and industry groups mobilized to produce them. These research efforts relied heavily on business networks and local affiliates maintained by the industry associations. The MPAA, representing Hollywood studios, and the IFPI, a London-based association of record labels, had the most far-reaching international networks, with local affiliates or partners in most national markets. The BSA was founded in 1988 and quickly developed its own extensive network of affiliates. The ESA was founded in 1994 and has a comparatively small international presence but nonetheless produced studies in ten to twelve countries per year between the late 1990s and the mid-2000s.

---

by the International Chamber of Commerce—comes out well by this standard. None of the work produced by the copyright industry groups makes a comparable effort.

IIPA reports tend to focus on qualitative accounts of enforcement efforts and on prescriptions for legislative and administrative reform. They detail successes and failures from the previous year and evaluate them as signs of progress, good faith, or backsliding in the fight against piracy. From the outset, they also introduced two quantitative benchmarks for piracy that acquired tremendous importance in policy debates: (1) estimates of the rates of piracy in different national markets and (2) estimates of the financial losses suffered by US industry in those markets. Consistently, these numbers headlined Special 301 submissions and wider debates about copyright and enforcement. They also acted as a universal solvent for widely differing industry research inputs and methods—creating a perception of consistency and confidence in the loss figures, in particular, that the underlying research usually did not support. Where the IFPI was wary of drawing conclusions about losses, for example, the RIAA—drawing on the same data provided by local affiliates—did calculate losses for countries it considered high-priority targets for enforcement. Although the ESA avoids the language of losses in its reports,[8] its estimates of pirated street sales—totaling some $3 billion in 2007—found their way into the industry-loss column in IIPA reports.

How Much Piracy Is There?

We have not made our own estimates of rates of piracy. Piracy is clearly ubiquitous in the developing world, and we see little prospect of (or benefit from) establishing more precise figures. Although we have doubts about the reliability of industry methods and—in many cases—the definitions of piracy used, we view the IIPA-cited rates as at least plausible and very possibly as understating the actual prevalence of pirated goods. When pressed, we find that this is often the view of industry representatives themselves.

In our view, understatement of the numbers is especially likely in developed countries, where capacities for digital distribution, storage, and sharing of media files have exploded in recent years. We see no clear strategy for measuring this wider culture of the copy in most sectors of the media market (with a partial exception for software). Although all the industry groups have invested heavily in online tracking and surveillance—including, but not limited to, P2P (peer-to-peer) networks—these simply do not account for the many ways in which digital files are now shared. P2P services, while prevalent, represent a diminishing share of these available channels. Increasingly, P2P is complemented by "file locker" sites like RapidShare or Megaupload, by unauthorized streaming services, and by the growing ease of more direct personal sharing of media files, currently measured in terabyte-sized portable hard drives. We have seen no studies that explore this evolving high-end personal-media ecology in any detail. Consumer surveys, which the MPAA and the IFPI have used to track the multiple channels of distribution affecting their goods, begin to run up against the problem of media collections so large that they are no longer actively managed—or manageable—by consumers. The emergence of cloud-based media services

---

8   Though not, at present, on its website: http://www.theesa.com/policy/antipiracy_faq.asp

and their fusion with local storage promises to accelerate this decline of the personal collection.

For the past four to five years, industry research has struggled with this changing landscape. The shift from point-of-sale or production-side observations to consumer-survey methods was intended to address the transition from optical disc piracy to a mixed economy of discs and downloads. In the case of film, in particular, it was an attempt to develop better models of how consumers respond to complex industry windowing strategies as films pass from theatrical release, to pay-per-view, to DVD release, to commercial broadcast, and so on down the line. The shift toward online monitoring, in turn, reflects the increasing irrelevance of optical disc piracy in high-value markets, such as the United States and Western Europe, where retail-level piracy has all but vanished and the informal street trade has diminished significantly. In 2007, the ESA became the first industry organization to decide that the optical disc channel was no longer worth tracking. Its new online monitoring tools debuted in the 2009 Special 301 submission by the IIPA.

Despite the tone of certainty that accompanies industry press releases about piracy, most of the industry researchers we spoke with showed considerable circumspection about their ability to accurately measure either rates or losses. Increasingly, industry researchers and representatives talk in more general terms about the magnitude of piracy, rather than about precise numbers. The USTR, for its part, appears to share this reticence and no longer includes top-line estimates for rates or losses in its Special 301 reports.

Efforts to encourage more independent research organizations to validate industry findings have also been problematic. When the International Chamber of Commerce (ICC) sponsored the OECD (Organisation for Economic Co-operation and Development) to conduct a study on *The Economic Impact of Piracy and Counterfeiting*, the resulting 2007 report endorsed the notion of major economic harms and cited industry estimates of losses but also concluded that "the overall degree to which products are being counterfeited and pirated is unknown, and there do not appear to be any methodologies that could be employed to develop an acceptable overall estimate." When the OECD followed up with its *Piracy of Digital Content* report in 2009, it relied on narrow studies of particular products or channels and qualitative claims about the scope of piracy. When the US Government Accountability Office (GAO) released its report on piracy losses in March 2010, it broadly followed the OECD line—repeating the "consensus" about losses, but without endorsing any particular account of them or method for determining them. When the World Intellectual Property Organization (WIPO) opened its Advisory Committee on Enforcement meeting in November 2009, it spent three days discussing the need for more research.

OECD and GAO hedging, in our view, is a sign that the golden age of big piracy numbers is past. Industry groups haven't had much success exporting their claims into more independent research bodies, and they don't appear willing—yet—to pull back the curtain from their own research practices in a way that would allow them to engage critics. This is a recipe for diminishing political returns. But the returns to date have, by all accounts, been considerable. Across a wide range of interviews, industry representatives and researchers appeared relatively comfortable acknowledging uncertainty in their research results—in our view, because they are still enjoying the advantages of earlier, uncontested discursive authority. As several representatives indicated, the case for massive losses has been made.

Absent new data, it is less clear what happens over time to the narratives of progress and backsliding on piracy that inform the enforcement conversation outside the United States. The conventional wisdom, supported by several studies (Thallam 2008; Varian 2004), is that international rates of piracy inversely (and loosely) track wider measures of socioeconomic development, such as per capita GDP (gross domestic product).

Table 1.1 Most Recent Industry-Cited Rates of Piracy (% of the market)

	Software *	Film	Music	Games **
Russia	67	81	58	79
Brazil	56	22	48	91
India	65	29 (90)***	55	89
United States	20	7	—	—
United Kingdom	27	19	8	—

* PC game piracy is modeled in BSA software piracy rates.
** ESA rates for game piracy include console games and other formats.
***MPAA number (recent Moser Baer estimate)

Source: Author based on BSA/IDC (2010b), IIPA (2010a), MPAA (2005) data, and interviews.

Given the relatively uniform global pricing for most media goods, a loose correlation is not surprising: the first determinant of access to media markets is income. Nor is the general assumption that countries "grow" themselves out of high piracy levels as the number of high-income consumers increases (and, correspondingly, as formal markets crowd out informal ones). Beyond this general tendency, however, we are skeptical of efforts to draw more precise trend lines from year to year or to establish cause-and-effect relationships with enforcement efforts. We think that industry research methods simply do not permit reliable estimates of change at this level of detail. Our work suggests that the scale of piracy has, rather, been determined primarily by shifts in technology and associated cultural practices, from the rise of CDs and VCDs (video compact discs) in the 1990s, to the explosive growth of DVDs in the early 2000s, to the more recent growth of broadband Internet connections. The movie piracy business, for instance, was transformed by the wave of cheap Chinese DVD players and burners that hit the market in 2003–4,[9] which increased both the supply of and the demand for pirated DVDs. Those DVD players, in turn, were often able to play MP3, MP4, and other digital formats, creating an infrastructure for the next

---

9   "In 2000, some 3.5 million players were produced [in China], of which nearly 2 million were for export. By 2003, China's DVD player output had soared to 70 million units—about three-quarters of worldwide output—of which some 5 million were sold domestically" (Linden 2004). Total production peaked in 2006 at 172 million players, of which a little over 19 million were sold domestically (CCID Consulting 2008).

wave of digital distribution. Enforcement, in our view, has played only a minor role in comparison to these larger structural factors.

Our reservations about measurement extend to the BSA's comparatively robust model of "rates" of piracy, which underpins the organization's very precise claims about changes in levels of piracy from one year to the next. The BSA studies rely on the relatively small and stable (and therefore predictable) number of packaged software applications installed on an average computer—what it calls "average software load," or ASL. ASL allows the BSA to estimate the total installed software base in a country and to compare that number to legal sales. The difference between the two is attributed to piracy. The model has no counterpart in music or film, where the size of personal libraries is subject to huge and growing variation. While solid in principle, however, the model is still very dependent on complicated inputs that the BSA's research vendor, the IDC (International Data Corporation), does not share. Conflicting estimates of the size of retail markets, for example, are relatively common outside the United States and Europe, as is difficulty in establishing how many computers are in use in different countries. In the case of Russia, for example, where the BSA prominently cites a 16% decrease in the piracy rate between 2005 and 2009 as evidence of effective enforcement strategies, we were unable to independently reproduce those inputs.

What Is a Loss?

Because the primary audiences for piracy research have been the USTR and US Congress, most industry research has focused on establishing the scale of US losses rather than losses to non-US businesses or other national economies. Although nearly all of these efforts involve the participation of global networks of industry affiliates, data flows up and only occasionally results in independently released studies of local impact. With few exceptions, local rights-holder groups have conducted very little research outside this framework.

In the last three to four years, however, the international associations have begun to make stronger efforts to localize anti-piracy discourse by establishing loss figures for domestic economies. The BSA, in particular, has worked to introduce the concept of domestic losses associated with what is, invariably, the piracy of mostly US-produced software. By the same token, in countries where distinct domestic stakeholders have emerged, governmental and industry groups have begun to develop their own research capacities to assert more control over the evidentiary basis of enforcement discussions. Recent studies in Russia, India, Mexico, and China point in this direction and intermittently part ways with the US-industry narrative.[10] Some of the assistance we received from local industry and governmental sources reflects growing recognition of the

---

10   See, for example, the *2008 Survey on Chinese Software Piracy Rate*, which somewhat disingenuously tries to shift the emphasis from overall rates of piracy to the street value of pirated software within the larger market—a calculation that yields a 15% share of revenues, rather than the 80% share of the market claimed by the BSA in 2009. The survey also claims that operating system piracy dropped from 68% in 2006 to 29% in 2008 (Chinese State Intellectual Property Office 2009).

importance of research in setting the terms of the enforcement dialogue. Inevitably, power in trade negotiations is partly a matter of who shapes the evidentiary basis on which claims and counterclaims are made.

So far, these local efforts have been, at most, skirmishes around the main story of rapidly rising global losses. And for most of the past decade, this story has belonged to the BSA. Through 2010, BSA-reported losses were an order of magnitude larger than those of any other copyright industry, and they accordingly dominated discussions of the economic impact of piracy. In 2003, the BSA claimed $29 billion in global losses.[11] By 2008, it claimed $53 billion. Much of this growth was attributed to rapid computer adoption in emerging economies. Rates of adoption in Russia, for example, averaged 50% per year between 2003 and 2008—and provide some context for the claim that Russian piracy losses rose from $1.1 billion in 2003 to $4.2 billion in 2008. Overall rates of piracy have nonetheless hovered around 40% since this round of studies began in 2003— stability the BSA attributes to offsetting decreases in software piracy in developed countries.

The MPAA, for its part, claimed $6.1 billion in US studio losses in 2005—the last year in which it reported. The RIAA came next with a claim of $5 billion in global losses to record companies handling US acts. The entertainment software industry made the less direct claim that the street value of pirated games in 2007 totaled $3 billion (a number that did not include Internet downloads and that certainly could have approached BSA levels had they used retail value).

Large industrializing and middle-income countries almost always place highly on these lists. Russian software piracy losses in 2008 ($4.2 billion) were edged out only by China ($6.6 billion) and the United States itself ($9.1 billion); Brazil trailed by roughly two and a half billion dollars ($1.64 billion).[12] The United States also led the way in film losses according to the MPAA's 2005 report, totaling some $1.2 billion, followed by Mexico at $480 million (three through six were the United Kingdom, France, Russia, and Spain). The appearance of high-income countries in the rankings generally reflects their much larger domestic markets, in which comparatively low rates of piracy can still generate high monetary losses.

Increasingly, direct losses are only the starting point of this conversation. Recent studies have also begun to estimate the wider impact of piracy on national economies, based on losses to the

---

11   BSA reported losses remained roughly steady throughout the late 1990s and early 2000s, and stood at $11 billion in 2002. In 2003, the BSA revised its list of tracked software to include Microsoft Windows and a number of consumer applications—effectively doubling the size of its baseline software market and making comparison with the earlier studies difficult. Reported losses took an immediate jump, and also began increasing at a roughly 30% annual rate, approximating the rate of growth of the global software market. In 2009, in the context of the global recession, the reported value of pirated software declined slightly to $51.4 billion (BSA/IDC 2010b).

12   The uneven impact of the global recession and the BSA's change in its definition of losses (discussed later) make 2008 a better representative of the trends of the past decade than 2009. In 2009, for example, reported Russian losses fell from $4.2 billion to $2.6 billion and Indian losses from $2.7 billion to $2.0 billion, while Brazilian losses climbed from $1.6 billion to $2.2 billion, and Mexican losses from $820 million to $1 billion.

secondary and tertiary businesses that rely on copyright, from music stores to security services for film production. This approach was consolidated in a series of studies conducted by Stephen Siwek in 2006–7 on behalf of several of the major industry associations. By using official US economic multipliers (RIMS II) for different industrial sectors, Siwek argued that $5 billion in losses to the US record industry actually represented a loss of $12.5 billion to the US economy (Siwek 2007a). Direct losses of $6 billion to the movie industry meant an overall economic loss of $20.5 billion (Siwek 2006). The total lost output to the US economy from piracy, Siwek argued, was approximately $58 billion in 2007 (Siwek 2007b).[13]

Most studies now also translate such numbers into job losses. This practice was pioneered by the BSA in 2007 when it developed a formula for converting future decreases in the rate of piracy into anticipated job growth—numbers that it calculated per country in an attempt to promote stronger local commitments to enforcement. Using his own version of this approach, Siwek calculated that global piracy cost the United States some 373,000 jobs in 2005 alone. Putting the Siwek method to work in the European Union in 2010, an ICC-funded study projected a cumulative loss due to piracy of between 611,000 and 1,217,000 jobs in Europe between 2008 and 2015 (BASCAP/TERA Consultants 2010).

Studies of economic effects are important but raise serious methodological challenges, of which we will highlight two:

- the difficulty of determining the *substitution effects* associated with piracy—that is, the likelihood that a pirated copy substitutes for a legal sale—and the importance of the price/income effects in that determination; and

- the importance of the *countervailing benefits* of piracy to both industry and consumers in any model of total economic impact and, consequently, the importance of treating piracy as part of the economy rather than simply as a drain on it.

Although a variety of studies now model substitution effects,[14] we are aware of only one that has attempted to model countervailing benefits: "Ups and Downs: Economic and Cultural Effects of File Sharing on Music, Film, and Games" (Huygen et al. 2009), commissioned by the Dutch government. Among the industry studies, all now acknowledge that substitution rates are less than one, but none offer any account or even acknowledgment of countervailing benefits. Consistently, they model only one side of the market—the industry losses but not the corresponding consumer surplus.

---

13   In arriving at this number, Siwek sidestepped the BSA's de facto one-to-one replacement ratio between pirated software and lost sales and instead appears to have discounted BSA loss estimates by 50%–60%. TERA Consultants did the same in a similar study of Europe in 2010.

14   For longer treatments of the substitution-effects literature, see Huygen et al. (2009) and Oberholzer-Gee and Strumpf (2009).

Substitution Effects

Claims of a one-to-one correspondence between pirated goods and lost sales are increasingly rare and are no longer part of the official methodologies of any of the largest industry groups. At best, they are an artifact of a period when industry research was based mostly on observations of retail supply rather than consumer behavior. Such assumptions had their political uses, however. One-to-one correspondence made for the highest possible loss estimates and a simple case against unauthorized use in all its forms. Problems with this assumption were flagged as early as 1992, when the Italian government objected to MPAA efforts to put it on the Special 301 "Priority Watch List" for an alleged $250 million annual loss in theatrical revenues due to video cassette piracy (Drahos and Braithwaite 2007). But such objections were isolated and generally ignored.

The MPAA held to a one-to-one equivalence in its research until 2004, when it shifted from retail observation to a consumer-survey-based methodology. The RIAA's practices are not public, but research staff indicated in 2009 that they take substitution rates into account when estimating losses for Special 301 reports (they do not reveal which rates). The ESA and the IFPI have never relied on the one-to-one claim.

The BSA position is often described as a claim of one-to-one correspondence because it calculates losses (or, beginning in 2010, what it calls the "commercial value of unlicensed software") by multiplying the estimated number of pirated copies of tracked products by a "blended average price" of those products across the different distribution channels (retail, volume licensing, "free" open-source distribution, and so on). Although functionally one-to-one, the BSA insists that its reasoning is more complex and reflects the assumption that although less piracy would not directly produce an equivalent increase in sales, it would do so indirectly by expanding economic activity, which would lead to increased sales. According to the BSA, "The two countervailing forces seem to cancel each other out" (BSA/IDC 2003).[15] As recently as 2009, the IDC argued that this effect "might even underrepresent" true losses to the industry (BSA/IDC 2009). In practice, they offer no account of substitution effects and, consequently, no account of consumer behavior.

In music and film markets, in contrast, substitution effects have become central to the debate about losses and changing market structure. Here, studies have tried to weigh substitution effects against possible sampling effects that describe additional purchases that follow from greater exposure to new goods. With respect to music, nearly all independent studies acknowledge the presence of both effects, albeit with significant variation in the findings, from alleged positive net effects on sales due to piracy (Anderson and Frenz 2008), to negligible impact (Huygen et al. 2009; Oberholzer-Gee and Strumpf 2007), to estimates of up to 30% displacement of legal digital downloads (Zentner 2006). Several studies also identify a correspondence between piracy

---

15   In correspondence with us in 2010, the BSA described this more broadly as a "linear relationship" be-
       tween lower piracy rates and larger software markets—an approach that could, at least, admit ratios of
       less than one-to-one but that in practice doesn't.

and increased media consumption in general, suggesting that piracy is most common among avid media consumers and reinforces or complements those habits. There are fewer studies of substitution effects for film, but a number of these show a stronger negative impact on theatrical visits and DVD sales (Peitz and Waelbroeck 2006; Bounie, Waelbroeck, and Bourreau 2006). Since Siwek's studies are arguably a bellwether of what the industry is prepared to think about this question, circa 2007, it is worth noting that he adopts a 65% substitution rate for physical piracy of music (that is, of pirated CDs replacing legal sales) and a 20% rate for downloads—both within the ballpark of existing studies.

We have no particular contribution to this debate and tend to view substitution and sampling rates as moving targets tied to changing gaps in convenience, quality, and price between licit and illicit services. With low-cost, high-quality, Internet-based music and video services emerging, moreover, the direction of the substitution becomes increasingly unclear. Do CD or DVD purchases compete with P2P downloads or with legal streaming services? Or with rentals, as Smith and Telang (2009) have tried to model? Does file sharing also displace secondary services around music and film, such as specialty stores or fan communities organized around print and web journals? The problem is far from new and has been at the center of long-standing tensions between record companies and radio stations over the direction of the benefits of radio airplay (Liebowitz 2004). As distribution channels proliferate, it will become still more complex.

We do note that such studies are conducted almost entirely in high-income countries and that the price/income ratios in most parts of the world dictate very different outcomes. The 65% physical-substitution rate and the 20% download rate simply make no sense in reference to Brazil or India, where purchasing power is far lower. The MPAA's 2005 movie piracy study is said to have explored substitution effects in the countries it surveyed—suggesting a potential wealth of data on price and income effects—but the MPAA has not released its findings or shared them privately (with us or, more surprisingly, with either the OECD or the GAO, both of which conducted their studies in the context of new enforcement initiatives). Other data points on this question remain scarce. One recent study of the relationship between file sharing and movie ticket sales in Hungary, a country with per capita GDP well below US and Western European levels, finds no measurable relationship between the two (Balázs and Lakatos 2010). When John Gantz, research director at the IDC, was asked about the impact of high Western software prices on piracy in developing countries, he suggested that possibly only one in ten unauthorized copies represented a lost sale. Absent clearer data, we would call this a plausible guess—and one that would have dramatically reduced the $29 billion loss that the BSA claimed in 2003. As Gantz observed, "I would have preferred to call it [the $29 billion] the retail value of pirated software" (Lohr 2004). In 2010, Gantz got his wish when the IDC started referring to these numbers as "the commercial value of unlicensed software" (BSA/IDC 2010b). This seemingly minor shift is, in fact, quite consequential: it salvages the one-to-one correspondence at the heart of the IDC method, putting it on firmer methodological ground. But any claims about losses are now gone.

Countervailing Benefits

Since 2006, industry-loss claims have been recycled into a wide range of broader estimates of the social and economic impacts of piracy. In our view, the current generation of economic-impact studies, including those of Stephen Siwek, the IDC, and TERA Consultants, simply does not provide a basis for understanding these wider impacts. Many of the problems we discussed earlier are repeated at this level, such as the lack of disclosure of the underlying datasets and key assumptions. But in extrapolating losses beyond the affected industries, these studies also introduce new problems. Fundamentally, they all misrepresent the relationship between piracy, national economies, and international trade. Consistently, none of them model the other side of the transaction—the consumer surplus—in describing overall economic impact. Two basic accounting problems have become emblematic of this approach.

First, domestic piracy may well impose losses on specific industrial sectors, but these are not losses to the larger national economy. Within a given country, the piracy of domestic goods is a transfer of income, not a loss. Money saved by consumers or businesses on CDs, DVDs, or software will not disappear but rather be spent on other things—housing, food, other entertainment, other business expenses, and so on. These expenditures, in turn, will generate tax revenue, new jobs, infrastructural investments, and the range of other goods that are typically cited in the loss column of industry analyses.

To make a case for national economic harms rather than narrower sectoral ones, the potential uses of lost revenue need to be compared: the foregone investment in the affected industries needs to represent a better potential economic outcome than the consumer surplus generated by piracy (Sanchez 2008). The net impact on the economy, properly understood, is the difference between the value of the two investments. Such comparisons lead into very complicated territory as marginal investments in different industries generate different contributions to growth and productivity. There has been no serious analysis of this issue, however, because the industry studies have ignored the consumer surplus, maintaining the fiction that domestic piracy represents an undiluted national economic loss. For our part, we take seriously the possibility that the consumer surplus from piracy might be more productive, socially valuable, and/or job creating than additional investment in the software and media sectors. We think this likelihood increases in markets for entertainment goods, which contribute to growth but add little to productivity, and still further in countries that import most of their audiovisual goods and software—in short, virtually everywhere outside the United States.

Second, and relatedly, the direction of trade matters greatly in calculating where losses (and benefits) fall. The global footprints of many software and media companies make the breakdown of revenue streams complicated, but the larger dynamic is relatively simple: With regard to imported IP goods, legal sales represent an outflow of revenue from the national economy. The piracy of IP imports, conversely, represents a welfare gain in the form of expanded "free" access to valuable goods. Because of US dominance of global film and software markets, the piracy of these goods in other countries falls overwhelmingly into this category—with revenue "lost" to US companies but "gained" by consumers on the receiving end.

Both Siwek and TERA have problems with this distinction. Siwek's estimate for film piracy, for example, starts with the $6.1 billion MPAA estimate of studio losses and applies a multiplier of roughly three (drawing on US Bureau of Labor Statistics sectoral models) to arrive at an estimate of total economic losses. Even accepting the MPAA numbers, however, this isn't the right starting point. The MPAA attributes some 20% of losses ($1.3 billion) to US-based piracy, which is not lost to the national economy but simply spent in other ways. The remaining $4.8 billion in overseas losses, in contrast, is "lost" to the United States in the first instance, but even this sum will continue to be spent and circulate in ways that will be partially recouped by US firms.

The closely related TERA study, for its part, assumes that losses fall solely on EU companies. For movies, music, and software markets in Europe, however, this is manifestly untrue. Hollywood films account for 67% of the EU market (European Audiovisual Observatory 2010), with ticket revenues roughly equally split between distributors (the studios) and local exhibitors (Squire 2004). Microsoft, Adobe, and other US-based companies have market shares well over 90% in many of the core business software categories.[16] For film and software, consequently, European countries are IP importers, and any comparison of domestic costs and benefits should first include the outflow of revenue. Under these circumstances, Europe might well realize a net welfare benefit from audiovisual and software piracy.[17]

A recent Dutch study of piracy makes a good case for exactly that in the case of music. Music is a more complicated sector to disaggregate due to the strong presence of local repertoire in most countries—a factor that should weigh in favor of real domestic losses. Nonetheless, Huygen et al. (2009) estimate the net welfare impact of music piracy in the Netherlands—industry losses compared to the consumer surplus—to be a positive €100 million per year.[18]

---

16   The IDC claims that roughly 80% of software revenues remain in Europe (BSA/IDC 2010a). Presumably this includes software produced in Europe, which would imply a less favorable split for foreign products. The IDC does not explain how it arrives at these numbers.

17   The TERA study buries these issues in the very last paragraph of its final appendix: "To be fully consistent, we should have considered the proportion of local/foreign pirated products (for all the covered creative products), but such data were not available." In our view, this omission fatally compromises the study. It makes a big difference, in the end, whose goods are pirated.

18   The 2009 Dutch study provides a strong set of reference points despite its narrow focus on the file sharing of movies and film in the Netherlands. Huygens et al. examine the impact of file sharing on both domestic and international producers, explore substitution effects in depth, and conclude that Dutch consumers enjoyed a net consumer welfare gain of around €100 million per year (in a country of sixteen million). In our view, a version of this analysis applied to developing countries would find substantially higher net benefits, based on much lower substitution rates due to lower income, the generally smaller scale of domestic culture industries, and the employment generated by the informal economy. The inclusion of business software, with its massive contribution to economic productivity, would push it higher still.

Among the industry consultants, only the IDC has shown much interest in determining how revenues are apportioned between domestic and foreign economies—interest we read at a general level as pushback against the perception that pirating foreign vendors has no domestic costs and more narrowly as pushback against local-development-based arguments for open-source software adoption. These estimates are the basis of the IDC's various papers on the domestic economic impact of reductions in piracy, which argue that $1 recouped from piracy generates $3–$4 of secondary domestic economic activity (BSA/IDC 2010a).[19] When the IDC, in a study prepared for Microsoft, tried to characterize the value of the Microsoft "software ecology" outside the United States, it argued that $1 in Microsoft revenues generates $5.50 in local business revenues (IDC 2009).

As usual, we must ask: compared to what? We see no reason to assume that the use of pirated software contributes less to economic growth than the use of licit software: a pirated copy of Windows or Photoshop will generally serve as well as a legal one. Relatedly, we see no reason to assume that pirated use does not also contribute to the growth of secondary markets for software services. To the best of our knowledge, no secondary applications or services require validated copies of the primary software platforms.

In contrast, we see a plausible case that Microsoft products have added value because of the positive network effects associated with Microsoft's dominance of the desktop (well over 90% in developing markets), which make Windows and related products de facto standards. But as the IDC's numbers indicate, this dominance in low- and middle-income countries is attributable almost entirely to software piracy, rather than legal licensing. As we will argue later, such network effects make piracy a key feature of software business models in emerging economies.

Rich software environments—such as the Windows environment—are basic infrastructure in modern economies and have a large positive impact on productivity. But the IDC studies offer no help in explaining why these benefits depend on legally licensed software or, for that matter, on Windows rather than its competitors. Instead, the IDC leaves readers to infer that other products add less or, potentially, nothing to local economies. By modeling only part of the market, the IDC studies limit themselves to a promotional role and do little to illuminate the relationship between piracy, jobs, and economic growth. It is this underlying complexity (and the unwillingness of the industry groups to address it) that led the US Government Accountability Office to discount all current estimates and conclude that "it is difficult, if not impossible, to quantify the net effect of counterfeiting and piracy on the economy as a whole" (GAO 2010).

---

19   The IDC breaks this number down by country, finding the domestic share of benefits from software purchases to be 76% in India, 73% in Brazil, 61% in Russia, and 68% in South Africa (BSA/IDC 2010a).

How Is Enforcement Organized?

The copyright industries invest heavily in enforcement advocacy and anti-piracy campaigns, from legislative lobbying, to police efforts to protect theatrical release windows for new films, to software legalization programs for governments and businesses. These efforts involve a wide range of actors operating at different geographical and political levels, including industry associations; local, national, and international law enforcement; licensing agencies; multilateral organizations like the WTO and WIPO; US government agencies; US and international chambers of commerce; and many others.

Such networks expanded dramatically in the past decade as countries implemented national enforcement plans. Both the number of groups involved and the level of financing of anti-piracy efforts rose significantly in the period, before tailing off in the wake of the recent global financial crisis. Predictably, budget numbers documenting this trend are hard to come by. Industry groups are reluctant to discuss enforcement budgets—especially in relation to their efforts in developing countries, where local associations and enforcement efforts are often funded by multinationals. Our rough estimate of the scale of operations of the top-level industry groups is in the low hundreds of millions of dollars per year. In 2009, CEO John Kennedy put the IFPI's enforcement budget at around 75 million British pounds ($120 million) (enigmax 2009)—a sum representing roughly half of the IFPI's estimated total budget of $250–300 million. The RIAA, for its part, has had a budget of $45–55 million per year in the last decade—much of it devoted to anti-piracy lobbying and enforcement efforts. Prior to cutbacks in 2009, the MPAA's anti-piracy budget was described as roughly $60–75 million per year—again approximately half its total budget (DiOrio 2009). The BSA is a $70-million-per-year organization, a large portion of which is self-financed through anti-piracy settlements (some $55 million in 2007, with roughly $10 million coming from member dues). The ESA is a $30-million-per-year organization with a comparatively small enforcement footprint (its primary responsibility is the annual E3 Expo tradeshow). The US Chamber of Commerce plays a significant role in both anti-piracy research, lobbying and educational initiatives, as do its many international franchises and analogs, including the International Chamber of Commerce and the 115 American Chambers of Commerce located around the world. We were unable to determine how much of the US Chamber's $150 million budget (2008) is devoted to IP issues. A number of the larger corporate sponsors of these groups, including Microsoft and Nintendo, also maintain anti-piracy operations and finance others. Microsoft's anti-piracy legal team in Redmond alone reportedly has a staff of around seventy-five (Hachman 2010).

Growth has not been without its challenges. The perception of low returns on investment has been a problem for all the organizations involved, and all except the BSA have faced significant budget cuts and/or challenges from membership in the past three years (Di Orio 2009).

The relatively modest size of the core industry groups compared to the scale of the pirate economy is an indicator of why stronger public enforcement is viewed as a top industry priority.

Buying Enforcement

In Brazil, police and government-agency units specializing in copyright enforcement depend on industry groups for logistical and financial support. According to one recent report on São Paulo's "Immaterial Property Police Department," these gifts range from printer cartridges, to car repair, to a refrigerator and new floor for the police department building. In Rio de Janeiro, we documented Association for the Protection of Movies and Music (Associação Anti-Pirataria de Cinema e Música—APCM) provision of police equipment, transportation for raids, locksmiths, and other support that make it unclear where the boundary between public and private policing lies. Because policing in Brazil is defined as a strictly public function, this private subsidization raises questions about the independence and impartiality of the police and has begun to attract scrutiny. In São Paulo, the APCM's gifts to police are being investigated by the public prosecutor. The APCM claims its donations are legal. As of late 2010 the matter remains unresolved.

The 2008 Pro-IP (Prioritizing Resources and Organization for Intellectual Property) Act, now coming into effect in the United States, called for $429 million in additional expenditures on enforcement between 2009 and 2013, with the sum rising each year (Congressional Budget Office 2008). Overall public expenditure, unfortunately, is almost impossible to determine in the United States because budgets for anti-piracy efforts are rarely broken out from more general law enforcement activities. We have seen no such estimates elsewhere either, though our country studies of Russia, Brazil, and South Africa documented comparable increases in police and other enforcement funding as new national-level enforcement plans went into effect in the past five or six years.

The primary goal of industry activism has been to shift enforcement responsibilities onto public agencies. Outside the United States, the USTR and industry groups have worked consistently to expand public investment in enforcement and increase private oversight of those efforts. Public-private partnerships already structure every stage of the enforcement business, from international policy formation to local policing. This model was visible (and highly controversial) in recent negotiations over a new international treaty on enforcement called ACTA (the Anti-Counterfeiting Trade Agreement), which was developed through private consultations between industry stakeholders and trade officials from friendly states.

Within countries, this model has given rise to webs of interlocking enforcement efforts and advisory groups that blur lines between public and private power. At the local level, industry groups both subsidize and participate in investigations, evidence collection, and raids. Inevitably, the increasing scale and complexity of such efforts brings coordination costs, which has led to the creation of new layers of bureaucratic intermediaries—liaison officials, "IP Czars," and other officials charged with managing the new cross-agency, public-private enforcement agendas.

Closer public-private coordination is almost always accompanied by industry calls for expanded police powers and the wider application of criminal law to copyright infringement. The IIPA has a list of standard demands for reengineering law enforcement around the needs of copyright holders, including provision for ex officio police powers (which empower police to act directly against suspected infringement without a complaint); greater use of ex parte hearings (which drop requirements to have the defendant present) and ex parte searches (which empower industry to conduct raids with lower police or judicial oversight); the application of anti-organized-crime statutes to commercial infringement (often modeled on US RICO laws); dedicated IP courts; longer prison sentences; higher fines; and diminished evidentiary requirements.[20]

Many of these measures are responses to the inefficiency of civil procedures in developing countries, which makes infringement lawsuits cumbersome and expensive. Our India, Russia, and South Africa studies document these problems in some detail. But expanded police power and diminished judicial safeguards are viewed in many countries as recipes for abuse—especially in contexts where police forces have been deliberately decentralized or subjected to sharp judicial checks on power, as in Mexico and Brazil. The private direction of public enforcement is also problematic on a number of levels, and raises concerns about accountability, fairness, and due process.

The lack of a clear enforcement endgame contributes to these concerns. The moral framework of anti-piracy campaigns makes it difficult to articulate an acceptable level of piracy that would set a boundary against the erosion of civil liberties. In this environment, enforcement policies have a strong tendency to fail up. Measures that do little more than inconvenience pirates will tend to be portrayed as insufficient rather than misguided, creating pressure for stronger, more pervasive, more expensive enforcement. Although greater public capacity to enforce might, in theory, diminish the incentives for private involvement, we have found no examples of private-sector pullback from this role in any of the countries examined in this report. Indeed the opposite is often true: greater public-sector buy-in on enforcement signals compliance, which spurs greater private sector involvement and investment. Although industry association members have shown signs of balking at the high costs of enforcement, they have expanded efforts to shift costs to other actors, including governments and ISPs.

Our country studies document these tensions between public and private power in considerable detail. Close relationships between industry and public officials are a large part of this story, most visible at the policymaking and administrative levels (see chapter 2). But these tensions also play out in less visible ways on the ground—in some cases with remarkable consistency from one country to another. Anti-piracy efforts at this level are not just about policing and courts but are arguably better understood in terms of confiscation and selective enforcement.

---

20   Such as permission to destroy seized goods on the spot rather than hold them as evidence and the right to bring charges based on the "sampling" of seized goods rather than a full inventory. In Brazil, informants described the last two points as the highest priorities for enforcement organizations—above even "three-strikes" legislation for Internet-based infringement. As one informant noted, the benefits of a three-strikes law remain hypothetical, but the evidence-storage costs incurred under current law are concrete.

The Confiscation Regime

Predictably, raids scale more easily than due process. Although no consistent or overall numbers are available, industry organizations and government agencies track and occasionally report the numbers of raids, arrests, and convictions in which they play a part. Most of the time, these numbers tell a striking story. In 2008, the Mexican Association for the Protection of Film and Music initiated 3,170 raids, resulting in 120 arrests and 7 convictions. In a single weeklong campaign during Russia's major anti-piracy crackdown of 2006–7, the Ministry of the Interior reported 29,670 "actions," generating 73 criminal cases and an unspecified number of convictions. The Russian BSA, in 2007, initiated 589 raids on local businesses for "end-user infringement," obtaining convictions in 83 cases. The Brazilian APCM reported 3,942 raids in 2008, leading to 195 convictions, most of which resulted in suspended sentences. Between 2000 and 2007 in India, there were 6 convictions for piracy (in 2008, the Indian Music Industry—IMI—reported 60).

Figure 1.1 Raids and Convictions in Brazil

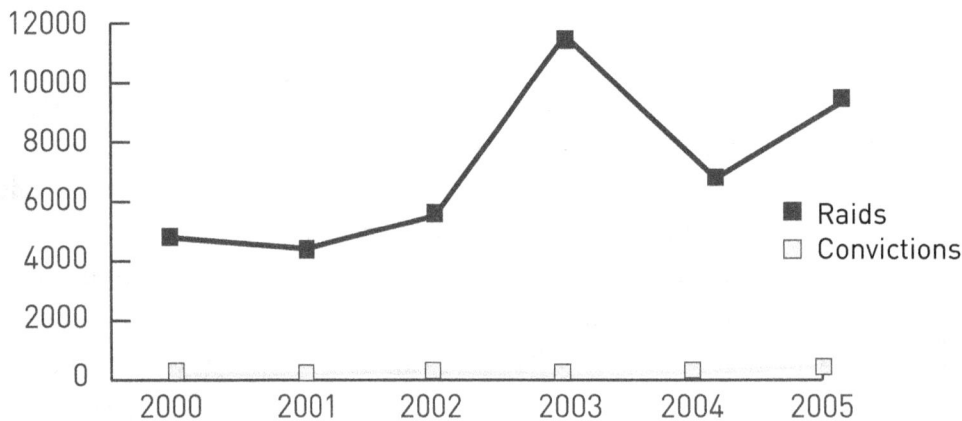

Source: Author based on IIPA-reported data.

Exceptions to this lopsided record generally come in the context of major campaigns against street vendors, in which investigative procedures have been streamlined or cases routed through the most accommodating courts. The South African film industry group SAFACT (Southern African Federation Against Copyright Theft), for example, reported 973 raids during its anti-vendor push in 2008, leading to 617 arrests and 447 convictions—a nearly tenfold increase in convictions over 2007. Nearly all resulted, however, in small fines or suspended sentences.

There are a variety of explanations for this disproportion—none mutually exclusive in our view. Due process in all the countries examined here is slow and inefficient, often in the extreme. Criminal cases can take several years to resolve and civil cases even longer. The cost of bringing charges in criminal and civil court is accordingly very high, and the prospect of significant fines or other penalties that can act as wider "deterrents" correspondingly low.

In such contexts, IIPA and other industry reports routinely present judges as obstacles to stronger enforcement outcomes. Unlike the dedicated police and administrative units at the center of anti-piracy efforts, judges have been much less reliable allies in the effort to scale up the number of convictions and increase the severity of penalties. Industry groups often attribute such resistance to ignorance of IP law or to a failure to grasp the severity and social costs of copyright infringement. Industry requests for the maximum allowable penalties are routinely ignored in favor of fines more commensurate with the (often very limited) ability of offenders to pay. Judges also frequently suspend fines or jail terms after sentencing, signaling that many do not view street-level vending, in particular, as a serious crime.

The training and "sensitization" of prosecutors and judges has, accordingly, been a top priority for stakeholder groups in the past decade. A full spectrum of corporate, government, and international actors fund and organize such efforts, from WIPO, to Microsoft, to the US Department of Justice and the US Patent and Trademark Office. These efforts have, by several accounts, improved coordination and procedures among the various law enforcement units needed to bring cases to court. As one Russian enforcement specialist noted, "We learned how to successfully combat piracy in its traditional form," referring to the police procedures and legal tools used to combat the optical disc retail trade in the early 2000s. But our interviews suggest that such programs have been less successful with the judiciary. Dismissively low conviction rates and penalties can also be read in part as judicial pushback against local enforcement drives—a view supported by a number of our interviews in South Africa, India, and Brazil.

The context for such resistance is obvious to anyone looking at the day-to-day activity of the criminal courts. In countries where judges routinely confront the consequences of extreme poverty and high rates of violent crime (see figure 1.2), the application of heavy fines and extended prison terms for street vending has proved a difficult sell. Chronic overcrowding of prisons means that judges are often forced to triage lower-level crimes. Efforts to characterize street piracy as commensurate with more dangerous forms of crime routinely fail this commonsense test. Street vendor tactics also play a part in this dynamic. In high-enforcement settings, such as major urban flea markets in Russia, South Africa, and India, vendors have adopted labor practices that shield them from direct exposure to police, including the use of foreign and underage sellers in kiosks and on the street. Judges have often been reluctant to use the full power of criminal sanctions in such cases.

Figure 1.2 Murders per 100,000 Inhabitants (2007/2008)

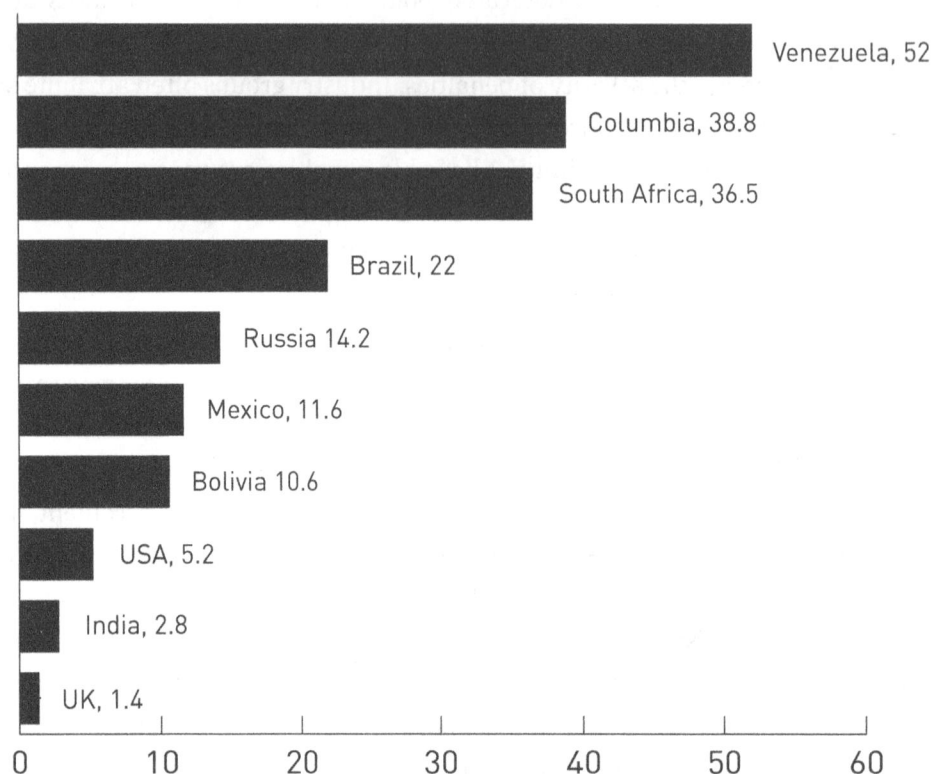

Source: UNODC (2009).

Slow due process and judicial recalcitrance also provide the context for other common IIPA demands, such as the creation of high statutory penalties for infringement that limit judicial discretion, or the creation of special IP courts that can process cases more quickly and decisively, or the application of a variety of extrajudicial forms of punishment, such as the use of pre-trial detention in cases of piracy arrests. In South Africa and parts of India, for example, such detention can last up to a year.

In the absence of an easy path through the courts, however, the main tool of dissuasion is the raid. Thousands of raids are carried out each year in the large middle-income countries, with optical disc vendors and suspected software-infringing businesses topping the list of targets. IIPA reports routinely complain about the lack of follow-through in these operations, which produce a great many confiscations but very few subsequent arrests or prosecutions. But the consistency of these outcomes suggests that this imbalance is a feature, not a defect, of the ramp up of enforcement efforts. Raids scale much more easily than due process, pushing police and industry representatives toward the fastest, most summary procedures at their disposal. The prominence of trivial-sounding disputes over obligations to pay for the storage of confiscated goods becomes clearer in this context. The churn of raids generates a lot of confiscated material. The slow pace of court cases means that the resulting responsibilities for storage are usually long term.

Raids, of course, are their own form of punishment. Although the pirate disc trade has evolved strategies to minimize the disruption from raids, they can be devastating to legitimate businesses. Stock or computers can be impounded for weeks while investigations play out, effectively shutting down businesses for the duration. Because licit software and discs are often hard to distinguish from illicit or unlicensed versions, the range of goods confiscated during raids is often indiscriminate, leading to the loss or impoundment of legitimate property. In Russia, for example, enforcement agents suggested that up to 30% of confiscated discs are legitimate—a number that reflects the broad interpenetration of licit and illicit markets. Software piracy investigations also pose problems due to typically very limited administrative capacities to evaluate installed programs. In these scenarios, the factors that make the court system so costly and slow for enforcement organizations also sharply limit opportunities for redress.

In countries where the costs of raids have fallen on politically connected domestic groups—the local business community or street vendor organizations, for example—enforcement efforts have met with political resistance. When the major Russian enforcement push in 2006–7 exacerbated problems of police shakedowns and commercially motivated harassment, local businesses successfully lobbied the federal government to curtail police authority to conduct raids. The relationships between Mexican street vendor organizations and police are marked by negotiated truces that reflect the integration of these organizations into the political system. Raid-based enforcement is inherently fragile and subject to a political calculus that weighs external pressure from the USTR and multinational groups against internal pressure from domestic business constituencies.

Selective Enforcement

Enforcement is, at all levels, a selective practice that picks and chooses targets from the ocean of infringing activity. This is inevitable in a context in which scarce enforcement resources confront ubiquitous piracy and is a source of many of the structural problems in its application. Enforcement, under these circumstances, has a strongly arbitrary character. At its worst, it is theatrical, politicized, and a tool of competitive advantage among businesses.

The counterpart to raid-based enforcement is the push for spectacular punishments in the handful of cases that do result in convictions. The punishment phase in such cases is often treated as an occasion for public education rather than proportional justice. High statutory penalties for individual acts of infringement in many countries mean that nearly any case can result in crushing penalties. In the United States, Joel Tenenbaum and Jammie Thomas-Rasset were sued by the RIAA for trivially minor acts of file sharing and fined $675,000 and $1.92 million, respectively.[21]

---

21   In 2008 and 2009. Thomas-Rasset was accused of sharing twenty-four songs; Tenenbaum, thirty. The US Department of Justice went on record that the Thomas-Rasset penalty was appropriate, indicating that such damages against individuals were intended in the 1999 Digital Theft Deterrence and Copyright Damages Improvement Act. In 2010, the Thomas-Rasset fine was reduced to $54,000 by a judge and then raised to $1.5 million when, at Thomas-Rasset's request, the case was heard again by a jury.

In Russia, a school principal, Aleksandr Ponosov, faced five years in prison when police discovered infringing software on twelve school computers in 2006. Cases against low-level suppliers or commercial intermediaries increasingly result in criminal charges and are periodically turned into media events by the industry groups themselves. In South Africa, the 2005 case against Johannesburg vendor Marcus Mocke became such an event. Mocke faced eight years in prison after the police seized four hundred pirated DVDs and PlayStation games in his home.

These high-visibility cases demonstrate the willingness of industry groups and at least some prosecutors to make use of the stronger penalties afforded by recent changes in national copyright laws. Ponosov and Mocke faced serious criminal charges for activities that a few years before would have been treated as misdemeanors at most and in all likelihood ignored. The Tenenbaum and Thomas-Rasset cases, for their part, were part of a larger industry experiment in shifting enforcement from commercial intermediaries (for whom such penalties were conceived) to the individual consumers who now represent the lion's share of infringing activity.

Whether such publicity does more good than harm for the industry enforcement effort is a matter of debate. Most observers view the Ponosov, Tenenbaum, and Thomas-Rasset cases as public-relations disasters for industry—with the former catalyzing a major open-source software movement in Russia and the latter two grounded in a mass-lawsuit strategy that has since been disavowed by all the major industry groups, including the RIAA.[22] Although infringement is routinely found in such cases, the push for disproportionate penalties has made adjudication very difficult. The charges against Ponosov were eventually dismissed. Thomas-Rasset's penalty was dramatically reduced by the judge (and then raised again in a retrial). Mocke received a fine rather than a prison sentence, which was later suspended. None of these penalties, so far, has been applied. None provide much evidence of achieving the "deterrent" penalty standard required by TRIPS—and if the continued prevalence of piracy is the criterion, then no countries meet the standard.

Samuelson and Wheatland (2009) have analyzed the increasingly arbitrary and extreme character of statutory damages in the United States. The current range of damages runs from $200 (in the case of "innocent" infringement) to $150,000 per work infringed.

22   Between 2003 and 2008, the RIAA threatened some 27,000 individuals with lawsuits, typically resulting in settlements in the low thousands of dollars. The retreat of the major players from this model has not dissuaded smaller groups from pursuing mass lawsuits. A handful of European law firms have refined the RIAA strategy into a business model based on Internet monitoring and the automated sending of letters demanding payment from alleged infringers (Masnick 2009). In 2010, the US Copyright Group—according to most reports a front for DC law firm Dunlap, Grubb, & Weaver (Anderson 2010) brought this practice to the United States, filing cases against alleged P2P infringers of individual films—including obscure low-budget films like *Far Cry* (2008) and *Smile Pretty* (2009) and more recently, and prominently, Oscar-winner *The Hurt Locker* (2008). By mid-2010, cases had been filed against some 14,000 "John Does," with the ISPs under pressure to identify users based on IP (Internet protocol) addresses. Like the RIAA lawsuits, the new suits are designed to produce quick monetary settlements rather than lengthy court cases. In the current round, the US Copyright Group's go-away price begins at $1,500 and escalates in the event of non-payment. By late 2010, the mass-lawsuit strategy appeared to be in jeopardy due to the slow handling of IP look-up requests by ISPs.

The Irony of Fate 2

Hit Russian movies *Day Watch*, *Night Watch*, and *The Irony of Fate 2* all benefitted from dedicated enforcement campaigns. A representative of Channel One, the Russian TV broadcaster that controlled the distribution of *The Irony of Fate 2*, observed:

> We simply scared them off. We asked the OBEP [police] to pass the word that our reaction [to pirated copies] will be harsh . . . Our access

to "administrative resources" undoubtedly helped. They would be unlikely to listen to anyone smaller than us. (Vershinin 2008)

"Administrative resources," in Russian business parlance, means political influence, which can be converted into raids, favorable attention from prosecutors, and even—in this case—preemptive notices from ISPs warning users not to pirate the film.

Predictably, corporate involvement in public enforcement also creates competition for enforcement resources and competitive advantages for companies that can make effective use of them. On one end of this spectrum are the various enforcement business strategies that become available in contexts of widespread illegality. These range from cases of borderline racketeering on the part of business or rights-holder groups, such as the OKO case documented in our Russia chapter, to the mass "John Doe" lawsuits underway in the United States and Europe, to the more common practices of the BSA and other software enforcement groups who self-finance through settlements. In the software arena, it is generally assumed that enforcement falls most heavily on small businesses, which have less sophisticated IT (information technology) management, limited influence with vendors and local authorities, and—above all—less capacity to contest legal threats. As with the suits against individuals, this is not a defect of the model—it is the model.

At the other end of the spectrum are the forms of commercial advantage that flow from influence with government agencies. Perhaps the most overt among these are the dedicated policing campaigns on behalf of particular products or brands. Dedicated enforcement campaigns have become relatively common sights in developing countries as part of the release strategies for major domestic films, with notable examples including *The Irony of Fate 2* (2007) in Russia, *Tsotsi* (2005) in South Africa, *Tropa de Elite 2* (2010) in Brazil, and *Lagaan* (2001) in India. Police mobilization in these situations is generally geared toward the suppression of street piracy during the initial release window for the film, when the majority of profits are made.[23]

---

23   The lack of dedicated protection for big Hollywood releases in the United States has been a long-standing source of annoyance for the MPAA, which recently argued that "the planned release of a blockbuster motion picture should be acknowledged as an event that attracts the focused efforts of copyright thieves, who will seek to obtain and distribute pre-release versions and/or to undermine legitimate release by unauthorized distribution through other channels. Enforcement agencies (notably within DOJ and DHS) should plan a similarly focused preventive and responsive strategy. An interagency task force should work with industry to coordinate and make advance plans to try to interdict these most damaging forms of copyright theft" (AFTRA et al. 2010).

Working with the Pirate Union

In 2006, in Bolivia, the La Paz city government brokered a deal between the Union of Cinema Workers and the National Federation of Small-Scale Audio-Visual and Music Merchants (an organization of street and kiosk vendors, dubbed a "pirate union" in many press reports) to limit the street piracy of new movies. The agreement required vendors to refrain from selling VCDs or DVDs of new films until after their exhibition in La Paz cinemas—typically a three-month period following first release. According to union officials, the agreement also stipulated protection in perpetuity for nationally produced films. The city police were assigned responsibility for enforcement.

Implementation of the agreement, however, broke down almost immediately. Press articles condemned the mayor's office for giving "a green light to piracy." Musicians' rights organizations condemned the lack of respect for the rights of international artists. But the real damage was done by non-unionized vendors and members of other vendor unions, who were not bound by the agreement and who undercut its control of the market. The agreement quickly fell apart, leaving street vendors and Bolivian rights groups back at square one. Since 2000, there has been one conviction for piracy in Bolivia (on behalf of Microsoft). The resulting one-year prison sentence was suspended.

Naturally, not all companies enjoy equal access to enforcement resources. As in other contexts, the power to deploy public resources tracks with—and reinforces—influence and size. Among the multinational firms, Microsoft, by nearly all accounts, operates in a league of its own, reflecting its market dominance, coherent developing-market strategy, and nearly bottomless wallet. The company figures centrally in most software enforcement efforts against large institutions, including public agencies, schools, large businesses, and computer-equipment manufacturers, and in the eventual negotiation of volume licensing agreements that bring those institutions into longer term contractual relationships.

Anecdotally, however, our work suggests that domestic companies and artists are often better able to mobilize attention from local authorities—even when representing products embedded in global circuits of investment and distribution, such as most high-end films. For obvious reasons, the politics of copyright enforcement on behalf of domestic producers are more attractive to local and national governments than enforcing Microsoft or Disney licenses. These preferences translate into a variety of formal efforts and informal norms to protect goods with strong local identities, often in ways that capitalize on protectionist sentiment among consumers. De facto deals between pirate vendors and authorities around local content have been common in India, for example, where regional cinema, especially, enjoys preferential treatment from local police. Film and recording artists in both India and South Africa have organized street-level enforcement efforts that focus exclusively on local materials (and sometimes shade into vigilantism). In Russia,

1C, a producer of accounting software and distributor of foreign titles, accounted for 126 of the 207 criminal indictments for software piracy between 2002 and 2008. Microsoft was second with 21.

How Effective Is Enforcement?

We see considerable evidence that raid-based enforcement can suppress the more organized forms of optical disc piracy at the retail level. Established stores are vulnerable to raids, and raids are now a regular feature of street life in most high-piracy countries. The result, however, is not the disappearance of the optical disc trade but its deformalization: its reduction to more mobile street vending with less stock, more transient labor practices, and—consequently—greater resilience to police pressure.

The deformalization of piracy is a common thread in our account and arguably the main achievement of enforcement efforts in developing countries. We see no evidence, however, that these efforts have significantly reduced the overall supply of pirated goods—and indeed quite a bit of evidence to the contrary. Optical disc prices have plummeted in most countries, indicating expanded supply and—often—sharper competition in the pirate marketplace. Increasingly, this competition comes from the growth in file sharing and other forms of non-commercial Internet distribution. Pirates, too, must now compete with free. But the underlying story is broader and involves the spread of cheap hardware throughout the media ecosystem, fueling the small-scale, local production of optical discs.

What about online? Lawsuits and injunctions against online intermediaries have become common in the past decade, directed against both non-commercial P2P sites and illicit or under-licensed commercial download sites like the Russian AllofMP3, which sold music at the unusual price of $0.01 per megabyte until its closure in 2008. Despite occasional friction between trading partners, TRIPS-era IP law is well suited to dealing with the latter category of commercial pirates, which generally involves direct, large-scale infringement and clear financial gain—both triggers for criminal prosecution under the TRIPS standard. But commercial websites of this kind have played a very small role in the growth of online copy culture. The current environment is built around an array of intermediary services, including P2P services, file locker sites, streaming services, social networking sites, and search engines. These have been more difficult to target, in part because the nature of their liability is harder to establish. Sites using BitTorrent—currently the dominant P2P protocol—are little more than specialized search engines that overlap the functionality of larger, general-purpose search sites like Google. Like Google, they can point to infringing content, but they neither host it nor directly participate in file exchanges. "Cyberlocker" sites like RapidShare or Megaupload are little more than online storage providers.

Since the Napster era in 1999–2000, rights-holder groups have filed suit against dozens of P2P sites and have generally succeeded in shutting them down.[24] Jurisprudence clarifying the secondary liability of site owners and administrators has been a different story, however, with some countries (such as the United States) developing relatively encompassing standards of contributory infringement,[25] while others (such as the United Kingdom and Germany) maintain more traditional requirements of proof of commercial gain.

Despite the stream of lawsuits and site closures, we see no evidence—and indeed very few claims—that these efforts have had any measurable impact on online piracy.[26] The costs and technical requirements of running a torrent tracker or indexing site are modest, and new sites have quickly emerged to replace old ones. P2P continues to account for a high percentage of total bandwidth utilization in most parts of the world, and infringing files represent, by most accounts, a very high percentage of P2P content (Felton 2010; IFPI 2006). ISP-traffic-monitoring firm ipoque put P2P use in 2009 at roughly 70% of total bandwidth in Eastern Europe, 60% in South America, and slightly lower percentages in northern and southern Europe (Schulze and Mochalski 2009).[27] US rates are generally estimated at 25%–30%, reflecting not so much lower utilization of P2P as higher utilization of streaming video services such as YouTube and Hulu. Rates of use of cyberlocker sites like RapidShare have grown rapidly, leading to pressure on those companies to monitor file uploads and sign deals with content providers. The IFPI, for its part, claims that some forty billion songs were shared on P2P networks in 2008, up from twenty billion in 2006, and that

---

24   Major BitTorrent site closures due to industry pressure include SuprNova (Slovenia, 2004), Finreactor (Finland, 2004), LokiTorrent (US, 2004), Grokster (US, 2005), EliteTorrents (US, 2005), TorrentSpy (US, 2006), OiNK (UK, 2007), The Pirate Bay (Sweden, 2009), and Mininova (Netherlands, 2009). Civil damages against site administrators have been common in these cases.

25   Principally as a result of MGM v. Grokster (2005), which introduced the concept of "inducement" to infringement as a basis for liability. Although the case fell short of setting a clear standard, it did establish a precedent for finding P2P services liable for secondary infringement.

26   The effects of the roughly 27,000 RIAA lawsuits brought against P2P users between 2003 and 2008 are occasionally debated in this context. The evidence for a deterrent effect on P2P use in the RIAA case is limited to a Pew Internet and American Life Project survey conducted in the wake of the first RIAA announcement. This survey showed a 50% drop in the percentage of users acknowledging use of P2P services, from 29% to 14%. By the time of Pew's 2005 survey, this number had reverted to 24% and Pew was drawing attention to the importance of other emerging digital distribution channels (Madden and Rainie 2005). For more analysis of the impact of the suits, see EFF (2008).

27   Ipoque relies on small sample sizes, and there is very little wider agreement about these estimates. ISPs rarely provide public data about traffic—either type or volume. Definitions of a "unit" of file sharing vary, and accurate measurement requires intrusive content monitoring. Ipoque's study is based on a handful of ISPs with which it has agreements. The IFPI, nonetheless, uses the ipoque study to claim that up to 80% of all Internet traffic is P2P (IFPI 2009)—a number found nowhere in the study itself. Cisco Systems put the figure at 55% in 2008 (2009). Zhang (2008) compared some sixty-eight studies and concludes that there is no basis for a reliable estimate.

legal downloads represent only 5% of the total circulation of digital music (IFPI 2009).[28]

Internet service providers have long been viewed as the logical choke points for monitoring, blocking, and punishing infringing behavior, and the next generation of enforcement activism focuses on exploiting the contractual links between individuals and ISPs. All the major industry groups support stronger ISP liability for infringing activity on their networks. All support either a direct ISP role in monitoring and enforcing copyright or an indirect role in forwarding industry warnings, leading to eventual cutoff of service. These are the so-called graduated-response, or three-strikes, laws, several of which are coming into effect in 2011.[29]

Three-strikes laws face a variety of legal and practical challenges—among them, the household-level organization of most consumer Internet service, which makes it difficult to identify and impossible to isolate individuals behind IP addresses. Collective punishment of families for the acts of individual members will be an inevitable (and legally very controversial) outcome. High courts in Spain, Finland, and France, for example, have declared Internet access a fundamental right, reflecting its growing role in social, cultural, and economic life. A 2010 BBC survey in twenty-six countries found that 79% of respondents shared this view. US law has not yet characterized access in these terms, but it is clearly the direction signaled by the FCC (Federal Communications Commission) in its recent National Broadband Plan.

Over the longer term, stronger consumer-directed enforcement is certain to produce an arms race between encrypted, anonymized services and industry detection techniques. Although the industry currently presents graduated response as an effective response to consumer piracy, it is far from clear that it will prove legally or politically viable, or do more than shift users to other forms of distribution. As recent MPAA and RIAA comments on enforcement submitted to the US government make clear, however, three-strikes is not the end of the digital enforcement fight but the beginning. The next steps down the path include preemptive content-filtering by ISPs, the inclusion of home-based monitoring software in ISP contracts, and the amendment of customs forms "to require the disclosure of pirate or counterfeit items being brought into the United States" (AFTRA et al. 2010). For the average 14- to 24-year-old with over eight hundred pirated songs in his or her collection in 2008 (Bahanovich and Collopy 2009), this would represent a serious dilemma.

---

28   As usual, the provenance of these numbers is unclear. The IFPI indicates that they are compiled from sixteen other unnamed studies.

29   Notably in France, the United Kingdom, New Zealand, Korea, and Japan. France started issuing warnings to alleged infringers in late 2010.

Does Education Work?

Access Copyright, Canada

Nearly all formal plans for IP protection, from the US Chamber of Commerce's "Campaign to Protect America" to the Brazilian government's "National Plan on Combating Piracy" to WIPO's Development Agenda stress that "repressive measures" are not enough—that enforcement also requires building a stronger "culture of intellectual property" through education and public awareness campaigns. Education efforts are accordingly widespread, ranging from anti-piracy curricula in public schools, to print and video campaigns, to technical seminars designed to "sensitize" judges and law enforcement officers to the severity of IP crime.[30] Because public awareness is an area where coordination between industry groups is relatively easy, local campaigns tend to look very similar from country to country and reinforce the same simple messages: equivalence between intellectual and material property, fear of being caught, and anxiety about buying dangerous or socially harmful goods. Distinctions between piracy and counterfeiting are almost always erased in these contexts, and alarming associations with organized crime, immorality, and disastrous

---

30   Efforts directed at children and students are quite common. Of the 202 campaigns listed in a WIPO enforcement database since 2000, 52 target "kids and teenagers." These include the BSA's "Define the Line" campaign and the ESA's "Join the ©Team" in the United States, the "Children Against Piracy" and "Change Starts with an Idea . . . It Can be Yours!" campaigns in Mexico, and the "Projeto Escola Legal," a Brazilian school-based curriculum examined in detail in chapter 5. These efforts have also produced a subgenre of comics, ranging from the MPAA's "Escape from Terror Byte City" (2009) to the short-lived Canadian hero, "Captain Copyright" (2006).

personal consequences are emphasized. As the teaching manual for the "Projeto Escola Legal" curriculum used in Brazilian elementary schools puts it: "It is no exaggeration to say that by buying a pirated product, an individual is worsening his own chances of getting a job, or even provoking the unemployment of a relative or friend" (Amcham-Brasil 2010). In a widely circulated Brazilian video spot, criminals address the pirate DVD consumer: "Thank you ma'am, for helping us to buy weapons!"

The effort to shape public discourse around piracy extends to the management of the print and broadcast news. Several of our country studies document the extent to which copyright industry messaging dominates print and broadcast coverage of piracy. Our South Africa team documented some eight hundred print and broadcast stories over a four-year period in a country with just three major media markets. A similar examination in Brazil collected roughly five hundred stories over a three-year period. The vast majority of this coverage reproduces a few standard templates: the raid or big arrest, the new piracy report, the aggrieved artist. Many of them report from industry press events or simply quote verbatim from industry press releases.

Despite the ubiquity of media piracy, contrasting or critical perspectives in this coverage are rare. Especially when the subject is enforcement action or research, there are few "other points of view" to feed the journalistic reflex for balance. A variety of factors contribute to this discursive dominance, from the professional press management strategies practiced by industry groups, to overstretched journalists in need of easily packaged stories, to the lack of civil-society engagement with enforcement.[31] This homogeneity stands in sharp contrast to the many online venues that harbor a wider range of positions on piracy and enforcement, and that collectively offer a much closer approximation, in our view, of the actual diversity of consumer attitudes.

What do these efforts to shape public discourse achieve? If dissuading consumers is the primary goal, the answer appears to be: very little. Our inquiries (mixing survey, focus group, and interview methods) found a remarkably consistent cluster of attitudes on piracy: (1) that it is often regarded with ambivalence by consumers, (2) that pragmatic issues of price and availability nearly always win out over moral considerations, and (3) that consumers know what they are buying. The classic scene of developing-world piracy—the kiosk or street vendor selling DVDs—produces very little misunderstanding on the part of consumers about the nature of the transaction. Consumers weigh tradeoffs between price and expectations of quality, but within a context of explicit black-market negotiation in which notions of fraud or deception—often borrowed from anti-counterfeiting discourse—generally don't apply. The price gap between licit and pirated media provides a clear signal of the origins of goods.

The legibility of this scene for consumers, in our view, provides a benchmark for other scenes of copying and infringement that are more commonly the subjects of uncertain or confused legal status—especially around practices of ripping, sharing, uploading, and downloading digital

---

31   We heard ample support for all three views from print journalists. A plausible—though here undocu-mented—fourth factor would be the control of the print and broadcast media by many of the same media conglomerates involved in enforcement advocacy.

material. Clarifying for students that the file sharing of copyrighted music is piracy seems entirely possible, for example, but we see no evidence that this knowledge will have any impact on practices. We see no real "education" of the consumer to be done.

This finding is consistent, we believe, with the preponderance of consumer-opinion surveys conducted in this area, including those by Pew in the United States, the BPI (British Recorded Music Industry) in the United Kingdom, PROFECO (the Attorney General for Consumer Affairs) in Mexico, IBOPE (Brazilian Institute of Public Opinion and Statistics) and Ipsos in Brazil, and many others. The most comprehensive comparative analysis of these issues to date is a 2009 StrategyOne study commissioned by the International Chamber of Commerce. StrategyOne examined some 176 consumer surveys and conducted new ones in Russia, India, Mexico, South Korea, and the United Kingdom. Like nearly all other surveys, StrategyOne's work showed high levels of acceptance of physical and digital piracy, with digital media practices among young adults always at the top of the distribution. The group concluded that "hear no evil, see no evil, speak no evil' has become the norm" (BASCAP/StrategyOne 2009).[32] At this point, such findings should come as no surprise. In the contexts in which we have worked, we can say with some confidence that efforts to stigmatize piracy have failed.

There is little room to maneuver here, we would argue, because consumer attitudes are, for the most part, not *unformed*—not awaiting definition by a clear anti-piracy message. On the contrary, we consistently found strong views. The consumer surplus generated by piracy is not just popular but also widely understood in economic-justice terms, mapped to perceptions of greedy US and multinational corporations and to the broader structural inequalities of globalization in which most developing-world consumers live. Enforcement efforts, in turn, are widely associated with US pressure on national governments and are met with indifference or hostility by large majorities of respondents. The reluctance of many governments to adopt stronger enforcement measures needs to be understood in light of these potentially high domestic political costs.

Although education is generally presented as a long-term investment in counteracting these attitudes, the lack of evidence for their effectiveness is striking. There have, after all, been a lot of campaigns in the past decade—StrategyOne counted some 333 in developed countries alone as of 2009. It would be reasonable to expect some benchmarks and tentative conclusions. But such follow-up appears to be almost universally avoided. We are unaware of any campaigns that have included subsequent evaluation. This also appears to be the conclusion reached by StrategyOne in its examination of 202 separate campaigns.

---

32   The BASCAP/StrategyOne study is an important but conflicted contribution to the literature. Consistently, it portrays the near-total failure of industry messaging on piracy in developing countries. It finds that the main drivers of piracy are price and availability and links these factors to widespread support for media piracy and general resentment of anti-piracy efforts, especially in developing countries. And it disaggregates findings for medicines and media products—in notable contrast to the usual industry practice of conflating health and safety risks associated with some categories of counterfeit goods to essentially harmless practices of media consumption. Yet, StrategyOne appears compelled to find that these structural factors are actually communication problems and that education efforts can (or more precisely, must) work given better messaging.

The proliferation of campaigns and the avoidance of bad news, in this context, strongly suggest the presence of other motives. Much of the continuing investment in education and public awareness, in our view, is attributable to strongly felt but ultimately wishful thinking about the future, as when StrategyOne describes the failure of education efforts, despite the evidence, as simply "unacceptable for us as individuals, for the companies and industries we work in and for society as a whole" (BASCAP/StrategyOne 2010). In other contexts, it is clear that educational initiatives provide useful political cover for governments publicly committed to enforcement but wary of further "repressive measures" and for industry groups looking to soften their agendas as they turn toward more direct ways of penalizing consumer infringement.

As we discuss at some length in the Brazil chapter of this report, educational campaigns can provide a path of least resistance between these contending interests and result in commitments to the most naive versions of these programs by public officials. Such compromises are why 22,000 Brazilian school children are now part of the "Projeto Escola Legal"—the flagship educational project of Brazil's National Plan to Combat Piracy—which, in a typical passage, advises teachers to address student concerns about affordable access to media with this logic: "The production of movies, music, books, etc., is vast, and therefore, if we cannot buy a ticket to watch a movie, we can't say that we do not have access to culture, but only to that specific movie, in that specific place, and that specific moment." We think it exceedingly unlikely that a culture of intellectual property will be built on such sophism and disconnection from consumer realities.[33]

What Is Consumption?

Traditionally, the high costs of media production and distribution dictated relatively sharp distinctions between producers, distributors, and consumers of media. The consumer sat at the end of a commodity chain that delivered finished goods and structured experiences—records played on stereos, movies shown in the theatres, and so on. Consumers' perspectives were valuable and eagerly solicited, but the opportunities for creative engagement with or appropriation of the work were generally marginal. This model has, of course, come under pressure as falling costs of production and distribution democratize those core functions of the media economy and as new technologies privilege forms of commentary, appropriation, and reuse. Such practices have arguably become the main tropes for thinking about digital media in general.

Our work generally validates and expands on this perspective. We see these shifts clearly in the emergence of new production and distribution chains at the very low end of media markets— almost always illicit at the outset but later evolving into mixed markets that include new, legalized competition. And we see it in a range of creative appropriations of goods that test the boundary between authorized and unauthorized use—often triggering charges of piracy.

---

33   As this report was going to press, representatives of Brazil's National Council on Combating Piracy indicated to the author that "Projeto Escola Legal" had recently been rejected by the government. There has been, as yet, no public announcement of this change in policy.

With regard to recorded media, however, our work highlights a more specific transformation in the organization of consumption: the decline of the collector and of the intentional, managed acquisition that traditionally defined his or her relationship to media. In our view, this notional consumer still organizes a large part of the cultural field and a significant share of the business models and supply chains for audiovisual media. But it is also clearly a shrinking cultural role, defined by income effects and legacy cultural practices.

The collector, our work suggests, is giving ground at both the high end and low end of the income spectrum. Among privileged, technically literate consumers, the issue is one of manageable scale: the growing size of personal media libraries is disconnecting recorded media from traditional notions of the collection—and even from strong assumptions of intentionality in its acquisition. A 2009 survey of 1,800 young people in the United Kingdom found that the average digital library contained 8,000 songs, with 1,800 on the average iPod (Bahanovich and Collopy 2009). Most of these songs—up to two-thirds in another recent study—have never been listened to (Lamer 2006).

Such numbers describe music and, increasingly, video communities that share content by the tens or hundreds of gigabytes—sizes that diminish consumers' abilities to organize or even grasp the full extent of their collections. Community-based libraries, such as those constituted through invitation-only P2P sites, carry this reformulation of norms further, structured around still more diffuse principles of ownership and organization. On such scales, many of the classic functions of collecting become impersonal, no longer individually managed or manageable. A related effect is that personal ownership becomes harder to specify and measure: consumer surveys are poorly adapted to mapping terrain where respondent knowledge is unreliable. Studies based on specific devices or media services (such as the handful of studies that use iTunes data) may only capture a portion of the media resources that consumers engage with. Increasingly, we live in an ocean of media that has no clear provenance or boundaries.

Several of our studies document the tension between the collecting model, which still has practical and affective connections to physical discs, and the "native" digital model, which generally does not. Inevitably this tension maps onto income effects, broadband availability, and age and consequently bears on relatively small portions of the populations of middle- and low-income countries. Original goods continue to play a variety of high-status roles in these contexts, as signals of wealth or—as our Russia study suggests—as the polite form for gifts.[34] But even in the short span of years covered in this study, the transformation of these practices is visible and striking. The relevant metric in middle-income countries is not the slow growth in average incomes but the fast decline in the price of technology.

The second and, in many countries, more significant consumer shift is the growth of mass markets for recorded media among the very poor and—in many cases—mass production of recorded media by the very poor. The contours of this revolution can be traced back to the

---

34   See also Wang (2003) on these distinctions.

profoundly democratizing and piracy-enabling recorded media technologies of the 1980s—the audio cassette and the cassette player (Manuel 1993). The much larger current wave of digital media production is built on the proliferation of a cheap VCD and DVD infrastructure in the past decade, including multiformat players, computers, burners, and discs—both fueling and fueled by the availability of cheap pirated content. Consumer practices at this level are organized differently, with less attachment to CDs or DVDs as elements of a private collection than as goods shared within extended families and communities. Collective consumption—viewing and listening—is more common in this context, reflecting the lower numbers of TVs, computers, and DVD players in poor households.

Neither the high-income nor low-income version of this shift has much currency in enforcement debates, which continue to be shaped, we would argue, by a nostalgic view of the consumer as collector—of people making deliberate choices to purchase, or pirate, specific goods for personal use. And despite the evidence of the collector's diminishing hold on digital cultural practices, we do not expect this to change: real or not, the collector is an important construct that anchors personal responsibility—and liability—in the copyright economy. As enforcement efforts shift from commercial intermediaries toward consumers, such anachronism takes on greater, not lesser, importance.

Does Crime Pay?

Claims of connections between media piracy and narcotrafficking, arms smuggling, and other "hard" forms of organized crime have been part of enforcement discourse since the late 1990s, when the IFPI began to raise concerns about the transborder smuggling of pirated CDs (IFPI 2001). Claimed connections between piracy and terrorism are a more recent addition. In 2003, the secretary general of Interpol, Ronald Noble, "sound[ed] the alarm that Intellectual Property Crime is becoming the preferred method of funding for a number of terrorist groups" (Noble 2003). In 2008, the US attorney general, Michael Mukasey, declared that "criminal syndicates, and in some cases even terrorist groups, view IP crime as a lucrative business, and see it as a low-risk way to fund other activities" (Mukasey 2008). In 2009, the RAND Corporation published what is to date the most exhaustive statement on this subject: a 150-page, MPAA-funded report on film piracy's links to organized crime and terrorism (Treverton et al. 2009).

Commercial-scale piracy is illegal, and its clandestine production and supply chains invariably require organization. It meets, in this respect, a minimal definition of organized crime. Pirated CD and DVD vending, moreover, is often concentrated in poor neighborhoods and informal markets where other types of illegal activity are common. Such contexts create points of intersection between the pirate economy and wider illegal and quasi-legal arrangements of the informal economy. It would be remarkable if they did not. But we found no evidence of systematic links between media piracy and more serious forms of organized crime, much less terrorism, in any of our country studies. What explains this result?

Thugs and Criminals

"With rare exceptions, the people procuring, producing, and distributing this pirated material are affiliated with large and dangerous international criminal syndicates." Film piracy is not being operated by "mom-and-pop operations." . . . "It is being done by business-minded thugs who fund this activity through money raised from other illicit activity such as drug dealing, gun running, and human trafficking (utilizing the same distribution networks), and who, in turn, fund these other activities through the money they raise through piracy." Consequently, "the odds are high that every dollar, pound, peso, euro or rupee spent on them is put into the pockets of bad people who will spend it in a way which is not consonant with our safety and security." Most alarmingly, these groups "have no qualms whatsoever about resorting to violence or bribery to conduct their operations, and they play for keeps."

–John Malcolm, senior vice-president and director of worldwide anti-piracy operations for the MPAA (quoted in McIllwain 2005)

Invariably, the rationale offered for criminal-syndicate and terrorist involvement is that piracy is a highly profitable business. The RAND report, for example, states (without explanation) that "DVD piracy . . . has a higher profit margin than narcotics" (Treverton et al. 2009:xii)—an implausible claim that has circulated in industry literature since at least 2004.[35] We think the record is clear that piracy was a highly profitable business through the early 2000s, when optical disc production facilities were expensive, industrial in scale, and relatively scarce. The concentration of production capacity in a few countries created an international pirate economy in which some countries emerged as exporters of optical discs (for example, Malaysia, Bulgaria, and the Ukraine), while others became primarily importers or transshipment points. International distribution, in these circumstances, involved the smuggling of physical goods and consequently mirrored—and sometimes shared—the distribution infrastructure for other counterfeit and contraband products. In our India and South Africa studies, in particular, we see evidence that this structure of piracy persists in regional trade networks connecting South Asia, the Middle East, South Africa, and parts of East Asia. But it is also clear that such networks are marginal to the larger pirate economy and rapidly waning—driven into unprofitability by expanded local production and free digital distribution. We see no evidence that piracy, outside a few niche markets, is still a high-margin

35   The initial version of this claim appears to come from a 2001 story in the French newsweekly Marianne, which stated that a kilogram of pirated CDs was worth more than a kilogram of hashish. The claim was picked up by Interpol in its 2003 report to the US Congress on "The Links Between Intellectual Property Crime and Terrorist Financing" and from there began a long life of circular citation in industry reports. This claim has been challenged before (Piracy Is Not A Crime.com 2006), but to update and reiterate the point: according to US customs authorities, a kilogram of hashish in New York sells for around $30,000. A kilogram of pirated DVDs (amounting to 60–65 discs averaging 16 grams each) has a street value of about $300 in New York, at the going rate of $5 per DVD. The IIPA repeats a version of this claim in its 2010 submission to the USTR.

business.

These trends have dominated pirate production since the early 2000s. Production costs and profit margins on optical discs have plummeted, leading to a collapse in prices. In 2001, quality DVDs typically cost five dollars or more on the street. In 2010, they are under a dollar at retail in many parts of the world. Burners and blank discs are now commodity items, and their greater availability has led to a massive expansion of local production, the displacement of smuggling, and—in many countries—a reorganization of production around small-scale, often family-based, cottage industry. Pressure on profit margins has increased, too, due to the rise of the massive non-commercial sphere of copying and distribution on the Internet, which has all but eliminated commercial optical disc piracy in high-income countries and appears poised to do so further down the GDP ladder. Increasingly, commercial pirates face the same dilemma as the legal industry: how to compete with free.

This decline in costs is, in our view, the primary factor shaping pirate markets and a growing disincentive for traditional organized-criminal involvement. Yet, to the best of our knowledge, no industry or law enforcement statements about alleged criminal connections have thought this worth mention. As in other contexts, the issue is avoided by conflating piracy and counterfeiting under the rubric of what Interpol calls "IP crimes." IP crimes include the counterfeiting of cigarettes, medicines, machine parts, and a variety of other industrial goods. Nearly all are high-margin goods distributed through transnational smuggling networks—indeed they are smuggled because they are high margin. Smuggling, in turn, creates opportunities for criminal groups to organize or tax the transit of these goods. Terrorist connections are possible in such contexts, and there is evidence that tobacco smuggling in particular—incentivized by high European and US taxes on cigarettes and abetted by major tobacco companies—is a significant revenue source for the Taliban, the Columbian FARC, and the PKK (Willson 2009).

Arguing that piracy is integral to such networks means ignoring the dramatic changes in the technology and organizational structure of the pirate market over the past decade. By necessity, evidentiary standards become very loose. Decades-old stories are recycled as proof of contemporary terrorist connections, anecdotes stand in as evidence of wider systemic linkages, and the threshold for what counts as organized crime is set very low. The RAND study, which reprises and builds on earlier IFPI and Interpol reporting, is constructed almost entirely around such practices. Prominent stories about IRA involvement in movie piracy and Hezbollah involvement in DVD and software piracy date, respectively, to the 1980s and 1990s. Street vendor networks in Mexico City—a subject we treat at length in the Mexico chapter—are mischaracterized as criminal gangs connected with the drug trade. Piracy in Russia is attributed to criminal mafias rather than to the chronically porous boundary between licit and illicit enterprise. The Pakistani criminal gang D-Company, far from "forging a clear pirate monopoly" in Bollywood, in RAND's words, plays a small and diminishing part in Indian DVD piracy—its smuggling networks dwarfed by local production.

The US record isn't more convincing in this regard. Jeffrey McIllwain examined the Department of Justice's IP-related prosecutions between 2000 and 2004 and found that only 49 out of the 105 cases alleged that the defendant operated within larger, organized networks. Nearly all of these were "warez" distribution groups for pirated software—hacker communities that are explicitly and often fiercely non-commercial in orientation. McIllwain found "no overt references to professional organized crime groups" in any of the DOJ's criminal charges (McIllwain 2005:27). If organized crime is a serious problem in these contexts, it should not be difficult to produce a stronger evidentiary record.

Disaggregating Industry Exposure

Piracy is generally presented as a uniform threat to the copyright industries, but in practice these industries have widely varying exposure to piracy, reflecting differences in how music, film, and software are consumed and how different business strategies and consumer expectations have shaped markets for those goods. The core copyright industries are also internally diverse, with a variety of revenue flows and business models that contribute to the bottom line.

We see considerable evidence that the digital transition is changing the mix of business models in the music, film, and software businesses—and undermining some very profitable ones, such as the markets for CDs and DVDs. But we see no evidence that the industries overall have diminished capacities to innovate or commercialize new work. By most measures this has been a very prosperous decade for the US copyright industries—up to and, in some sectors, including the current economic crisis. All of the US copyright industries—film, business software, entertainment software, book publishing, and even music (including live performance)—grew in total revenues through 2008.

Insofar as the quantity of new products is an indicator of the health of a cultural sector, the first decade of the new millennium was a veritable golden age in the United States. The number of new albums released more than doubled in the period, from 35,516 in 2000 to 79,695 in 2007 (Oberholzer-Gee and Strumpf 2009). The number of Hollywood films released ranged between 370 and 460 in the 1990s and between 450 and 928 in the 2000s, with the peak year in 2006 and some 677 produced in 2009 (MPAA 2006, 2010).[36] Software industry growth has been dramatic, averaging 20%–30% annually until 2009. The video-game sector averaged nearly 17% growth between 2005 and 2008, with growth rates in 2007 and 2008 of 28% and 23% (Siweck 2010).[37] According to the IIPA, the core copyright industries in the United States averaged 5.8% growth

---

36   We use these numbers with reservations. The European Audiovisual Observatory relies on MPAA numbers, and the MPAA appears to have revised its counting method in 2010, leading to different (and generally higher) numbers of films reported between 2005-2009 and a somewhat sharper decline in recent production (MPAA 2010). How this impacts numbers before 2005 is unclear.

37   The economic crisis produced a 10% contraction in the market in 2009.

between 2003 and 2007—well above the roughly 3% annual US growth rate in the period (Siwek 2009). According to the World Association of Newspapers and News Publishers, total media and entertainment spending posted an annual growth rate of 5.3% in the United States between 2002 and 2008 and 6.4% globally (WAN-IFRA 2008). Losses to piracy need to be placed in this context of overall industry growth—and in some cases remarkably rapid growth.

Our work reinforces the view, however, that business models built around the sale of high-priced recorded media—CDs, DVDs, and stand-alone software products—are becoming less viable. This is especially true in environments where consumer expectations are oriented around ownership, rather than licensing or rental, and above all in countries where price/income ratios remain high. Piracy is a major source of pressure on recorded media markets, but by no means the only one. In particular, it is increasingly difficult to separate the impact of piracy on CD or DVD markets from the impact of low-cost legal competitors that have emerged in the past several years—streaming music and video services like Spotify in the United Kingdom and Hulu in the United States, very low-priced video rental services like redbox in the United States or BigFlix in India, and "debundled" products like digital music singles that are supplanting the higher-cost album as the main unit of sale. Piracy has undoubtedly been a catalyst for the emergence of these low-cost models, insofar as it resets consumer expectations around cheaper, on-demand availability. But increasingly, the pressure on the high-end market comes from legal innovators at the low end.

Music

Our study adds relatively little to the volumes that have been written about the digital transition in the music industry—often held up as the "canary in the coal mine" for other media markets. We share the increasingly consensual view that the situation is better understood as a crisis of the high-margin CD business—and of the "big four" record labels (EMI, Sony Music Entertainment, the Universal Music Group, and Warner Music Group), which have relied nearly exclusively on it for their profits—rather than a crisis of the music business in general. The decline in this side of the business has, without doubt, been precipitous (see figure 1.3). According to the IFPI, global recorded music sales dropped from $33.7 billion in 2001 to $18.4 billion in 2008—almost entirely attributable to the decline of CD sales. In the United States, CD sales fell from $7 billion in 2004 to $3.1 billion in 2008—a situation somewhat mitigated by the rise in digital sales from zero to $1.8 billion in that period. Recorded music sales in most other countries have been in similar free fall. Between 2004 and 2008, Brazilian recorded music sales shrank from $399 million to $179 million; Russian sales dropped from $352 million to $221 million; sales in Mexico from $237 million to $145 million. In South Africa, considered a bright spot in international sales, sales grew through 2007—peaking at $129 million before falling to $119 million in 2008.

Figure 1.3 Recorded Music Sales (trade value, in billions of US dollars)

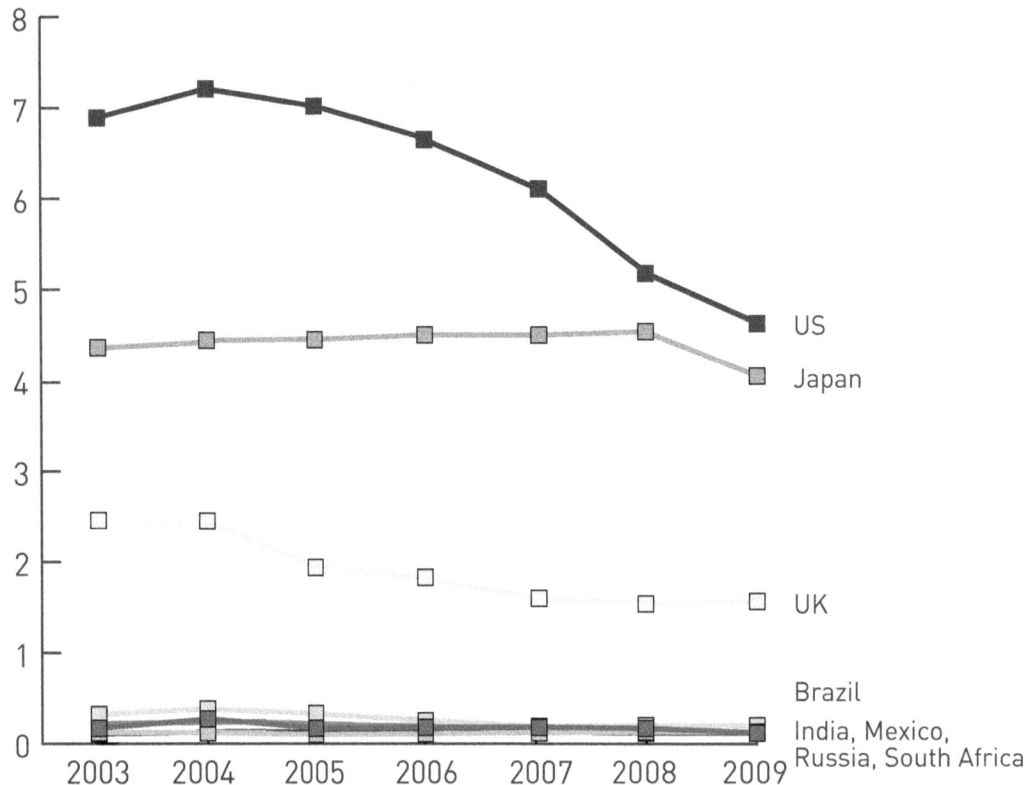

Source: Author based on IFPI data, 2004–10.

Industry representatives tend to attribute this decline to piracy—and in high-income countries to the boom in Internet piracy inaugurated by the launch of Napster in 1999.[38] Most recent histories of the music business, in contrast, cite a broader range of factors that pushed the CD market into decline in the early 2000s, including the maturation of the market in the late 1990s as customers replaced their LP collections,[39] the proliferation of other media goods and services (DVDs, video games, cell-phone services) competing for the same pool of disposable income, and the debundling of the album format as customers cherry-picked lower-priced digital singles, to cite only a few. As we have discussed, the contribution of piracy to this decline is hard to specify and is a matter of considerable disagreement in the research literature.

---

38   Cary Sherman, RIAA president, said in 2003, "The root cause for this drastic decline in record sales is the astronomical rate of music piracy on the Internet." IFPI CEO Jay Berman similarly claimed, in 2001, that "the industry's problems reflect no fall in the popularity of recorded music: rather, they reflect the fact that the commercial value of music is being widely devalued by mass copying and piracy" (Hu 2002).

39   CD sales were a massive growth engine for the recording industry in the United States in the 1990s, rising from $3.4 billion in 1990 to a peak of $13.2 billion in 2000. Much of this growth is attributed to "replacement costs" as customers repurchased their LP collections in CD format.

Nonetheless, total expenditures on music in the period—including concerts and digital formats—have been stable or slightly increasing. The CD's sharp decline in the United States has been offset by the growth in digital sales and concert revenues: the latter more than tripled, from $1.3 billion in 1998 to $4.2 billion in 2008. Such numbers point to a shift from a high-margin industry dominated by CD sales, the album format, and the big four labels to a lower-margin business with more emphasis on performance and related rights.[40] They do not, in our view, point to an existential threat to the music business, much less to music culture.

Developing countries share in these trends, including the fall in CD sales and the growth of the live-performance market. But the structure of the global marketplace also creates important points of divergence. In broad terms, this structure is relatively simple, marked by (1) the near complete dominance of the big four labels in most developing markets—some 84% of the market in Brazil, 82% in Mexico, and 78% in South Africa, for example;[41] (2) the concentration of 80%–85% of revenues in the United States, Western Europe, Japan, Australia, and Canada; and (3) the absence, in most developing countries, of strong domestic competitors capable of building viable alternative distribution strategies, such as Apple and other digital distributors are doing in the United States.

In practice, these factors reinforce the high-price, very-small-market dynamic visible in most developing countries. They create a context in which the big four labels have every incentive to protect high-income markets but little incentive to change their pricing strategies in low- and middle-income markets. Compared to high-value markets like the United States, the United Kingdom, and Japan, the emerging markets are simply inconsequential. Price cuts to expand the market in Brazil, South Africa, or Mexico would have a very limited upside in this context and a potentially serious downside if they began to undermine pricing conventions in the high-income markets. The majors' evaluation of this tradeoff is clear: none have significantly lowered prices in emerging markets.

The dominance of the majors means, too, that there are fewer local actors capable of developing business models at price points below the retail CD market. Competitors in the digital market, who have driven the shift in business models in the high-income countries, are still nascent in most developing countries: legal services have emerged only in the last few years and major players like the iTunes Store are generally absent from the music and video market.[42] Consequently, the

---

40   There is a dearth of empirical work on the impact of this shift in revenue streams. It will undoubtedly be bad for some artists, but whether it is good or bad in general is not something we can clarify here. Proportionally, a much larger percentage of concert revenues than of CD sales remain in the hands of artists, reflecting more direct artist control over concert deals.

41   Compared to a 70%–75% share globally.

42   The iTunes App Store is widely available due to Apple's global marketing of the iPhone, but music and video sales are much less widely supported. An iTunes Music Store launched in India in 2008 and Mexico in late 2009, for example, but is unavailable in many other countries, including South Africa, Russia, and Brazil.

market is still "stuck" on the CD model in ways that have widened the gap on price, convenience, and variety with the pirate market. The continued decline of CD sales and the massive growth of piracy are predictable consumer-driven results. Recent IIPA reports cite rates of music piracy in excess of 90% in China, India, Mexico, and Brazil. Less and less of this traffic takes place on the street, as physical piracy shifts toward the narrower stock and higher margins of DVDs.

Most of our data points to the continued erosion of this model. Pressure on the majors is coming from all sides of the business. Diminished costs of production and the growing ease of digital distribution have produced a wave of new entrants into the low end of music markets. Digital distribution is just beginning to break the majors' lock on access to international markets. Telecom providers are beginning to push on the local pricing conventions for mobile-music sales.

Our country studies explore this shift from the perspective of vendors, consumers, and industry actors and, overall, demonstrate the advantages of local industry control in developing markets. In countries where local recording labels and local repertoire are especially powerful—among our reports, in Russia and India—the reconfiguration of music business models is a given. Indian companies like T-Series compete fiercely with the pirates on price and have hugely expanded the market for recorded music; Russian music labels, which never had a stable CD market to rely upon, have increasingly consigned the CD to a promotional role for live performance and set prices well below those of licensed international albums.

The limit case, in our studies, is Bolivia, where the impasse of high prices, low incomes, and ubiquitous piracy shuttered all but one local label in the early 2000s and drove the majors out altogether. The tiny Bolivian legal market, worth only $20 million at its peak, was destroyed. But Bolivian music culture was not. Below the depleted high-end commercial landscape, our work documents the emergence of a generation of new producers, artists, and commercial practices— much of it rooted in indigenous communities and distributed through informal markets. The resulting mix of pirated goods, promotional CDs, and low-priced recordings has created, for the first time in that country, a popular market for recorded music. For the vast majority of Bolivians, recorded music has never been so prolific or affordable.

The resulting global picture is complex and unresolved. The significance of the cheap CD model pioneered by T-Series and other vendors in India is not that it eliminated or even marginalized piracy—it did not. The point is that competition and technological innovation in the Indian music business drove prices to a much lower level, expanded access beyond the commercial elite, and proved viable as a local business model. In other countries, the dominant international labels have not followed suit: the Bolivian case illustrates not so much the failure of a market as the lack of interest, on the part of the incumbents, in reinventing it.

In developing countries where the majors dominate, the legal CD market was never a mass market, and at this stage, never will be. The format is headed for obsolescence, and with it the high-price/small-market dynamic it anchored. The current pirate market, in contrast, is a mass market, but it remains to be seen how many legal publishers can—or will—offer competitive pricing and availability. In a period of both unprecedented access to music and unprecedented levels of production of new music, this is a subject of intense interest but not, in our view, a cause for general alarm.

Movies and TV Shows

Claims about movie piracy drive much of the industry enforcement agenda. In 2009, the MPAA Chairman Dan Glickman called piracy a "dagger in the heart" of the movie industry. When Senator Patrick Leahy, fresh from his cameo in 2008's *The Dark Knight*, unveiled the 2008 PRO-IP Act, he called piracy a threat to "all of the value" created by the film. And in many respects he picked a good example. *The Dark Knight* appeared on BitTorrent sites well before its theatrical release and became the most pirated film of 2008. It also broke all box office records and earned over $1 billion worldwide.

The message from Hollywood consequently has a schizophrenic quality: the movie business is in crisis; the movie business is thriving. Since 2002, the US movie industry has been a $9–10.5 billion business in domestic box office revenues, with successive record-setting years in 2007, 2008, and 2009. International distribution brought in some $16.6 billion in 2007, $18.1 billion in 2008, and $19.3 billion in 2009 (MPAA 2009). DVD sales are a separate, massive revenue stream: global sales peaked at $23.4 billion in 2007 before dropping to $22.4 billion in 2008 and falling further in 2009. Licensing of movie-related merchandise is a third revenue stream, estimated at roughly $16 billion per year (Oberholzer-Gee and Strumpf 2009).[43] This success is not limited to Hollywood. The Indian movie industry—second in global revenues—has also boomed in recent years and registered 13% growth in 2008, with up to $2.2 billion in box office revenues (see figure 1.4). Revenues in 2009 dropped slightly to $1.86 billion (Kohi-Khandekar 2010).

---

43   The cross-marketing of key movie "properties" makes revenue figures hard to disaggregate: movies, games, books, and other products are increasingly part of an integrated media mix that generates revenues—and audiences—across sectors. According to Disney, licensed merchandise alone generated $30 billion in 2008, including $3.7 billion from the 2006 film Cars and $2.7 billion from Hannah Montana tie-ins (Walt Disney Company 2010).

Figure 1.4 Domestic Box Office (in billions of US dollars)

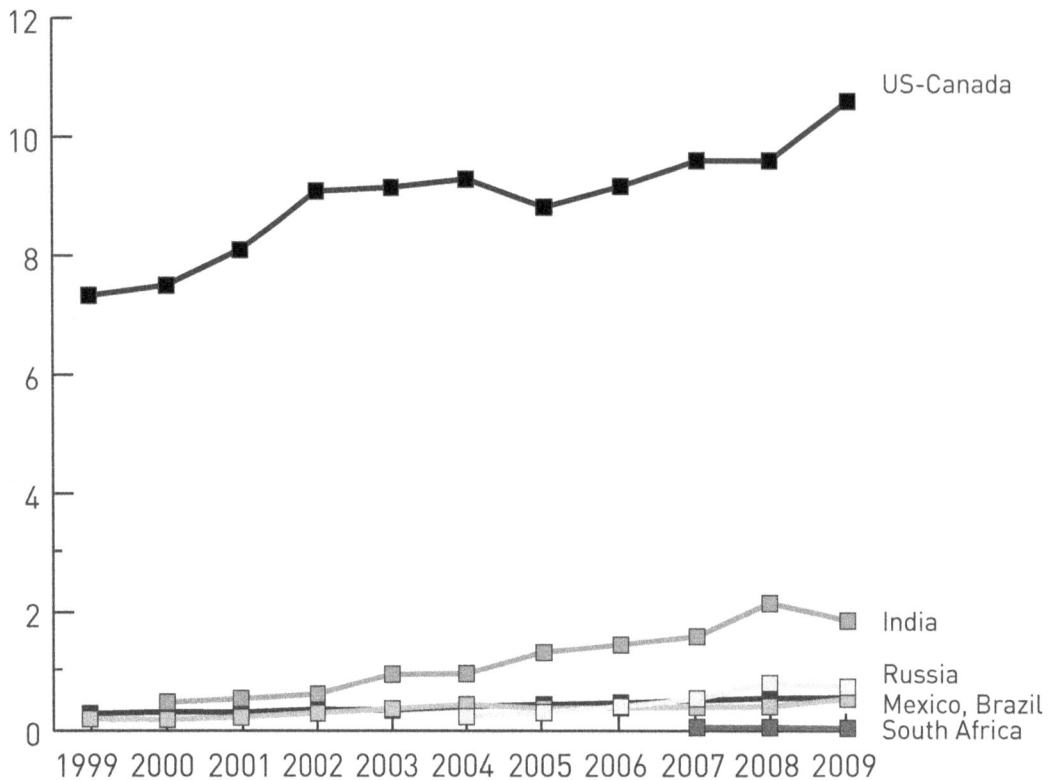

Source: Author based on data from European Audiovisual Observatory (2001–10).

As bandwidth and computing power catch up to the higher demands of video piracy, industry representatives fear that the studios will follow in the footsteps of the record companies. We think this is likely, but insist on carrying the analogy through. The high-priced DVD market is clearly vulnerable to piracy and to the growing range of low-cost legal alternatives, such as streaming services like Netflix and Hulu and automated rental kiosks like redbox in the United States. The displacement effects between these different channels of distribution and consumption will be increasingly difficult to isolate. But theatrical revenues, like live performance in the music business, appear remarkably solid even in a period of sharp cutbacks in consumer spending. Merchandising, cross-media franchising, and other sources of income are also largely independent of these changes in the distribution channel. Unlike the major music labels, the studios control these other revenue streams, leaving them in a far better position to maintain their core business model. If the DVD market collapses as quickly as the CD market, Americans may one day face a $50–60 billion domestic movie industry rather than a $60–70 billion one.

Hit movies nearly always top the list of most-pirated media (table 1.2)—though torrents can also fuel viral hits like the 2008 British gangster film *RocknRolla*, which received minimal distribution in the United States. But overall, American TV series dominate the P2P channel.

Table 1.2 Top Downloads for 2009 by Category

Movies

	Number of Downloads	(Worldwide Gross)
Star Trek	10,960,000	$385,459,000
Transformers: Revenge of the Fallen	10,600,000	$834,969,000
RocknRolla	9,430,000	$25,728,000
The Hangover	9,180,000	$459,422,000
Twilight	8,720,000	$384,997,000

Television Series (top single episode)   (US TV audience)

Heroes	6,580,000	5,900,000
Lost	6,310,000	11,050,000
Prison Break	3,450,000	5,300,000
Dexter	2,780,000	2,300,000
House	2,590,000	15,600,000

PC Games (sales figures generally unavailable)

Call of Duty: Modern Warfare 2	4,100,000
The Sims 3	3,200,000
Prototype	2,350,000
Need for Speed: Shift	2,100,000

Console Games

New Super Mario Bros. (Wii)	1,150,000
Call of Duty: Modern Warfare 2 (XBox 360)	970,000
Punch-Out!! (Wii)	950,000
Wii Sports Resort (Wii)	920,000
Street Fighter IV (XBox 360)	840,000

Source: Author based on TorrentFreak data.

Much of this traffic comes from outside the United States, where local distribution of hit series is usually delayed by months and sometimes years. Television networks have been very slow to adopt the global simultaneous release practices of the major studios. Until recently, even major English-speaking markets like Australia waited a year or more for the broadcast of American hits. The international premier of *Lost* one week after the US broadcast in February 2010 represents the most radical compression of TV windowing practices to date.

Mom and Dad in Munich

Question: Your business is watching file sharing. So is it spreading to the mainstream? Are Mom and Dad from Sheboygan pirating content?

Eric Garland: Oh yes, particularly Mom and Dad in Munich; Mom and Dad in Seville; Mom and Dad in Paris. When we talk about video the reason I single out the European cities is because that's where people are forced to wait a long time to see content legally. In the digital world, we don't want to wait three months, six months. We're just not accepting that anymore . . . we want it all, we want it right now and even Mom and Pa Kettle are getting to the point where they say if it's not on, let's just fire up the computer and watch it. If they want me to wait six months, I've got other options. And people don't really have a conscious or qualms about that, or at least it's mitigated by their feeling that they are entitled to keep up with the Jones'. It is the Twitter, real-time Internet expectations.

–Interview with Eric Garland, CEO of BigChampagne (Sandoval 2009)

The underground distribution of American TV shows is an example of the incomplete globalization of media documented throughout this report, in which global media cultures and global marketing efforts outstrip nationally bounded, time-delayed distribution channels. The role of "national" P2P sites specialized in local media reflects the same breakdown of industry time-management. These sites—DesiTorrents in India, Torrents.ru in Russia, and many others—cater to much smaller publics than the most visible global torrent sites like The Pirate Bay and Mininova. They also disproportionately serve diasporic communities, who often live in high-bandwidth countries with limited access to music, television, and movies from home. Over 20% of the DesiTorrents user base is in the United States and the United Kingdom. Fan-based subtitling communities have also played a role in the circumvention of slow—or sometimes nonexistent—exportation and localization of media products. Anime fan communities began subtitling series available only in Japan in the early 2000s, signaling market demand that distribution companies eventually recognized and moved to meet. Bollywood films are commonly subtitled for African and Asian pirate distribution. The enormously popular Brazilian site, Legendas.tv, distributes only subtitle files for video downloaded through other means. It delivered a complete Portuguese version of *Lost* four hours after the US premiere.

Entertainment Software

According to the ESA, entertainment software sales in the United States reached $11.7 billion in 2008, registering a 28% jump over 2007's already record-setting numbers and surpassing revenues from both movie tickets and CD sales. The global market for games, including those played on personal computers, consoles, and mobile devices, reached $46.5 billion in 2009 (Wu 2010).

Console games account for the lion's share of this revenue—some 39% of the total in 2007, according to the Interactive Software Federation of Europe (ISFE 2009). Sales numbers for PC games are harder to characterize because they are generally split between stand-alone games, which have been in slow decline for a decade and currently represent a $4 billion market, and online games such as *World of Warcraft,* which represent a $7–8-billion market on the PC and a $15-billion market across all platforms. Mobile and handheld games account for another $13 billion.

By most industry accounts, video-game piracy is concentrated within the traditional stand-alone PC-game market, resulting in pressure on developers to abandon the PC in favor of console-only titles. PC games with cracked serial numbers or activation codes are widely available online and in pirate optical disc markets. Unlike record companies or film studios, PC-game developers and publishers have a variety of ways of estimating the prevalence of pirated copies of their games, such as tracking the percentage of calls to technical support from gamers playing with pirated copies (Ghazi 2009). For popular games, reported ratios of ten pirated copies for every purchased copy are routine.

Console games have traditionally been less vulnerable to piracy because of the technical knowledge needed to install a "mod chip" or patch a console's operating system. Among current-generation consoles, both the Wii and the Xbox 360 can be "soft-hacked"—that is, modified without replacing chips. The PlayStation 3 has proved a much tougher nut to crack, with usable hacks available only in late 2010. Complete "modded" consoles can be purchased through retail in many parts of the world, and simple-to-install, mass-produced mod chips have been introduced for a number of systems—with much of the attention falling on the Nintendo DS.

"Anti-circumvention" rules criminalizing the modding of systems are a major feature of the WIPO Internet Treaties, but courts in several countries, including Canada, Spain, France, and Australia, have found wide latitude for modding under existing copyright law, primarily on the grounds that the circumvention of protection measures is not itself an act of copyright infringement and has substantial non-infringing uses.[44] In the United States, the Digital Millennium Copyright Act incorporates strong anti-circumvention language and industry groups have been successful in pushing law enforcement to bring criminal complaints against both modders and mod-chip vendors.

---

44   Such as opening up traditionally closed systems like the Nintendo DS to the developers of "homebrew" software that expands the functionality of the device (without paying Nintendo licensing fees). This is the subject of an ongoing battle between Nintendo and video-game-accessory distributor Divineo over the sale of "linkers," which allow access to the otherwise closed Nintendo DS operating system. Nintendo won two favorable verdicts against Divineo in the United States and in Hong Kong in 2006 (the latter in absentia) but lost a 2009 verdict in the Paris High Court, where the judge found Divineo to be operating legally under French and European law that privileges interoperability between systems. The ruling over whether Nintendo has a right to maintain a closed system is emerging as a flashpoint in the larger battle over consumer and corporate rights over devices. Nintendo has appealed.

Despite the prominence of modding in enforcement conversations, we are aware of no research on the prevalence of mod chips or modded systems and cannot find a credible estimate of how far the practice goes beyond tech-hobbyist communities. In 2007, Nintendo claimed that some seven million DS handhelds had been modded via a widely available Chinese-produced chip, contributing what Nintendo characterized as losses of $975 million across platforms (Nintendo 2009). Nintendo's USTR submission for 2009 singled out Mexico, Brazil, China, Paraguay, and South Korea as hot spots for game piracy. The major US enforcement action against modding in recent years—Operation Tangled Web in 2007—netted just 61,000 mod chips, however, suggesting

Xbox Live in Brazil

In many countries, it can be difficult to be a legal gamer. Although game culture has become global in the past decade, game markets, in many instances, have not. In Brazil—by all accounts a high-piracy country for video games—Sony has withheld release of PlayStation 3 despite its relative immunity to hacking.* Microsoft and Nintendo market current-generation consoles and games in Brazil, but most third-party game publishers do not, resulting in a very diminished legal retail market. Brazilian customers have been locked out of many of the newer digital services, such as Xbox Live, a popular online portal that enables Internet play of Xbox 360 games, which was not launched in Brazil until late 2010. Adding to the difficulty, prices for consoles and most games are higher than in the United States and Europe. An Xbox 360 that costs $299 in the United States retails for over $700 in Brazil—a premium attributable to high taxes on foreign-software imports and complicated local certification requirements.

Brazilian use of Xbox Live exemplifies the complex geography of gaming markets. The service, which costs $60 per year, is for many gamers a primary reason to buy an Xbox 360. The subscription model also ties the Xbox into much stronger server-based authentication of hardware and games. Although the console has been successfully hacked, the Xbox Live service has not, enabling Microsoft to effectively exclude users of modded machines. Before the service was legally available in their country, Brazilian gamers got around this by subscribing under false addresses, and—according to our sources—they mostly still do: a recent spot-check found that the Brazilian version of the service had only a few games available.

The in-service economy based on Live Points had also been closed to Brazilians, but there are many sources of unofficial currency exchange that enable residents of unsupported national markets to pay and play. Microsoft can identify player location by IP address but has a variety of reasons for tolerating these practices and their associated informal markets—among them, the intense customer loyalty demonstrated by the effort to access the service. Among hard-core gamers, high game prices and the high value of the XBox Live service can justify having two Xbox 360s: one modded for pirated games and one reserved for Xbox Live use. Similar strategies allow Brazilians access to Sony's online portal, the PlayStation Network, which is still unavailable through legal channels.

* Correction: Sony launched the PS3 in Brazil in August 2010, for the modest price of $1225.

a problem on a much smaller scale, at least in the United States (Associated Press 2007a).[45] The ESA, for its part, indicates that its analysis of online distribution finds comparable numbers of pirated console and PC games—challenging conventional wisdom on this point and pointing to a mass-market phenomenon. Clearly, this is a subject requiring more detailed study.

A large part of the game business operates on a model that is, for all intents and purposes, immune to end-user piracy. Online PC games such as *World of Warcraft*—a category worth some $7 billion in 2007—operate on monthly subscriptions, which makes unauthorized use for any length of time virtually impossible.[46] A wide variety of other game types are increasingly tethered to publishers' servers and require online authentication to play. The relative ease with which game producers can incrementally add value to games in return for the validation of copies also represents a powerful anti-piracy tool, contributing to the game industry's robust position among the copyright industries. Unlike film, music, or business software, games often entail a relationship between the consumer and the developers or publishers that extends beyond the initial sale and often feeds back into game development. Developers are solicitous of these relationships, and the game community often responds with considerable loyalty. Online game forums host heated debates between consumers about the ethics and practice of piracy that, in our experience, are unique among the copyright industries.

Why Is Business Software Piracy Different?

The business software market is unique to an extent that warrants a very different understanding of piracy. As we noted earlier, the BSA has simultaneously the most robust model for estimating rates of piracy and—prior to 2010—the most exaggerated model of actual losses. The now defunct assumption of a one-to-one ratio of piracy to lost sales has been only part of the problem, however. More significant, in our view, is the elective blindness of the BSA and many industry representatives to the value of the network effects generated by piracy in emerging software markets.

In software markets, network effects refer to contexts in which the value of software rises with the size of the installed base. The more widely used a piece of software or software service, the more it becomes a de facto standard that shapes user decisions about adoption and investment. Platform technologies such as operating systems exhibit strong network effects because a popular platform will foster a rich secondary market in applications and services, which in turn increases the platform's value. "Lock-in" occurs when the costs of leaving a particular software environment are high—whether because switching would require significant repurchasing of software, or because the use of less common standards is disadvantageous, or simply due to costs of retraining.

---

45 In 2008, some thirty-five million game consoles were sold in the United States.

46 Both the ESA and the IIPA have reported the growth of fraud in the online-game sector—generally in the form of copycat servers hosting subscription-based games. Whether this is a serious problem or not is unclear. We have seen no estimates of the scale of this practice, and it appears to be entirely addressable in a commercial infringement and consumer fraud context.

For near-monopolies such as Microsoft in the operating systems and office software markets, network effects reinforce market power and increase the value of their products. Lock-in effects, in turn, ensure that customers are less likely to switch to competitors.[47]

As BSA piracy figures indicate, these dynamics in emerging economies are primarily (and sometimes overwhelmingly) a function of pirated-software adoption, not legal adoption.[48] Piracy, in effect, has allowed the major vendors to dominate low- and middle-income markets (or, as they develop, market segments within them) that they have little financial incentive to serve. Perhaps most important for market-dominating firms, piracy acts as a barrier to entry for competition, especially "free" open-source alternatives that have no upfront licensing costs. When these emerging markets begin to grow, as most did in the last decade, piracy ensures they do so along paths shaped by the powerful network and lock-in effects associated with the market leaders.

In our view, these factors should figure in any full accounting of the costs and benefits of software piracy. Top-tier vendors have established and maintained their dominant positions in emerging markets through piracy, often prior to or in the absence of significant local investment. Any losses they incur at the margins of the consumer and business markets in those countries should be weighed against the value of maintaining their dominant positions. For near-monopolies, we would argue that this value is very high. For vendors working in highly competitive markets or selling products that do not function as standards or platforms, that value is clearly lower. We have seen no work that empirically measures or distinguishes these effects and so can only speculate here as to their relative worth.

Enforcement representatives interviewed for this project generally disagreed with this view of how software markets work and held to the notion that piracy is first and foremost a loss of revenue and a disincentive for investment—both foreign and local. We call this elective blindness because the relationship between piracy and network effects appears to be well understood elsewhere in these firms—including among such industry leaders as Bill Gates, who has referred repeatedly to the importance of piracy in securing market share and undercutting Linux adoption in China.[49] As

---

47   There is an extensive and—for the most part—highly speculative business literature on network effects that has attempted to model the decision points that shape policies of tolerance and enforcement toward software piracy (for an overview, see Katz [2005]). The actual estimation is highly complex, and we are unaware of any compelling estimates across different software lines or in developing countries.

48   BSA-derived rates of software piracy in Russia hovered around 90% through the early 2000s. China was at 90% as recently as 2008. India has spent most of the past decade around 70%; Brazil, 60%–70%.

49   Microsoft Chairman Bill Gates to students at the University of Washington, in 1998: "And as long as they're going to steal it, we want them to steal ours. They'll get sort of addicted, and then we'll somehow figure out how to collect sometime in the next decade" (Grice and Junnarkar 1998). Or more recently: "It's easier for our software to compete with Linux when there's piracy than when there's not. . . . You can get the real thing, and you get the same price" (Kirkpatrick 2007). The same logic also holds for smaller companies seeking to establish a presence in developing markets, such as LogMeIn, a $320 million vendor of remote access software. As CEO Michael Simon observed, echoing Gates, "If people are going to steal something, we sure as hell want them to steal our stuff" (Vance 2010).

Microsoft executive Jeff Raikes observed: "In the long run the fundamental asset is the installed base of people who are using our products. What you hope to do over time is convert them to licensing the software" (Mondok 2007).

The major vendors have done just that in the past decade in the institutional sectors of emerging markets, through a combination of price discrimination and enforcement. This strategy has focused on computer manufacturers and vendors, large businesses, school systems, and other public-sector institutions because they combine two things the software companies like—relatively high ability to pay and vulnerability to enforcement—with two things that they don't like but must confront: sufficient market and/or political power to extract pricing concessions and sufficient technological capacity to make credible threats of open-source adoption. In 2007, the Russian government played this game with a consortium of commercial vendors to obtain a 95% discount on Windows and a bundle of productivity applications for Russian schools. Chinese municipalities did so in 2008, following a Chinese edict requiring legal software in government use. The Indian state of Karnakata did so in 2009 for its government agencies, and so on. In both the Russia and China cases, the BSA cited the licensing of public institutions as a major factor in reported drops in the local rates of piracy (BSA/IDC 2009). When these licenses come up for renewal (in the Russian case, at the end of 2010), network effects and lock-in costs will factor on the side of commercial vendors in any renegotiation.

In the retail channel, in contrast, prices remain very high relative to local incomes—usually matching and sometimes exceeding US or European levels. One might reasonably ask why. It is no secret, including among vendors, that very few Indian or Brazilian customers will pay $300 for Windows or $1,000 or more for Adobe's Creative Suite. There is no significant market at that price level. In practice, however, vendor strategies don't require one. The retail channel plays a very small part in the marketing strategies of the major vendors even in developed countries and far less in developing ones where price/income ratios are several multiples higher.[50] The institutional channel is the revenue generator.

Retail prices, in these contexts, can remain high because the retail market is not needed to build market share. Piracy does that. High retail prices are, nonetheless, valuable for two reasons: they prevent arbitraging of low-priced goods across borders,[51] and they set expectations about how much software should cost—and accordingly set a baseline for licensing deals. Some vendors have made efforts to "complete" these underserved markets through price discrimination in the retail sector, but without notable success. Efforts to sell stripped-down versions of Windows—the various "Starter" packages announced over the past decade—are perhaps the best-known example, widely distributed but doomed in markets where full versions are available at little or no cost. As an Indian respondent observed: free software in India means Microsoft Windows.

---

50   According to quarterly earnings reports, Microsoft's consumer market—here including retail purchases and (often discounted) sales through manufacturers—represents around 20% of total business software revenue.

51   Even of local-language software, which generally sells at no more than a slight discount.

The BSA's valuation (until 2010) of every pirated copy as a lost sale is worth returning to in this context because we can now see that it answers the wrong question. In a market dominated by volume-licensing deals, the question is not "how many legitimate copies does piracy displace," regardless of whether the answer is 90% or 10%, but rather: "given the high market share already achieved by vendors in high-piracy markets, for which segments of the market are price discrimination and enforcement profitable strategies?" Here, vendors face the downside of economies of scale: the smaller the customers, the higher the costs of engaging them in contracts or threatening them with enforcement. Completing a market, in this context, is an expensive proposition with diminishing returns. In our view, the BSA piracy rates are descriptions of this decision point.

Small business is the main enforcement frontier, actively contested by the BSA and local affiliates, the major vendors, and police. Small and medium-sized businesses face sharp dilemmas insofar as they are vulnerable to enforcement, lacking in leverage with software vendors, and often unable to afford operating fully within the licit economy. A software-compliance audit or raid can be a business-threatening experience in such circumstances, as we document in our Russia study. The BSA, for its part, is regularly criticized for its small-business enforcement tactics, which include unrealistic proof of licensing requirements and a practice of basing settlements on the unbundled, highest-possible-retail-price of infringing software rather than the actual purchase cost (Associated Press 2007b). Such practices are in notable contrast to the accommodations and discounts made for large institutional infringers and are part of a dynamic in which enforcement does not so much dissuade piracy as enable price discrimination—down or, occasionally, up in the form of settlements—based on the power relations between the two parties.

The acceptability and even optimality of this approach can be weighed against the various alternatives available to business software vendors. All the major companies could adopt stronger online authentication measures, making it more difficult to use and maintain pirated software. All of them could create obstacles to the over-installation of licensed copies within businesses, which is routinely cited as the most prevalent form of infringement. But strong versions of these options go unexercised for a variety of reasons, including fear of alienating paying customers, fragmenting the installed-code base (which could increase security risks for licensed users), and diminishing the other positive network effects of widespread use.[52] The anti-piracy strategies of

---

52   As Bradford Smith, deputy general counsel for Microsoft, characterized it in 2001: "By the late 1980s every single company abandoned that approach [copy protection] for the simple reason that legitimate customers did not like it. They found that there were times when they needed to make additional copies: they sold the computer and bought a new one and wanted to move their software, or their hard discs crashed and they needed to reinstall it. And even though at the time worldwide piracy rates for software were in excess of 80% the need to take care of the legitimate 20% of the market place took precedence over trying to deal with the rest. And that same bias very much exists today, I see it all the time when these issues are debated inside Microsoft" (Katz 2005). In late 2010, Microsoft abandoned its Windows Genuine Advantage program, which tied Windows and Office updates to regular authentication on Microsoft servers.

An Investment in Friendship

"Piracy helped the young generation discover computers. It set off the development of the IT industry in Romania. It helped Romanians improve their creative capacity in the IT industry, which has become famous around the world . . . Ten years ago, it was an investment in Romania's friendship with Microsoft and with Bill Gates."

– *Romanian President Traian Basescu* (during a press conference with Bill Gates) (Reuters 2007)

PC-game publishers in the past few years offer an informative contrast. Because games rarely function as platform technologies or standards, publishers have less to gain from the network effects associated with piracy and have moved much more quickly toward strong forms of online authentication. Despite a number of controversial missteps and botched launches (for example, *Spore* in 2008 and most of the Ubisoft lineup in 2010 when its authentication servers crashed), the lock down of the PC-gaming environment is well underway.

Credible threats of open-source software adoption in Brazil, Russia, India, South Africa, and many other countries also place a sharp upper bound on business software enforcement strategies. Once again, the logic is simple but rarely acknowledged: the most likely consequence of the widespread enforcement of licenses in Russia or China would be the widespread adoption of open-source alternatives—and very possibly a spur to development of alternatives where no open-source equivalents yet exist, as in the case of Autodesk's specialized AutoCAD tools. As we detail in our Russia and India chapters, these risks are not hypothetical: Microsoft and other vendors go to great lengths to underbid open-source providers in institutional contexts to ensure that open-source adoption does not reach the point where it generates comparable network effects.[53] Where the institutional or symbolic stakes are unusually high, this competitive dynamic can push licensing fees to zero.

Given the rules of this game, open-source adoption policies have become targets of IIPA criticism, despite the irrelevance of this issue to IP protection. The government of Indonesia, for example, characterized its recently announced open-source procurement policy, plausibly, as a measure to combat the use of infringing software. Rather than applaud the measure, the IIPA's 2010 report criticized Indonesia for establishing a trade barrier that "does not give due consideration to the value of intellectual creations" and, as such, "fails to build respect for intellectual property rights" (IIPA 2010b).[54] Whether such procurement policies represent a trade

---

53 For a textbook example, see Volker Grassmuck's study of Linux adoption in Munich in Karaganis and Latham (2005).

54 Similar complaints appear in the 2010 IIPA reports on India, Brazil, Thailand, Vietnam, and the Philippines.

barrier—unjustifiable or not—is a worthwhile question that has been debated within the open-source community (O'Reilly 2002). But the implication that open source undermines IP rights is tendentious. Quite the contrary, open-source licensing derives from and depends on strong copyright.

The BSA continues to push the enforcement envelope by calling for stronger penalties and audit powers, including the criminalization of "organizational end-user piracy" to increase pressure on businesses. End-user criminal provisions have been implemented in a handful of countries, mostly through US-driven bilateral agreements (for example, in Australia and Singapore), but they go significantly beyond international IP obligations under TRIPS and remain controversial. This is true in the United States as well, where end-user criminal liability is implied in the sweeping No Electronic Theft Act (1997) but has never been targeted. Given the viability of the institutional-legalization strategy and the balancing act between enforcement and open-source adoption, we see little incentive for the major commercial vendors to upset the status quo.

In the end, with growth rates around 30% and high-value network effects structuring key software markets, we see no strong evidence that there are *any* real losses to market leaders from business software piracy. But the enforcement effort does play an important role in defining the boundaries of vendor institutional-licensing strategies. With the massive subsidization of local IT infrastructures through pirated software and—to date—very inconsistent adoption strategies for open-source alternatives, it appears that most governments are also willing to play this slow game of legalization with vendors, with cooperation on enforcement and open-source adoption as the carrot and stick.

Pricing

Price comparisons between pirated and licit goods in different countries offer a simple but powerful lens on the organization of national media markets. To illustrate these differences, we compared the most common legal prices of a range of media goods to the most common pirate prices and then translated those numbers into a "comparative purchasing power" (CPP) price that reflects how expensive the item would be for Americans if priced at an equivalent percentage of US per capita GDP (table 1.3).[55]

Prices were collected in late 2008 and 2009 and should be treated as approximations. The prices of goods vary according to a range of factors, including the location of sale, perceptions of demand, and—in the pirate market—differences in quality, packaging, and degree of bundling of goods on a single disc. Currency fluctuations also have a large impact on price comparisons. To facilitate comparison, we focused on single-title, high-quality CD and DVD equivalents of licit goods.

---

55   IMF 2009 estimates for exchange rate (OER) GDP per capita are used here rather than the more common purchasing power parity (PPP) numbers—United States: $46,857; Russia: $8,694; Brazil: $8,200; Mexico: $8,135; South Africa: $5,824; India: $1,031.

Table 1.3 Comparative Prices: International Hits, 2008–9
Parentheses indicate lowest observed price (generally wholesale).

Coldplay: *Viva la Vida* (CD)

	Legal Price	CPP Price	Pirate Price	Pirate CPP Price
United States	$17	—	—	—
Russia	$11	$55	$5	$25
Brazil	$14	$80	$2.5	$14
South Africa	$20.5	$164	$2.7	$22
India	$8.5	$385	$1.2	$54
Mexico	$14	$80.50	$(.4) 1	$5.75

The Dark Knight (DVD)

	Legal Price	CPP Price	Pirate Price	Pirate CPP Price
United States	$24	—	—	—
Russia	$15	$75	$5	$25
Brazil	$15	$85.50	$3.50	$20
South Africa	$14	$112	$(.4) 2.8	$22.40
India	$14.25	$641	$(.3) 1.2	$54
Mexico	$27	$154	$(.4) .75	$4.25

Source: Author.

Coldplay's album *Viva la Vida* and Warner Bros.' movie *The Dark Knight* were blockbuster international hits in 2008. Coldplay has sold over nine million *Viva la Vida* CDs since its release, and the album topped the charts for digital downloads for months. *The Dark Knight* brought in over $1 billion in global box office receipts and broke all records for DVD sales when it was released toward the end of 2008.

Although quintessential global goods in many respects, these products are not trade goods in the same sense as cars or electronics or other manufactured items. Albums and films are licensed separately in each country in which they are sold. The license generally permits the reproduction of a specific number of copies, which are almost always produced locally. The parallel importation of copyrighted goods is restricted in many countries, ensuring that differences in pricing cannot be easily arbitraged.

Licensing costs are controlled by the rights holders—nearly always the major labels, software publishers, or studios. In the case of music licensing, the final retail price is often the result of deals between the labels and other players in the distribution chain—distributors, retailers, and radio stations. This introduces variability in pricing, with prices for the same album differing significantly from one country to the next.

Film studios demonstrate stronger consistency around pricing. The DVD for a major recent release starts at $14–$15 in most markets, with higher prices the norm in some countries. With the exception of some brief experiments with cheap DVDs, notably in China and Eastern Europe, the major studios have made very few efforts to cater to differences in local incomes or to price goods at levels that compete with the pirated goods. In neither the film nor the music market are goods priced at levels that serve more than a niche customer base. CDs and DVDs remain luxury items in most middle- and low-income countries. Price/income ratios roughly comparable to those of US and European media markets are found only in the pirate markets in these countries.

This dynamic extends to locally produced music and film (table 1.4). Local record labels are not as constrained by the norms of major-label return on investment and generally have a stronger interest in promoting live performance. Local CDs, consequently, demonstrates more variability in price. This flexibility is not present for most domestically produced films, however, which generally only range "up" from the high floor set by distributors. Unlike local music labels, local film studios are already tightly integrated into international networks of film production, distribution, and anti-piracy enforcement and follow their pricing conventions.

Table 1.4 Price Comparison: Domestic Hits, 2008–9

Domestically Produced Hit Albums

	Legal Price	CPP Price
Krematorium: Amsterdam (Russia)	$6.50	$32.5
Thermal and a Quarter: first album (India)	$7	$315
Victor and Leo: Borboletas (Brazil)	$9.50	$54
Thalia: Primera Fila (Mexico)	$15	$86

Domestically Produced Hit Films

	Legal Price	CPP Price
Tropa de Elite (Brazil)	$10	$57
The Inhabited Island (Obitaemiy Ostrov) (Russia)	$15	$75
Mr. Bones 2 (South Africa)	$18	$144
Arráncame la Vida (Mexico)	$17.6	$100
Jaane Tu . . . Ya Jaane Na (India)	$3.8	$171
Oye Lucky! Lucky Oye! (India)	$2	$90
Mission to Nowhere (Nigeria)	$3	$123

Source: Author.

The Movies Go Upmarket

Although US domestic box office revenues have grown 40% since 2000, the real growth markets have been overseas. Box office revenues have roughly quadrupled in India since 2000 and tripled in Brazil; they have tripled in Russia since 2004.[1] Because this growth took place against low baselines, however, these markets remain very small compared to their US and European counterparts. Nearly all of it, moreover, is attributable to rising prices rather than increased attendance (figure 1.5 and figure 1.6). Movies have moved rapidly upscale in the past decade, tracking growing middle-class affluence in industrializing countries. (In the United States, attendance has slightly fallen since 2000.) The push toward 3D and IMAX theatres is the next round of this premium-pricing strategy.

In countries where domestic companies actively compete for audiences, average price is usually hard to calculate: exhibitors practice elaborate price-discrimination strategies, including different tiers of theatres, various club or subscription discounts, and—in India—differential pricing for better seats and services within the same theatre. India stands out for its extraordinarily wide range of prices, from $0.60–$0.70 seats in the older cinema halls to a $10-12 high-end reserved for Hollywood and top-tier

Bollywood films. This mix of exhibition spaces ensures that Indian film remains available to a broad audience: Indian per capita annual movie attendance hovered around three throughout the last decade—a number that dwarfs attendance rates in other developing countries. But even in India, average prices have risen dramatically.

Price increases may well be a valid revenue-maximizing strategy: audiences have proved relatively insensitive to price so far, though it is not hard to imagine an eventual tipping point in this relationship. But the price increases have also significantly expanded the gap between the licit and illicit markets and narrowed the margin in which the pirate and licit markets can be plausibly said to compete.

The only country to buck this trend, in our study, is South Africa, where the dominant distributor, Ster-Kinekor, lowered prices in 2005 in a bid to attract the growing black middle class. In so doing, it triggered a price war that sent tickets below $2. The result was a bifurcated market in which the two major exhibitors established premium and budget cinema chains, each showing Hollywood films at dramatically different prices. This model survived the end of the price war in 2007 and is this report's only example of a price-cut driven effort to expand film audiences.

i   Mexico and South Africa saw much lower growth in the period, and we were unable to find numbers for Bolivia.

Figure 1.5 Average Ticket Prices (in dollars)

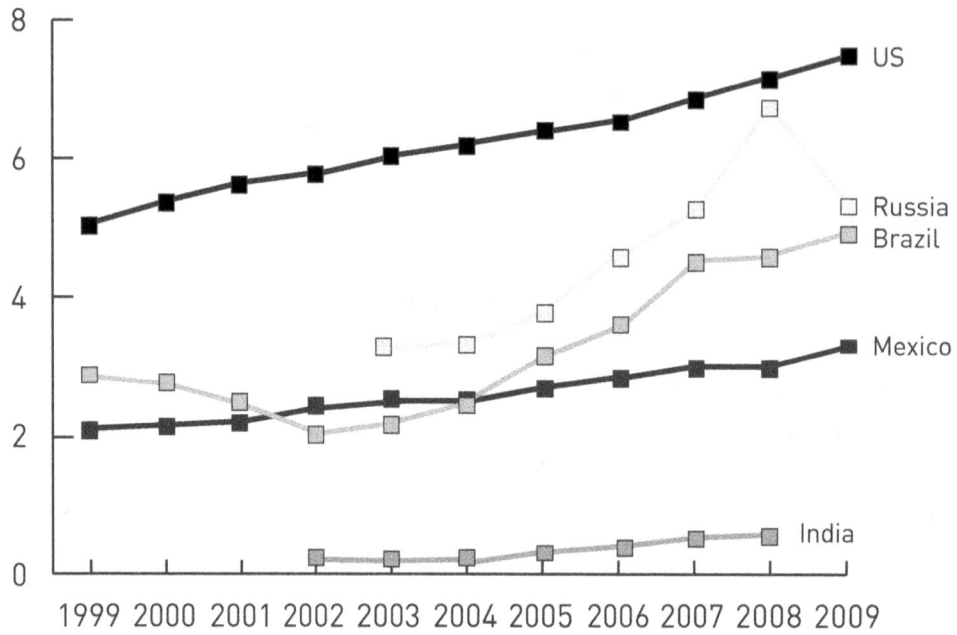

Source: Author based on data from European Audiovisual Observatory (2001–10).

Figure 1.6 Movie Admissions Per Capita Per Year

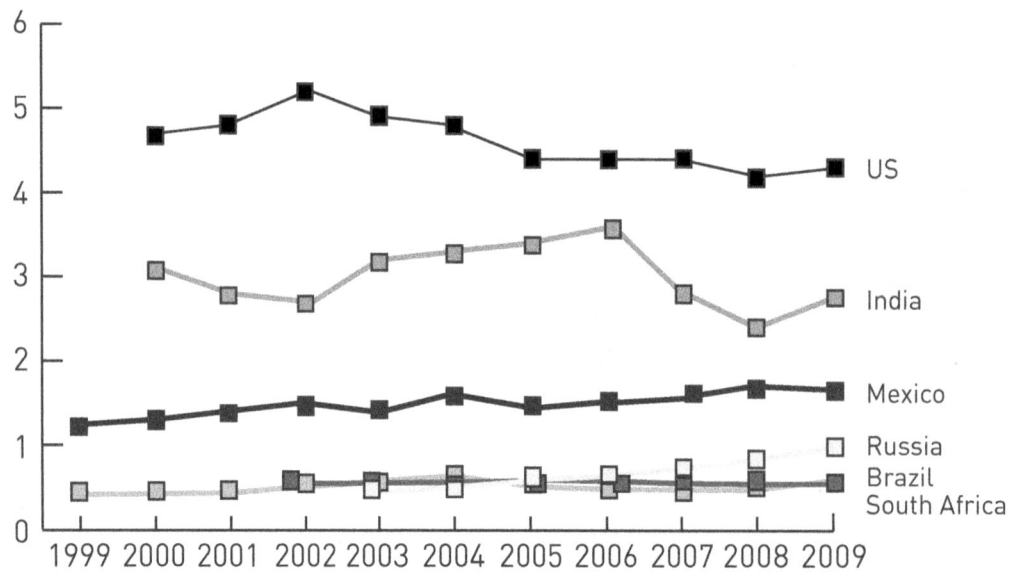

Source: Author based on data from European Audiovisual Observatory (2001–10).

The most notable exceptions to this rule are India and Nigeria—both of which host large domestic film industries that compete for local audiences. The ticket and DVD pricing structure in India is strongly bifurcated between Hollywood films exhibited at close to Western prices and Indian films extending into much lower price tiers. A number of the major DVD distributors—notably Moser Baer, the largest Indian distributor—have gone further in upsetting licensing conventions in the DVD market, creating a mass market for Indian home video that competes with the high end of the pirate market. The Nigerian home-video market—built largely on the piracy of Bollywood films and still very reliant on informal vendor networks for distribution and sales—also operates at a price level that competes with pirated DVDs (Larkin 2004). The main lesson of this price comparison is relatively simple: in countries where domestic companies dominate production and distribution, those companies compete on price for local audiences. In countries where domestic production and distribution is controlled by the multinationals, they generally don't.

Pirate CD and DVD pricing also indirectly illustrates the different structures of pirate markets in these countries. In the early 2000s, the retail price of a pirated DVD in all these countries was in the vicinity of $5. By 2009, the price had dropped to $1 in many countries, with wholesale and lower-quality retail discs often available for significantly less. Our work suggests that $1 is the current retail floor for decent-quality DVDs in competitive pirate markets—including competition from other vendors and, increasingly, from the Internet. Anything above $1 reflects a constraint on trade, whether due to enforcement, higher-priced inputs, or collusion between vendors. The country studies in our larger report offer examples of all three.

The high prices of pirated goods in Russia and the United States stand out in this context. In the United States, the pirate optical disc market has all but disappeared—displaced by P2P and other digital services. Pirated goods at the organized-retail level are virtually nonexistent. Street vendors can still be found in major US cities but fill only niche markets, such as the markets for "camcordered" copies of new movie releases or specialty genres like reggae. High prices in the United States reflect this niche-market status and, more generally, a higher ability to pay.

High prices for pirated goods in Russia, on the other hand, appear to reflect the successful consolidation of production in the hands of large-scale and—by many accounts—state-protected pirates, who have acquired enough market power to prop up prices. A key component of this consolidation was the crackdown against small-scale retail and local producers that began in 2006, which swept away the middle tier responsible, in other countries, for the strongest competition on price and volume production.

The China Syndrome

The rule of high multinational DVD pricing has a number of minor exceptions, several of which are documented in this report, and to date, one major one: China. Between 2003 and 2007, DVD prices in China dropped from an average of 100 renminbi ($15) for international titles and 50 renminbi ($7.50) for domestic ones to as little as $1.50 for foreign titles and $1.20 for many domestic ones. Today, most high-quality foreign titles are sold for around 20–30 renminbi ($3–$4.50), with cheaper versions generally available in the lower-quality DVD-5 format.

Price-cutting was initiated by the domestic studios in response to the growing divide between legal and pirate prices. Because the Chinese market is overwhelmingly dominated by domestic, state-controlled studios and distributors, state studios wielded enough market power to compel the marginal foreign players to follow. Warner Bros. and Paramount Pictures, eager to maintain their positions in the Chinese market, made their own price cuts in 2007—led by Warner Bros.' "10-Renminbi Blockbuster" collection of popular movies in DVD-5 format, available for roughly $1.50. Because this low-cost initiative put no measurable dents in the pirate marketplace, it was discontinued the following year.[1] Licit prices subsequently rebounded slightly, and volume retailers like Walmart now sell new-release DVDs for around 22 renminbi ($3.20).

Today, DVD prices in China are sufficiently compressed that the important market differentiator is more often the quality of the copy than the price. Here, the studio's attempt to differentiate cheap DVD-5 copies from higher-priced DVD-9 copies maintained the value window for pirated goods, and indeed a common criticism of the Warner Bros. effort was the low-quality of its recordings, jacket, and materials. This price compression is now moving into the emerging Blu-ray market, where both pirate and legal prices have dropped to as little as 30 renminbi ($4.50).

In our view, Warner Bros. and Paramount's tolerance of lower prices in the Chinese market is part of a larger pattern of Chinese exceptionalism in the global media marketplace, in which the potential future size of the market (and the very aggressive present-day government intervention in it) dictates a short-term drive for market share and accommodation of the domestic incumbents rather than profits. Among other admittedly narrow data points, Microsoft sold Office 2007: Home and Student Edition in China for as little as $26 in 2010, dramatically undercutting the prices documented in India, Brazil, and other developing countries. Consistent with this exceptional status, we see little inclination among the multinationals to extend these pricing practices to other countries.[2]

---

i   Interview with Warner Bros. representative.

ii   Findings on the Chinese DVD market draw on work conducted by Jinying Li.

Table 1.5 Comparative Prices: Software, 2009

Microsoft Office 2007: Home and Student Edition

	Legal Price	CPP Price	Pirate Price	Pirate CPP Price
United States	$149	—	—	—
Russia	$149	$745	—	—
Brazil	$109	$621	—	—
South Africa	$114	$912	—	—
Mexico	$155	$883	$1	$4
India	$100	$4500	$2	$90

Halo 3 (Xbox 360)

	Legal Price	CPP Price	Pirate Price	Pirate CPP Price
United States	$40	—	—	—
Russia	$101	$505	—	—
Brazil	$60	$342	—	—
South Africa	$53	$424	$30	$240
Mexico	$54	$308	$2	$11
India	$36	$1620	—	—

Source: Author.

Software offers few surprises in this broader context. The retail prices for most productivity software in developing countries are at or near Western prices—with small discounts for local-language versions that have less export value. Such prices demonstrate the general irrelevance of the retail software market in these countries and provide a context for the different developing-market strategies proper to the two major software sectors: (1) volume licensing and institutional enforcement by the business software sector, and (2) technological lock down and underinvestment by the entertainment software sector. For the vast majority of consumers, the market remains split between exorbitant retail and cheap pirate goods (table 1.5).

Distribution

In middle- and low-income countries, the counterpart to high prices is weak distribution. Movie theatres, DVD and CD retailers, bookstores, and software vendors are scarce and typically clustered in the capital cities, in proximity to wealthy elites. Smaller cities and the provinces are chronically underserved—sometimes entirely so. In Brazil, the cities of São Paulo and Rio de Janeiro contain roughly 9% of the population but have 41% of the movie screens (Funarte 2009).

In Russia, Moscow and St. Petersburg represent 11% of the population but have a third of the screens (Berezin and Leontieva 2009). In South Africa, the first multiplex in a black township opened in 2007. The number of screens per capita in most countries is a fraction that of the United States, with density only slowly rising in the past decade (see figure 1.7). The quality of copies and exhibition infrastructure also falls off with distance. Despite the move toward global simultaneous release as a strategy to deter piracy, the circulation of new releases to the provinces often takes weeks as exhibitors wait for copies to rotate through their towns.

Figure 1.7 Screens per 100,000 Inhabitants

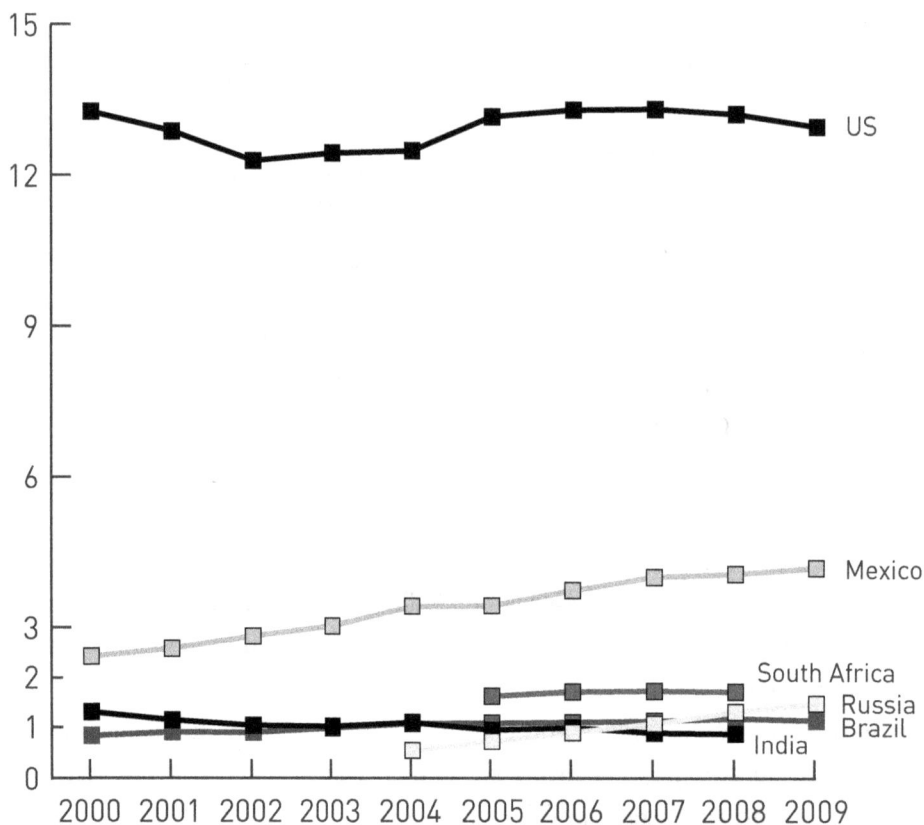

Source: Author based on data from European Audiovisual Observatory (2001–10).

Much the same is true in the optical disc market, where the status of discs as luxury goods generally ensures that they are only carried in a handful of retail chains. This has begun to change in several of the markets we examined as distributors try to combat the massive convenience advantage of pirate vendors, who simply sell where people congregate. In India, T-Series pioneered this approach with cassettes in the 1980s, distributing to a much wider array of vendors and retail outlets than other distributors had tried to reach. We document a number of cases in South Africa, Brazil, India, Russia, and Bolivia in which the superiority of the informal sector as a distribution channel has led legal distributors to try to adopt its methods and approach its price points, in some cases co-opting pirate networks to distribute competitively priced legal goods. The Nigerian home-video industry—the second-largest film industry in the world in terms of the number of features released—was built primarily on such practices and is extending them throughout Africa (Larkin 2004).

The fate of such efforts, our work suggests, depends heavily on access to sufficient capital and market power to build new distribution channels over time—and in particular to prevail in conflicts with incumbent distributors for access to content. Such initiatives have proved viable for large firms in India and the United States but very fragile in emerging markets where multinationals dominate the production and distribution channels. The dilemma is a profound one for local artists especially, and it relegates most of the innovation in media access in developing markets to the legally contested or illegal margins of the media economy.

Where there is no meaningful legal distribution, the pirate market cannot be said to compete with legal sales or generate losses for industry. At the low end of the socioeconomic ladder where such distribution gaps are common, piracy often simply is the market. The notion of a moral choice between pirated and licit goods—the basis of anti-piracy campaigns—is simply inoperative in such contexts, an impractical narrative of self-denial overwhelmed by industry marketing campaigns for the same goods.

Looking Forward

Despite the rapid growth of broadband connectivity, the pirate optical disc trade remains the main form of access to recorded music and film in emerging markets. Enforcement efforts in these markets, accordingly, continue to focus on the links in this commodity chain, from optical disc producers, to distributors, to retailers, to street vendors. Online versions of these businesses—pay-MP3 sites and "download-to-burn" services—have also been targeted, and a handful of prominent cases have become points of contention in trade negotiations between the United States, Russia, and China. Business software enforcement, for its part, continues to focus on private companies and public institutions. Enforcement, in other words, is still directed at the commercial and institutional contexts of infringement, where policing and private settlements have relatively high returns.

As broadband connections and cheap digital storage become more common, however, the focus of enforcement is shifting toward non-commercial activity and the consumer space. The crowding out of the industrial-retail-disc pirate chain by non-commercial digital piracy is largely complete in high-income countries and underway in the middle-income countries we examined. The targeting of BitTorrent sites and other P2P services is part of this shift, and courts have generally been receptive to industry arguments about third-party liability in such contexts, even when these sites do little more than replicate the functionality of search engines. But developing countries are ill equipped and, so far, disinclined to bring enforcement to bear against consumers—

especially stronger criminal procedures. Despite significant pressure from industry, none of the countries examined here have tried. The push for three-strikes laws will be a significant test of this position in the next years.

Not all content-industry positions point in the direction of stronger enforcement, however. Industry positions are evolving as conventional wisdom begins to assimilate the breakdown of the older commodity chains and as businesses conceived as responses to that breakdown become incumbent players in their own right. Shifts in the way industry representatives speak about piracy and technological change provide a good indicator. From the early 1980s through the early 2000s, Jack Valenti of the MPAA arguably set the industry tone regarding the control of new consumer media technologies: a completely uncompromising one expressed most famously in his 1982 comparison of the VCR to a serial killer (Valenti 1982). The same hard line was still visible twenty years later, when Jamie Kellner of Turner Broadcasting claimed that "any time you skip a commercial . . . you're actually stealing the programming" (Kramer 2002).[56]

By 2009, however, it was possible to find even MPAA representatives with less Manichean views of unauthorized use and strikingly different accounts of piracy's relationship to the licit market. In interviews in 2009, the director of special projects, Robert Bauer, sketched out a different agenda for the industry group: "to isolate the forms of piracy that compete with legitimate sales, treat those as a proxy for unmet consumer demand, and then find a way to meet that demand."

The conceptual distance traveled between Valenti's attacks on consumer copying and Bauer's view of piracy as a signal of unmet consumer demand is considerable and, in our view, describes a split in the current debate about piracy and intellectual property within the various affected industries. For the past half decade, industry conversations have had a schizophrenic quality, marked by an enforcement debate organized around the hard line of Valenti and others and a business-model debate organized around the soft line articulated by Bauer.

Our work generally validates Bauer's path as the only practical way forward for the media industries—and one well underway in countries with competitive media sectors. But it is not the only short-term path, and our studies raise concerns that it may be a long time before such accommodations to reality reach the international policy arena. Hard-line enforcement positions may be futile at stemming the tide of piracy, but the United States bears few of the costs of such efforts, and US companies reap most of the modest benefits. This is a recipe for continued US pressure on developing countries, very possibly long after media business models in the United States and other high-income countries have changed. This international policymaking landscape—and its drift toward stalemate—is the subject of the next chapter.

---

56   Or Joe Biden, US vice president, in 2010, announcing the release of the US Joint Strategic Plan on Intellectual Property Enforcement: "Piracy is theft. Clean and simple. It's smash and grab. It ain't no different than smashing a window at Tiffany's" (Sandoval 2010).

About the Chapter

Chapter 1 synthesizes and extends arguments developed throughout the report. It relies heavily on the research conducted for the other chapters, as well as on a range of contributions from team members and other researchers, including Jinying Li, Jaewon Chung, Emmanuel Neisa, Nathaniel Poor, Sam Howard Spink, and Pedro Mizukami. The chapter also draws on correspondence and interviews with roughly thirty experts in the piracy research and enforcement fields, including staff at the IIPA, the BSA, the ESA, the RIAA, the IFPI, and the MPAA. This input was invaluable on many levels and kept the chapter grounded, whenever possible, in the details of business practices and empirical cases.

Synthetic work of this kind presents a variety of difficulties, most immediately in the pricing study where shifting exchange rates, especially, make comparisons approximate and unstable over time. Access to reliable data is another problem in this area, with many sources on media markets either proprietary, not comparable across countries, exorbitantly priced, or some combination of the three. We have done our best to cobble together market structure data from authoritative sources—which are often themselves cobbled together from other sources.

Since there is no overall acknowledgments section for this report, I will give special thanks here to Alyson Metzger, who improved the report in innumerable ways as an editor and copy editor, and to Jaewon Chung, who wore many research and management hats over the life of the project. The report layout and design are the work of Rosten Woo. And we have enjoyed the constant and very patient support of our funders at the Ford Foundation and the IDRC, with thanks especially to Alan Divack, Ana Toni, Jenny Toomey, Phet Sayo, and Khaled Fourati.

References

AFTRA (American Federation of Television and Radio Artists), Directors Guild of America (DGA), International Alliance of Theatrical and Stage Employees (IATSE), Motion Picture Association of America (MPAA), National Music Publishers' Association (NMPA), Recording Industry Association of America (RIAA), and Screen Actor's Guild (SAG). 2010. Letter to the Intellectual Property Enforcement Coordinator (IPEC) in response to request for written submissions, March 24.

Amcham-Brasil. 2010. ABC do PEL. http://www.projetoescolalegal.org.br/wp-content/uploads/2010/02/ABC-do-PEL-2010.pdf.

Anderson, Birgitte, and Marion Frenz. 2008. "The Impact of Music Downloads and P2P File-Sharing on the Purchase of Music: A Study for Industry Canada." Ottawa: Industry Canada.

Anderson, Nate. 2010. "The RIAA? Amateurs: Here's How You Sue 14,000+ P2P Users." Ars Technica, June 3. http://arstechnica.com/tech-policy/news/2010/06/the-riaa-amateurs-heres-how-you-sue-p2p-users.ars.

Associated Press. 2007a. "Raids in 16 States Seek to Thwart Video Game Piracy." *New York Times*, August 2. http://www.nytimes.com/2007/08/02/business/02raid.html.

_____. 2007b. "Software 'Police' Accused of Targeting Small Businesses."

Bahanovich, David, and Dennis Collopy. 2009. *Music Experience and Behaviour in Young People.* Hertfordshire, UK: UK Music and the University of Hertfordshire.

Balázs, Bodó, and Z. Lakatos. 2010. "A filmek online feketepiaca és a moziforgalmazás" [Online black market of films and the movie distribution]. *Szociológiai Szemle* [Review of Sociology of the Hungarian Sociological Association].

BASCAP (Business Action to Stop Counterfeiting and Piracy)/StrategyOne. 2009. *Research Report on Consumer Attitudes and Perceptions on Counterfeiting and Piracy.* Paris: International Chamber of Commerce. http://www.internationalcourtofarbitration.biz/uploadedFiles/BASCAP/Pages/BASCAP-Consumer%20Research%20Report_Final.pdf.

BASCAP (Business Action to Stop Counterfeiting and Piracy)/TERA Consultants. 2010. *Building a Digital Economy: The Importance of Saving Jobs in the EU's Creative Industries.* Paris: International Chamber of Commerce.

BBC World Service. 2010. *Four in Five Regard Internet Access as a Fundamental Right: Global Poll.* http://news.bbc.co.uk/2/shared/bsp/hi/pdfs/08_03_10_BBC_internet_poll.pdf.

Berezin, Oleg, and Ksenia Leontieva. 2009. *Russian Cinema Market: Results of 2008.* St. Petersburg: Nevafilm Research.

Bounie, David, Patrick Waelbroeck, and Marc Bourreau. 2006. "Piracy and the Demand for Films: Analysis of Piracy Behavior in French Universities." *Review of Economic Research on Copyright Issues 3* (2): 15–27. Accessed January 11, 2010. http://papers.ssrn.com/sol3/papers.cfm?abstract_id=1144313.

BSA/IDC (Business Software Alliance and International Data Corporation). 2003. *The Economic Benefits of Reducing PC Software Piracy.* Washington, DC: BSA.

_____. 2009. *Sixth Annual BSA-IDC Global Software Piracy Study.* Washington, DC: BSA. http://global.bsa.org/globalpiracy2008/studies/globalpiracy2008.pdf.

_____. 2010a. *The Economic Benefits of Reducing PC Software Piracy.* Washington, DC: BSA. http://portal.bsa.org/piracyimpact2010/index.html.

_____. 2010b. *2009 Global Software Piracy Study.* Washington, DC: BSA.

CCID Consulting. 2008. "Review and Forecast of China's DVD Market in 2008." December 26. http://www.digitaltvnews.net/content/?p=6125.

Chinese State Intellectual Property Office. 2009. *2008 Survey on Chinese Software Piracy Rate.* http://www.chinaipr.gov.cn/news/government/262954.shtml.

Cisco Systems. 2009. *Cisco Visual Networking Index.* http://www.cisco.com/en/US/solutions/collateral/ns341/ns525/ns537/ns705/ns827/white_paper_c11-481360.pdf.

Congressional Budget Office. 2008. CBO Estimate: S.3325 *Enforcement of Intellectual Property Rights Act of 2008.* Washington, DC: CBO.

DiOrio, Carl. 2009. "Hollywood Studios' Trade Group Faces Leaner Budget." *Hollywood Reporter*, February 3.

Drahos, Peter, and John Braithwaite. 2007. *Information Feudalism: Who Owns the Knowledge Economy?* New York: New Press.

EFF (Electronic Frontier Foundation). 2008. *RIAA v. The People: Five Years Later.* http://www.eff.org/wp/riaa-v-people-years-later#4.

enigmax. 2009. "Pirate Bay Trial Day 8: Pirates Kill the Music Biz." *Torrent Freak* (blog), February 25. http://torrentfreak.com/the-pirate-bay-trial-day-8-090225/.

European Audiovisual Observatory. 2001–10. Focus: *World Film Market Trends.* Annual reports. Paris: Marché du Film.

Felton, Edward. 2010. "Census of Files Available via BitTorrent." *Freedom to Tinker* (blog), January 10. http://www.freedom-to-tinker.com/blog/felten/census-files-available-bittorrent.

Funarte. 2009. *Cultura em Numeros.* Brazil Ministry of Culture.

GAO (US Government Accountability Office). 2010. *Intellectual Property: Observations on Efforts to Quantify the Economic Effects of Counterfeit and Pirated Goods.* GAO-10-423. Washington, DC: GAO. http://www.gao.gov/new.items/d10423.pdf.

Goldstein, Paul. 1994. *Copyright's Highway: From Gutenberg to the Celestial Jukebox.* New York: Hill and Wang.

Ghazi, Koroush. 2009. "PC Game Piracy Examined." *TweakGuides*, June. Accessed August 15, 2009. http://www.tweakguides.com/Piracy_1.html.

Grice, Corey and Sandeep Junnarkar. 1998. "Gates, Buffett a Bit Bearish." *CNET News*, July 2. http://news.cnet.com/2100-1023-212942.html.

Hachman, Mark. 2010. "CSI Redmond: How Microsoft Tracks Down Pirates." *PCMag.com*, April 26. http://www.pcmag.com/article2/0,2817,2363041,00.asp.

Huygen, Annelies et al. 2009. "Ups and Downs: Economic and Cultural Effects of File Sharing on Music, Film and Games." TNO-rapport, February 18, TNO Information and Communication Technology, Delft.

Hu, Jim. 2002. "Music Sales Dip; Net Seen as Culprit." *CNET News*, April 16. http://news.cnet.com/Music-sales-dip-Net-seen-as-culprit/2100-1023_3-883761.html?tag=mncol;txt.

IDC (International Data Corporation). 2009. *Aid to Recovery: The Economic Impact of IT,* Software, and the Microsoft Ecosystem on the Economy. Framingham, MA: IDC.

IFPI (International Federation of the Phonographic Industry). 2001. *IFPI Music Piracy Report.* London: IFPI.

———. 2006. *The Recording Industry 2006 Piracy Report: Protecting Creativity in Music.* London: IFPI.

_____. 2009. Digital Music Report 2009. London: IFPI.

IIPA (International Intellectual Property Alliance). 2010a. *2010 Special 301 Report on Copyright Protection and Enforcement: India*. Washington, DC: IIPA.

_____. *Indonesia: 2010 Special 301 Report on Copyright Protection and Enforcement*. Washington, DC: IIPA. http://www.iipa.com/rbc/2010/2010SPEC301INDONESIA.pdf.

ISFE (Interactive Software Federation of Europe). 2009. "The Economics of Gaming." http://www.isfe-eu.org/index.php?PHPSESSID=d4f6r9d0oap3sdeulekc4ore92&oidit=T001:8ca835a1574ad46296a34393b4e28c57.

Johns, Adrian. 2010. *Piracy: The Intellectual Property Wars from Gutenberg to Gates*. Chicago, IL: University of Chicago Press.

Karaganis, Joe, and Robert Latham. 2005. *The Politics of Open Source Adoption*. Social Science Research Council, New York. http://wikis.ssrc.org/posa/index.php/Main_Page.

Katz, Ariel. 2005. "A Network Effects Perspective on Software Piracy." *University of Toronto Law Journal 55*.

Kirkpatrick, David. 2007. "How Microsoft Conquered China." *Fortune*, July 17. http://money.cnn.com/magazines/fortune/fortune_archive/2007/07/23/100134488/.

Kohi-Khandekar, Vanita. 2010. *The Indian Media Business*. Delhi: Sage Publications.

Kramer, Staci D. 2002. "Content's King." *Cable World,* April 29.

Lamer, Paul. 2006. "What's On Your iPod?" *Duke Listens!* (blog), May 22. http://blogs.sun.com/plamere/entry/what_s_on_your_ipod.

Larkin, Brian. 2004. "Degraded Images, Distorted Sounds: Nigerian Video and the Infrastructure of Piracy." *Public Culture* 16:289–314.

Liebowitz, Stan J. 2004. "The Elusive Symbiosis: The Impact of Radio on the Record Industry." *Review of Economic Research on Copyright Issues 1* (1). Accessed February 26, 2010. http://papers.ssrn.com/sol3/papers.cfm?abstract_id=1146196.

Linden, Greg. 2004. "China Standard Time: A Study in Strategic Industrial Policy." *Business and Politics 6*.

Lohr, Steve. 2004. "Software Group Enters Fray Over Proposed Piracy Law." *New York Times*, July 19. http://www.nytimes.com/2004/07/19/technology/19piracy.html.

Madden, Mary, and Lee Rainie. 2005. "Music and Video Downloading Moves Beyond P2P." Project memo, March, Pew Internet and American Life Project, Washington, DC. http://www.pewinternet.org/~/media//Files/Reports/2005/PIP_Filesharing_March05.pdf.pdf.

Manuel, Peter Lamarche. 1993. *Cassette Culture: Popular Music and Technology in North India*. Chicago, IL: University of Chicago Press.

Masnick, Mike. 2009. "Profitable 'Pay Us Or We'll Sue You For File Sharing' Scheme About To Send 30,000

More Letters." *Techdirt*, November 25. http://www.techdirt.com/articles/20091125/1047377088. shtml.

McIllwain, Jeffrey. 2005. "Intellectual Property Theft and Organized Crime: The Case of Film Piracy." *Trends in Organized Crime 8*:15–39.

Mondok, Matt. 2007. "Microsoft Executive: Pirating Software? Choose Microsoft!" *Ars Technica,* March 12. http://arstechnica.com/microsoft/news/2007/03/microsoft-executive-pirating-software-choose-microsoft.ars.

MPAA (Motion Picture Association of America). 2005. *The Cost of Movie Piracy*. Washington, DC: MPAA.

———. 2006. U.S. Theatrical Market Statistics. Washington, DC: MPAA.

———. 2009. Theatrical Market Statistics. Washington, DC: MPAA. http://www.mpaa.org/2008%20 MPAA%20Theatrical%20Market%20Statistics.pdf.

Mukasey, Michael. 2008. Speech at the Tech Museum of Innovation, San Jose, CA, March 28.

Nintendo. 2009. *Nintendo Anti-Piracy Training Manual*. Nintendo Anti-Piracy Program. ap.nintendo. com/_pdf/Nintendo_Antipiracy_Training_Manual.pdf.

Noble, Ronald K. 2003. "The Links Between Intellectual Property Crime and Terrorist Financing." Statement of the secretary general of Interpol to the US House Committee on International Relations, 108th Cong., July 16. http://www.interpol.int/public/ICPO/speeches/SG20030716. asp.

Oberholzer-Gee, Felix, and Koleman Strumpf. 2007. "The Effect of File Sharing on Record Sales: An Empirical Analysis." *Journal of Political Economy 115:*1–42.

———. 2009. "File-Sharing and Copyright." Working Paper 09-132, Harvard Business School, Boston, MA.

OECD (Organisation for Economic Co-operation and Development). 2007. *The Economic Impact of Counterfeiting and Piracy*. Paris: OECD. http://www.oecd.org/document/4/0,3343, en_2649_33703_40876868_1_1_1_1,00.html.

———. 2009. *Piracy of Digital Content*. Paris: OECD. http://www.oecd.org/document/35/0,3343, en_2649_34223_43394531_1_1_1_1,00.html.

O'Reilly, Tim. 2002. "Software Choice vs. Sincere Choice." O'Reilly on Lamp.com, September 27. http://www.oreillynet.com/pub/wlg/2066.

Peitz, M., and P. Waelbroeck. 2006. "Why the Music Industry May Gain from Free Downloading: The Role of Sampling." *International Journal of Industrial Organization 24*:907–13.

Piracy Is Not A Crime.com. 2006. "Terrorist Involvement." http://www.piracyisnotacrime.com/stats-terror.php.

Reuters. 2007. "Piracy Worked for Us, Romania President Tells Gates." *Washington Post,* February 1.

RIAA (Recording Industry of America). 2010. "Piracy: Online and on the Street." http://www.riaa. com/physicalpiracy.php.

Rose, Mark. 1993. Authors and Owners: *The Invention of Copyright.* Cambridge, MA: Harvard University Press.

Samuelson, Pamela, and Tara Wheatland. 2009. "Statutory Damages in Copyright Law: A Remedy in Need of Reform." *William & Mary Law Review.*

Sanchez, Julian. 2008. "750,000 Lost Jobs? The Dodgy Digits Behind the War on Piracy." *Ars Technica.* http://arstechnica.com/tech-policy/news/2008/10/dodgy-digits-behind-the-war-on-piracy.ars.

Sandoval, Greg. 2009. "Q&A: A Front-Row Seat for Media's Meltdown." *CNET News,* October 27. http://news.cnet.com/8301-31001_3-10383572-261.html.

_____. 2010. "Biden to File Sharers: 'Piracy Is Theft.'" *CNET News,* June 22. http://news.cnet. com/8301-31001_3-20008432-261.html.

Schulze, Hendrik, and Klaus Mochalski. 2009. *Internet Study 2008/2009.* Leipzig: ipoque.

Sherman, Cary. 2003. Statement of the RIAA president and general counsel before the US Senate Committee on the Judiciary, 108th Cong., September 9. http://judiciary.senate.gov/hearings/ testimony.cfm?id=902&wit_id=2562.

Siwek, Stephen. 2006. *The True Cost of Motion Picture Piracy to the U.S. Economy.* Lewisville, TX: Institute for Policy Innovation.

_____. 2007a. *The True Cost of Sound Recording Piracy to the U.S. Economy.* Lewisville, TX: Institute for Policy Innovation.

_____. 2007b. *The True Cost of Copyright Industry Piracy to the U.S. Economy.* Lewisville, TX: Institute for Policy Innovation.

_____. 2009. *Copyright Industries in the US Economy.* Washington, DC: IIPA.

_____. 2010. *Video Games in the 21st Century: The 2010 Report.* Washington, DC: Entertainment Software Association.

Smith, Michael, and Rahul Telang. 2009. "Competing with Free: The Impact of Movie Broadbases on DVD Sales and Internet Piracy." *MIS Quarterly 33:*321–38.

Squire, Jason E. 2004. *The Movie Business Book.* New York: Simon & Schuster.

Sundaram, Ravi. 2007. "Other Networks: Media Urbanism and the Culture of the Copy in South Asia." In *Structures of Participation in Digital Culture,* edited by Joe Karaganis. New York: Social Science Research Council.

Thallam, Satya. 2008. *The 2008 International Property Rights Index.* Washington, DC: Property Rights Alliance.

Treverton, Gregory F. et al. 2009. *Film Piracy, Organized Crime, and Terrorism.* Santa Monica, CA: RAND Corporation.

UNODC (United Nations Office on Drugs and Crime). 2009. "The Eleventh United Nations Survey of Crime Trends and Operations of Criminal Justice Systems: 2007-2008." http://www.unodc.org/unodc/en/data-and-analysis/crime_survey_eleventh.html.

USIBC (U.S.-India Business Council)/Ernst & Young. 2008. *The Effects of Counterfeiting and Piracy on India's Entertainment Industry.* Washington, DC: USIBC.

Valenti, Jack. 1982. Testimony of the MPAA president to the US House Committee on the Judiciary, Subcommittee on Courts, Civil Liberties, and the Administration of Justice, 97th Cong., April 12. http://cryptome.org/hrcw-hear.htm.

Vance, Ashlee. 2010. "Chasing Pirates: Inside Microsoft's War Room." *New York Times,* November 11. http://www.nytimes.com/2010/11/07/technology/07piracy.html.

Varian, H. 2004. *Copying and Copyright.* Berkeley: University of California.

Vershinen, Alexander. 2008. "Vzyali na ispoug." *Smart Money/Vedomasti,* February 18. http://www.vedomosti.ru/smartmoney/article/2008/02/18/4937.

Walt Disney Company. 2010. *Fiscal Year 2009 Annual Financial Report and Shareholder Letter.* http://amedia.disney.go.com/investorrelations/annual_reports/WDC-10kwrap-2009.pdf.

Wang, Shujen. 2003. *Framing Piracy: Globalization and Film Distribution in Greater China.* Lanham, MD: Rowman & Littlefield.

WAN-IFRA (World Association of Newspapers and News Publishers). 2008. *World Digital Media Trends.* Darmstadt: WAN-IFRA.

Willson, Kate. 2009. "Terrorism and Tobacco." *Tobacco Underground,* June 28. http://www.publicintegrity.org/investigations/tobacco/articles/entry/1441/.

Woodmansee, Martha, and Peter Jaszi. 1993. *The Construction of Authorship: Textual Appropriation in Law and Literature.* Durham, NC: Duke University Press.

Wu, Jia. 2010. *Global Video Game Market Forecast.* Boston: Strategy Analytics.

Zentner, Alejandro. 2006. "Measuring the Effect of File Sharing on Music Purchases." *Journal of Law & Economics 49*:63–90.

Zhang, Mia. 2008. "Internet Traffic Classification." Cooperative Association for Internet Data Analysis. http://www.caida.org/research/traffic-analysis/classification-overview/#discussion.

Chapter 2: Networked Governance and the USTR

Joe Karaganis and Sean Flynn

Contributors: Susan Sell, Parva Fattahi, and Mike Palmedo

Introduction

Intellectual property and enforcement policy was once a fairly narrow area of law and practice, administered by a handful of government agencies. Copyright and patent offices played a role, as did the customs office and a mix of national and municipal police agencies that targeted commercial infringers. Substantive responsibilities were set through domestic law and guided— at the international level—by voluntary treaties managed by the World Intellectual Property Organization (WIPO).

This situation has changed dramatically in the past quarter century. IP policy has become the subject of a proliferating array of international treaties and agreements, involving many more international actors, from the World Trade Organization (WTO), to the World Health Organization (WHO), to the European Commission, to—perhaps most prominently—the Office of the US Trade Representative (USTR), which has reshaped global IP policy through bilateral and multilateral trade agreements.

Enforcement efforts have become correspondingly complex. In the United States, domestic enforcement responsibilities are now shared among half a dozen major agencies, including the Departments of Commerce, Homeland Security, Justice, and State; the FBI (Federal Bureau of Investigation); and Customs and Border Protection (CPB)—as well as the copyright and patent offices. In Russia, responsibility is centered on the Ministry of the Interior but includes the Prosecutor's Office, the Federal Security Bureau, the Federal Customs Service, the Ministry of Culture, the Russian Cultural Protection Agency, the Federal Anti-Monopoly Bureau, and the Ministry of Commerce and Industry.

In Brazil, enforcement efforts cut across an array of decentralized police forces, including the Federal Police, the Federal Highway Police, state civil and military police, customs agents, and the Municipal Guards, as well as federal and state prosecutors. Coordination at the federal level is the responsibility of the National Council on Combating Piracy (CNCP), which pulls members from government ministries and the major industry associations. Some individual states and municipalities are also creating their own versions of the CNCP.

Inevitably, new agencies and initiatives have emerged to coordinate these growing enforcement networks. The United States recently launched its second such effort in the past decade, replacing the National Intellectual Property Law Enforcement Coordinating Council (NIPLECC) with a

Chapter Contents

75 Introduction

77 TRIPS and TRIPS+

80 Toward ACTA

84 The USTR and Special 301

85   The History of Special 301

86   Sanctions and Bilateral Agreements

87   The Warnings Regime

90 Participation and Influence

91   A Symbiotic Relationship

94   Foreign Countries

95   A Transitional Regime

97 About the Study

97 References

new "IP Czar" position in the White House. Russian enforcement has also gone through two major revisions in the period, first with the creation of the Governmental Committee for the Prosecution of Intellectual Property Violations, Its Legal Protection and Usage in 2002 (run by then vice–prime minister Dmitri Medvedev), and later with a major administrative reorganization in 2006. Brazil's CNCP was created in 2004, published its anti-piracy plan in 2005, and published a revamped plan in 2009.

All these coordinating bodies work closely with industry groups and ensure that industry plays a role in directing enforcement efforts. The industry groups, for their part, coordinate research, policy positions, and activism across the continuum of national and international venues in which they work, as well as with each other through higher-level coordinators such as the IIPA and BASCAP.[1] International organizations such as WIPO also provide connective tissue—both top-down in the form of treaties and bottom-up via technical training for lawyers, judges, customs officers, and other actors in the enforcement business. Money flows across these networks as international industry groups subsidize their local counterparts. The result is dense "networked governance," to use Peter Drahos' phrase, in which relevant policymaking and jurisdiction are spread across overlapping public institutions and corporate networks (Drahos and Braithwaite 2007).

This proliferation is the result of several decades of IP policy activism on the part of industry groups and aligned states—especially the United States, the European Union, and Japan (Sell 2003; Drahos and Braithwaite 2007). At the national level, it reflects a process of accumulation of public resources for the enforcement effort, marked

---

[1] Respectively, the International Intellectual Property Alliance, based in Washington, DC, and Business Action to Stop Counterfeiting and Piracy, a program of the International Chamber of Commerce.

by periodic attempts to organize and consolidate the resulting alphabet soup of participating agencies. Internationally, it reflects a process of "forum shopping" by industry representatives for venues in which stronger and more binding IP measures can be passed. Where existing venues prove insufficiently accommodating, as with WIPO and now the WTO, industry lobbyists have pushed for new institutions with less representation of the obstructing parties. Over time, protection ratchets up as new international agreements create pressure for changes in national law and changes in national law set new baselines for international agreements.

The history of IP policy over the last three decades is largely a history of such maneuvers, undertaken whenever international or domestic institutions proved unwilling to adopt stronger protection measures. The resulting policymaking process can be very difficult to follow as it moves between venues, and indeed this has been an important advantage for well-coordinated state and industry actors, who have the resources to manage—and exploit—this complexity.

TRIPS and TRIPS+

Arguably the defining example of such forum shopping was the 1994 Trade-Related Aspects of Intellectual Property Rights (or TRIPS) agreement, which culminated a decade-long effort to move responsibility for global IP norms out of WIPO and into the new WTO. Much of TRIPS deals with trademarks, patents, and other forms of industrial property and with the extension of protection to biotechnology, software, semiconductors, and industrial designs. With respect to copyright, TRIPS was first and foremost an exercise in harmonizing national law around the Berne Convention for the Protection of Literary and Artistic Works—the 1886 agreement that established international recognition for minimum copyright terms, author's and performance rights, the "automatic" establishment of copyright upon the creation

Acronyms and Abbreviations

ACTA	Anti-Counterfeiting Trade Agreement
BASCAP	Business Coalition to Stop Counterfeiting and Piracy
CNCP	(Brazil) National Council on Combating Piracy
FTA	free trade agreement
GATT	General Agreement on Tariffs and Trade
GSP	Generalized System of Preferences
IIPA	International Intellectual Property Alliance
IP	intellectual property
IPR	intellectual property rights
ISP	Internet service provider
IT	information technology
MPAA	Motion Picture Association of America
NAFTA	North American Free Trade Agreement
NET Act	(US) No Electronic Theft Act
NGO	non-governmental organization
P2P	peer-to-peer
PhRMA	Pharmaceutical Research and Manufacturers of America
RIAA	Recording Industry Association of America
SECURE	Standards to be Employed by Customs for Uniform Rights Enforcement
TRIPS	Agreement on Trade-Related Aspects of Intellectual Property Rights
USTR	Office of the United States Trade Representative
WCO	World Customs Organization
WHO	World Health Organization
WIPO	World Intellectual Property Organization
WTO	World Trade Organization

of a work (rather than via formal registration), and a variety of other features of modern copyright law. TRIPS was viewed as a more powerful instrument than Berne for universalizing those norms because, unlike WIPO, the WTO had a strong dispute resolution process, which could result in the loss of trade privileges when a member successfully pursued a complaint.

Developing countries had two expectations for TRIPS. First, they believed that the harmonization of IP rules would end long-standing disagreements with high-income countries over the appropriate levels of IP protection. Second, they believed that the dispute resolution process would end the strong-arming of developing countries through bilateral and regional trade negotiations—contexts in which the United States exercised obvious advantages (Sell 2003; Bayard and Elliot 1994). Both assumptions proved incorrect. Instead, TRIPS inaugurated a period of intense policy activism in other venues. WIPO, the administrator of the Berne Convention, immediately launched a round of negotiations to extend protection beyond the TRIPS baseline. These negotiations resulted in the Copyright Treaty (1996) and the Performance and Phonograms Treaty (1996)—collectively known as the Internet Treaties because of their provisions regarding the protection of digital works. The United States and the European Union also continued to negotiate regional and bilateral agreements that included higher levels of IP protection and, especially, stronger provisions for enforcement. These requirements are generally described as TRIPS+.

In contrast to its many formal provisions for protecting creative work, the Berne Convention is largely silent on the question of enforcement, specifying only that "infringing copies of a work shall be liable to seizure" in member countries (Art. 16). Here, TRIPS went significantly beyond Berne in specifying how norms should be enforced, notably by requiring that countries "provide for criminal procedures and penalties to be applied at least in cases of willful trademark counterfeiting or copyright piracy on a commercial scale" (Art. 61).

As with many TRIPS provisions, the language provides an explicit path toward stronger measures but also affords countries discretion in defining key terms, such as the meaning of "commercial scale" or the nature of the penalties "sufficient to provide a deterrent" to infringement. In the past two decades, the commercial scale standard, in particular, has become a source of considerable tension in trade negotiations. In Berne-era national copyright laws, variations on commercial scale provided the most widely used threshold for criminal liability in cases of infringement. Consistently, the standard evoked not the number of infringing copies per se but rather the for-profit purposes of the infringement, signaled through such phrases as "commercial advantage" (EU and US law) and "financial gain" (US) (Harms 2007). Such language traditionally exempted non-commercial and personal infringement from criminal liability—though not civil liability. As digital distribution allowed nonprofit and individual copying to scale to levels once reserved for commercial entities, the assumptions underlying the standard began to fray, and industry groups began to press for an expansion of criminal liability to all infringing acts.

Concerted industry pressure to eliminate the commercial scale threshold for criminal liability dates back to the 1980s, when computer companies started to take notice of non-commercial software hacking. In the United States, revisions to copyright law in the 1990s progressively expanded the scope for criminal liability and increased its associated penalties. This push culminated in the 1997 passage of the No Electronic Theft Act (NET Act), which effectively gutted the commercial scale standard by redefining financial gain to include "the receipt, or expectation of receipt, of anything of value, including the receipt of other copyrighted works" (Sec. 2a).[2] Under the new definition, all infringement is subject to criminal prosecution, with penalties of up to three years in prison for a first offense and six for a second. In practice—and despite continuing industry pressure—the law has been invoked only a handful of times, primarily against distributors of pirated software (and never against users of peer-to-peer services). In most high-income countries, including the United States, law enforcement agencies have been wary of expanding criminal prosecution in this direction and have instead favored the use of civil liability.[3] In most middle- and low-income countries, in contrast, criminal prosecution has become the norm (Harms 2007; Correa and Fink 2009).

The allocation of public resources is another area of explicitly national discretion in TRIPS, and another area that has come under significant pressure since its passage. As countries adopt broader criminal liability for infringement, demands by rights-holder groups for additional public resources to enforce the law have increased. As we discussed in chapter 1, the expansion of liability and the relatively modest growth in actual capacity to enforce have led to a divergence between the law and its application and to a related set of conflicts around TRIPS regarding the extent to which states must prioritize enforcement. Article 41.5 puts the matter clearly:

> Nothing in this Part creates any obligation with respect to the distribution
> of resources as between enforcement of intellectual property rights and the
> enforcement of law in general.

The passage reaffirms the right of states to set their own priorities regarding the use of public resources and recognizes that difficulties associated with IP enforcement are often inseparable from larger problems with judicial process and the rule of law. A related passage in Article 61 strikes a more complicated balance, affirming that "remedies available shall include imprisonment

---

2   The NET Act also calls into doubt the other traditional requirement for criminal liability: intention to infringe. On this and related issues see Bailey (2002).

3   Even in the civil arena, file sharing cases against individuals have been extremely rare. In the United States, civil suits against Jammie Thomas-Rasset and Joel Tenenbaum are the only examples to have gone to trial—both brought by the Recording Industry Association of America. In both cases, juries supported the RIAA's demands for extremely high penalties, leveling a $1.92 million fine on Thomas-Rasset for sharing twenty-four songs over a P2P network (later reduced to $1.5 million) and a $675,000 fine on Tenenbaum for thirty songs. In 2009, the US Department of Justice went on record that the Thomas-Rasset penalty was appropriate under the 1999 Digital Theft Deterrence and Copyright Damages Improvement Act, which had raised statutory penalties for infringement.

and/or monetary fines sufficient to provide a deterrent, consistently with the level of penalties applied for crimes of a corresponding gravity."

As usual, the language here is indirect enough to support a wide range of interpretations about the nature of member obligations—partly turning on the status of the "available" deterrent remedies. The main problem with this formulation from the rights-holder perspective is that copyright infringement has traditionally been categorized as a minor offense in the laws of most developing, IP-importing countries. Penalties for commercial infringement have generally been low, and evidentiary standards for conviction relatively high. Law enforcement has usually lacked ex officio authority to investigate infringement or make arrests, instead requiring complaints by rights holders or their representatives before taking action.[4] Most countries, in this context, have also resisted pressure to establish special courts or "fast-track" procedures for prosecuting copyright violations independent of wider civil and criminal procedures—a fact that has kept enforcement at the slow, often multi-year pace of other civil and criminal actions.

Pushing copyright infringement up the ladder of priorities for law enforcement has, consequently, been a top goal for industry groups and their government partners. TRIPS+ measures routinely include provisions for special IP courts, dedicated police and prosecutorial units, and lower evidentiary standards. Securing ex officio authority for police to open investigations, conduct raids, seize goods, and prosecute IPR (intellectual property rights) cases is a standard demand. Such provisions are key to transforming copyright infringement from a primarily civil matter into a criminal one, in which the burden of enforcement is borne by public agencies. Collectively, these measures make the principle that intellectual property is a private right rather than a public one—a core principle strongly reaffirmed in the TRIPS preamble—an all-but-dead letter.

Toward ACTA

In 2007, the United States filed the first WTO complaint regarding the enforcement practices of another country—China—for what it described as inadequate action against commercial vendors of pirated software. The case turned in part on a dispute over the meaning of the Chinese words for "large," "huge," and "severe" as applied to the vendor's activity—a strange linguistic quarrel widely viewed as an attempt by the United States to dictate the interpretation of commercial scale infringement. In January 2009, the WTO ruled in favor of Chinese discretion on this point, frustrating US hopes that it would override national prerogatives in resolving TRIPS ambiguities.[5]

The case also highlighted the broader difficulty of securing agreement among WTO members for stronger enforcement language in TRIPS. A bruising battle over access to medicines in the

---

4   This requirement explains the relatively common sight of police standing by in markets that openly sell pirated goods.

5   The ruling was a split decision on the two other points in the complaint, with China winning an important point about border-control compliance with TRIPS and the United States winning a less important point about the range of works accorded copyright protection (WTO 2009).

early 2000s had established that developed countries would resist any significant loosening of IP standards, even in the face of public health emergencies. In the following years, the split between developed and developing countries on these issues widened and moved beyond the WTO. By the middle of the decade, it was clear that stronger enforcement efforts faced serious and potentially insurmountable obstacles in all the broadly representative international bodies. WIPO emerged as the focal point for this resistance. In 2007, a coalition of developing countries (including Brazil, South Africa, and India) succeeded in pushing the Development Agenda through the WIPO General Assembly, which requires that new IP policy prioritize social and economic development goals.[6]

Industry plans to circumvent the emerging stalemate were laid in a series of meetings in 2004 and 2005—initially at the inaugural Global Congress on Combating Counterfeiting, sponsored by the Global Business Leaders' Alliance Against Counterfeiting[7] and hosted by Interpol and WIPO. These plans gained steam at the July 2005 G8 meeting, where Japanese representatives proposed the creation of a new enforcement regime to battle piracy and counterfeiting. They received a further push from the launch of BASCAP, a program of the International Chamber of Commerce, which has become a leading voice for more aggressive enforcement policy.

The first fruit of this effort emerged in 2006, when G8 members opened negotiations at the Brussels-based World Customs Organization (WCO) to strengthen international customs enforcement standards via a new agreement called SECURE (Standards to be Employed by Customs for Uniform Rights Enforcement). The WCO was viewed as a congenial forum for IP rights holders because corporations operated on equal footing with governments and did not have to publicly disclose their negotiations—or admit the participation of advocacy or consumer groups.

SECURE envisioned new responsibilities for customs agencies and included a number of "TRIPS-Plus-Plus" provisions that went beyond incremental extension of TRIPS protections. It expanded the scope of customs enforcement from its traditional point of application during the importation of goods to the full set of activities defining cross-border trade, including export, transit, warehousing, and transshipment. It diminished the obligations of rights holders to provide evidence of infringement in order to initiate a search or seizure, and it empowered customs authorities to impose deterrent penalties in the case of violation of any IPR laws—not just traditional issues of counterfeiting and piracy. Critics feared that the latter would create presumptions of guilt on complex issues that customs officials had little ability to adjudicate, such as patent infringement—an issue of particular concern to manufacturers of generic medicines— and violations of anti-circumvention laws governing technical protection measures (TPMs) used in electronics and software (Li 2008).

---

6   WIPO Development Agenda, http://www.wipo.int/ip-development/en/agenda/.

7   GBLAAC members include Coca-Cola, Chrysler, Pfizer, Proctor & Gamble, American Tobacco, Phillip Morris, Swiss Watch, Nike, and Canon (Shaw 2008).

Because the WCO, like the WTO, is a multilateral organization with representation from some 170 countries, it provided a forum for developing countries to raise concerns about these expanded powers and their costs of implementation. When it became clear in 2008 that this coalition was large enough to block the adoption of SECURE, the process was slowed and eventually abandoned in mid-2009.

SECURE was not the only enforcement initiative underway, however. Two weeks after WIPO's September 2007 adoption of the Development Agenda, US, European, and Japanese officials announced that they would negotiate an agreement to "set a new, higher benchmark for enforcement that countries can join on a voluntary basis" (USTR 2007). This proposal became the Anti-Counterfeiting Trade Agreement (ACTA), which was negotiated over three years and declared finalized in October 2010 (despite some obvious points on which it remained incomplete). The primary partners in the negotiation were the United States, the European Commission, Japan, Germany, Switzerland, Australia, Korea, Canada, New Zealand, Jordan, Morocco, Singapore, and the United Arab Emirates. Notably absent were the industrialized middle-income countries that have been the principal obstacles to stronger enforcement at WIPO and the WTO—and the principal targets of US and European enforcement pressure in the past decade.[8]

Although initially billed as an effort to strengthen anti-counterfeiting coordination among customs agencies, the various drafts leaked (and later, released) throughout 2009 and 2010 described a much broader agreement, designed to create higher TRIPS+ enforcement standards on a full spectrum of patent, trademark, and copyright issues. Some of the most controversial provisions dealt with Internet regulation, including provisions for strong secondary liability and US-style "notice and takedown" procedures for infringing content—measures that would give rights holders much more control over Internet service providers (ISPs) and other web services. Drafts also included strong anti-circumvention language intended to criminalize workarounds for technical protection measures used to protect digital content, such as the copy protection used on DVDs.

In the background of these discussions—and intermittently visible in the drafts—was the shift in industry attention from traditional commercial infringement to consumer-based infringement, conducted largely over the Internet. Although the draft language never mentioned consumers or individuals in this context, it created a framework for requiring ISPs and other web services to take whatever measures were deemed appropriate by rights holders to stop infringement. Much of the current debate about Internet enforcement—and the only specific measure mentioned in the drafts—concerns the implementation of "three-strikes" laws designed to punish consumers accused of multiple infringements. But ACTA's language is flexible regarding what constitutes

---

8   As Hisamitsu Arai, an advisor to Japanese prime minister Koizumi, reportedly argued to US Tokyo Embassy staff: "the intent of the [proposed ACTA] agreement is to address the IPR problems of nations such as China, Russia, and Brazil, not to negotiate the different interests of like-minded countries" (Wikileaks cable 06TOKYO4025; July 20, 2006).

adequate compliance and thereby sets the stage for upward revisions of the standard.[9]

The final draft released in October 2010[10] differs in some significant ways from earlier ones, and it will be some time before the implications of the specific changes in the language are fully understood. Accounts to date suggest that the United States abandoned much of the Internet agenda late in the negotiations, under pressure from European negotiators and in an effort to bring the process to a rapid close. Little of the strong language on secondary liability survives in the final draft. Also gone are the "notice and takedown" provisions designed to universalize the US standard on this issue. The surviving language on anti-circumvention has been watered down and now differs little from earlier WIPO treaties.

But in other respects, the finalized document takes a large step toward a more pervasive culture of copyright liability and enforcement. The final text notably broadens the definition of commercial scale infringement, with "direct or indirect economic or commercial advantage" (Sec. 4) superseding the narrower definitions and exceptions present in much national law. The fact that no one can say with certainty what "indirect economic advantage" means, in this context, is not an accident. Rather, it creates a framework in which nothing is clearly excluded from the criminal standard—a position much closer to the US NET Act assertion that the receipt of anything of value constitutes financial gain.

Elsewhere, the text endorses (and may mandate) the use of statutory penalties for infringement, as opposed to the common practice, outside the United States, of allowing judges discretion over damages. It endorses the use of any "reasonable" rights-holder-submitted claims of damages, specifically including the retail price of the good (despite the fact that all the copyright industry groups have abandoned the use of retail price in describing losses).

To the extent that there is flexibility on these issues, it lies in the convoluted language of the text, which permits multiple interpretations but which also sets up a familiar dynamic in which power relations between states will determine whose interpretations matter. Evidence of concern for the exceptions and limitations that make copyright a broader social compact between creators and users, in contrast, is completely absent.

Other aspects of ACTA have also generated controversy—notably the secrecy of the negotiations, which while broken on several occasions by leaks, was lifted only briefly in April 2010 with the release of a public draft.[11] The method of adoption is also controversial. In the United States,

---

9   The RIAA and the MPAA (Motion Picture Association of America) have previewed their next steps in comments to the US "IP Czar," Victoria Espinel in 2010. These include the development of new international norms requiring preemptive content-filtering by ISPs and the inclusion of home-based monitoring software (AFTRA et al. 2010).

10 The consolidated draft text is available on the European Commission website: http://ec.europa.eu/trade/creating-opportunities/trade-topics/intellectual-property/anti-counterfeiting/.

11 This secrecy extended to the elected officials of governments involved in the negotiations and has produced some striking internal tensions. In March 2010, the European Parliament voted to demand disclosure of ACTA documents from the European Commission, which is a party to the negotiations. The Canadian government—also a party to the negotiations—has advocated for fuller disclosure. A US

the Obama administration has signaled that it would treat ACTA as a "sole executive agreement" rather than a treaty—a very unusual maneuver for an agreement of this scope, intended to sidestep the usual requirement of Senate ratification and the resulting public debate (Lessig and Goldsmith 2010).[12] Internationally, the battle lines are beginning to be drawn, with India—not a party to the negotiations—signaling that the proper venue for enforcement discussions is the WTO.

India's efforts to bring the WTO to bear highlight the most important aspect of ACTA: not the specific measures in the treaty, but rather the emergence of a new institution designed to co-opt the existing international framework for enforcement. Even in its watered-down form, ACTA consolidates a variety of TRIPS+ measures into a new baseline for IP enforcement. And as with TRIPS, this will almost certainly not be the end of the story. ACTA will provide a venue for further ratcheting up of protection and enforcement measures. If the past is any indicator, the job of building ACTA+ will fall primarily to the USTR.

The USTR and Special 301

It is hard to overstate the importance of the Office of the US Trade Representative to the process of forum shopping and leveling-up of international norms. The USTR's annual Special 301 reports are the stick in the US carrot-and-stick approach to international IP policy. The reports weigh countries' compliance with IP and enforcement standards—both those in existing treaties and those the United States would like to see adopted. They convey recommendations for changes in domestic law and signal US conditions for accession to international agreements, such as the WTO. They threaten and reward countries with inclusion on or delisting from the annual "Watch List" and "Priority Watch List" and signal US intentions to pursue sanctions or other measures when demands are not met.

US copyright industries and the USTR have, in key respects, a symbiotic relationship. The IIPA was instrumental in the creation of the Special 301 process, and annual IIPA country submissions furnish the primary and often only evidence on copyright issues cited in the Special 301 reports. In all but a few cases in any given year, the USTR closely follows IIPA recommendations in assigning countries to the watch lists. In 2008, the USTR accepted forty-six of the IIPA's fifty-four recommendations (84%). In 2010, it accepted all the Priority Watch List recommendations and twenty-one of twenty-four for the Watch List (an acceptance rate of 91%). For the most part, IIPA findings and recommendations simply pass through into USTR reporting.[13]

Senator petitioned the USTR for more public information. Appeals for transparency were made by the governments of Spain, France, and Sweden.

12   Sole executive agreements are considered minor agreements that do not require congressional ratification or even a presidential signature. Several scholars, including Lessig and Goldsmith, have argued that the use of this model for ACTA is unconstitutional because the setting of intellectual property policy is clearly defined as a legislative function, not an executive one. See also Flynn (2010).

13   The Pharmaceutical Research and Manufacturers of America (PhRMA) is the second largest submitter.

This close relationship is not an accident. The USTR was created in 1974 to explicitly strengthen the ties between industry and government in trade negotiations. Its mandate was revised repeatedly in the 1970s and 1980s to make the USTR more responsive to business needs and revised further to ensure that it would not be limited or constrained by the provisions of existing trade agreements, such as the GATT and later the WTO. While other countries assumed that the multilateral dispute resolution process of the WTO would make unilateral USTR efforts redundant, in practice, the role of Special 301 expanded throughout the last two decades.

The History of Special 301

Special 301 builds on an earlier trade policy mechanism known as Section 301, established via the US Trade Act of 1974. Section 301 was created to address the lack of effective enforcement tools in existing international trade agreements—notably the General Agreement on Tariffs and Trade (or GATT), the set of global trade rules that preceded the WTO. Section 301 authorized the president to take economic measures against countries that "burden or restrict United States commerce," including the suspension of trade agreements, the imposition of tariffs or restrictions on imported goods, and the withdrawal of special trade benefits for developing countries (known as the Generalized System of Preferences, or GSP).[14]

Section 301 findings are, by definition, unilateral findings by the United States and subject to its own standards. At the time, this meant that foreign practices and policies did not have to contravene the GATT (or later, the WTO) to be found "unreasonable." Nor did it require the United States to take into account a country's level of economic development in determining what was fair or unfair—a sharp departure from GATT rules that favored differential treatment for developing countries.

Section 301 was strongly supported by US exporters, who wanted greater access to foreign markets during a period of widespread foreign protection of domestic industries.[15] Initially, a wide range of exporting industries participated in these efforts—especially automobile manufacturers and electronics companies worried by the rise in Japanese exports in the 1970s. But the initiative soon shifted to IP-based industries—drug companies, semiconductor manufacturers, and the entertainment industry especially—which increasingly viewed IP protection as the key to expansion into global markets. The push for a stronger international regime for intellectual property rights soon became the main front in this effort.

The USTR has historically adopted around 75% of PhRMA's recommendations.

14   GSP programs provide exemptions for developing countries from the equal treatment requirements of the WTO, generally in the form of more favorable tariffs or other terms of trade. The US and the EU maintain GSP programs with most developing countries.

15   For a general history of this process and its eventual outcome in the WTO, see Sell (2003), Harris (2006, 2008), and Drahos and Braithwaite (2007). On GATT enforcement rules, see Lowenfeld (2002).

Section 301 was amended several times in the next decade in response to industry lobbying. An initial amendment in 1979 transferred its functions from the Department of Commerce to the Office of the USTR and increased its responsiveness to private-industry complaints. A second amendment in 1984 established "adequate and effective protection of intellectual property rights" as grounds for 301 investigation and sanctions. This change reflected the growing coordination and assertiveness of IP industry lobbying efforts—also signaled that year by the founding of the IIPA.

At the urging of the IIPA, and especially Jack Valenti of the MPAA, the Section 301 statute was amended again in 1988 to create the new IP-focused Special 301 reporting and sanctions process. Under Special 301, the USTR is required to identify foreign countries that "deny adequate and effective protection of intellectual property rights" or "fair and equitable market access to United States persons that rely upon intellectual property protection"[16] and subject them to expedited investigation. These requirements resulted in the well-known Watch List and Priority Watch List, which serve as warning mechanisms to countries out of compliance with the USTR's preferences on IP policy. A third status, Priority Foreign Country (PFC), represents a final warning stage. PFC designation triggers a thirty-day countdown during which the targeted country must "enter into good faith negotiations" or "make significant progress in bilateral or multilateral negotiations" or else face sanctions.[17]

Sanctions and Bilateral Agreements

The bilateral approach to building a stronger global IP regime came together in the mid-1980s. After the 1984 revision to the US Trade Act, the Reagan administration made quick use of Section 301's new IP provisions to launch investigations of Korea and Brazil—both countries with histories of domestic-industry protection.

Actions against Brazil were designed to end Brazilian protection of its domestic pharmaceutical and computer sectors. A 1985 case targeted the lack of copyright protection in Brazil for computer software—an innovation adopted in US law only in 1980. Brazil gave in to US demands, and the case was settled without sanctions in 1988. A second action in 1987 targeted Brazil's distinction between pharmaceutical processes (which were accorded patent protection) and final pharmaceutical products (which were not). This distinction had been widely and, under international law prior to TRIPS, legally employed to encourage the "reverse engineering" of important drugs and, relatedly, the development of local pharmaceutical industries. Because the public health benefits of the Brazilian position were clear and popular and the domestic commercial interests more entrenched than in the IT field, the Brazilian government refused to amend its law. As the dispute escalated in 1988 and 1989, the United States imposed duties on imported Brazilian goods, worth

---

16   19 USC 2242(a).

17   19 USC 2242(b)(1).

some $39 million.[18] Brazil responded with a suit under the GATT, challenging the legality of the retaliation. The United States, in turn, blocked the formation of a dispute settlement panel, making adjudication of the complaint impossible. Sanctions were eventually lifted in 1990 when the new Brazilian president, anticipating fuller patent protection requirements in TRIPS, agreed to adopt pharmaceutical end-product patents.

The 1985 case against Korea, also primarily on pharmaceutical patents, established what one negotiator described as a "blueprint" for the resolution of Special 301 disputes: bilateral treaties, or side agreements, that committed the targeted country to higher levels of patent and copyright protection (Drahos and Braithwaite 2007:103; US/Korea 1986).

In both cases, USTR actions eventually led to the passage of stronger Korean and Brazilian IP laws designed to bring the countries into closer compliance with US wishes. From the US industry perspective, these outcomes validated the 301 process and encouraged efforts to pass the still-stronger Special 301 provisions in 1988.

The strategic dimension of these actions grew more explicit in the late 1980s as the Uruguay Round of GATT negotiations neared its conclusion and set the stage for a new international trade agreement—the eventual WTO. Developing countries, led by India and Brazil, supported the strengthening of existing provisions on counterfeiting but opposed the inclusion of broader IP rules in the form of TRIPS. Such inclusion had no precedent in earlier trade agreements and duplicated existing international forums such as WIPO, which already managed a wide range of copyright and patent treaties.

The United States placed five of the ten "hard-liners" opposing TRIPS in the first Special 301 Report in 1989—Brazil, India, Argentina, Yugoslavia, and Egypt. Two years later, India, China, and Thailand became the first Priority Foreign Countries, triggering Section 301 investigations. Brazil lost its GSP benefits in 1988, Thailand in 1989, and India in 1992—all on matters related to pharmaceutical patents. US pressure, combined with assurances that TRIPS would end such unilateral action, eventually broke the anti-TRIPS coalition.

The Warnings Regime

With the passage of TRIPS, the USTR secured most of the IP goals it had pursued in the 1980s. But success did not end the program. Instead, Congress amended the trade statute in 1994 to specify that even countries fully compliant with TRIPS might lack "adequate and effective" IP protection. The amended statute authorized the use of Special 301 to promote IP and enforcement policy beyond what was required by TRIPS. The USTR quickly took up the task.

---

18   The number reflected a Section 301 requirement that sanctions "be devised so as to affect goods or services of the foreign country in an amount that is equivalent in value to the burden or restriction being imposed by that country on United States commerce."

Figure 2.1 Special 301: Number of Cited Countries

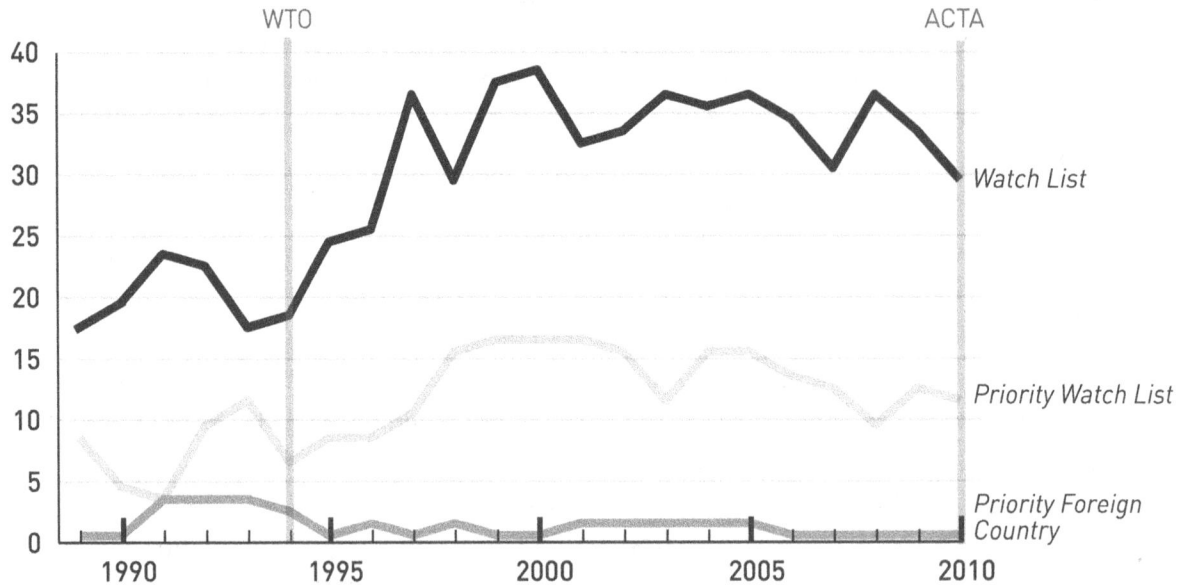

Source: Authors.

Figure 2.1 shows the number of countries placed on the US watch lists since the creation of Special 301 in 1989. As it suggests, the launch of the WTO had a powerful effect on USTR strategy: after 1994, the use of sanctions dropped off sharply, and the Special 301 process became predominantly a surveillance and warnings regime.

This change reflected concerns about the legality of Special 301 within the WTO framework. The WTO had remedied the GATT's deficiencies on enforcement by allowing dispute panels to be formed without the consent of both parties and by establishing a strong sanctions mechanism for members found in violation of WTO rules. The resulting Dispute Settlement Understanding[19] required WTO members to use the prescribed process to settle any questions about TRIPS compliance.[20] In 1999, a dispute settlement panel initiated by the European Commission reviewed the use of Section 301 in non-IP cases and held that the United States could not use the process to impose trade sanctions outside the dispute settlement process (WTO 1999). The ruling was taken as a signal that sanctions imposed under Special 301 regarding conduct covered by TRIPS would themselves be in violation of the WTO agreement. The USTR's subsequent actions appear

---

19   WTO Understanding on Rules and Procedures Governing the Settlement of Disputes, http://www.wto. org/english/tratop_e/dispu_e/dsu_e.htm.

20   Article 23.2 states: "Members shall not make a determination to the effect that a violation has occurred, that benefits have been nullified or impaired or that the attainment of any objective of the covered agreements has been impeded, except through recourse to dispute settlement in accordance with the rules and procedures of this understanding."

to reflect this concern. Since 1994, the USTR has initiated sanctions through Special 301 only once against a WTO member—Argentina in 1997, for alleged violation of pharmaceutical patents. Argentina capitulated quickly, and the United States avoided a direct challenge to the legality of the Special 301 program.

There is also some basis in WTO jurisprudence for seeing the watch lists as a violation of WTO rules. In the same 1999 dispute settlement, the WTO panel explained that the "threat alone" of unilateral sanctions outside the dispute settlement process risks undermining the basic principle of WTO legitimacy, the "equal protection of both large and small":

> Members faced with a threat of unilateral action, especially when it emanates from an economically powerful Member, may in effect be forced to give in to the demands imposed by the Member exerting the threat . . . To put it differently, merely carrying a big stick is, in many cases, as effective a means to having one's way as actually using the stick. The threat alone of conduct prohibited by the WTO would enable the Member concerned to exert undue leverage on other Members. It would disrupt the very stability and equilibrium which multilateral dispute resolution was meant to foster and consequently establish, namely equal protection of both large and small, powerful and less powerful Members through the consistent application of a set of rules and procedures." (WTO 1999: Para. 7.89)

Many WTO observers interpreted the ruling as a shot across the bow of the US watch lists, but the matter has never been pursued.

In the WTO era, the punitive power of Special 301 has consequently become more indirect. Watch list status still signals US displeasure, but that displeasure no longer leads to sanctions for WTO members. The United States has other ways to advance its positions in trade disputes, including through the WTO dispute settlement process itself, of which it is the most frequent user.[21] Due to requirements that aspiring WTO members negotiate accession agreements with major trading partners, entry into the WTO has also become a bottleneck where Special 301 demands are brought to bear. This has been the case, notably, for Russia and several other post-Soviet republics that have adopted TRIPS+ standards in an effort to secure US approval for accession.[22]

---

21   Of the WTO's 402 dispute settlements to date, the US is the complainant in 92 cases. The EU is second with 80 disputes initiated. Among other lead filers: Canada, 31; Brazil, 23; Mexico, 17; India, 16; Korea, 13; China, 6 (since joining in 2001).

22   Adoption of the WIPO Internet Treaties has been a key benchmark of compliance with USTR demands. After twelve years on the Priority Watch List, Russia signed the Internet Treaties in February 2009. Kazakhstan (Watch List, 2000–2005) signed the treaties in 2004, Azerbaijan (Watch List 2000–2005) in 2006. The Ukraine, which was sanctioned for copyright piracy between 2001 and 2005, signed in 2002 (and acceded to the WTO in 2008).

Free trade agreements are also frequently part of the settlement process around Special 301 disputes, and US officials have acknowledged that inclusion in the annual Special 301 report can hinge on a country's attitude toward such negotiations. FTAs almost always include IP obligations stricter than those found in TRIPS, including—in the copyright realm—accession to the WIPO Internet Treaties and strengthened enforcement procedures by police, courts, and border officials. Signing an FTA, however, does not ensure a free pass from the USTR. Israel, Canada, Mexico, and Chile have all maintained their places on the watch lists after signing FTAs, including on grounds of poor implementation of those agreements. Chile was moved from the Watch List to the Priority Watch List years after signing its FTA with the United States.[23]

Participation and Influence

The USTR has direct ties to industry through various advisory committees. The Industrial Functional Advisory Committee on Intellectual Property Rights for Trade Policy Matters (or IFAC-3) plays the leading role where Special 301 is concerned and includes the IIPA, PhRMA, Time Warner, the RIAA, and a long list of other companies and industry organizations.[24] Such formal linkages are complemented by the long-standing revolving door between the USTR and its industry clients, which creates a reward system for USTR officials who cater to industry requests. The USTR and other federal agencies with IP and enforcement responsibilities have been regular way stations for lucrative industry careers in the past three decades—and, importantly, bipartisan way stations.

The list of former senior USTR officials working for organizations that lobby the USTR is a long one and includes Harvey Bale (now at the International Federation of Pharmaceutical Manufacturers and Associations—IFPMA), Joe Papovich (now at the RIAA), and former head Mickey Kantor (currently at PhRMA). The Copyright Office also has its share of influential alumni, including former general counsel for policy and international affairs Shira Perlmutter (now at the IFPI) and former liaison to the ACTA negotiations Steven Tepp (now at the US Chamber of Commerce). The revolving door extends to the Department of Justice and DC law firms that represent media and technology companies and includes an assortment of other politically connected former office holders, including Dan Glickman, former secretary of agriculture and head of the MPAA until 2010, and Bruce Lehman, former head of the US Patent and Trademark

---

23   Israel signed an FTA in 1985 but has made frequent Watch List (1997, 2003, 2004) and Priority Watch List (1998–2002; 2005–9) appearances. Canada signed the North American Free Trade Agreement (NAFTA) in 1994 but has maintained a place on the Watch List every year since 1989 (with the exception of 1994) and graduated to the Priority Watch List in 2009. Mexico has been on the watch lists repeatedly since signing NAFTA (1999; 2003–9), mostly in regard to film piracy. The US-Chile FTA became effective in January 2004, and Chile has spent some fifteen years on the Watch List (1994–2006) and Priority Watch List (2007–9).

24   The full list is available on the USTR website: http://ustraderep.gov/Who_We_Are/List_of_USTR_Advisory_Committees.html.

Office and current director of the International Institute for Intellectual Property, an industry-supported think tank.[25]

Until recently, there was little pressure for greater participation or procedural transparency in relation to Special 301. The institutional culture discouraged it; the most obvious affected parties—other countries—had no meaningful standing, and the traditional obscurity of trade policy sheltered it from the public attention directed at policymaking bodies in related areas, such as the Federal Communications Commission. The legal status of Special 301 reinforces these tendencies. The Special 301 process is an "informal adjudication" as opposed to a formal adjudication or rulemaking process. As described by the US Administrative Procedure Act, adjudication is a technical determination of rights and responsibilities based on existing rules and past conduct, whereas rulemaking is forward looking. In our view, this distinction misses the primary function of Special 301 as an instrument for pushing foreign and American IP commitments beyond existing obligations without the inconvenience of a strong public comment process (as required in rulemaking) or a structured adversarial process (as required in formal adjudication).

Informality plays an important part here and creates considerable leeway with respect to procedures. Notably, informal adjudications do not have to be "on the record after the opportunity for an agency hearing." Even this lax requirement has been ignored for most of Special 301's history: the first hearing took place in February 2010.

Nonetheless, the term has been subject to a variety of legal interpretations regarding what constitutes due process in such contexts, with strong consensus in the courts that "a minimum procedure must include at least some form of notice and an opportunity to be heard at a meaningful time and in a meaningful manner."[26] In our view, the Special 301 process has been out of compliance with a reasonable understanding of this standard. Minimal and still inadequate notice was made possible only in 2008. The first meaningful opportunity to be heard was the hearing in 2010.

A Symbiotic Relationship

Despite this obscurity, the USTR has to meet certain basic requirements to justify its findings, including acting on the basis of evidence collected during the Special 301 process. With some fifty to sixty countries placed annually on the watch lists, the research requirements of the Special 301 process are considerable. The USTR's role in this process was never clearly defined by statute and quickly defaulted to industry, which ramped up its research capacities throughout the 1990s

---

25   For a doubtless incomplete list, see "The Revolving Door," IP Enforcement Database, https://sites.google.com/site/iipenforcement/the-revolving-door.

26   32 Fed. Prac. & Proc. Judicial Review § 8136 (1st ed.), stating: "Generally, all informal adjudications have some form of three elements—notice, some opportunity to participate and reasons." See also 32 Fed. Prac. & Proc. Judicial Review § 8201 (1st ed.), stating: "Several courts have said that a minimum procedure must include at least some form of notice and an opportunity to be heard at a meaningful time and in a meaningful manner."

to meet the new demand. This division of labor quickly became reflected in the USTR's internal organization: in 2009, only eight of the roughly two hundred USTR staff worked on IP issues. Most of the findings, legal recommendations, and country detail discussed in the Special 301 reports simply recapitulate IIPA (and other industry) work. For nearly two decades, the IIPA and the USTR have been the research and policy sides, respectively, of a larger collective enterprise.[27]

The Special 301 process begins each year with a public comment period designed to gather information for the report. This is, in principle, a fact-finding exercise that takes into account "any information . . . as may be available to the Trade Representative and . . . as may be submitted . . . by interested persons."[28] Interested persons can include other countries, non-US industry groups, non-governmental organizations (NGOs), and—in principle—individuals. In practice, it has overwhelmingly meant US industry. The USTR's interest in hearing from other parties has generally been viewed as negligible, and this perception was reinforced by the unusual restrictions on the comment process itself. Until 2008, all comments from all parties were due on the same day—a requirement that made the notification of countries about complaints and same-year replies to them impossible. Under these circumstances, only a handful of countries (and typically no civil society groups) bothered to submit comments at all, and the few that did generally responded to the previous year's comments.

Under new rules that went into effect in 2008, countries (but not NGOs or other parties) were permitted two additional weeks to submit comments after industry submissions were received. This small opening had a dramatic effect on participation (see table 2.1): the number of countries submitting comments jumped from three to twenty-four. In 2010, the country participation held steady, but the number of individual and nonprofit comments exploded, following efforts by legal advocacy and public interest groups to draw attention to ACTA and to IP policymaking more generally.

Table 2.1 Special 301 Comments

	2007	2008	2009	2010
Companies and Industry Groups	21	19	30	37
Countries on Previous Watch Lists	4	3	24	25

---

27   The handful of cases in which the USTR has departed from IIPA recommendations (in 2009, Sweden, Nigeria, Kazakhstan, Lithuania, and Brunei) are suggestive of the somewhat broader field of political inputs that affect USTR decisions, including geopolitical goals, conflicting industry requests, and other factors shaping bilateral relationships. In South Africa, for example, the controversial dispute over patent protection for AIDS medicines in the late 1990s cast a shadow over subsequent IIPA requests for South African inclusion on the watch lists. The USTR ignored these requests between 2000 and 2006, and South Africa has since fallen off the list of countries targeted by the IIPA.

28   19 USC 2242(b)(2)(B).

	2007	2008	2009	2010
Individuals	0	2	1	441
Nonprofits	1	0	0	26

Source: Authors.

USTR sensitivity about the Special 301 process has grown in recent years as the scope of trade negotiations has expanded and their public profile has risen. Like other government agencies, the USTR is also subject to new requirements to adopt higher evidentiary standards and more transparency about the research it uses in policymaking. Much of this pressure, ironically, originated with industry groups looking for tools to head off unwanted regulatory action resulting from federally funded scientific research. This is the background, notably, of the 2000 Data Quality Act, which established procedures for complainants to challenge data used in policymaking.[29] While many view the act as a victory of lobbying over science, the interesting question for agencies like the USTR is what the act implies in contexts where there is no scientific research culture to undermine.

In 2005, the Office of Management and Budget issued an interpretation of the Data Quality Act that requires peer review whenever the federal government disseminates "scientific information [that has] a clear and substantial impact on important public policies or private sector decisions" worth more than $500 million (OMB 2005). The OMB did not limit this requirement to the natural sciences—in fact, it specifically included economic and other policy-relevant research. It noted further that an adversarial comment process, in which contending parties submit and challenge each other's comments, is not an adequate substitute for peer review. When the Department of Commerce implemented the OMB directive in 2006, it placed emphasis on "transparency—and ultimately reproducibility" as the crucial standard in policy research and clarified that transparency "is a matter of showing how you got the results you got" (Department of Commerce 2006).

The de facto outsourcing of research to the IIPA and other industry groups allows the USTR to exempt Special 301 from such quality-control efforts. Nothing in the Data Quality Act or the OMB bulletin addresses transparency requirements for privately produced research or discusses how to improve policymaking processes that depend entirely on it. The absence of hearings or a reasonably structured comment process ensures, further, that Special 301 undershoots even the lower evidentiary standards of an adversarial process, in which the stakeholders comment and respond to each other. The USTR does, nonetheless, set two modest requirements for submitted comments. It specifies that (1) comments should "provide all necessary information for assessing the effect of the acts, policies, and practices"; and (2) "any comments that include quantitative loss

---

29   The brainchild of tobacco-industry lobbyists, the Data Quality Act has been used to challenge federally funded research on a variety of health and environmental issues, from the effects of exposure to pesticides like Atrazine to studies of animal habitats used to restrict logging permits on federal land.

claims should be accompanied by the methodology used in calculating such estimated losses."[30] As we have argued in chapter 1, by any reasonable standard, these requirements go unmet.

None of this is particularly surprising given the history and goals of Special 301. The program has very ably performed its mission of translating US industry positions into trade policy. But as trade negotiations assume greater importance in national and international politics, standards of procedural fairness and credibility can and should change. Although the USTR bears no direct responsibility for industry claims, it does have statutory responsibility for the information it presents as factual, and it can discount or reject material that fails to meet its own evidentiary standards. Although peer review is difficult to reconcile with third-party comment submission, the USTR could do much more to ensure a credible and—in our view—more effective policymaking process. These steps would start with (1) taking its own evidentiary requirements seriously; (2) creating a more dynamic and open comment process; and (3) building more diverse representation into the layer of advisory and coordinating committees that set the USTR's agenda, including consumer groups.

Foreign Countries

The spike in country comments that began in 2009 was also marked by a perceptible change in tone. Traditionally, foreign countries have been deferential in their dialogues with the USTR—often highly so. Country comments typically catalogue the actions taken in the previous year to meet American wishes and on that basis request removal from the watch lists. Local policy and enforcement activities in targeted countries also often follow the seasonal rhythm of the Special 301 process, as governments seek to head off placement on the lists.

Occasionally, countries have filed more pointed objections to USTR claims and the industry research underlying them. In 1992, Italy challenged the $224 million estimate of losses by the MPAA to pirated movie cassettes—focusing in particular on the assumption that pirated video cassette sales represented a one-to-one loss with respect to ticket sales (Drahos and Brathwaite 2007:97). But such comments and, especially, research-related comments have been rare. Countries have ignored, acquiesced to, or tried to finesse the Special 301 process. They have seldom contested it.

There are signs that this politics of avoidance is beginning to change. Country comments from 2009 and 2010 include a number of unusually blunt rebuttals, including criticism of the Special 301 process, of IIPA claims, and of USTR complaints about policies that comply with TRIPS.[31] Most of these comments take note of the lack of consistency in the evidence and standards that underpin the various warnings. Israel—on the Priority Watch List in 2008 and 2009—responded sharply in 2009 to IIPA and USTR criticism of its recently revised copyright law. In considerable

---

30   19 USC 2242(b)(2)(B).

31   The comments can be found at regulations.gov: http://www.regulations.gov/search/Regs/home. html#home (Dockett ID # USTR-2009-0001).

detail, it objected to the unilateralism of US demands that it go beyond existing international obligations on issues such as the copyright term for sound recordings, the scope of fair use provisions, the legal protection accorded technical protection measures (anti-circumvention), takedown procedures for ISPs, the liability of end-users of pirated software, compensation for the accidental seizure of licit goods, and much of the rest of the TRIPS+ playbook (Israel 2009). Turkey, which appeared on the Priority Watch List from 2004 to 2007 and on the Watch List in 2008, 2009, and 2010, offered similar criticism of US unilateralism in reference to ongoing disputes over pharmaceutical patents.[32]

Spain, which the IIPA described in 2009 as having "the worst per capita Internet piracy problem in Europe and one of the worst overall Internet piracy rates in the world," also made an active rebuttal of IIPA claims, arguing that "numerous assertions in the report are not based at all on data contained in the report or on coherent arguments" (Jordan 2009). Drawing on its own consumer-survey data, the Spanish government challenged the rates of music piracy cited in the IIPA report, drew attention to gaps in the IIPA data, underscored its own solid ranking in the Business Software Alliance's piracy reports, and reminded the USTR of its commitment to enforcement through its participation in the ACTA negotiations.

A Transitional Regime

For our part, the details or even accuracy of the recent country rebuttals are less interesting than what they suggest about the evolution of the Special 301 process. The more accessible comment window and the obvious inclination of countries to use it marks a step toward openness and accountability of a kind the USTR has generally avoided since its beginning.[33]

These small steps also bring into relief the tensions in what appears to be a transitional moment in the global IP policy and enforcement regime. Since the inauguration of the WTO in 1994, the USTR has operated in a space of ambiguous legality and soft power—able to threaten countries but largely unable to make good on those threats for fear of generating an adverse WTO ruling. The stability of this position, in our view, was the product of a number of factors, including the industry's virtual monopoly on the evidentiary discourse around piracy; the disorganization of developing-country coalitions on IP policy; and the general obscurity of copyright and enforcement issues, which allowed IP policymaking to fly under the radar of most consumer and public interest groups. Where all these factors held true six or seven years ago, it is difficult to make a strong case for any of them today. Industry research has been delegitimized by its opacity and the excesses of

---

32   "There is no rule obligating the members to apply patent linkage in the TRIPS Agreement . . . [A] patent linkage process as stipulated in the US legislation is not a global rule and the lack of such a linkage cannot be interpreted as a weakness in protecting IPR" (Turkey 2009).

33   There is also some circumstantial evidence that the growth in participation in 2010 influenced that year's Special 301 report. Notably, USTR fidelity to PhRMA recommendations fell from 75% in 2009 to around 60% in 2010, following a major push for delisting countries by health advocates.

its advocacy campaigns, developing countries are more organized and assertive with regard to IP policy, and enforcement has begun its "consumer turn" toward measures that are likely to make obscure IP policy venues like the USTR much more visible and controversial in the public eye. A more transparent USTR may be the only possible way forward.

The waning of this fifteen-year Special 301 interregnum is also visible in the USTR's leadership of the current round of IP enforcement forum shopping. Signs of openness at the agency come at a moment when some of the USTR's major accomplishments have been folded into ACTA. ACTA's eventual jurisdiction is unclear at this point and may remain so long after the agreement is ratified—if it is ratified. At a minimum, however, ACTA seems likely to undercut the WTO's preeminence in enforcement and push the bilateral and multilateral regimes back into closer alignment, at least temporarily. Special 301 is unlikely to disappear in this context. More likely, it will become a mechanism for pressuring other countries to adopt the new multilateral regime, and eventually for new ACTA+ policies. The continuation of this state of affairs would, in our view, be a mistake. In an era in which trade and IP agreements shape basic questions of social welfare, from health to taxes to broader prospects for economic growth, the process needs sunlight, wider participation, and greater legitimacy. The USTR and Special 301 are too powerful to remain insiders' games.

About the Study

This chapter draws on work by Joe Karaganis, Sean Flynn, and Susan Sell. Mike Palmedo and Parva Fattahi provided valuable research assistance.

References

AFTRA (American Federation of Television and Radio Artists), Directors Guild of America (DGA), International Alliance of Theatrical and Stage Employees (IATSE), Motion Picture Association of America (MPAA), National Music Publishers' Association (NMPA), Recording Industry Association of America (RIAA), and Screen Actor's Guild (SAG). 2010. Letter to the Intellectual Property Enforcement Coordinator (IPEC) in response to request for written submissions, March 24.

Bailey, Aaron. 2002. "A Nation of Felons? Napster, the NET Act, and the Criminal Prosecution of File-Sharing." *American University Law Review 51*.

Bayard, Thomas O., and Kimberly Ann Elliot. 1994. *Reciprocity and Retaliation in US Trade Policy*. Washington, DC: Peterson Institute for International Economics.

Berne Convention. 1886. Berne Convention for the Protection of Literary and Artistic Works. http://www.wipo.int/treaties/en/ip/berne/trtdocs_wo001.html#P192_37445.

Correa, Carlos, and Carsten Fink. 2009. "The Global Debate on the Enforcement of Intellectual Property Rights and Developing Countries." Issue Paper No. 22, Programme on Intellectual Property Rights and Sustainable Development, International Centre for Trade and Sustainable Development, Geneva, February. http://ictsd.net/i/publications/42762/.

Department of Commerce. 2006. "Department of Commerce: Information Quality Guidelines." http://ocio.os.doc.gov/ITPolicyandPrograms/Information_Quality/dev01_003914.

Drahos, Peter, and John Braithwaite. 2007. *Information Feudalism: Who Owns the Knowledge Economy?* New York: New Press.

Flynn, Sean. 2010. "Over 75 Law Profs Call for Halt of ACTA." American University College of Law, Program on Information Justice and Intellectual Property. http://www.wcl.american.edu/pijip/go/blog-post/over-75-law-profs-call-for-halt-of-acta.

Harms, Louis. 2007. The Enforcement of Intellectual Property Rights by Means of Criminal Sanctions: An Assessment. Geneva: WIPO Advisory Committee on Enforcement. http://www.wipo.int/edocs/mdocs/enforcement/en/wipo_ace_4/wipo_ace_4_3.pdf.

Harris, Donald. 2006. "Carrying a Good Joke Too Far: TRIPS and Treaties of Adhesion." *Journal of International Law 27* (3): 681–755.

———. 2008. "The Honeymoon is Over: The US-China WTO Intellectual Property Complaint." *Fordham International Law Journal 32* (1): 96–187.

IIPA (International Intellectual Property Alliance). 2009. *2009 Special 301 Report on Copyright Infringement and Enforcement*: Spain. Washington, DC: IIPA. http://www.iipa.com/

rbc/2009/2009SPEC301INDIA.pdf.

Israel. 2009. "2009 Submission of the Government of Israel to the United States Trade Representative with Respect to the 2009 'Special 301 Review.'" http://www.regulations.gov/search/Regs/home.html#documentDetail?R=09000064808e9bc5.

Jordan, Carmen. 2009. "Special 301 2009 Review: Comments from the Spanish Government." http://www.regulations.gov/search/Regs/home.html#documentDetail?D=USTR-2009-0001-0050.1.

Lessig, Lawrence, and Jack Goldsmith. 2010. "Anti-Counterfeiting Agreement Raises Constitutional Concerns." Washington Post, March 26.

Li, Xuan. 2008. "SECURE: A Critical Analysis and Call for Action." South Bulletin, no. 15.

Lowenfeld, Andreas F. 2002. International Economic Law. London: Oxford University Press.

OMB (Office of Management and Budget). 2005. "Final Information Quality Bulletin for Peer Review." Federal Register 2664.

Sell, Susan K. 2003. Private Power, *Public Law: The Globalization of Intellectual Property Rights*. Cambridge, UK: Cambridge University Press.

Shaw, Aaron. 2008. "The Problem with the Anti-Counterfeiting Trade Agreement (and What to Do About It)." *Knowledge Ecology Studies 2*.

Turkey. 2009. "2009 Submission of the Government of Turkey to the United States Trade Representative with Respect to the 2009 'Special 301 Review.'"

US/Korea. 1986. "US-Korea Intellectual Property Rights and Insurance Understandings (1985-1986)." Bilaterals.org. http://www.bilaterals.org/spip.php?article388.

USTR (Office of the US Trade Representative). 2007. "Schwab Announces U.S. Will Seek New Trade Agreement to Fight Fakes." News release, October 23. http://www.ustr.gov/ambassador-schwab-announces-us-will-seek-new-trade-agreement-fight-fakes.

WTO (World Trade Organization). 1999. "DS152: European Communities v. US, Sections 301–310 of the Trade Act 1974." Dispute Settlements. http://www.wto.org/english/tratop_e/dispu_e/cases_e/ds152_e.htm.

———. 2009. DS362: China—Measures Affecting the Protection and Enforcement of Intellectual Property Rights. Geneva: WTO. http://www.wto.org/english/news_e/news09_e/362r_e.htm.

Chapter 3: South Africa

Natasha Primo and Libby Lloyd

Contributors: Natalie Brown, Adam Haupt, Tanja Bosch, Julian Jonker, and Nixon Kariithi

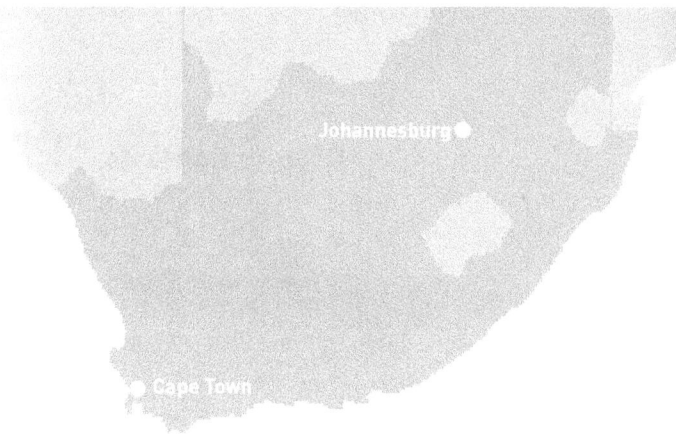

Introduction

As in many other countries, media piracy in South Africa is shaped by poverty and social inequality. Low incomes—some one-third of the population lives on less than $US1 a day—high media prices, and a pervasive advertising culture create high demand for media goods but very limited legal access for the great majority of South Africans. Inevitably, pirated cassettes, books, discs, and now digital formats fill the gap.

Although this dynamic is commonplace in low- and middle-income countries, piracy in South Africa is also the product of a distinctive history of repression, political contestation, and diplomatic tension, reaching back to the apartheid era. The cultural economy of South Africa under apartheid was marked by illicit flows of many kinds, including books, video cassettes, and audio cassettes. The economic boycotts of South Africa in the 1980s and early 1990s made cultural goods expensive and often unavailable, leading to widespread and widely tolerated copying—perhaps most prominently of school textbooks. Government censorship and book bans made illegal copying an act of political resistance and gave rise to an array of clandestine distribution networks that enabled the circulation of dissident views. Apartheid's restrictions on the movement of blacks and the geographical concentration of services in white communities further skewed media access, ensuring that the majority black population had almost no access to legal cultural markets. Vastly unequal purchasing power between blacks and whites meant that geographical barriers to access were, in most cases, redundant.

Sixteen years after the country's first democratic elections, the formal restrictions on movement are gone, but the racial and economic geography of media access remains largely unchanged. Movie theatres, bookstores, and music retailers continue to be located almost exclusively in the (formerly whites-only) suburbs, while most black South Africans still work, live, and seek entertainment in townships situated at the peripheries of major cities. Large chains dominate the market, supplanting the older array of independent theatres and retailers. There are, today, less than a hundred movie complexes in a country of forty-seven million, with all but a handful located in expensive shopping malls and districts.

Trends in media consumption over the past decade are mixed. Although the global market for music CDs peaked in 2004, the South African market continued to expand through 2007—reaching $126 million before falling slightly in 2008 and 2009. The percentage of the population attending the cinema at least once a year has dropped slightly, from 9.7% in 2001 to 8.7% in 2008—led by

Chapter Contents

99   Introduction

101   The Medicines Dispute and Its Shadow

103   Enforcement Autopilot

105   Sectoral Effects

106   Books

107   Software

108   Music

109   Games

111   Movies

114   Piracy and South African Film

117   Piracy on the Internet

120   Attitudes toward Piracy

120   Piracy Stories

122   Software Piracy in Music Production

124   Law and Enforcement

125   TRIPS Compliance and Enforcement Legislation

126   Anti-piracy under the Counterfeit Goods Act and the Copyright Act

127   Judicial Pushback

129   Public-Private Justice

131   How Piracy Works in South Africa

132   Optical Disc Piracy

133   Flea Markets and Street Vendors

134   Market Differentiators

135   Bruma

136   Noord/Plein

137   Cultural Nationalism

137   The Deformalization of the Trade

138   Hanover Park

138   Neighborhood Vending

139   Consumption

140   What People Watch

141   Conclusion

144   About the Study

144   References

a sharp decline in attendance by whites (SAARF 2008).[1] This drop is usually attributed to rising ticket prices, which average 35 rand ($5) despite a brief price war from 2005 to 2007, and to the growth in the use of DVD players, which surged from 3% of total households in 2003 to 48.8% by the end of 2008 (Euromonitor International 2009). Much of the resulting demand for DVDs is served by pirate suppliers—up to 80% according to the industry group South African Federation Against Copyright Theft (SAFACT). Of the remaining legal market, rentals make up nearly 50% of revenues, leaving a very small home-video retail market.

The end of apartheid and economic sanctions in the mid-1990s produced a rapid flow of cultural goods into the South African marketplace, including movies, books, audio and video cassettes, and music CDs. The high prices and underdeveloped retail sector for these goods, however, meant that existing grey- and black-market practices for acquiring, copying, and circulating media retained their place in South African life—especially in poor communities. Textbook piracy remained ubiquitous and, according to the International Intellectual Property Alliance (IIPA), was responsible for larger total losses than either film or music piracy throughout the 1990s and early 2000s. As a global trade in pirated cassettes and discs emerged in the 1990s, South Africa became both a destination market and a transit point for CD and, later, DVD smuggling into other African countries. Industry accounts attribute much of this traffic to disc production in Southeast Asia—especially Malaysia—but South Asian networks also played a role as Pakistani immigrants began to cater to, and cultivate, South African tastes for Bollywood films and music.

---

1   Among the major racial groups, only blacks showed an increase in attendance, from 4.2% to 5.2% (OMD South Africa 2009, 2002). This trend may be changing, with the major chains reporting increased attendance figures for 2009, driven by Hollywood blockbusters like *Avatar* and *Iron Man*.

Despite the large informal economy, claims of losses to piracy in South Africa were never very high. The IIPA—calculating only losses to US companies—put the figure at $129 million in 2000, of which two-thirds was attributed to business software. Brazil, Mexico, and Russia, by comparison, regularly approached $1 billion in the late 1990s in the same reports, representing two to four times the per capita loss. Nonetheless, South Africa loomed very large in wider debates about the intellectual property (IP) obligations of middle-income and low-income countries under the Agreement on Trade-Related Aspects of Intellectual Property Rights (TRIPS), which sets minimum standards for IP protection for members of the World Trade Organization (WTO).[2]

The Medicines Dispute and Its Shadow

In the late 1990s, South Africa faced a massive HIV/AIDS crisis, with infection rates approaching 20% of the adult population. New antiretroviral drug "cocktails" had proven effective in controlling the disease but were inaccessible to the vast majority of South Africans: the standard treatments cost, on average, $12,000 per patient per year. In 1997, President Nelson Mandela signed the South African Medicines and Medical Devices Regulatory Act, which legalized compulsory licensing for HIV/AIDS drugs and—more importantly—the parallel importation of drugs from lower-priced sources, notably India. Although many observers viewed the act as compliant with relatively vague TRIPS

---

2   South Africa faced the challenges of compliance earlier than many other comparable countries because it agreed to comply with TRIPS as a developed country—a status based on its leadership of the Southern African Customs Union (SACU), which the WTO classified as a developed-country region. As a result, South Africa began to implement TRIPS in 1995. India, in contrast, negotiated a ten-year TRIPS implementation schedule that gave them until 2005 to comply. Signatories identified by the UN as Least Developed Countries (LDCs), such as Lesotho, also a member of the SACU, have until 2016.

Acronyms and Abbreviations

BACSA	Business Against Crime South Africa
BSA	Business Software Alliance
CGA	Counterfeit Goods Act
DTI	Department of Trade and Industry
ECT Act	Electronic Communications and Transactions Act
ESA	Entertainment Software Association
FIFA	Fédération Internationale de Football Association (International Federation of Association Football)
IFPI	International Federation of the Phonographic Industry
IIPA	International Intellectual Property Alliance
IP	intellectual property
ISP	Internet service provider
ISPA	Internet Service Providers' Association
P2P	peer-to-peer
MPAA	Motion Picture Association of America
PICC/ SABDC	Print Industries Cluster Council/ South African Book Development Council
RIAA	Recording Industry Association of America
RiSA	Recording Industry of South Africa
SACU	Southern African Customs Union
SAFACT	Southern African Federation Against Copyright Theft
SAPS	South African Police Service
SARS	South African Revenue Service (including the South African Customs Administration)
TRIPS	Agreement on Trade-Related Aspects of Intellectual Property Rights
USAID	United States Agency for International Development
USTR	Office of the United States Trade Representative
WIPO	World Intellectual Property Organization
WTO	World Trade Organization

provisions on parallel importation, it set the Mandela government on a collision course with US pharmaceutical companies, which filed suit in South African court to overturn the law.[3]

The US government took the side of the pharmaceutical companies and brought diplomatic pressure to bear. In 1999, the Office of the US Trade Representative (USTR) placed South Africa on its Special 301 "Priority Watch List" of countries that do not provide "adequate and effective" protection of intellectual property rights, signaling strong disapproval and raising the threat of sanctions (Sell 2003). The subsequent public backlash from groups such as Oxfam and Médecins Sans Frontières, however, proved damaging to the pharmaceutical companies and embarrassing to the Clinton administration, leading to pressure for a settlement.

The conflict was defused later that year when South Africa and the United States came to a written "understanding" regarding intellectual property protection, in which South Africa backed off threats of parallel importation in return for promises of more favorable arrangements with pharmaceutical companies and an end to US trade and diplomatic pressure. The pharmaceutical companies, for their part, dropped their lawsuit. The agreement also set in motion broader discussions at the WTO about strengthening access to medicines in poor countries facing public health crises—a debate framed initially by the Doha Declaration on the TRIPS Agreement and Public Health in 2001 and later provisionally resolved in 2003 by the "Medicines Decision," which established parallel importation rules to combat health emergencies.

For much of the next decade, the medicines dispute and its resolution overshadowed other IP conflicts with South Africa. Throughout the early 2000s, the IIPA admonished South Africa for its enforcement failures in the copyright area—notably in the context of the flood of pirated CDs and DVDs that were arriving on the market. The IIPA recommended that South Africa be placed on the Special 301 Watch List in 2001and 2002 and the Priority Watch List in 2003. It made "special mention" of South African enforcement problems in 2005, 2006, and 2007. Wary of further conflict, the USTR refused to implement these recommendations, making South Africa the main exception in recent years to the IIPA's otherwise considerable influence on the Special 301 process.[4]

Throughout the period, the IIPA's major concern was optical disc piracy. In 2003, it reported a leap in DVD piracy from 10–15% to 30–35% of the home-video market. The IIPA also reported that South Africa had become "one of the world's largest breeding grounds for DVD retail piracy"— an improbable statement given that claimed losses never exceeded $35 million in the period.

Despite IIPA efforts to push South Africa onto the watch lists, the country figured only intermittently in industry statistical reporting. As in other countries, the Business Software Alliance (BSA) was the most reliable engine of piracy statistics, though it is unclear whether South

---

3   For a summary of the dispute and the larger public health context, see Fisher and Rigamonti (2005).

4   In 2008, the USTR accepted 84% of the IIPA's recommendations. In 2009, it accepted 91%. For more on this, see chapter 2.

Africa's software piracy rate was determined from direct samples or extrapolated from regional estimates. The Motion Picture Association of America (MPAA), for its part, did not include South Africa in its 2005 survey due to the insignificant size of the local movie market, which hovered around $50 million per year. The Entertainment Software Association (ESA) conducted several consumer surveys in the early 2000s but stopped in 2003. The Recording Industry Association of America (RIAA) has made periodic reports to the IIPA that draw on estimates provided by its local counterpart RiSA (the Recording Industry of South Africa) but has never claimed more than $8.5 million in losses in a given year.

In the absence of statistical inputs, the IIPA's South Africa reports relied heavily on reports of seizures and police activity, on local industry groups' accounts, and on a more or less stock array of criticisms of South African law and institutional arrangements. Few of the IIPA's complaints, consequently, were unique to South Africa. Through 2007 it continued to criticize the prevalence of book and business software piracy. It reiterated the need to address delays in the court system and the financial burden placed on plaintiffs in cases involving the seizure of goods.[5] It objected to the scope of "fair dealing" provisions in South African copyright law, the lack of criminal penalties for "end-user piracy" of software by businesses, the burden of proof on complainants to demonstrate ownership in cases against alleged copyright infringement, and various failures to "cure remaining TRIPS deficiencies"—particularly around the implementation of South Africa's 1997 Counterfeit Goods Act. The 2007 IIPA report included this warning:

> The impact of piracy in South Africa is devastating for legitimate right holders, legitimate distributors, and retail businesses (sale as well as rental), so much so that local copyright owners are mobilizing to take a stand against piracy. Legitimate distributors have reduced employment levels, some rental outlets have reported year-on-year decreases in business in the region of 30%, and many rental outlets have actually closed. (IIPA 2007)

Enforcement Autopilot

Much of this rhetoric is the IIPA on autopilot. By 2008, the level of industry complaints had fallen below the IIPA's radar, and no new reports had been filed. Local industry groups present a generally positive view of the enforcement effort and government cooperation, especially regarding efforts to suppress the optical disc trade. Connections between the Department of Trade and Industry (DTI), local and national police, and industry groups have strengthened over the past decade, and

---

5   Notably, the obligation of the complainant to pay the storage bill for seized goods. Such obligations are a common feature of TRIPS-era customs law and are designed to protect the rights and property of the accused.

the DTI in particular has become a powerful coordinator of enforcement activities and advocate for legislative change.

The 2010 World Cup, which South Africa hosted in June and July, provided the rationale for a further enforcement push, resulting in a boom in arrests and convictions of street vendors. Much of this effort took place in the context of the wider emphasis on public order and safety surrounding the event. But it also reflected agreements with FIFA and other corporate sponsors to protect World Cup–related merchandising from the inevitable knockoffs and copies. The South African Revenue Service (SARS), which includes the customs agency, significantly increased the number of raids on areas where counterfeit sports apparel were traded and created a new unit specifically to coordinate enforcement actions with FIFA. Other enforcement agencies also stepped up their activities, putting vendors of pirate DVDs and CDs under new pressure.

Despite this uptick in government involvement, anti-piracy efforts in South Africa remain, in key respects, a US-directed enterprise, only partially taken up by local networks of cultural producers and law enforcement agencies—with the DTI a strong standout in this regard. Although there has been some mobilization on the part of South African musicians, and although US-industry proxies, like SAFACT, also represent South African film studios and publishers, enforcement discourse and policy at a national level continue to be driven by US- and multinational-funded industry associations. The evidentiary discourse is similarly one-sided. Industry reports remain the primary—and in some cases the only—documents in the South African conversation on IP policy and enforcement.

By most official accounts, South Africa has become an enforcement success story. Industry reports have documented the decline of piracy on the street and in major flea markets, especially in the main urban centers. Cooperation between law enforcement and industry is strong, and agencies like the DTI have become activists within the government for still stronger enforcement policy and practices. Our work broadly confirms this ramp up, notably against the more organized forms of pirate vending in flea markets.

We were unable to gauge, however, whether these enforcement efforts have had any impact on the overall availability of pirated media. The weight of our evidence suggests not. Street vending networks, for instance, show considerable resilience in the face of police pressure. In our surveys of several major flea markets described as recently raided or shut down, pirated goods were widely available. And as in other countries where police pressure has been brought to bear against the organized street vendor sector, we found abundant signs of the deformalization and geographical dispersion of the trade, especially—in the South African case—into low-income areas, such as the black townships. These forms of distribution are organized predominantly around neighborhood networks and house-to-house vending, which makes them very difficult to police. Geographical differences also play an important role in enforcement, with a strong concentration of police activity in the major media markets of Johannesburg and Cape Town and less attention to piracy in more remote areas.

As elsewhere, the real problem for enforcement in South Africa is the proliferation of distribution channels. To date, South African rights holders have benefited from an unusual grace period with respect to Internet piracy. Prior to 2010, connectivity in and out of the country was limited to a single undersea cable running down the west coast of Africa (the SAT3 cable), resulting in very limited bandwidth and high prices. According to the Organisation for Economic Co-operation and Development, in 2008, South Africa had only 1.35 million broadband connections, representing a 2.8% penetration rate (Muller 2009).[6] Quality of service has been poor, and low-end broadband services have generally included sharp bandwidth caps. The uptake of peer-to-peer (P2P) services in South Africa has consequently been limited, and the few notable efforts to create local P2P services have been aggressively blocked by the recording industry association, RiSA.

Over the next few years, however, this bandwidth shortage is expected to end as new undersea cables come into service. With computers and other digital storage and playback technologies also becoming more widely available, South Africa is likely to play rapid catch-up in the global digital-media economy—in both its licit and illicit forms.

The balance of this chapter examines these dynamics in more detail, with a focus on the relationship between licit and illicit media markets in South Africa, attitudes toward piracy, the legal framework for and practices of enforcement, and the complex interactions between inequality, de facto segregation, high prices, and an increasingly globalized media environment that shape the social organization of piracy. Like the other contributions to this report, this chapter explores these issues along two primary axes: (1) the role of piracy within different media sectors, including the markets for books, movies, music, and software; and (2) the relationship of different consuming publics to piracy.

We take up the second question through a series of snapshots of the organization of piracy within different South African communities, including a survey of software use among musicians, a social geography of South African pirate markets, and a study of pirate media practices in Hanover Park, a low-income township outside Cape Town.

Sectoral Effects

Collectively, these snapshots portray an incomplete process of media globalization, in which millions of South Africans have been integrated into a globalized media culture without a corresponding expansion of access. As elsewhere in this report, the "problem" of piracy in South Africa is also the problem of high-priced, anemic legal markets, which open the door to cheap, convenient, illicit alternatives. This is not a uniform dilemma, however. It varies considerably across the copyright sector, as do its consequences, modes of enforcement, and possible solutions.

---

6   The roughly comparable US rate was 65% (NTIA 2010).

Books

In the late 1990s and early 2000s, book piracy in South Africa was a primary concern of the IIPA and domestic publishing industry groups, with the latter represented primarily by the Print Industries Cluster Council (PICC, now the South African Book Development Council—SABDC[7]), consisting of publishers, retailers, and other stakeholders in the book value chain. In 2001, the IIPA argued that "at least 30–50% of text[books] used countrywide are pirate photocopies." In 2004, the PICC observed that:

> copyright infringement in South Africa is not a matter—at least not yet—of the
> mass piracy of trade books, like the pirated editions of Harry Potter titles that
> have appeared internationally, but of systematic copying of various kinds in the
> educational sector, public sector and businesses. (PICC 2004:55–56)

Book piracy in South Africa is a legacy of the academic and cultural boycotts of the apartheid era, when large-scale copying of academic texts was condoned on university campuses (PICC 2004:55–56). Photocopying provided access to educational materials that were otherwise unaffordable due to the combination of academic sanctions, which restricted access to overseas publications, and economic sanctions, which resulted in a poor exchange rate for the South African rand that doubled or tripled prices at the local level (Haricombe and Lancaster 1995:89). State censorship also played a role. The numerous books and articles banned by the apartheid government circulated widely via photocopies and private desktop publishing (Berger 2002:532). For many opponents of the government, book copying was an act of political opposition rather than a crime.

Reported levels of book piracy dropped precipitously in the early 2000s, from $21 million in 2000 to $2–3 million by 2006. This was attributed to greater "copyright consciousness" among educational institutions—especially universities, which began to assert more control over the copying of course materials. IIPA reporting nonetheless kept the issue alive in its Special 301 submissions through 2007.

Local partners, such as the PICC/SABDC, provide statistics on book piracy to the IIPA, but the PICC/SABDC does not undertake systematic data collection comparable to the consumer surveys of other industry groups. The PICC's data on South African book piracy is composed of estimates made by "local representatives and consultants" (PICC 2004). In such cases, local affiliates—book publishers and vendors—are asked to provide estimates of the scope of piracy as a percentage of the total market. Such "supply-side" methodologies have a strongly subjective dimension and have been replaced by consumer surveys as the primary research methods of other industry groups. As

---

7   PICC became the SABDC in 2007. The SABDC remains a notable South African example of public-private partnership in the enforcement area, as it is jointly funded by the Department of Arts and Culture and industry stakeholders.

the practice of book copying shifts from copy machines to digital files, the number of distribution channels and the scale of the resulting collections are likely to explode, rendering both methods obsolete (see chapter 1).

From an enforcement perspective, book piracy is difficult to prosecute: students and academic staff are generally shielded from liability by exemptions in the Copyright Law that allow copying for personal use or study (at least with respect to text and images).[8] Copy shops do have liability as commercial ventures but can be hard to prosecute because they provide an on-demand service and do not hold stock. While the copying of course packs by lecturers is being brought under tighter university control, we see no evidence of a wider impact on book copying nor any research that would provide a reasonable measure of its scope. The days of course-pack and book photocopying, in any event, are clearly numbered. The sharing of large-scale digital libraries among students promises to dwarf the practice, fueled by the coming wave of cheap digital readers.

Software

Of all the industry groups, the Business Software Alliance is arguably the most entrepreneurial in its approach to reporting and local enforcement. In South Africa, as elsewhere, this role includes software-licensing audits of businesses, which frequently result in large monetary settlements. It includes the use of monetary incentives for informers, who in theory can earn up to R100,000 ($13,000) for information that results in the successful settlement or prosecution of a company using unlicensed software (though we are aware of no actual collections of that amount). For cases related to the retailing of pirated software, the BSA has sought penalties under the 1997 Counterfeit Goods Act (CGA), which specifies up to a R5,000 ($700) fine per illegal copy and/or imprisonment for up to three years for a first conviction (with penalties scaling upward for subsequent convictions).

After reporting steady decreases in software piracy between 1997 and 2002, the BSA changed its methodology in 2003 to include Microsoft Windows and a variety of consumer applications. It registered a fourfold increase in losses to $119 million for that year. For 2008, the BSA claimed some $335 million in losses—an order of magnitude larger than any other industry estimate.

Through 2009, the BSA's loss claims were based on a formula equating pirated copies with lost sales. After years of criticism of this position, the BSA in 2010 dropped its references to losses in favor of a more general assertion about the "commercial value" of unlicensed copies. This was long overdue: high software prices in South Africa combined with the availability of open-source alternatives in many software categories always made this rough equivalence implausible. As we discuss in chapter 1, software piracy plays an important role in the price-discrimination and market-building strategies of the major vendors. Vendors weigh the strict enforcement of licenses

---

8   Section 12(1)(a) of South Africa's Copyright Act provides an exception for literary and artistic works for, inter alia, personal use and study provided that it is consistent with "fair dealing." Current law does not appear to provide a basis for extending fair dealing to film and sound recordings.

against the very real possibility of large-scale open-source software adoption if end-user costs rise. This prospect is made more likely in South Africa by the government's very favorable open-source policy, adopted in 2007, which tilts public procurement decisions toward open-source solutions "unless proprietary software proves to be significantly superior" (DPSA 2009). In practice, the major software vendors step very carefully around these issues and do not make full use of their abilities to enforce licenses.

In contrast to its claims of losses, BSA-reported rates of software piracy in South Africa have held nearly steady since 2002 at around 35% (after falling dramatically in the late 1990s and early 2000s). A 1% uptick in 2008 generated a number of BSA-sourced stories about a rise in piracy, but as we discussed in the first chapter of this report, we are skeptical that the BSA can reliably measure trends at this level of detail. Minor year-to-year changes are likely to be statistical noise, outweighed by uncertainty at other levels, such as the difficulty of measuring the size of the open-source market or the number of computers being used in the country.

The BSA numbers do nonetheless tell a compelling, if different, story. At 35%, South Africa has one of the lowest reported business software piracy rates in the developing world. Russia, according to the same reports, had a 67% piracy rate in 2009, Mexico 60%, and Brazil 56%. South Africa's rate was also lower than many European countries: Greece, for example, had a 58% rate, Italy 49%, Spain 42%, and France 40% (BSA/IDC 2010). In other African countries, software piracy rates routinely exceed 80%.

Still, BSA representatives continue to press home the anti-piracy message. According to former BSA-South Africa chairperson Stephan Le Roux: "Software piracy is rampant in [South Africa] and found across all business sectors, including financial services, technology and manufacturing companies. This is having an impact on their efficiency and data security and ultimately on our economy" (Mabuza 2007). Even the global economic downturn becomes a rationale for expensive software licensing in this context. As current BSA chair Alaistar de Wet creatively rationalized: "In these uncertain economic times it is vital that companies do not skip corners and use unlicensed software, as this would increase the detrimental impact on those businesses and consumers as well as the local and global economy" (Manners 2009). This is the BSA on autopilot.

Music

The South African music market is tiny, registering $120 million in wholesale revenues in 2009, with a roughly equal split between international product and local repertoire from South Africa's vibrant music scene (IFPI 2010). It is also an unusual market in that the transition to CDs happened slowly, reflecting the impact of the country's sharp socioeconomic and racial divide on consumer technology adoption. As a result, South Africa had a sizable cassette tape market through the early 2000s, built on the legacy infrastructure of cheap radio/cassette players. According to RiSA, the South African affiliate of the International Federation of the Phonographic Industry (IFPI), cassette sales accounted for R35 million ($5 million) as recently as 2007—down 75% from their

1998 level but still representing a significant distribution channel in poor and rural communities because of their low price (Durbach 2008). CD sales, for their part, continued to grow through 2007, providing a rare exception to the global decline of the format since 2004 (IFPI 2010).

According to the most recent numbers on South Africa reported by the RIAA (from 2006), US losses represent a tiny percentage of the music market. RIAA claimed $8 million in losses in 2004 and $8.5 million in 2005. RIAA representatives note that these losses are not equivalent to retail sales but, rather, are discounted to reflect a lower-than-one "substitution rate" between pirate and licit purchases. Consistent with the other industry groups, RIAA does not reveal how it calculates this number; nor has RiSA released information about how it conducts its consumer surveys.

Newer figures supplied by RiSA reflect a shift in methods and an apparent effort to escape the difficulties of substitution effects. The new RiSA data distinguishes between the street and retail values of pirated materials. For 2009, RiSA estimated the street value of pirated music at $6.2 million and the equivalent retail value at $30.2 million, without speculating about where "losses" fall within that range.[9] Of the physical product seized, RiSA reports that some 65% was local repertoire, providing one indicator of a pirate market dominated by local production rather than imports.

Although there has been very little other published information on the overall scale or impact of music piracy in South Africa, the topic has consistently attracted public attention—not least through the outreach efforts of local musicians. Following a string of news reports in 2004 and 2005 about former music greats dying in poverty, both RiSA and musicians' groups launched new campaigns against music piracy. The most prominent of these was Operation Dudula—a short-lived anti-piracy "movement" organized by local recording artist and poet Mzwakhe Mbuli, which attracted industry support before disintegrating amidst charges of vigilantism (see later in this chapter for a fuller discussion).

Games

The Southern African Federation Against Copyright Theft represents the interests of both movie and video-game companies and is one of the few international affiliates of the US-based Entertainment Software Association. With the support of the ESA, SAFACT engaged in a relatively broad range of consumer survey research and enforcement activities in the late 1990s and early 2000s, dating back to the first reported confiscation of pirated PlayStation discs in 1998.

In 1999, SAFACT launched a widely reported "war" against video-game pirates, consisting of a series of raids on pirate vendors and distributors. Cape Town and Mitchells Plain (a low-income township) were the focus of much of this effort, with the former described as a "hub of piracy" in South Africa and the latter as the "pirate capital of South Africa" (IOL 1999). This war, at least, appears to have ended. Despite IIPA statements as recently as 2007 that "pirated entertainment

---

9   Estimates supplied by RiSA in e-mail communications, December 14, 2009.

software products continue to be imported from Southeast Asia, particularly Malaysia," (IIPA 2007) no new data on entertainment software piracy has been reported for South Africa since 2001—the last year in which the ESA conducted a survey. In 2009, we could not find any pirated video games in the Noord/Plein, Bruma, or Fordsburg markets—all major flea markets in Johannesburg hosting vendors of pirated CDs and DVDs. Our conclusion—largely shared by ESA but still surprising in bandwidth-poor South Africa—is that video-game piracy has moved online and, secondarily, into informal distribution networks among friends and local gaming communities.

As in other developing countries, gamers in South Africa appear split about the ethics of such behavior. Our informal survey of discussions of game piracy on the forums of MyBroadband. co.za (a popular Internet service provider) and other sites suggests that game companies are held in significantly higher esteem by their customers than the music and movie companies, whose products, market practices, and enforcement tactics generate more consistent criticism. The combination of rapid innovation in gaming and the very slow rollout of premium consoles and gaming services in South Africa supports a more easily understood case for commercial investment, the importance of legal markets, and consequently, the protection of content. Most of the major console vendors and games publishers have simply ignored the South African market over the past two decades, treating its small size and remoteness (and the dominance of low-cost PlayStation 1 and 2 systems and games) as disincentives to entry.[10] Distinct from other more fully globalized markets, such as film, this deprivation makes itself felt among high-income consumers who could otherwise afford such goods, and fuels a variety of informal forms of parallel importation. As South Africa begins to be better integrated into global gaming markets, this dynamic is producing a comparatively complex debate about piracy within gaming communities, with arguments on both sides.

Nonetheless, the realities of international pricing and domestic incomes place a practical limit on this ambivalence. As of late 2009, *Gods of War* (PlayStation 2) and *Halo 3* (Xbox 360) retailed online at R274 ($39) and R400 ($57) on Kalahari.net—both well above their US retail levels. Like gamers in other countries with weak domestic publishing and distribution networks, South African consumers routinely order from foreign websites that offer wider stock and lower prices than local retailers—even after international shipping charges. This consumer-driven parallel-import sector in game software has no real equivalent in film or music.

---

10   Among earlier-generation systems, the PlayStation 1 and PlayStation 2 dominated the South African (and wider African) market. None of the Sony systems' major competitors, including the Nintendo 64, the Sega Saturn, the Sega Dreamcast, and the original Xbox, were launched in South Africa. The Nintendo GameCube was picked up by local distributors but inadequately supported. The Xbox 360, launched in South Africa in 2006, was the first major system to be strongly marketed contemporary with its release in high-income countries. The popular Xbox LIVE service, however, is not available at the time of writing (it is scheduled for launch in November 2010). And in contrast to the PlayStation 1 and 2, Sony has not made a serious bid for the South African market with the PlayStation 3, which was priced at $852 on its launch in 2007.

Movies

There are, in total, under a hundred cinemas in South Africa—an absurdly small number for a country with forty-seven million inhabitants, though not an unusual ratio.[11] Over the past fifteen years, two companies, Ster-Kinekor and Nu Metro, have emerged as the dominant players in film exhibition, as well as home-video and video-game licensing. In 2009, the two companies owned seventy-eight of South Africa's multiplexes, with a combined total of over seven hundred screens. The remaining cinemas are owned by a handful of much smaller companies, such as the Avalon Group, and by the few remaining single-theatre owners.

Nearly all South African multiplexes are located in casino complexes or shopping malls in the formerly whites-only suburbs of the major cities. Prior to Ster-Kinekor's launch of a new multiplex in Soweto in 2007, there were no major cinemas in the predominantly black township areas. Both price and ease of access skew audience composition toward the wealthy, white minority.

Throughout the late 1990s and early 2000s, the Ster-Kinekor/Nu Metro duopoly progressively raised movie ticket prices, with average prices reaching R35 ($5.75) by 2005.[12] Although cinema attendance by the new black middle class grew in that period, overall, audiences slightly declined—led by a sharp 30% drop in white audiences. In 2005, competition for this shifting audience triggered a price war between the two distributors. Ster-Kinekor converted some 70% of its screens into a budget cinema chain called Ster-Kinekor Junction, dropping prices to R14 ($2.30). Shortly after, Nu Metro dropped prices across its theater chain to R12 (roughly $2.00) (Worsdale 2005). The price war proved costly to both firms, forcing the sale and closure of several cinemas. When it was declared over in 2007, it left a transformed market—split between high-end multiplexes charging R35–R40 ($5–$5.70) and budget cinemas charging R17–R18 ($2.50). The high end is now undergoing further segmentation as Ster-Kinekor and Nu Metro convert theatres to digital and 3D projection, hoping that the new features can justify higher prices and differentiate the theatrical experience further from the growing home-video market. The handful of new 3D theatres command a premium ticket price of R60 ($8.50).

South Africa also has a small but viable market for Bollywood movies and music, and consequently for pirated Bollywood CDs and DVDs. In the early 2000s, Bollywood piracy was widely ascribed to Pakistani syndicates, who smuggled discs into South Africa for sale in the large South Asian immigrant communities. As this report's India work has documented, these syndicates moved into the DVD trade because of opportunities created by the conflict in

---

11   South Africa has more screens per capita, for example, than the other main countries in this report: Brazil, Russia, and India (see chapter 1). The United States, by comparison, has nearly six thousand cinemas, or roughly eight times the number of screens per capita.

12   The value of the rand against the dollar fluctuated significantly in the latter half of the last decade, from under 6:1 in 2005 to nearly 8:1 in 2010. Where possible, we have used values appropriate to the year cited.

Kashmir, which blocked legal trade between India and Pakistan and created a large Pakistani black market for Indian film. Their production and organizational capacities, in turn, leveraged the large Pakistani diaspora, creating an infrastructure for DVD exports to South Asian communities overseas—including the roughly 1.2 million South Asians living in South Africa.[13]

Legal Bollywood exhibition, retail, and rental infrastructure emerged in South Africa in the early 2000s, notably with the growth of the small Avalon Group cinema chain and Global Bollywood Music—the licensee for Indian music distributor T-Series. But Bollywood exhibition remains a very small niche market within the already small South African film market. The most successful Bollywood release for 2008 and 2009 was *Rab Ne Bana Di Jodi*, a romantic comedy and the third-highest-grossing film in Bollywood history. In 2009, it earned just $151,000 in South Africa, placing it 101st in revenues among releases for the year.

The Bollywood case illustrates the "chicken or the egg" dilemma for new entrants into the South African (or any other) cultural market. In the absence of legal distribution, piracy creates the audience—in this case an audience grounded in the South Asian community but growing well beyond it. But the same process undercuts opportunities for legal distribution, reinforcing the low equilibrium of South Africa's media market. Legal distribution channels, in this context, become a poor indicator of popularity. If the widespread availability of Bollywood films in the pirate marketplace is any guide, film revenue numbers significantly understate Bollywood's presence in South African media culture.

After years of ignoring piracy in their overseas markets, Bollywood studios have become more active in enforcing rights in the United States and the United Kingdom, where large South Asian populations and poor local distribution ensure active markets in pirated Indian media. But the major Bollywood studios remain disorganized in South Africa. Few have registered trademarks for protection; none (as of mid-2010) have appointed local representatives who can collaborate with South African enforcement agencies in identifying infringing goods. SAFACT, the MPAA's representative in South Africa and the primary enforcer of film rights, has no Indian membership. This lack of on-the-ground presence has a direct effect on enforcement efforts: South African police and customs routinely ignore pirated Bollywood CDs and DVDs because there is no complainant to pursue a case or—more materially—to cover the storage costs associated with seized goods as required under the Counterfeit Goods Act. For all intents and purposes, there is no enforcement of Bollywood film rights in South Africa.[14]

In general, pirate DVDs become available on or near a film's official release date. Prices vary dramatically by location, reflecting socioeconomic differences in the regular clientele of the main

---

13   Malaysia is the other frequently named foreign source for Bollywood (and Hollywood) DVDs—home, like South Africa, to a large South Asian population eager for new Bollywood films.

14   In the United Kingdom and the rest of Europe, the Bollywood distribution infrastructure is more developed, and Bollywood pirates have been subject to slightly more pressure. Yash Raj Films, one of the largest Bollywood studios, has played a particularly active role in both raids and lobbying (Agence France Press 2008) and has successfully brought charges in a handful of cases.

markets. The suburban Bruma Lake market, which primarily serves a middle-class Johannesburg public, occupies the high end, with prices typically ranging between R20 and R40 ($2.50–$5.00) for a new release. The Noord/Plein market in the center of town serves mostly black public-transit commuters and sells DVDs within a much lower price range—generally R10 to R20 ($1.25–$2.50).

The range of products available at these markets also reflects differences in clientele. At the Noord/Plein market, local music, South African movies, and older international action movies predominate. At Bruma Lake, the latest international movies and not-yet-broadcast television series are the ubiquitous items. In Fordsburg, a predominantly South Asian immigrant community, Bollywood movies and music are the norm. It is easy, though possibly too simplistic, to see these differences in terms of relative privileges in access to media and in an accompanying structure of preferences. Whereas the city center offers local favorites and old standbys to the mostly working-class commuters, the Bruma market is oriented toward consumers who have more access to a globalized media culture and who anticipate products not yet released in South Africa due to the windowing strategies of film and television studios.

As elsewhere, windowing strategies are giving way to simultaneous international release dates for movies and shorter delays between theatrical and DVD releases. Such efforts undercut one of the two basic advantages of the pirate marketplace. As James Lennox, current head of SAFACT, observes:

> The delayed release of movies provides an opening for pirates to sell movies and DVDs that haven't yet been released. Delayed release and window periods between cinema and video release undoubtedly create a gap, but it must be emphasized that more and more films are released theatrically in South Africa within a day or two of release in a major territory, such as the US and UK.[15]

For the most part, television re-broadcasts of hit American series—a favorite genre in the Bruma market—have not yet followed suit. New seasons of *Lost* and *24*, both major hits in South Africa, have typically been re-broadcast a year or more after their premieres in the United States. Subscription television channels, like M-Net, offer shorter delays—indeed this is a large part of their added value. The last two seasons of *Lost* aired on M-Net only two months after their US premieres. But for many South African consumers, even such shortened delays are sharply felt: although national release windows can be managed and staggered, corresponding efforts to time popular anticipation surrounding global media goods have largely collapsed, driven by global advertising campaigns and the Internet. Demand in South Africa, consequently, nearly always runs ahead of supply—in terms both of price and availability. In this context, pirate distribution sets the consumer standard.

---

15   Interview with James Lennox, CEO of SAFACT, 2009.

Lennox and other industry representatives are aware that the inadequacy of the local distribution infrastructure cedes access and convenience to the pirates, and he points to industry efforts to address the issue:

> There are too few places to access legitimate products—cinemas, video shops, and retailers—in South Africa, and this creates a space for pirates. Many people who buy from street vendors are impulse buyers, and the legitimate outlets don't always cater to them. I don't think people think, "I am going off to buy, say, *Mama Mia* from a street vendor today." They pass a vendor, browse, see it, and buy. Our members are continually endeavoring to make genuine product more accessible, including selling DVDs at garages, clothing shops, and workplace vending machines.

But Lennox is unsympathetic to the other obvious differentiator—price:

> We are aware that the cost is also used as an excuse by consumers to justify buying pirated goods . . . Prices at first release date have shown a steady rate of decline [and] . . . prices of DVDs come down after the first six months so consumers need to wait six months and then can buy the original—with all the extras—at anything from R50 to R90.[16]

Incremental price decreases reflect the hold of the Hollywood studios on local DVD pricing but bear little connection to consumer expectations or the de facto norms of media access in South Africa. Although there will probably always be a public disinterested or patient enough to balance waiting against discounts in the market for licit goods, this model clearly doesn't address the population that has been most effectively captured by Hollywood marketing. And therefore it offers no serious alternative to the pirate market. As in other countries, the growth of broadband and other digital media infrastructure will almost certainly widen this gap between the legal and illicit models.

Piracy and South African Film

Local film industries often feel these dilemmas acutely, but it is important to put them in perspective. The South African film industry has many problems, but it is far from clear that piracy is foremost among them. Its immediate difficulty is that Hollywood movies completely dominate the South African box office: only one South African film (*District 9*) appeared in the top-fifty box office earners in 2009 and only three in 2008.[17] The massive production and advertising leverage

16   Ibid.

17   One through ten in 2009 were: *Avatar, Ice Age: Dawn of the Dinosaurs, The Proposal, Transformers, 2012, Couples Retreat, The Hangover, Up, Fast and Furious,* and *Harry Potter and the Half Blood*

of Hollywood and the scarcity of screens means that many South African films never get picked up for theatrical distribution. The South African film industry, consequently, is small and fragile. Despite recent high-profile hits like *District 9*, its fate has hinged largely on publicly funded television production—and in 2009 this market collapsed when the largest source of funding, the South African Broadcasting Corporation, suffered a financial crisis that undermined its ability to commission new work or even pay existing bills. As a result, the South African film industry shrank dramatically, from 25,000 employees in 2008 to around 8,000 in 2010.

At its peak, the domestic film industry produced only a handful of feature-length films per year, and few of these reached the mainstream film circuit. *Tsotsi* and *District 9* have been the notable international successes in recent years, with *Tsotsi* winning an Oscar for best foreign language film in 2006. The great majority of South African films are lower-budget productions aimed at South African audiences. *White Wedding*, *Jerusalema*, and the popular slapstick comedies of Leon Schuster are characteristic examples. Like most Hollywood and Bollywood blockbusters, these are generally available in pirate markets at the same time as their theatrical release. *Tsotsi*, notably, was widely reported as the most pirated DVD in South Africa in 2006 (Maggs 2006).

Indignation at the piracy of South African movies has been a key ingredient of SAFACT-sponsored radio and TV anti-piracy campaigns. These campaigns commonly rely on and reinforce a nationalist approach to IP rights, blaming "unpatriotic" consumers for buying pirated South African films. The campaigns appear to have had some localized success in highly policed markets, like Bruma, where pirate DVD vendors are reluctant to display South African movies (while showing no such qualms about Hollywood movies).[18]

Predictably, concern about the wider failure of the market for South African film has sparked interest in other distribution models. Here, the goal of broadening the audience points less to high-tech streaming solutions—like the US-based Hulu and Netflix, which are still years away from significant penetration of the South African market—than to the low-tech appropriation of street vendor networks as distributors and retailers for DVDs. This latter model has a notable success story: the Nigerian movie industry, which has a growing presence in the South African (and more broadly, African) DVD market.[19] Now one of the largest movie industries in the world in terms of the total numbers of productions (with 1,200 movies released in 2008), the Nigerian movie industry emerged in the 1990s from a context of economic collapse and, arguably, very successful cultural protectionism—factors that combined to sharply constrain the supply of foreign films

---

Prince. The first Bollywood films on the list appear at number 99 (*Love Aaj Kal*) and number 100 (*Kambakkt Ishq*). See Box Office Mojo, http://boxofficemojo.com/intl/southafrica/yearly/.

18   This is not universal. Local films were on prominent display at the downscale Noord/Plein market in the center of Johannesburg. The enforcement agents we spoke with do not credit street pirates with much restraint when it comes to pirating local productions, and more than one indicated that South African films are pirated as flagrantly as Hollywood and Bollywood blockbusters.

19   Nigerian films have also been extensively taken up by the South African–owned Multichoice pay-television platform and are carried via satellite across Africa on its African Magic channel.

and TV during an extended period in the 1970s and 1980s.[20] Because there was almost no formal distribution or exhibition infrastructure in Nigeria, early domestic producers co-opted the large informal sector for distribution, selling video cassettes at very low wholesale prices to vendors for resale (Ogbor 2009; Larkin 2004).

The possible use of informal vendors as distributors of licit goods has been a topic of repeated discussion in South Africa. The major distributors of Hollywood films, represented by SAFACT, have generally rejected this approach as too complicated. According to Lennox:

> There are some obstacles to this for legitimate industry players, including the need to comply with Street Trading Bylaws, paying of minimum wages, acceptable working conditions, all of which if adhered to make the "informal" trading environment ineffective. The Road Traffic Act, for instance, prohibits selling by licensed street vendors within five meters of a junction. All pirated product sold by street vendors is sold within five meters of a junction, thus exposing the employers of "legitimate" traders operating in this zone to possible prosecution.[21]

A number of local industry players, however, have stronger incentives to test the model and have attempted to overcome the obstacles cited by Lennox. Bliksem DVDs, launched in December 2009, was one such effort. Bliksem sought to create a middle ground within street vendor culture that would allow legal, affordable DVDs to be sold to Johannesburg commuters. The Bliksem model depended on arrangements with local producers and distributors who were willing to break with the pricing conventions of the international studios—a list that eventually included most of the major South African production companies. Bliksem's prices, in turn, were set not in relation to international DVD prices or existing licensing conventions but in relation to the chief competitors for disposable income: the pirate DVD market and, Bliksem founder Ben Horowitz notes, the pre-paid cell phone "air-time" market, which sets a de facto price benchmark:

> [Consumers are] buying DVDs and CDs from pirates at R10 [$1.25] and R20 [$2.50]. I think the commuter market, the working class market . . . walking down the street and buying from vendors, their currency is based on buying air-time for pre-paid cell phone usage (which most commuters would purchase for between R35 and R50, on average). You're either buying air-time or you're buying some other luxury item.

---

20   Nigerian cultural protectionism continues to this day. The Nigerian Broadcasting Commission recently prohibited the screening of foreign movies during prime-time television slots. See Oxford Business Group (2010:190).

21   Interview with James Lennox, CEO of SAFACT, 2009. From a legal perspective this is a weak claim: if the distributors sell wholesale to street vendors, they are not legally responsible for the street vendors' compliance with bylaws.

Anything R50 [$6.25] is super luxury. Anything over R100 is out of reach. Nobody in this market is buying home entertainment for more than R100.[22]

Bliksem DVDs sold legitimate copies of South African films for between R20 ($2.50) and R60 ($7.50)—prices higher than those of the low-end pirate vendors but significantly cheaper than retail outlets. And it delivered a strong anti-piracy appeals to consumers in the process—a posture that won it praise from Sony Pictures and allowed it to make deals with major local distributors, such as Nu Metro. But in contrast to its major counterparts in the US and Europe (and in India), the initiative was undercapitalized, could not secure significant discounts on Hollywood content, and had no opportunity scale or refine its model. Bliksem DVDs closed in July 2010. The two structural problems associated with Hollywood dominance of developing markets—high prices and poor distribution, especially for local film—remain unaddressed. As Ferdie Gazendam, CEO of Ster-Kinekor, observed:

It's not just about putting local content onto the shelves. It's very tough to find the shops. . . . The dilemma is that we are set in our ways because of the studios we represent. When it comes to local content we need to learn different ways to get the product to the market. (Smith 2006)

Piracy on the Internet

Until the summer of 2009, South Africa was linked to the wider global Internet by a single undersea cable, with a very modest total capacity of .8 gigabytes per second. Broadband services have been correspondingly expensive and low speed. The vast majority of consumer services come with bandwidth caps in the vicinity of 3 gigabytes per month—a level easily surpassed by streaming a single high-definition movie. The quality of South African broadband services also ranks near the bottom of international surveys (Muller 2009), with most ISPs unable to guarantee consistent performance for even common video services, such as YouTube. In this context, the uptake of bandwidth-intensive P2P services in South Africa has been very limited.

The bandwidth shortage, however, is expected to ease as new undersea cables come into service. Total bandwidth in and out of South Africa should improve to 2.58 gigabytes per second in 2010 and—according to plans—to 10.5 terabytes per second in 2013 (World Wide Worx 2009). Internet stakeholders on all sides expect that this increase will bring cheaper, better broadband services. As computers, storage, and playback technologies also become more widely available, South African participation in both the licit and the illicit global digital media economies is likely to expand dramatically.

The bandwidth shortage has given most rights holders a respite from Internet piracy in South Africa and has kept the rate of return on optical disc enforcement relatively high. Also as

---

22   Interview with Ben Horowitz of Bliksem DVDs, 2009.

a result, existing jurisprudence on Internet-based infringement is thin—as is, by most accounts, enforcement and policing expertise regarding online activity. Industry groups, however, are preparing for the inevitable. RiSA, in particular, has laid the groundwork for wider online surveillance and enforcement, including attempts to establish ISP (Internet service provider) responsibility for blocking infringing third-party content and, ultimately, infringing users. These initiatives have met with mixed reactions from ISPs, and none have been directly tested in court.

Much of this effort is framed by the 2002 Electronic Communications and Transactions (ECT) Act, which introduced a standard set of immunities and liabilities for Internet and web-based-software service providers in regard to third-party infringement. Under the act, service providers have no obligation to monitor content on their service and bear no direct responsibility for infringement facilitated by or exposed through their service. In contrast, they are required to respond to take-down notices from rights holders when those parties provide a detailed account of infringing activity utilizing their service.

In South Africa, as elsewhere, these guidelines have left a great deal of room for uncertainty about the scope of the "safe harbor" for service providers, and to date there is very little clarifying South African jurisprudence. In particular, there is currently no basis for the "contributory infringement" standard established in the United States as an outcome of the suit against the Grokster P2P service.[23] Rights-holder organizations are, however, pushing strongly in that direction. In 2008, RiSA sent notices to South Africa's Internet Service Providers' Association (ISPA), demanding that they block access to two recently established local BitTorrent sites, Bitfarm and Newshost, which were accused of linking to infringing content. The ISPA cooperated, but the legal basis for the request was unclear and produced a challenge. Reinhardt Buys, the lawyer representing the site owners in the two cases, argued:

> There is no legal precedent, whether in case law or legislation, in South Africa or
> elsewhere, confirming that the hosting of torrents and NZBs [a format for retrieving
> Usenet posts] and the indexing of such files is unlawful or illegal. (MyBroadband
> 2008)

This is a narrow but accurate reading of the international record. Although a large number of BitTorrent sites outside South Africa have been shut down through legal action, none of these suits have challenged the legality of torrent search engines per se: the underlying functionality is common to all search engines (including Google, which also returns infringing torrent files in searches). Cases have turned, instead, on other criteria for establishing liability, such as the proportion of infringing activity on the site, profits made in the course of the activity, or the flaunting of take-down requests—all factors, for example, in the high-profile 2009 guilty verdict in the Pirate Bay case in Sweden.

---

23   MGM Studios, Inc. v. Grokster, Ltd. 545 U.S. 913 (2005).

But such outcomes are not clear cut. In early 2010, a similar suit against the administrator of the UK-based P2P site OiNK failed due to the lack of a clear principle of contributory infringement in UK law. Such criteria would have been similarly challenging to establish in the Bitfarm and Newshost cases, and the lenient record of most South African judges on infringement probably did not provide much reassurance. Having succeeded in blocking the sites, RiSA did not pursue further action against the site owners. In a July 2009 interview with the authors, Buys continued to argue for the non-commercial status of the sites and their inclusion under existing safe harbor provisions. The case was like "punishing the photocopying machine in a library because it might be used for photocopying books"—a resonant example in South Africa. Responding to RiSA's claims that the sites were organized by piracy syndicates, Buys observed that "they were just kids."

In July 2009, RiSA served a number of ISPs with blocking notices regarding two foreign websites, www.gomusic.ru and www.soundlike.com, both of which sell MP3s at prices well below those of local online music vendors.[24] In this case, the ISPA resisted the request and informed RiSA that the notices fell outside the scope of the ECT Act. As ISPA General Manager Ant Brooks argued:

> The law does not make provision for blocking a website, especially ones that are located overseas. ISPs are not the police and cannot just block access to a website on the very thin premise that at least some of the content is infringing copyright. Otherwise this will lead to anyone from anywhere giving any reason to block access to websites. (Vecchiato 2009)

Like most safe harbor clauses, the ECT Act limits the take-down procedure to content or services hosted on the network of the service provider (MyBroadband 2009). Broader requests to block criminally infringing activity of the kind alleged against gomusic.ru must still pass through a criminal investigation and a court order.

Discussions between the ISPA, RiSA, and other rights-holder groups about curbing Internet piracy continue. RiSA has started to track downloads using content-monitoring techniques, and there are reports of ISPs threatening users with account suspension and blacklisting on the basis of alleged infringement (MyBroadband 2009). Such practices are in line with wider industry efforts to make ISPs play a stronger role in enforcing rights-holder claims, but the law surrounding such practices remains unclear and may face serious challenges in the context of South Africa's constitutionally protected right to privacy.

---

24   The two services priced tracks at R1.20 ($0.15) and R0.72 ($0.09), respectively—far less than the online music services licensed in South Africa: the Nokia Music Store, which charges R10 per track ($1.30) and R100 per album, and RhythmMusicOnline, which charges R7 per track for local music (the iTunes Store is not available in South Africa). The sites are the latest Russian variations on the popular AllofMP3 music store, which made use of a loophole in Russian law that allowed downloads to be characterized as broadcasts, subject to a modest compulsory license fee. Recent revisions to Russian copyright law have closed this loophole, but enforcement against the services has been inconsistent.

Attitudes toward Piracy

In the anti-piracy struggle, legal efforts and police actions are complemented by extensive "hearts-and-minds" campaigns waged through the media. South African newspaper readers and TV audiences, in particular, are exposed to a flood of industry-generated stories about the negative effects of piracy on local artists and producers. Between 2005 and 2008, the major broadcast and newspaper outlets carried some 846 stories on media piracy—a huge number in a country with just three major media markets (see table 3.1).

Table 3.1 Media Piracy Stories, 2005–2008

	2005	2006	2007	2008
Broadcast	113	215	108	33
Print	7	34	190	146
Total	120	249	298	179

Source: Authors based on data provided by Monitoring South Africa.

Piracy Stories

Raids, seizures, the release of new industry studies, and anti-piracy activities by local artists and celebrities all generate stories and—accounting for differences in time, place, and context—overwhelmingly present the same story of artists and local entrepreneurs victimized by piracy. Metaphors of bodily harm—of piracy killing and strangling artists, of pirates as bloodsuckers and parasites—are common and constitute a virtual template for media coverage (Naidu 2007).

SAFACT, RiSA, the BSA, and other industry associations have been remarkably successful at managing this media attention and the larger media narrative on piracy. As in other countries, however, we see a significant disconnect between the media narrative and actual consumer attitudes, which show much greater diversity and widespread tolerance of piracy in many circumstances. This diversity of opinion becomes immediately apparent in the various online forums that discuss IP policy and enforcement in South Africa—such as MyBroadband.co.za—where consumer perspectives, moral ambiguities, and industry narratives around piracy are routinely discussed. It is also visible in our interviews and limited survey work, which echo other South African studies in finding pirate practices completely normalized and integrated into daily life (Van Belle, Macdonald, and Wilson 2007). As one source observed, "You can walk into any house in Soweto and find a stack of pirate CDs and DVDs."[25]

---

25   Interview with filmmaker Peter Ndebele, 2009.

Empirical data on South African attitudes toward piracy remains thin. The most recent study, a DTI-funded consumer survey on counterfeiting and piracy from 2006 (Martins and van Wyk), found that personal copying was widespread: 58.1% of respondents indicated that they copied CDs, DVDs, computer software, and/or other material for themselves; 53.5% said they did so for friends; and 40.1% said they knew of people who copied discs for the purpose of resale.[26]

Such surveys are challenging in communities where the informal economy plays a large role. In Hanover Park, a poor neighborhood where we investigated disc piracy (discussed in detail later in this chapter), residents often drew no distinction between pirated and legal goods. Pirated movies were simply referred to as movies, or sometimes as kwaai (good) movies. The remoteness—both geographic and economic—of the licit markets for these goods makes the distinction literally meaningless for many residents.

Anti-piracy advertising and media campaigns do not penetrate very far in these contexts. Some respondents in our Hanover Park survey did not understand the concept of piracy, obliging the researcher to explain the term. Others justified piracy in terms of the inequality between poor South Africans and the American firms that dominate the global media trade:

> Morality doesn't really play a big factor because in America they don't feel the R80 [$10.00] that they're losing from me. They can afford it. They all live in big fancy mansions, whereas I live in a flat here in Hanover Park; and so it doesn't really matter to me what they think about me being a pirate because my circumstances are different from theirs. It's more economical to get that pirated DVD. No, I don't think piracy is a crime.

Others made reference to the powerful contradictions of living at the periphery of global film culture—subject to its attractions and social effects but largely excluded from legal access:

> I know piracy is a crime, but we will go for it . . . so we can say, "I am also part of the crowd. I also watched that movie."

Still others offered more practical rationales, such as the need to keep children off the streets and away from the ubiquitous gang culture.

Collectively, our Hanover Park interviews revealed the extent to which piracy is embedded in—and, to a degree, an enabler of—an array of social functions around media that extend well beyond entertainment. Unlike in high-income homes with more screens, computers, and living space, DVD piracy in poor neighborhoods does not typically serve private acts of consumption.

---

26   The study drew on a small sample of 604 respondents from metropolitan areas and large cities in four key provinces: KwaZulu-Natal, the Western Cape, Gauteng, and the Eastern Cape. The sampling was random.

Rather, it serves collective ones, organized around groups of family and friends. Pirated discs are not primarily destined for personal collections but are circulated within social networks. Access to pirated media, in these contexts, feeds practices of viewing and sharing that reinforce links across families and within communities. In a context, moreover, in which access to audiovisual media is usually limited to radio and a handful of broadcast TV channels, pirated optical discs are the closest approximation of the rich, on-demand media culture taken for granted by most high-income consumers. Piracy in this context is flexible, low risk, cheaper than a ticket to the cinema and often highly customer-oriented. As one resident noted, "people even come to my door with movies."

Software Piracy in Music Production

Because the struggling musician is arguably the iconic figure in anti-piracy campaigns, we decided to survey some of them about their attitudes toward piracy. We opted to focus not on music piracy, however, but on the use of the software tools that have become ubiquitous in recording and production. Increasingly, musicians and producers start their careers through self production, often at home, and consequently rely heavily on sound editing and mixing software. For genres that extensively sample or transform sounds, such as much contemporary pop and hip-hop, these tools are basic to production itself—not just to post-production. The high cost and, often, lack of availability of the most common software packages create a familiar dynamic in this context. As one independent-label CEO observed to us, South African hip-hop was built on home production and pirated software.

Among hip-hop artists and producers, the software toolbox revolves around a handful of widely used titles, including Fruity Loops Studio, Reason, Sony's ACID Pro, and Apple's Logic Studio. None of these products are designed or manufactured in South Africa, but demo versions with limited functionality can be downloaded for free from the official sites, and full versions are available online at prices ranging from $200 to $499. Because the production community is relatively small and interconnected, shared production techniques and training introduce strong network effects in the choice of products. Producers tend to use software that has been "vetted" in their communities, and these choices tend to self-reinforce as producers and musicians exchange knowledge. In our interviews, open-source alternatives, such as Audacity, did not even register.

Predictably, artists and producers are much more aware of and divided on the subject of piracy than residents of Hanover Park. Hopes for commercial success among the former still pass primarily through recording contracts and CD sales. Equally predictably, artists and producers at the margins of the business (for the purposes of our survey, those who produce professionally but who do not earn a living from it) rely heavily on pirated software for the basic infrastructure of their craft.

Twenty-eight musicians and producers responded to our questionnaire (distributed via e-mail and Facebook). Nineteen of the twenty-eight respondents described themselves as established artists—a term that reflects their own perceptions of their professionalization and notoriety in their communities. Several had recorded in professional studios. All had participated in less formal practices of recording and mixing, notably on computers at home. None of the respondents were signed to major music labels.

Because the surveyed group was small, we do not place much weight on statistical findings from the study. Nonetheless, the results were striking: of the twenty-eight artists, twenty indicated that they use at least some pirated software. Half (14) indicated that they have "obtained a cracked copy from a friend/associate for free," and just under half (13) have downloaded pirated software from the Internet. Just under half (13) said that they have purchased at least some software legally. When asked about their reasons for using unlicensed software, half (14) indicated that the genuine copies were too expensive, while six said that the software they needed was not available at local retailers.

Only two respondents stated that they were categorically opposed to software piracy. Others either described it as a necessary practice (12) or reported mixed views (13). In contrast with our consumer interviews in Hanover Park and elsewhere, few musicians offered blanket approval. Much more common was the articulation of an "ability to pay" principle, under which profiting through use of the software brings an obligation to purchase it legally. As one music producer argued:

> This is Africa! We do not have access to the latest and greatest software when they release it. When it eventually lands the prices are ridiculous. So for a beginner it is a waste of time and money. If you are using the software to obtain more knowledge, then I do not have a problem with it even though it is illegal. Once you start using this software in an environment where you profit, then I have a problem. Go and buy it!

Variations on this theme ran throughout the musicians' comments, combining respect for software developers, who are viewed as enablers of the new lower-cost production culture, with recognition that, in Africa, the cost of legal entry is prohibitive:

> I suppose it is either a good thing or a bad thing depending on one's perspective. The software companies and programmers will feel that cracked software will impact on their revenues, while cash-strapped music programmers or artists might feel that it enables them to do what they enjoy without having to lay out potentially large sums of money. From my personal perspective, if my only access to beat-making software was to pay for it, it is highly unlikely that I would ever have taken up this work at all.

Law and Enforcement

Despite the history of industry complaints, South Africa is now usually presented as an enforcement success story. Reported rates of piracy are the lowest in Africa and among the lowest for developing countries overall. The South African government cooperates closely with industry and, in the past decade, has significantly expanded public investment in the enforcement effort, including the creation of new courts and police units and the introduction of progressively stronger policing measures. IIPA reports—a reliable barometer of US-industry concern—repeatedly downgraded South Africa after 2002 and dropped reporting on the country altogether in 2007. The last IIPA report (2007) offered the usual denunciation: "The impact of piracy in South Africa is devastating for legitimate right holders, legitimate distributors, and retail businesses." But it is clear that the overall industry perception of enforcement in South Africa has improved—and continues to improve. SAFACT CEO James Lennox, for example, is very positive about enforcement efforts and even defends the state against industry demands for more support:

> Industry, in general, shouldn't complain about a lack of state action in South Africa
> as it is up to industry to use the law to combat piracy. We have found the state to be
> very willing to assist—in all aspects, from raids to asset seizure. Some critics seem to
> think that the state should do it all.[27]

Proof of commitment, for industry, is usually measured in raids, arrests, and convictions—and here the signs of government cooperation are tangible. In 2008, SAFACT participated in 853 piracy investigations and 973 raids, leading to some 617 arrests and 447 convictions. These figures represent a 59% increase in the number of raids over 2007 and a remarkable 936% increase in the number of convictions—a jump attributed to SAFACT's decision to pursue criminal prosecutions in all cases of alleged infringement and to its increasingly close relationships with investigators and prosecutors associated with the commercial crimes courts, where most IP cases are heard.

This level of cooperation represents a significant turnaround from the early 2000s, when the IIPA and other industry groups routinely complained about the lack of public commitment to enforcement. The change owes much to South African government actions to expand enforcement efforts in the wake of TRIPS and to involve private partners in the process. In 1997, the Counterfeit Goods Act introduced TRIPS-level border controls and created an administrative architecture for enforcement. In the wake of the CGA, several government agencies began to take on new or more active roles in IP protection and enforcement, including the Department of Arts and Culture and especially the Department of Trade and Industry.

The DTI was not a newcomer to these issues. It played the main coordinating role in bringing South African legislation into compliance with TRIPS in the late 1990s and subsequently in the

---

27   Interview with James Lennox, CEO of SAFACT, 2009.

implementation and coordination of anti-counterfeiting efforts under the CGA. These tasks included modernizing the system for the local registration of trademarks and patents, strengthening border enforcement, and coordinating the growing range of public and private entities involved in enforcement, including the national and metropolitan police services, the South African Revenue Service, and the Department of Justice and Constitutional Development. For over a decade, the DTI has been the main actor in South African enforcement efforts and the main conduit of industry pressure for policy change.

TRIPS Compliance and Enforcement Legislation

Three pieces of legislation currently structure copyright and enforcement efforts in South Africa: the Counterfeit Goods Act, the Copyright Act, and the Electronic Communications and Transactions Act. Nearly all the provisions for policing and judicial process are found in the CGA, and much of the industry struggle for stronger enforcement in the past decade has involved the application of—and controversy around—its main provisions.

The CGA was enacted in 1997 to meet South Africa's obligations under the TRIPS agreement, which entered into force on January 1, 1998. The act criminalized the import, export, manufacture, trade, distribution, and display of "counterfeit goods"—a term usually reserved for trademark infringement but which has been applied more broadly under the CGA to pirated optical discs and other copyright-infringing goods. In practice, the law sets up a dual regime for enforcement: the CGA for products that violate trademarks (that is, counterfeit goods) and the Copyright Act for goods that violate copyright but not trademark (that is, pirated goods that make no effort to pass as legitimate products).

As with most anti-counterfeiting legislation, the CGA targets the commercial middlemen in the counterfeiting trade rather than consumers or end-users. The possession of counterfeit goods is not, in itself, an offense under the CGA: the criminal standard requires intent to sell. The act also makes clear that the possession of counterfeit goods for private or domestic use is not prohibited. The interpretation of this provision has been a point of ongoing controversy with rights-holder groups, particularly the BSA in regard to the use of pirated software by businesses, which has remained exempt from the criminal statute. Calls for the criminalization of "organizational end-user" piracy are part of the standard list of IIPA and BSA demands, based on the view that business use of pirated software should be treated as commercial-scale activity regardless of the intent to sell. To date, only a handful of countries have taken this step, usually as part of bilateral trade agreements with the United States. South Africa has so far resisted such a dramatic expansion of criminal liability.

The Copyright Act, for its part, follows the international norms for film and sound recordings set out in the Berne Convention, including a copyright term, for most works, equal to the life of the author plus fifty years and no registration requirement (in contrast to trademarks and patents, which must be registered) (Dean 1989). Within this framework, two varieties of infringement can become the subject of civil litigation in South Africa: (1) direct, or primary, infringement; and (2) indirect, or secondary, infringement. Primary infringement refers to acts such as the unauthorized reproduction, adaptation, or other exploitation of a work (where such use falls outside the scope of the various exceptions and limitations to copyright, such as "fair dealing" provisions). Secondary infringement is committed by trading in infringing copies. The act thus establishes liability for street vending as well as dealing in so-called grey goods or parallel imports that violate customs or local-licensing agreements. There are a variety of possible remedies in such cases, including injunctions, damages, and the payment of royalties.[28]

Criminal penalties under the Copyright Act are subject to two conditions: (1) a vendor must know that a copy is an infringing copy, and (2) that copy must be destined for sale or other commercial purposes. This standard is stricter than in civil litigation and often turns on determinations of intention. Intention, in turn, has become a complicated issue in South African enforcement because many vendors have responded to police pressure by staffing stalls with legal minors or persons who do not speak a local language, making the standard difficult to apply. When such cases do meet the threshold, the maximum allowable criminal penalty is extremely punitive, consisting of fines of up to R5,000 ($625) and/or imprisonment for up to three years for each infringing article, in the case of a first conviction. These penalties rise to R10,000 ($1,250) and five years per article for any subsequent convictions.[29] There is, in practice, no such thing as an isolated act of infringement, and in most such cases infringements number in the hundreds or thousands.

Anti-piracy under the Counterfeit Goods Act and the Copyright Act

Under the CGA, industry lobby groups have gained a prominent role in public enforcement efforts. Industry representatives often guide the process at every stage, from the initial investigation, to the raids themselves, to their eventual roles as plaintiffs in court.

Most such cases begin with complaints of copyright or trademark infringement by industry representatives to the DTI. The DTI generally responds by organizing a raid under the auspices of the CGA. This typically involves at least nominal coordination between the national police service, the SARS, the DTI, and the complainants. If the raid results in the seizure of goods suspected of trademark infringement, prosecutors can initiate a criminal complaint under the CGA. If only copyright infringement is at stake, the case can be pursued under either the CGA or the Copyright Act.

Raids may also be organized under the Copyright Act, in conjunction with the more general Criminal Procedure Act. Unlike the CGA, these provide no role for private groups, though industry representatives are known to participate anyway. In cases limited to copyright infringement, the Copyright Act offers two notable advantages for industry groups: (1) its implementation involves

---

28   See section 24 of the Copyright Act.

29   See section 27(6) of the Copyright Act.

better-established and more familiar routines for both the police and the judiciary, and (2) the complainant does not bear the cost of storage of evidence seized during a raid or bear responsibility for indemnifying the police in the event of a successful countersuit for damages (in sharp contrast to the CGA). In the case of large raids in which tens or even hundreds of thousands of discs are seized, such storage costs can be significant. This exemption has made the Copyright Act the preferred tool of the RiSA, in particular.

The DTI and industry groups continue to bring new legal tools and interpretations to bear on enforcement. In the past year, the DTI and prosecutors have started to use the 1996 Proceeds of Crime Act—an organized crime statute—to apply pressure on market owners and other businesses higher up the distribution chain. In the most common example, flea market owners are told to evict vendors accused of illegal trading or else face charges of benefiting financially from criminal activities conducted in their market. Such methods circumvent the problems—and the due process—associated with convicting vendors of pirated goods. Even when the threat of charges is not explicit, the prospect of continual police raids raises incentives for market owners to comply with police and industry demands.

Judicial Pushback

When civil claims or criminal charges are filed, cases enter the overburdened South African justice system, whose chronic problems, including prohibitive bail and lengthy pretrial detention, have been amply documented by the South African government itself (van Zyl 2009). From the industry perspective, the system creates obstacles to enforcement at every stage, from the slow pace of prosecutions, to the low rate of convictions, to the suspended sentences applied in most cases.

The DTI has been active in training different professional sectors to streamline the process and make maximal use of the police and judicial powers available under the law. They have worked with Metro Police in Johannesburg, for example, to change the legal regime under which street vending is policed—ending the long-standing practice of simply confiscating the goods of street vendors found trading without a license in favor of more frequent arrests designed to build police records on repeat offenders. They have helped train the commercial crimes unit within the South African Police Service (SAPS) to investigate and prosecute copyright infringement. They have pushed prosecutors to demand the maximum allowable sentences for vendors rather than the traditional admission of guilt and modest fine. And they have worked on South African judges, in particular, whose resistance to imposing stronger penalties has been a point of contention in the push for stronger public-private cooperation.

Judicial resistance to the enforcement agenda has been widespread throughout the court system, up to and including the commercial crime courts, which specialize in IP cases. Industry requests for the maximum statutory penalties have generally been ignored in favor of fines more commensurate with the ability of offenders to pay. Judges also frequently suspend fines or jail terms after sentencing, suggesting that many do not view street-level vending, in particular, as a serious crime. In a few notable cases involving the confiscation of small quantities of infringing goods, judges have sent rebukes to prosecutors by imposing fines lower than those applicable in cases of admitted guilt.

Such pushback has been a persistent source of irritation for industry groups and commercial crimes prosecutors. Since the 1990s, the IIPA and other organizations have argued that sentences are too lenient to act as a deterrent (a point on which they are surely right) and that the lack of deterrence through the courts places South Africa out of compliance with TRIPS (a charge that could be brought against every other country in the WTO as well).[30] The first WTO ruling on this global issue came two years ago, in a case brought by the United States against China. On the key point of whether Chinese provisions for enforcement were adequate, the WTO found in favor of Chinese discretion, very likely setting precedent on the issue (WTO 2009).

The context of South African judicial resistance is obvious to anyone looking at the day-to-day activity of the criminal courts. South Africa averages 19,000 murders per year, according to police reports—among the highest in the world (see figure 3.1). It ranks similarly in other categories of violent crime. The court system, for its part, is hugely overburdened, with pretrial detentions of up to a year in many cases (van Zyl 2009).

Figure 3.1 Murders per 100,000 Inhabitants, 2007/2008

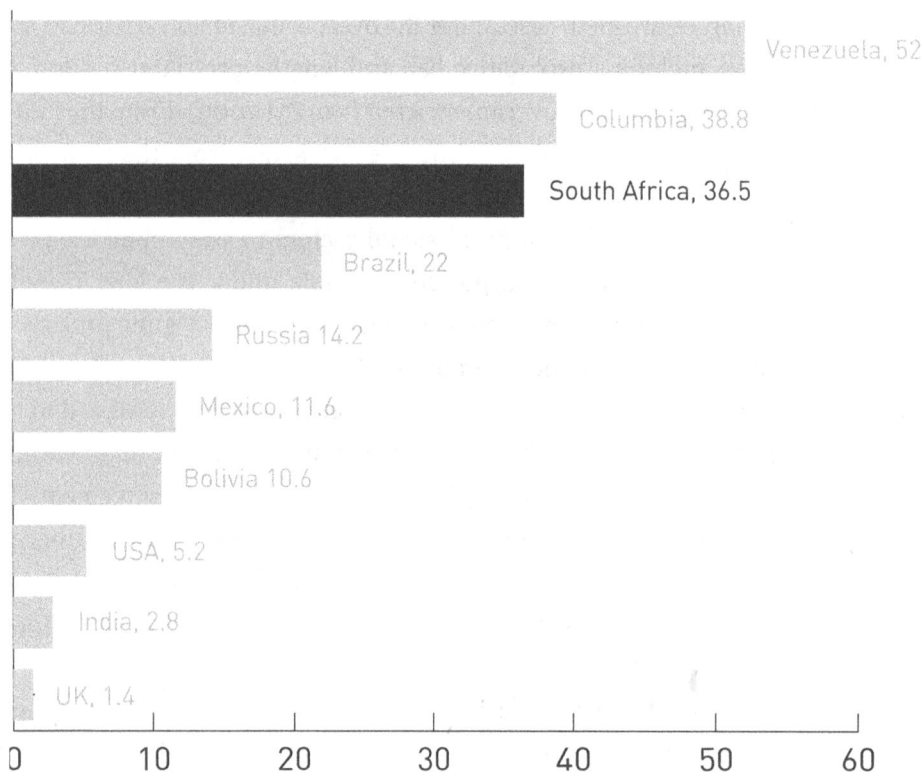

Source: UNODC (2009).

---

30 For our part, we have found no evidence of effective deterrent penalties in any of the countries examined in this report—including the United States. See chapter 1 for more on this issue.

As we repeatedly document in this report, there are inevitable tradeoffs in allocating scarce police, judicial, and penal resources among different types of crime. The scale of violent crime in South Africa makes these choices unusually stark, forcing judges to triage a variety of low-level offenses, including most forms of street-level piracy.

Judges and prosecutors face these tradeoffs in their daily activities. IP stakeholders, in contrast, do not and continue to push for greater prioritization of infringement in the courts. This pressure often takes the form of IP training programs for judges and prosecutors, which are run by the DTI and geared toward bringing judicial attitudes into closer alignment with industry enforcement goals. Industry groups are frequent sponsors of these programs, but funding and technical support also come from the array of foreign government proxies of those groups, notably the US Department of Justice, the US Department of Commerce, and the US Agency for International Development (USAID). The World Intellectual Property Organization (WIPO) has also played a long-term role in such technical assistance. Despite the prevalence of these events over the past decade, our work suggests that judicial culture in South Africa remains relatively insulated from industry pressure. In our interviews, neither prosecutors nor industry representatives felt that the training programs had accomplished much. Although we have no objective measures of this impact (or lack thereof), low conviction rates and weak penalties remain the norm.

The lack of strong sentences, however, does not imply a lack of punishment. Raids based on suspicion of piracy, confiscation of goods, arrests, and pretrial detention are the much more common, de facto forms of punishment in South Africa. The shift by industry groups toward criminal penalties for all infringement cases pushes vendors further into this extrajudicial punishment regime, providing more context for the resistance of judges to harsh sentencing once the case ends up in court. Conversely, even when significant fines are imposed, industry rights holders sometimes have difficulty collecting awards. The large role played in the optical disc retail trade by Pakistani and Chinese immigrants poses a particular challenge, in this regard, as such defendants can often transfer assets abroad or leave the country to avoid sentences.

Public-Private Justice

One anti-piracy measure routinely advocated by the IIPA and other industry groups is the creation of separate IP courts to expedite infringement cases. In South Africa, industry groups got much of what they wanted with the establishment of the commercial crime courts, which hear cases of fraud and other crime directed against businesses. The first commercial crime court was established in Pretoria in November 1999 through a partnership between the South African Police Service, the National Prosecuting Authority, the Department of Justice, and Business Against Crime South Africa (BACSA), a not-for-profit organization funded by South African businesses and USAID. BACSA played a critical role: it supplied the initial funding to hire prosecutors for the court and in so doing set the precedent for the quasi-privatization of the criminal justice system that has come to characterize the enforcement effort more generally. The perceived success of the Pretoria

Operation Dudula

Operation Dudula was an anti-piracy campaign organized by the recording artist, anti-apartheid activist, and poet Mzwakhe Mbuli, who had enjoyed considerable popularity in the 1990s. The campaign was built around Mbuli's organization, Concerned Musicians, and launched in 2006 with a series of marches in Johannesburg, Cape Town, Durban, Port Elizabeth, and Polokwane. Mbuli framed the marches as an effort to do what the industry groups and the state had failed to do: save artists "the millions" they were losing to pirates. In a typical interview, Mbuli declaimed:

> It is time for this criminal behavior to stop . . . This is not only infringement of copyright material, it is also economic theft and I am appealing to all proud and patriotic South Africans to stand with us in this fight for the very life of our industry. On your marks, get set, ready, GO. Run Criminals Run! (Biggar 2006)

With its strong populist overtones, the campaign quickly attracted the participation of large numbers of local artists, who joined Mbuli in police raids on "burner labs" and flea markets, starting a trend of direct participation by aggrieved musicians in enforcement activities. The musicians, in turn, were often accompanied by film crews, who captured emotional scene-of-the-crime testimonials.

Like many campaigns, Operation Dudula was a media-centered effort, designed to intimidate pirate vendors and shame the South African consumer. It ran into trouble almost immediately, however, when marchers began to violently confront street vendors and local retailers—mainly Pakistanis and Chinese. The resulting physical harassment and destruction of property led to counterclaims of theft and assault. The resulting bad publicity brought the marches to an end. The legality of musician-accompanied raids was also challenged: none of the musicians held the rights to their own work, and they therefore had no legal standing to accompany police as complainants. As a result, Operation Dudula produced a wave of cases that could not be prosecuted.[1]

Operation Dudula was initially supported by RiSA, the recording industry association. From RiSA's perspective, the movement provided welcome evidence of local support for the anti-piracy agenda and dramatized the role of artists, rather than corporations, as the main victims of piracy. But support from RiSA died amidst the charges of vigilantism and the controversy surrounding the marches. The break between RiSA and Operation Dudula went public in August 2006, with Mbuli calling for the resignation of RiSA chief Ken Lister (Coetzer 2006). Wider cooperation on raids between Dudula members and RiSA, the DTI, and the SAPS came to an end, and Operation Dudula eventually folded in 2008.

---

1   Interview with Advocate Nkebe Khanyane, National Prosecuting Authority of South Africa, 2009.

Commercial Crime Court led to the creation of a Johannesburg court in 2003, two additional courts in Durban and Cape Town, and new commercial crime units within the police service—all enjoying BACSA financial support.

Industry-sponsored staff training and education campaigns are routinely conducted for and through such public institutions and mark another side of public-private cooperation. BACSA continues to train prosecutors at the commercial crime courts. RiSA's Anti-Piracy Unit supports and trains officials in the South African Police Service, the National Prosecuting Authority, and the South African Revenue Service. Sony BMG funds customs and excise officials to deliver anti-piracy seminars to high school students. Microsoft South Africa funds local activities of the US-sponsored "STOP!" initiative—the Strategy Targeting Organized Piracy—which, in collaboration with the US Embassy and US Information Agency, also trains South African judges. As the other contributions to this report document, the expansion of enforcement in developing countries takes place through such blurring of public-private boundaries in law enforcement, policymaking, and the judicial system. The judiciary's central role and comparative insulation from stakeholder capture tend to foreground the resulting contradictions.

If these contradictions have a poster child in South Africa, it is Marcus Mocke, a small-scale Johannesburg distributor of pirated optical discs. In 2004, the Johannesburg Commercial Crime Task Force raided Mocke's residence and found four hundred pirated DVDs and PlayStation games. Mocke pled guilty to distributing and trading pirated goods in flea markets. In January 2005, the Johannesburg Commercial Crime Court handed down what was widely reported as the harshest piracy sentence in South African history: a choice between eight years in prison or a R400,000 fine ($65,000). Mocke agreed to pay the fine. Fred Potgieter, then head of SAFACT, heralded the sentence as "a landmark decision and a major breakthrough in the war against piracy" (Bizcommunity 2005). The presiding judge later suspended the fine, contingent on Mocke's good behavior.

How Piracy Works in South Africa

The copying and circulation of illicit books, tapes, and other media has a long history in South Africa, linked to practices of political and cultural resistance to apartheid. Inevitably, these practices overlapped a wider range of economic needs, entrepreneurial practices, and implicit or explicit acts of social disobedience. Economic sanctions against South Africa made the informal economy the primary form of access to many kinds of goods, from textbooks to electronics. A grey-market service economy also flourished in this context as skilled workers offered their services directly to customers, circumventing both white-owned businesses and state taxes. The widespread theft and resale of factory goods mapped racial divides between labor and management and blurred the lines between criminal and political behavior. In this fashion, the consumption of pirated goods was normalized and integrated into wider South African political and social practices. Piracy became part of everyday life and, as such, rarely needed justification.

Although the political valences of piracy have mostly dropped away in the post-apartheid era, the sharply racialized patterns of inequality and access to media have not—nor has the normalization of piracy and its role in a wider ethic of social disobedience. In this section, we explore this daily side of piracy and its complicated social geography. We draw on fieldwork conducted in three Johannesburg flea markets where vendors sell pirated music and film—Bruma, Noord/Plein, and Fordsburg Square—and from our investigation of the market for pirated goods in Hanover Park, a Cape Town township.

Optical Disc Piracy

The South African market for pirated media has two characteristics that distinguish it from other developing countries: (1) the very slow development of broadband services at the high end of the consumer spectrum and (2) the still active market for cassette tapes at the low end. In the middle lies the ocean of pirated CDs and DVDs.

As in other countries, the introduction of the cassette tape in South Africa in the 1970s gave rise both to industrial-scale cassette piracy and—notably—to the first large-scale consumer-based copy culture, built around "mix tapes." South Africa was an importer of tapes from the United Kingdom but also had significant local production capacity that served the domestic market and enabled exports to surrounding countries. As inexpensive CD players became available in the early 1990s, the format went into rapid decline. By 1996, 42% of music sold in South Africa was in cassette format, compared to 53% on CD (DACST 1998).[31]

But in South Africa, as in many other poor countries, decline did not mean extinction. Rather, the cassette tape moved downmarket, catering primarily to low-income, often rural consumers who still depended heavily on battery-operated radio/cassette players. The available content in the cassette market shifted accordingly, away from international hits and toward music in the local vernaculars, including Shangaan music, Maskandi and Tswana traditional songs, music from Lesotho, and gospel. Legal copies of these cassettes remain relatively inexpensive, ranging from R14 ($1.75) to R25 ($3.10), with prices practically unchanged for the last fifteen years. By 2003, the formal market had shrunk to $3.2 million, and by 2008 to less than $400,000 (Euromonitor International 2009). Such numbers do not include the large informal market for cassettes. The Noord/Plein flea market, which serves working-class commuters, still has cassette vendors. And outside the urban centers, the cassette remains a significant medium that anchors a variety of local media practices, from music listening to the recording and sharing of local performances and religious sermons.

By industry accounts, 2000 and 2001 were the watershed years in South Africa's shift to CD and DVD piracy. The IIPA's South Africa report for 2001 noted the rapid influx of pirate CDs and DVDs and on that basis called for South Africa's inclusion on the USTR's watch list. In our

---

31   In comparison, cassettes accounted for 24% of the US market in 1996.

view, this nervousness involved a certain amount of projection on the part of industry. Music CDs were indeed circulating in South Africa in larger numbers by the end of the 1990s, driven by a ramp up in East European and South Asian production. But the high price of DVD players in 1999 and 2000 meant that the market for pirated movies was miniscule. The real boom in movie piracy, globally, would have to wait for the wave of cheap Chinese DVD players that hit the market beginning in 2003 and 2004. By 2005, multiformat DVD players were commodity items, priced as low as R250 ($35). By 2008, DVD players were in 48% of South African households. By 2007, the IIPA had stopped bothering to report on South African piracy.

FLEA MARKETS AND STREET VENDORS

Optical disc piracy is part of the much larger informal economy in South Africa and shares much of its infrastructure. Traditionally, piracy has been associated with the flea markets that dot South African towns and cities. Vendors at flea markets typically operate from fixed stalls that they rent from market owners or managers, selling daily or on weekends, according to the market calendar. Street vending, in contrast, is a much more dispersed and transient practice. Vendors congregate around high-traffic intersections, trading in what they can physically carry during the day.

The main Johannesburg flea markets operate as both retail and wholesale sources for pirated goods—often, in the latter capacity, serving as supply hubs for wider networks of neighborhood, town, and rural vendors. Among the more established wholesalers, this system is highly organized: buyers place written orders, and sellers circulate lists of available titles, often outside the city.

Some buyers come from considerable distances to the main market hubs. During our interviews in Johannesburg, we encountered a buyer from Bloemfontein, a city some five hundred kilometers away, who said that he had come to buy the newest titles for "private clients." These buyers connect flea markets and production networks in the urban centers to the more remote town- and neighborhood-based vending networks, where limited access to broadband, especially, constrains the range of readily available goods. In the larger cities, where broadband Internet is accessible but still scarce, some suppliers run subscription services that provide access to the most popular downloads in a given month. Others operate cottage production facilities, taking orders and delivering discs to their regular buyers.

Both RiSA and SAFACT publish newsletters that list "hot spots" where pirated goods are retailed and detail the results of police surveillance, raids, and pressure brought to bear on market operators. Many of the named hot spots in Johannesburg (Bruma and Rosebank) and in Pretoria (the Montana flea market) are described as "closed down" following rights-holder and police efforts. But our investigations had no difficulty identifying pirated music and films on sale in the markets we visited. In South Africa, as elsewhere, strong street enforcement appears capable of suppressing the more organized forms of retail piracy but shows less evidence of impacting the lower strata of the informal economy, where more transient vending practices are the norm.

When fixed stalls are shut down or their vendors evicted, pirate vendors often move to adjacent spaces, like the parking areas of shopping malls, or into the townships. The distribution chain reconfigures itself to limit risk.

Historically, the storage and packaging of imported pirate discs from Malaysia, China, and Pakistan required large local intermediaries to manage warehousing and distribution. These operations reflected what was still a largely industrial organization of CD and DVD piracy, built around centralized production in foreign factories and large-scale smuggling operations. Such operations were also obvious targets for enforcement, and South African police and customs registered a number of high-profile raids, arrests, and prosecutions throughout the early 2000s. By the second half of the decade, seizures of large shipments of CDs and DVDs had begun to drop. The SARS reported 165 seizures in 2006/7, 50 in 2007/8, and 37 in 2008/9, in an environment of rising enforcement activity.[32]

As in the other countries documented in this report, we see little evidence that stronger enforcement is the determining factor in this decline. Rather, the industrial-scale, smuggling-based model is being supplanted by smaller, more distributed operators who produce discs locally, in closer proximity to their networks of vendors. According to Ben Horowitz, CEO of Bliksem DVDs, the new generation of small-scale pirate operations supplies "two or three areas; that's about as big as [an operation] gets in the [Johannesburg] city center."[33]

Production and vending are almost always separated in these contexts—a strategy that keeps producers at arm's length from the police. As Horowitz observes, this means that "the vendors have to take all the risk. They buy the DVDs probably for R5 [$0.65], and the suppliers are producing for about R1.50 [$0.20] per DVD." Such arrangements mitigate risk for the producers while keeping production in relatively close alignment with consumer demand—a crucial advantage in a pirate market dominated by local repertoire. But as elsewhere, the underlying shift is technological: the rapidly declining costs of burners and other copying technologies have obviated much of the need for large-scale factories and for high-cost/high-risk investments in cross-border smuggling. In turn, this cottage industry is facing its own technological obsolescence as non-commercial Internet distribution and personal copying and storage technologies shift the locus of activity to the consumer.

MARKET DIFFERENTIATORS

Pirate CD and DVD vendors practice a variety of forms of market segmentation, visible in the differences in disc titles, quality, and prices available in South African flea markets. Fordsburg Square's clientele is predominantly of South Indian descent and varied age and income. Vendors offer a wide selection of Bollywood film titles and music, supplemented by the usual mix of

---

32   Interview and follow-up e-mail communications with Sean Padiachy, head of the SARS FIFA World Cup customs unit, 2010.

33   Interview with Ben Horowitz, 2009.

Hollywood films and popular Western music. The quality is generally very good, and discs are available across a range of prices, depending on the packaging and included extras. Fordsburg vendors—mainly South Asians from Pakistan and Bangladesh—also compile music CDs "on demand, while you wait," pulling tracks together from different CD releases. Burning facilities are on site but generally kept in back rooms closed to the public.

While Fordsburg caters to the Bollywood niche market, the Bruma and Noord/Plein flea markets serve a broader range of consumer preferences, notably including local film and music. Bruma is frequented by a mostly middle-class and tourist clientele, Noord/Plein by predominantly black, working-class commuters traveling through the city center. Accordingly, Bruma is busiest on weekends, while the Noord/Plein market has a stronger daily presence.

BRUMA

Situated among the middle-class suburbs of Johannesburg, the Bruma flea market is one of the better-known hot spots for CD and DVD piracy, in part due to the ongoing lack of cooperation between market owners and police.[34] On the weekends, Bruma shoppers crowd the stalls selling pirated DVDs. Vendors generally cater to middle-class tastes shaped by global advertising cultures. The available stock tends toward US productions, such as television series like *Desperate Housewives* and *CSI*, currently exhibited movies (at the time of our visits in March 2009, *Slumdog Millionaire*, *Race to Witch Mountain*, and *Marley and Me* featured prominently), and a range of older Hollywood hits. There are no art-house movies on display, and classic titles are limited to old James Bond movies, Clint Eastwood films, and other assorted action films—though vendors can generally fill special requests on demand.

Staff are often young, immigrants, or both—signs of the disposable labor strategies that minimize liability and shelter owners from arrest. Many vendors, consequently, display very limited knowledge of the movies they sell. When asked about the availability of a classic, such as *Casablanca*, most vendors we questioned looked blank. Similar responses were received to queries about upcoming films that were not yet in theatres, such as *Duplicity* and *Monsters vs. Aliens*.

Bruma vendors price their goods for a clientele with middle-class disposable income and are generally prepared to meet the higher level of customer scrutiny that comes with those prices. Television sets with DVD players are set up on site to verify the quality of goods before purchase, though vendors are quick to volunteer which are "good" copies and which are poor. DVDs sell for R40–R50 ($5.20–$6.50) for "good-quality" copies and R20–R25 ($2.60–$3.20) for inferior copies. Only one stall that we visited had an extensive music collection, prominently identified as legitimate by signs on the shelves and walls.

---

34   Bruma is one of the "few flea markets who still refuse to take action against tenants committing illegal acts." Interview with James Lennox, CEO of SAFACT, 2009.

Noord/Plein

The Noord/Plein market surrounds a central Johannesburg taxi station where minibuses drop off and pick up thousands of commuters each day.[35] A wide variety of goods are sold, from secondhand clothing and shoes, to fruit and vegetables, to CDs and DVDs. The CD and DVD vendors are massed in a wide area around the taxi line, organized into two main zones often described by locals as separate markets. The zone immediately in front of the taxi line offers more costly goods with higher-quality packaging. A second zone behind the taxi line offers more obviously artisanal copies, mostly packaged in small plastic bags. The higher quality goods in front generally sell for R20 ($2.50), while those in the back sell for R10 ($1.25). Although Nollywood (Nigerian) films are not sold in the market proper, stalls specializing in the Nigerian "watch and buy" DVDs congregate just a few meters away, selling apparently legitimate copies of these movies for R20. Individual vendors also roam the area, selling the most popular recent films, both South African movies, such as *White Wedding* (a comedy in cinemas at the time of our visit) and *Jerusalema* (a drama released in 2008), and international hits, like *The Fast and the Furious 4* (released in South Africa in April 2009). On a visit during the week that Michael Jackson died, both zones were doing a brisk business in his CDs and DVDs.

Behind the taxi line, the low-end vendors stack DVDs and CDs in plastic sleeves on black crates. South African film and music—especially gospel—dominate the displays. The typical DVD pile includes most of the local hits of the past two decades: *Jerusalema*, *Madloputu*, *Sarafina*, *Swop*, *White Wedding*, the Schuster comedies *Mr. Bones 1* and *Mr. Bones 2* and *Mama Jack*, and others. Old but popular local television dramas, such as *Kwa Khala Nyonini* and *Bophelo Ke Semphego*, are also well represented. International films are available but tend toward older action movies, such as *The Fast and The Furious* and *The Good, The Bad and The Ugly*. Pornography is also widely available.

The market in front of the Noord/Plein taxi rank is smaller and caters to commuters with more money. DVDs are sold in packaging similar to that of the original products. South African gospel music and movies predominate here too, but there is a greater variety of titles and more of the current international hits, such as the recent *Transformers: Revenge of The Fallen* and albums by Beyoncé and Kanye West.

In interviews, consumers regularly voiced concerns about the quality of their purchases. As one woman admitted: "I know I'm buying at my own risk as [the discs] often do not play." Pirate vendors catering to the high end of the market are sensitive to this problem and often have TVs and DVD players available to demonstrate the quality of the product to prospective buyers. Even low-end vendors without such equipment value return customers, however, and many mark their discs with personal symbols, which allow the buyer to return them if they prove defective.

---

35   The Noord/Plein market is always busy. When asked what hours he worked, one vendor responded that business is always good for them as there are always people flocking to the area. Another vendor reported that on a "good day" he can earn as much as R3,000 ($375). Both of the vendors said they owned their own stalls, placing them in positions of relative privilege in this informal economy.

CULTURAL NATIONALISM

Although local content dominates markets like Noord/Plein, vendors are often circumspect about displaying pirated local goods. Musician-driven Operation Dudula and dedicated enforcement campaigns for major South African films like *Tsotsi* reflect—and have helped foster—a strain of cultural nationalism in South African anti-piracy efforts that appears to at least partially influence the behavior of vendors. Operation Dudula, in particular, framed piracy of local music and film as not merely wrong but "unpatriotic"—a dynamic we have seen at work in many locally grounded enforcement efforts and one that tends to operate to the benefit of a handful of high-profile local products. At the Noord/Plein market, for example, shelves are filled with South African gospel music, but many of the top stars, such as the "Queen of Gospel," Rebecca Malope, are conspicuously absent. Malope was an active and visible participant in Operation Dudula marches against vendors. Similarly, in the Bruma market, there are surprisingly few South African titles on the shelves, and no copies of the hugely popular Leon Schuster slapstick comedies in sight. The visible stock is generally limited to lesser-known Afrikaans comedies, such as *Vaatjie Sien Sy Gat*.

When asked about South African movies, many Bruma vendors refused to talk further. A few said that they had copies of *Tsotsi* (2005) and *Jerusalema* (2008) but that these were "old now" and no longer routinely stocked. Unlike the slapstick Afrikaans titles on display, both are "serious" films, often presented as representatives of the national cinema. When we asked about a copy of the hit 2008 Schuster comedy *Mr. Bones 2*, a runner was sent off and returned twenty minutes later with the disc and photocopied cover in hand. As he surreptitiously handed us the copy, he whispered that "you can get arrested for having this" and just laughed when we asked whether this was true of any of the other clearly pirated goods on display. He admitted they regularly suffered raids and confiscation of goods, at which point "we just have to start copying them all over again." He seemed concerned that openly displaying South African titles, like the Schuster movies, could have more serious repercussions, particularly if an angry customer elected to file a complaint. Unfortunately, our research was unable to explore this nationalist dimension in greater detail. It was unclear to us, for example, whether the *Tsotsi* campaign and Operation Dudula were the main sources of this anxiety or just leading indicators of wider enforcement bias for local goods. Vendors, at least, appear to believe the latter.

THE DEFORMALIZATION OF THE TRADE

Markets like Bruma and Noord/Plein are relatively exposed to pressure from police and industry groups. Such sites are easily raided, and the formalized rent and ownership structures in the markets give the DTI and the SARS leverage over market owners. As pressure on vendors has increased, we see evidence of a shift to less-exposed forms of trade, including more mobile street vending and the use of underage and/or immigrant labor to conduct most illegal sales. Pressure on the flea markets also appears to have reinforced long-standing practices of neighborhood vending, especially in poor neighborhoods with limited access to media and larger markets. In this respect, our work confirms recent findings by SAFACT, RiSA, and other anti-piracy groups, which also describe a process of deformalization of the pirate trade.

In many cases, neighborhood vending involves home-based production of CDs and DVDs, and often home-based sales. Although such vending practices are small-scale, they are much harder to police, and they remove the vendor from the precariousness of operating in the flea market or on the street. They also embed vendors in closer service relationships with regular networks of customers, providing them (typically) more prominent roles as opinion-leaders within those networks. Inevitably, such security and prominence comes at the expense of volume and, in some contexts, the ability to charge premiums to tourists or middle-class customers. Such factors make neighborhood vending a lower-risk but also more marginal business than stalls in high-traffic marketplaces.

Hanover Park

Hanover Park is on the Cape Flats, on the outskirts of Cape Town. It is one of the many urban ghettoes to which black South Africans were removed during apartheid and was referred to colloquially as the "dumping ground of apartheid." Today, it is a so-called "coloured," working-class township, located between a large industrial zone to the east and the middle-class coloured neighborhood of Lansdowne to the west.[36]

Neighborhood Vending

Piracy takes several forms in Hanover Park. Many people have opportunities to visit flea markets, but the most common form of access to recorded media is the purchase of pirated feature films from local neighborhood vendors, usually for between R5 ($0.65) and R10 ($1.30). Increasingly, these vendors have their own computers and burners and can create products on demand for customers—by either downloading from the Internet, copying from their existing stock of popular films and music, or acquiring goods from wider vendor networks. When production tools are unavailable, neighborhood vendors act as retailers, purchasing discs wholesale or placing custom orders with the more established vendors and distributors in the large urban markets. Business models vary and include rental models and resale back to the vendor.

In most cases, only a few gatekeepers within families or larger social networks maintain direct contact with the sellers. Most people in the community get their pirated media through a

---

36   The term "coloured" is an apartheid designation that refers to people of mixed race origin. Today it is controversial, if still widely used. Of South Africa's 47 million people, 9% are coloured, and most of these are concentrated in the Western Cape region, primarily Cape Town. Hanover Park is a coloured residential area, home to roughly 30,000 people, of whom some 11,000 are between the ages of 15 and 34. Only 1,700 possess a high school matriculation certificate. The overwhelming majority—nearly 80%—are native Afrikaans speakers. Unemployment is above 50% (Statistics South Africa 2010). The neighborhood is well known for gangsterism and by the late 1970s was home to nearly twenty gangs, each "owning" their own small patch of ghetto (Steinberg 2004).

family member or close friend. Such networks are informal and highly dependent on personal trust.[37] Vendors maintain a stock of goods based on their judgments of what will be popular; the buyers (often male heads of households) choose the titles they think will be most appropriate for themselves and their families.

In our visits with home vendors, the latest movies were usually kept in a disc flipbook that buyers could browse. In a few cases, vendors showed customers previews of films. The recommendations given by vendors are often personalized and based on long-term relationships with their customers. Unlike the stall vendors in flea markets, who are often selected for their expendability in the event of a raid, neighborhood vending places a premium on expertise—real or feigned. Consistently, in our fieldwork, vendors cultivated the impression that they have seen all the films they have available for sale and provided brief synopses and reviews for prospective buyers. In this way, they become local opinion-leaders, influencing which films circulate in the neighborhood and gaining some modest corresponding social status.

Although vending at the low end of the market is often associated with low-quality products, our Hanover Park respondents had decidedly mixed views on this subject. A perceived tradeoff between low price and the risk of low quality was clearly present, as was the general willingness of consumers to make that bet. But the majority of Hanover Park residents indicated that the quality of the discs they acquired was usually high. Discs worked and showed few of the typical signs of a poorly pirated copy, such as subtitles in another language or screener code. Although we do not treat such opinions as definitive, they are consistent with the larger compression of pirate markets, evident throughout this report, as cheap reproduction equipment and access to high-quality Internet distribution improves. The obsolescence of much of the audiovisual equipment in Hancock Park homes also likely plays a role in lowering expectations.

Consumption

In middle-class South African communities, viewing movies and listening to music is increasingly a private experience, mediated by the availability of multiple screens, playback devices, and headphones and by the broader expansion of personal media. According to one recent study of South African students, pirated video—here, the US television show *Grey's Anatomy*—is almost always viewed alone (McQueen 2008). In Hanover Park, in contrast, all thirty of our interviews described viewing practices organized around friends and family.

The collective dimension of viewership was not described as a necessity in our interviews but rather as a basic part of the media experience that anchors other forms of sociability. As described by one respondent:

---

37   As we found in trying to establish relationships with local vendors. Our successful contacts came only
      with local assistance and took considerable time and repeat visits.

> What I normally do is I invite my mother down for a nice cup of coffee, then I put her on the bed—she's an elderly lady—and I say to her, "You like Bollywood movies," and I put on a Bollywood movie. And all of us, we're seven children, we all come together and talk about that afterwards.

Respondents expressed little interest in collecting or holding on to the pirated discs, once viewed. In a handful of cases, respondents indicated that they had resold their discs to other buyers in order to raise money to buy new films. Most, however, gave their discs away to friends or family members.

Consumers generally indicated that the cost of a pirated disc is low enough to permit a free flow of "used" goods. Consequently, for many individuals, pirated movies arrive second- or thirdhand—circulating after the initial purchase or acquisition. Such consumers were rarely interested in viewing the latest titles but rather saw value in media that had filtered through the community.

What People Watch

It is a measure of both the naturalization of American cultural influence and the emergence of a newly globalized Indian culture that when Hanover Park respondents were asked whether they viewed "foreign" films, a substantial majority indicated Bollywood films. When explored further, actual viewing preferences almost always tended toward Hollywood, and to explorations of black culture and gang life in particular.

Genre preferences followed fairly stereotypical patterns in our interviews, with younger women often signaling preference for romantic comedies and teen movies (for example, *High School Musical*), younger men indicating a preference for action films, and older respondents adding dramas to the mix. One of our main informants in Hanover Park, a pirate vendor named Rafiek, reported that "nigga gangster movies" were his top-selling genre—a statement our interviews broadly confirmed, and one that resonates with the strong local hip-hop culture and interest in black American culture in general.[38] A large number of those interviewed also indicated that they often viewed pirated DVDs of standup comedy sets, usually by African American comedians, such as Dave Chapelle or Chris Rock, but also—and with some of the typical ambivalence toward acts of local piracy—South African comedians, like Joe Barber and Marc Lottering.

When asked about South African media, the response was remarkably thin. Among movies, only *Tsotsi* and *Jerusalema* were mentioned. Several respondents listed the TV programs *Generations* and *7de Laan* as among their favorites. Vendors—though not customers—indicated that local slapstick Afrikaans comedies like *Vaatjie Sien Sy Gat* and *Poena is Koning*, which caricature poor white South Africans, were popular items. Field observation also points to sales of

---

38   Hip-hop emerged on the Cape Flats in the early 1980s as one of many responses to apartheid. It was particularly powerful in Cape Town, where it became a vehicle for expressing the tensions of racial marginalization (Watkins 2001).

pornographic films and gospel-music videos, but there was little corroborating evidence for either in our interviews.

It is indicative of the poverty of media access in South Africa that these thin collections of pirated goods actually diversify the media environments in which people live. Very few South Africans inhabit the long-tail media environment characteristic of high-income, high-bandwidth countries.[39] Although enforcement pressure plays a role in forcing vendors to limit the breadth and quantity of stock, the selection of pirate goods is still surprisingly limited: a survey of titles in Noord/Plein turned up only forty-three distinct film and music discs, reproduced across dozens of stalls and tables. In Hanover Park, Rafiek casually sold our researcher a spindle of 120 European films that he had not been able to sell elsewhere—for R1 ($0.13) per movie. All were high-quality copies of screeners or original DVD releases, but such work had no local audience. For film, especially, the primary function of South African pirate networks is to make the larger, advertising-driven mass culture much more accessible.

One sign of this dynamic is that, even in communities like Hanover Park where moviegoing is rare, the first selling point for pirate vendors is almost always speed. A large majority of Hanover Park respondents indicated that they sought out the latest releases. The "latest," in this context, meant either a new release or—in some cases—an anticipated release that had been delayed in South Africa due to studio windowing strategies. Being able to view films at roughly the same time as more economically privileged consumers seems like a trivial concern, but our work suggests it is part of an increasingly powerful experience of inclusion in a globalized media community. Such forms of inclusion are especially significant in countries like South Africa, where real and perceived marginality—geographic, economic, racial, and other—are written into daily experience on many levels. Piracy is, in this limited sense, a means of bridging that marginality. Put differently, it is what happens when wildly successful marketing campaigns meet with wildly unsuccessful efforts to serve local audiences.

Conclusion

When asked whether the IIPA or the USTR play a role in DTI decisionmaking, Amanda Lotheringen, deputy director, responded with an unequivocal no. Given the diminishing interest of these US-based actors in South Africa in the last half decade, we see no reason to doubt this answer. But it seems equally clear that, where the DTI is concerned, no arm-twisting is needed. Like many other developing-country agencies responsible for trade policy, the DTI has taken the lead in pushing for stronger formal IP protection and enforcement measures. It has adopted the agenda—or perhaps more accurately, the underlying assumptions—of the foreign interests that

---

39   The baseline for (and often the limit of) media diversity in South Africa is the four "free-to-air" broad-cast television channels, which reach nearly all South Africans. SAARF (2009) data shows that the next most popular form of access, subscription satellite television, reaches only 20% of viewers.

dominate the copyright economy in South Africa, and it has been a very effective de facto advocate for those interests. Beginning with the 2010 World Cup, South Africa seems poised for a new round of enforcement activism, including more public investment in policing and an expected overhaul of intellectual property rules, also under the guidance of the DTI.[40]

The anti-piracy fight in South Africa has benefited considerably from association with wider security efforts, visible most prominently in the measures adopted for the World Cup. Police budgets have grown in recent years, with an anticipated further one-third growth in the national police budget between 2009 and 2012 (Parliamentary Monitoring Group 2009). Neither the DTI nor the SARS would reveal the size of their enforcement budgets, except to acknowledge that these too have increased.

As in other countries, however, the expansion of the enforcement agenda in South Africa faces a range of internal constraints, beginning with the heterogeneity of the different agencies and layers of government involved and the resistance of some of them to the enforcement maximalism of US-led industry groups. As elsewhere, there is an array of more pressing social problems that have their own claims, constituencies, and institutional centers of power. Copyright enforcement, under conditions of scarce policing resources and overburdened courts, is a zero-sum game that inevitably draws resources away from other issues. In the enforcement context, the obvious counterpoint is the crisis of public security, headlined by South Africa's high murder rate. But wider issues of access to knowledge and public health have also played important roles in shaping South Africa's IP politics.

Cooperation by the police and the DTI with industry groups has earned South Africa some respite from the continuous pressure placed on other countries in this report—most notably Russia, India, and Brazil, which often act as the geopolitical peers of South Africa in international forums like the WIPO and WTO. But our research was unable to answer one of the basic questions that should guide any expansion of public investment in this area: do enforcement efforts have any effect on the availability of pirated goods in South Africa? As in the other contributions to this report, we see evidence that enforcement can harass the more vulnerable parts of the retail pirate-disc channel, but we find no evidence that this represents any serious constraint on the consumer availability of pirated goods. In our view, availability in South Africa is shaped by factors that are largely exogenous to the enforcement effort: poverty, cheap consumer technologies, uniquely expensive broadband Internet service, the globalization of media culture, and the chronic weakness of legal distribution and exhibition channels. None of these seem likely to change in ways that will diminish the availability of pirated goods in the coming years.

Nonetheless, given the South African institutional landscape, the prospect of stronger enforcement policies—including consumer-directed enforcement—is a real one. The DTI and local industry groups are likely to push for the strongest available international norms, such as strict

---

40   Interview with Amanda Lotheringen, DTI deputy director, 2009.

ISP liability and Internet surveillance of file transfers. Planned copyright law revision—and, very possibly, the recently completed international Anti-Counterfeiting Trade Agreement—will create a context for this push.

From our perspective, however, the central question should be how to create vibrant, accessible media markets and how, in particular, to move South Africa out of the high-price, small-market equilibrium it shares with many other developing countries. The conventional wisdom among industry groups is that stronger enforcement spurs the growth of the legal media market and thereby improves access to media. While we understand this logic, we do not see it operating in South Africa. The more salient factor, in our view, is competition within the domestic media market. In countries where large domestic media industries compete for audiences, piracy has been a catalyst for new, legal lower-cost business models. In countries where media markets are dominated by foreign multinationals, competition on price and services within the legal market is rare, and piracy becomes the primary form of local compensation. In such contexts, enforcement answers none of the core problems of media access or market growth.

South Africa straddles these global issues. In many respects, it presents a typical case of a multinational-dominated media market and an exemplary case of government cooperation with those multinationals on enforcement. But it has also seen local movie exhibitors dramatically reduce prices as a means of building local markets (providing the sole exception in this report to soaring ticket prices around the world). The Ster-Kinekor/Nu Metro price war significantly expanded the South African movie audience and proved that exhibitors in Hollywood-dominated markets can, under some significant constraints, pursue alternative market strategies. In most other respects, however, the high-price structure of South African media markets remains intact, with only hints of change at the peripheries in the form of experiments like Bliksem DVDs and new external players, such as the increasing presence of low-cost media from Nigeria and India. The key question for South Africa is whether such peripheral actors can prosper—lowering prices, democratizing access, and creating a mass legal market. The alternative, in our view, is simply more of the same: slowly growing legal markets pegged to rising incomes, fast-growing pirate markets pegged to decreasing technology costs, and expanded public investment in an enforcement effort with little demonstrable impact on either.

About the Study

The primary research for this chapter was conducted in 2009 and early 2010. A number of researchers lent their expertise to the project. We are grateful to Adam Haupt and Tanja Bosch who conducted case study with marginal music producers and the residents of Hanover Park, respectively; Julian Jonker for producing a background paper on the legal framework in South Africa; and Nixon Kariithi, who conducted a discourse analysis drawing on more than eight hundred print, radio, and television reports (between 2006 and 2008) on piracy in South Africa. Joe Karaganis collaborated on the secondary research, writing, and editing. Natalie Brown provided competent, efficient, and intellectually engaged research assistance throughout the project.

We would also like to thank some of the many people who gave their time to be interviewed: Ootz (independent hip-hop music producer and DJ), Peter Ndebele (a young, independent black filmmaker from Soweto who self-finances his projects and sells them from the trunck of his car), Sean Padiachy (SARS), Amanda Lotheringen (DTI deputy director), Police Superintendent Mangaliso (Johannesburg Metro Police), Advocate Nkebe Khanyane (National Prosecuting Authority), Ben Horowitz (Bliksem DVDs), Dan Jawitz (independent film producer), Braam Schoeman (RiSA), James Lennox (SAFACT), Aifheli Dzebu (National Film and Video Fund), and Jacques Stoltz (Gauteng Film Commission). Not least we want to thank Melody Emmet for her support in setting up these interviews in a very short time frame.

References

Agence France Press. 2008. "Bollywood Piracy Fighters Take Battle to US Congress." April 21. http://afp.google.com/article/ALeqM5i5u8wX6GI_QtBpBoDg9FCnGZSvaw.

Berger, Guy. 2002. "Deepening Media Density: What South African Freedom Shows Us." *The Round Table* 91 (366): 533–43.

Biggar, Taryn-Lee. 2006. "Operation Dudula Anti-Piracy March!" *Music Industry Online*. http://www.mio.co.za/article/operation-dudula-anti-piracy-march-2006-04-17.

Bizcommunity.com. 2005. "Harshest Sentence in SA History Against DVD Pirate." January 25. http://www.biz-community.com/Article.aspx?c=87&l=196&ai=5583.

BSA/IDC (Business Software Alliance and International Data Corporation). 2010. *Seventh Annual BSA/IDC Global Software Piracy Study*. Washington, DC: BSA.

Coetzer, Diane. 2006. "South Africa Schism." Billboard, September 9. http://www.allbusiness.com/retail-trade/miscellaneous-retail-retail-stores-not/4565794-1.html.

DACST (Department of Arts, Culture, Science and Technology, South Africa). 1998. *The South African Music Industry: The Cultural Industries Growth Strategy; Final Report*. Pretoria: DACST. http://www.info.gov.za/view/DownloadFileAction?id=70494.

Dean, O. H. 1989. *The Application of the Copyright Act, 1978, to Works Made Prior to 1979*. PhD diss., University of Stellenbosch.

DPSA (Department of Public Service and Administration, South Africa). 2009. "FOSS Policy and Strategy." http://www.oss.gov.za/?page_id=485.

Durbach, Dave. 2008. "Not Digital Enough: Where is the Revolution?" *Levi's Original Music Magazine*, July 16. http://www.levi.co.za/MusicMag/Category/Detail/Detail.aspx?ID=552.

Euromonitor International. 2009. *Consumer Electronics in South Africa*. London: Euromonitor International. www.euromonitor.com/Consumer_Electronics_in_South_Africa.

Fisher, William W, III, and Cyril P. Rigamonti. 2005. *The South Africa AIDS Controversy: A Case Study in Patent Law and Policy*. Working paper, Harvard Law School.

Haricombe, Lorraine J., and F. W. Lancaster. 1995. *Out in the Cold: Academic Boycotts and the Isolation of South Africa*. Arlington, VA: Information Resources Press.

IFPI (International Federation of the Phonographic Industry). 2010. *Recording Industry in Numbers 2010*. London: IFPI.

IIPA (International Intellectual Property Alliance). 2001. 2001 *Special 301 Report: South Africa*. Washington, DC: IIPA. http://www.iipa.com/rbc/2001/2001SPEC301SOUTHAFRICA.pdf.

———. 2003. 2003 Special 301 Report: South Africa. Washington, DC: IIPA. http://www.iipa.com/rbc/2003/2003SPEC301SOUTHAFRICA.pdf.

———. 2007. *2007 Special 301 Report: Special Mention South Africa*. Washington, DC: IIPA. http://www.iipa.com/rbc/2007/2007SPEC301SOUTHAFRICA.pdf

IOL (Independent Online). 1999. "War Goes On against Video Game Pirates." October 9. http://www.iol.co.za/index.php?sf=136&set_id=7&click_id=81&art_id=ct19991005215227452P625715.

Larkin, B. 2004. "Degraded Images, Distorted Sounds: Nigerian Video and the Infrastructure of Piracy." *Public Culture 16* (2): 289–314. http://muse.jhu.edu/login?uri=/journals/public_culture/vo16/16.2larkin.html.

Mabuza, Ernest. 2007. "Crackdown Coming on Illegal Software Users." *Business Day*, April 25. http://mybroadband.co.za/nephp/6243.html.

Maggs, Jeremy. 2006. "Fake Fakes." *Financial Mail*, May 19. http://secure.financialmail.co.za/06/0519/admark/aam.htm.

Manners, Tom. 2009. "South Africans Pirate More Software." *t*, May 12. http://mybroadband.co.za/news/Software/8011.html.

Martins, Johan, and van Wyk, Helgard. 2006. "Investigation Into the Use of and Trade in Counterfeit Goods in South Africa." Study commissioned by the DTI, University of South Africa Bureau of Market Research.

McQueen, K. 2008. "An Anatomy of Grey's Anatomy." Unpublished research paper, University of Cape Town.

Muller, Rudolf. 2009. "Broadband: SA Versus the World." MyBroadband, May 27. http://mybroadband. co.za/news/Broadband/8208.html.

MyBroadband. 2008. "RISA and Torrent Website Truce?" November 24. http://mybroadband.co.za/ news/Internet/6101.html.

———. 2009. "Music Piracy and Local ISPs." August 11. http://mybroadband.co.za/news/ General/9138.html.

Naidu, E. 2007. "Local Music Fights Piracy for a Digital Future." *IOL Technology*, May 29. http:// www.ioltechnology.co.za/article_page.php?iSectionId=2892&iArticleId=3855863.

NTIA (US National Telecommunications and Information Administration). 2010. *Digital Nation: 21st Century America's Progress Toward Universal Broadband Internet Access.* Washington, DC: US Department of Commerce. http://www.ntia.doc.gov/reports/2010/NTIA_internet_use_report_ Feb2010.pdf.

Ogbor, J. 2009. *Entrepreneurship in Sun-Saharan Africa: A Strategic Management Perspective.* Bloomington, IN: AuthorHouse.

OMD South Africa. 2002. *South Africa Media Facts 2002.* Johannesburg: OMD. http://www.omd. co.za/samediafacts2002.pdf.

———. 2009. South Africa Media Facts 2009. Johannesburg: OMD. http://www.omd.co.za/ samediafacts2009.pdf.

Oxford Business Group. 2010. *The Report: Nigeria 2010.* London: OBG. http://www. oxfordbusinessgroup.com/publication.asp?country=70.

Parliamentary Monitoring Group. 2009. "South African Police Service: Strategic Plans & Budget 2009/12." http://www.pmg.org.za/report/20090630-south-african-police-service-strategic-plans-budget-200912.

PICC (Print Industries Cluster Council). 2004. *PICC Report on Intellectual Property Rights in the Print Industries Sector.* Cape Town: PICC. http://www.publishsa.co.za/downloads/intellectual_ property_report.pdf.

SAARF (South African Advertising Research Foundation). 2005. SAARF AMPS 2005. Presentation. http://www.saarf.co.za/AMPS/PPT/2005AfricanResponse23Nov.zip.

———. 2008. SAARF AMPS 2008. Presentation. http://www.saarf.co.za/AMPS/PPT/AMPS%20 2008B%20Industry%20Presentation.zip.

———. 2009. SAARF AMPS 2009AB. Presentation. http://www.saarf.co.za/AMPS/PPT/AMPS%20 2009AB%20-%20Industry%20Main.zip.

Statistics South Africa. 2010. StatsOnline. http://www.statssa.gov.za/.

Sell, Susan K. 2003. *Private Power, Public Law: The Globalization of Intellectual Property Rights.* Cambridge, UK: Cambridge University Press.

Smith, Theresa. 2006. "Pirates Sail Off with DVD Market." *Tonight*, January 23. http://www.tonight.co.za/index.php?fSectionId=358&fArticleId=3077775.

Steinberg, J. 2004. *The Number*. Cape Town: Johnathan Ball Publishers.

UNODC (United Nations Office on Drugs and Crime). 2009. "The Eleventh United Nations Survey of Crime Trends and Operations of Criminal Justice Systems: 2007-2008." http://www.unodc.org/unodc/en/data-and-analysis/crime_survey_eleventh.html.

Van Belle, J., B. Macdonald, and D. Wilson. 2007. "Determinants of Digital Piracy Among Youth in South Africa." *Communications of the IIMA 7* (3): 47–64.

van Zyl, Deon Hurter. 2009. Annual Report for the Period 1 April 2008 to 31 March 2009. *Cape Town: Judicial Inspectorate of Prisons*. http://judicialinsp.pwv.gov.za/Annualreports/Annual%20Report%202008%20-%202009.pdf.

Vecchiato, P. 2009. "ISPs Take a Stand for Internet Freedom." *ITWeb*, August 7. http://www.itweb.co.za/index.php?option=com_content&view=article&id=25180:isps-take-a-stand-for-internet-freedom&catid=182:legal-view.

Watkins, L. 2001. "Simunye, We Are Not One: Ethnicity, Difference and the Hip-hoppers of Cape Town." *Race & Class 43* (1).

World Wide Worx. 2009. *Internet Access in South Africa 2008*. Pinegowrie: World Wide Worx. http://www.worldwideworx.com/archives/107.

Worsdale, A. 2005. "Cinema Ticket Price War Erupts in South Africa." *ScreenDaily*, March 21. http://www.screendaily.com/cinema-ticket-price-war-erupts-in-south-africa/4022426.article/.

WTO (World Trade Organization). 2009. DS362: China—Measures Affecting the Protection and Enforcement of Intellectual Property Rights. Geneva: WTO. http://www.wto.org/english/news_e/news09_e/362r_e.htm.

Chapter 4: Russia

Olga Sezneva and Joe Karaganis

Investigators: Oleg Pachenkov, Irina Olimpieva, and
Anatoly Kozyrev

"The only way to kill piracy in Russia is strong copyright law with stern penalties and government resolve to enforce that law."

– Jack Valenti, president of the Motion Picture Association of America[1]

Introduction

Since 2000, Russia has held an unshakeable spot on the USTR's Special 301 "Priority Watch List," backed by industry claims of billions lost to US companies in software, music, and film piracy. For 2008, the IIPA reported US$2.3 billion in losses in business software alone. Had the motion picture, music, and entertainment software industry groups reported, the total would almost certainly have exceeded $3 billion. The only country with higher reported losses was China, where the IIPA cited some $3 billion in software losses in the same year (IIPA 2009).[2]

Despite these numbers, quantitative reporting on Russia has fallen off sharply in the past several years, with the Motion Picture Association of America (MPAA) dropping out after 2005, the Entertainment Software Association (ESA) after 2006, and the Recording Industry Association of America (RIAA) after 2008. With the recognition by the Business Software Alliance (BSA) in 2010 that its "losses" are better characterized as "the commercial value of unlicensed software," there is no longer any current industry reporting of piracy losses in Russia.

Reported rates of piracy in Russian media markets (as opposed to monetary losses) generally decreased throughout the 1995–2009 period, with sharper drops after 2006—a period of stepped-up Russian enforcement. For reasons discussed later in this chapter, it is unclear to us whether these trendlines reflect real differences in the availability of pirated goods. The size of the Russian market for all types of media goods except recorded music increased dramatically over the past decade, making higher total quantities of pirated goods and lower overall rates a plausible combination, and one validated by our experience with Russian consumers. Russian sources in enforcement, for their part, generally share these doubts. But the reported decreases have played

---

1   See Arvedlund (2004).

2   In 2009, in a flagging economy, reported software losses in Russia dipped to $1.86 billion. Losses in China remained stable, at just over $3 billion (IIPA 2010).

Chapter Contents

149 Introduction

154 A Brief History of Piracy in Russia

158 Disaggregating Piracy by Sector

158   Movies

159   Music

160   Software

161   Enter the Internet

162 How Piracy Works

163   Licenses Everywhere

166   Legal, Grey, and Illegal

167   Stamps and Stickers

170 The Social Organization of Production

171   Consolidated Production after the Crackdown

174   Geography

175 The Social Organization of Distribution

175   Warehouses

176   The Deformalization of Retail

179   Consumption

179   The Socioeconomics of Consumption

180   Patterns of Consumption

181   Comparative Purchasing Power

185   Author and Copyright

186   Counterfeiting

187   Local Effects

188 The Economic Function of Piracy

189   Survival in the Informal Economy

190   The Small-Business Dilemma

191 The Cultural Function of Piracy

192   Specialty Stores

194   Expanding the Market

195   Reproducing the Intelligentsia

196 Law and Enforcement

199   Governmental Actors

200   Non-governmental Actors

201   The Effectiveness of Enforcement

an important role in the domestic politics of enforcement and above all in the US-Russia dialogue as evidence of Russian compliance with US demands.

Lobbying by US copyright interests has made piracy a source of continuous tension in US-Russian relations over the past decade—at times placing it on diplomatic par with global security issues such as nuclear proliferation. Russian government attitudes toward this pressure have varied, with conspicuous efforts to comply with US demands emerging in 2005–7 as the prospects for Russian accession to the World Trade Organization (WTO) seemed to grow closer. A bilateral Russia-US agreement on trade and IP (intellectual property) enforcement was signed in November 2006 and set in motion a wide array of changes to Russian law, enforcement practices, and—ultimately—the organization of piracy, in which a relatively formalized pirate retail sector gave way to a range of more informal and less visible channels.

Domestic pressure for stronger enforcement also grew in the period, as Russian software and movie interests, especially, emerged from the economic turbulence of the late 1990s and began pushing for stronger local controls. As elsewhere, these local conversations have been shaped by (and fed back into) the wider context of international copyright lobbying, but they also are indicative of new locally grounded debates over the costs and benefits of enforcement. In our view, the experience of the past four to five years suggests that these domestic Russian conversations will have more impact on enforcement efforts than the USTR or other external forces—if not always at the level of state policies, then at the level of the actual practices of enforcement and the margins accorded piracy in Russian business and consumer life. A primary goal of this chapter is to contribute to that conversation.

IIPA reports continue to provide the dominant account of piracy in Russia, however, and debates tend to be framed by its claims and those of affiliated groups. In part, this dominance reflects the very effective use of the reports by Western-affiliated industry groups active in enforcement lobbying, notably the Russian Anti-Piracy Organization (RAPO) in film, the BSA in software, and the International Federation of the Phonographic Industry (IFPI) in music. In part, this discursive power simply reflects the lack of alternatives: as in other countries, the same industry affiliates provide the only broad-based infrastructure for research and reporting on piracy. Although local political interests have, on several occasions, challenged perceived overreach by the police and industry groups, this opposition has not articulated a clear alternative account and has produced little independent research or data.

The picture that emerges from IIPA and other industry research follows, for the most part, the well-established IIPA template for high-piracy countries. Failures of Russian political will, popular ignorance of the law, insufficient deterrents, corruption, and inadequate police powers all figure prominently and repeatedly in the reports of the past five years. Although our work lends support to some of these accusations, this narrative of failure provides a very incomplete perspective on the wider social, political, and economic contexts of piracy in Russia. Most important, it does very little to explain the prevalence and persistence of piracy in Russia despite more than a decade of international pressure, institutional growth, policy change, and expanded enforcement efforts.

Like the other contributions to this report, our account of media piracy in Russia emphasizes the relationship between pirate and legal media markets. As elsewhere, the dominance of domestic media markets by multinational companies means that media prices remain high and the variety of available goods low. As elsewhere, the growth of domestic media markets in the past decade is largely a function of rising incomes in major cities (in the case of film, matched by rising prices)—not of efforts by multinational companies to compete on price. (The Russian music market, with its unusually large percentage of locally owned labels and close ties to the live-performance market, offers a partial

Chapter Contents, cont.

206	Selective Enforcement
208	Business Community Pushback
209	The Ponosov Case
211	State Capture
212	Conclusion
214	About the Study
214	References

Acronyms and Abbreviations

CPP	comparative purchasing power
CRM	collective rights management
DRM	digital rights management
IP	intellectual property
IP address	Internet protocol address
ISP	Internet service provider
IT	information technology
P2P	peer-to-peer
BSA	Business Software Alliance
ESA	Entertainment Software Association
IFPI	International Federation of the Phonographic Industry
IIPA	International Intellectual Property Alliance
MPAA	Motion Picture Association of America
RIAA	Recording Industry Association of America
TRIPS	Agreement on Trade-Related Aspects of Intellectual Property Rights
USTR	Office of the United States Trade Representative
WTO	World Trade Organization

Acronyms and Abbreviations

This chapter preserves the original Russian-language system of acronyms and abbreviations through transliteration.

Department K	computer crime unit, Ministry of the Interior
FAS	Antitrust Service
FSB	Federal Security Bureau
FTS	Federal Customs Service
MILITIA	Department of Public Safety, also known as municipal police
MVD	Ministry of the Interior
NFPP	National Federation of Phonogram Producers
NPD	Nonprofit Partnership of Distributors
NP PPP	Nonprofit Partnership of Software Suppliers
OBEP	Department of Economic Crimes, Ministry of the Interior
Prokatura	Prosecutor's Office
RAO	Russian Author's Society
RAPO	Russian Anti-Piracy Organization
RARE	Restricted Access Regime Enterprise
RFA	Equal Rights Phonographic Alliance
ROSP	Russian Association of Allied Rights
VOIS	Russian Organization for Intellectual Property

exception, visible in lower CD prices.) Recurrently, our study suggests that the strongest competition on price and services in Russia takes place in the pirate market—in the optical disc channel, on the Internet, and in the various quasi-legal ventures that have exploited confusion around Russian copyright licensing laws. As in other middle-income and low-income countries, Russia's pirate market provides the only truly mass market for recorded media and often the only source of any kind for niche-market goods, such as non-Hollywood foreign films.

In key respects, Russia is also an outlier in our work. Many of the most common features of pirate markets—selective enforcement, conflicting official action on piracy, and inadequate domestic licensing regimes, to cite a few—are extravagantly present in Russia and represent virtual limit cases in this study. The chapter explores, for example, how operating fully within the licensed economy is a luxury reserved for large, well-capitalized businesses and how the capture and use of enforcement resources by those businesses conveys competitive advantages—tracking with and reinforcing influence and size.

It explores the surreal history of copyright licensing in Russia, which deserves a study of its own. Where other national licensing regimes for music or film are merely inefficient, leading to high prices and limited availability of goods in the legal market, the Russian media market is the product of a wild proliferation of licenses, of counterfeit licenses, and—most significantly—of licensing authorities, with the result that licensing has largely ceased to be a viable means of distinguishing licit from illicit goods. Efforts to address this through the consolidation of licensing authority in groups like the Russian Author's Society (RAO) have, so far, produced hyperactive rights enforcement and a massive (some would say, indiscriminate) expansion of rights claims and litigation but nothing yet resembling a transparent, credible basis for artist royalties.

Perhaps the most striking feature of Russian pirate markets, however, is the evidence of state protection of pirate optical disc production. A number of industry and government sources have drawn attention to the role of Restricted Access Regime Enterprise (RARE) sites—such as military industrial facilities and nuclear power plants under the protection of state security forces—as hosts of pirate production lines. Factories on such sites also are among the major "legal" suppliers of discs in Russia and contribute, in particular, to a DVD pirate market saturated by the "above-quota" production of high-quality discs. While the RARE sites have been targeted for closure in recent years by the Russian government—with some success according to industry sources—the larger web of police and security-service protection for the major factories appears to be mostly intact.

Conversely, we see no evidence of ties between pirate networks and broader "organized-crime syndicates" or the so-called ethnic mafia often alleged to dominate Russian black markets. Such allegations are relatively common in online forums and it would be surprising, in our view, if opportunistic connections between piracy and other forms of criminal activity were absent from the Russian market. But we have seen no evidence of systematic relationships and believe such activity to be, at most, peripheral to the larger dynamics of Russian piracy.

In 2006–7, Russian law enforcement agencies conducted a major crackdown on pirate producers and retailers, leading to praise from both domestic and foreign government officials and industry groups. But in an environment in which the major producers enjoyed relative immunity, the crackdown had a perverse outcome. It was, by most accounts, successful in sweeping out the most exposed middle-tier producers and retail vendors of pirated CDs and DVDs. But in so doing it consolidated the power of the large, protected factories and—above all—sheltered them from the mid- and low-level competition that has collapsed prices for pirated optical discs in many other countries. One consequence is that pirate disc prices in Russia remain unusually high—averaging $4–$6 for a high-quality DVD. In countries where cheap burners and raw materials have led to extensive small-scale, low-end competition among producers, DVD prices have fallen to $1–$2 at retail and often lower at wholesale and in the least formalized sectors of the market.[3]

The crackdown and its aftermath also provide evidence of the increasingly complex balance of forces that shapes the politics of IP and enforcement in Russia. Although the crackdown and related changes to Russian law have been framed by the IIPA and other international stakeholders as responses to US pressure—notably in the context of Russian efforts to join the WTO—a domestic reading is also possible in which local rights holders played the critical role in supporting the crackdown and later in limiting it as business-class discomfort with the raids grew. The two views are not irreconcilable, but the latter is largely invisible in the international arena. Over the past half decade, we see evidence of growing autonomy in Russian approaches to IP policy and enforcement, shaped by struggles between domestic stakeholders and by calculations of domestic costs and benefits.

---

3   See the India, Mexico, and Bolivia chapters in this report.

From the perspective of Russian consumers concerned above all with access to media, these cost-benefit calculations are relatively simple and unambiguously favor piracy. Our work finds near-universal participation by Russian consumers in the pirate economy, differentiated mostly by the frequency of pirated purchases or downloads and the degree of (mostly inconsequential) ambivalence toward the practice. Piracy, our work suggests, is not just a drain on the cultural economy in Russia—*it is one of the primary forms of that economy* and is woven into a wide range of licit practices, forms of enterprise, and patterns of consumer behavior. Much of what follows is an effort to understand enforcement in relation to this other side of piracy, this feature of everyday life in Russia.

A Brief History of Piracy in Russia

Since the transition from socialism in 1991, there have been three distinct phases in the organization of Russian piracy:

- An initial period, running from 1991 to roughly 1999, characterized by (1) the widespread smuggling of optical discs from other countries into Russia (especially from the former Eastern-bloc countries), (2) weak law enforcement, and (3) generally low public awareness of IP law. Cassettes—audio and video—were the first generation of pirate goods, complemented in the mid-1990s by increasing numbers of CDs. These markets grew and, by the late 1990s, had become widely embedded in small and medium scale retail. Enforcement was minimal in this context: criminal charges and civil lawsuits were rare, successful convictions rarer, and penalties negligible.

- A second period, running from the late 1990s to 2006, marked by a shift toward domestic production as manufacturing costs dropped and by the growth of a relatively diverse pirate economy with low barriers of entry and a wide range of producers, distributors, and vendors. A mixed international and domestic enforcement lobby also began to take shape in this period, resulting in several rounds of changes to IP law and the reorganization of enforcement operations, but with little visible impact on street piracy.

- A third period marked by (1) the emergence of much stronger cooperation between industry groups and the state, (2) a resulting sharp increase in enforcement pressure against local vendors and distributors, and (3) the growth of the Internet as a competitor to the pirate optical disc channel.

In the first years of the post-socialist era, the Russian government worked to build a free market economy by creating legal institutions based on US and European norms. Western-style intellectual property law was part of this first round of legal reform. When Russia passed its Copyright Statute in 1993, it closely followed the standards set by the Berne and Rome Conventions for the protection of author's rights and the range of neighboring rights in performance, recording, and broadcast. This included, notably, the concept of "transferable" copyright separate from the inalienable "moral" rights of a creator over his or her work—a distinction that remains important in Russian legal and popular understanding of intellectual property.

Despite these legal innovations, the concept of unauthorized copying as an illegal activity was slow to emerge in Russian public life. The unauthorized recording, sharing, and gifting of cassettes had been common behavior in Russia, especially in regard to the Western cultural goods that official censorship kept out of legal reach of Soviet audiences.[4] Such practices continued well after the fall of the Soviet Union, fostered by the lack of developed legal markets. When pirated CDs and video cassettes flooded into Russia after the transition, they were perceived by most Russians as the market—neither legal nor illegal but simply available, part of the consumer surplus promised by capitalism.

The first organized "pirate" networks emerged to meet this demand. Illegal copying on an industrial scale began in the early 1990s, as formerly state-run optical disc factories across the region lost their primary buyers and turned to production for the black market. Because of Soviet-era decisions about the placement of optical disc factories, much of this production took place outside Russia, in the former socialist countries of the Council for Mutual Economic Assistance (COMECON). Factories in Bulgaria and the Ukraine, in particular, became major suppliers of pirated CDs and later DVDs to the emerging Russian market (and to other Eastern and Western European markets). Discs from Bulgaria were transported in large quantities by truck across the southern border of Russia, often disguised as industrial waste to avoid customs. Piracy in this period was embedded in these wider regional networks and their complicated supply chains, which started with the acquisition of original CDs and film prints from Western distributors. According to several experts interviewed for this report, studio originals, not cruder "camcordered" copies, were the standard for pirated new movie releases. Because the optical disc factories were among the last high-tech industries to be built under socialist rule, the quality and quantity of output throughout the 1990s was generally high.

On the enforcement side, the 1990s were, by most accounts, a period of impunity for pirates in Russia. Although both the IFPI and the MPAA had active enforcement efforts underway by the mid-1990s, with numerous raids and seizures of infringing goods, the first successful prosecution for optical disc smuggling came only in 2001, in the so-called Bulgarian case.

The Bulgarian case was the first trial for music piracy in Russia—so named because it centered on CD shipments intercepted en route from Bulgarian CD factories. The case was the first to reveal the complexity of the international pirate trade in Russia but was otherwise notable mostly for its glacial pace. The investigation began in 1995. Two years passed before initial arrests were made in 1997. Court proceedings took another four years before guilty verdicts were handed down in 2001. The leader of the ring was sentenced to three years in prison, but because he had been detained throughout the period of investigation, his sentence was considered served. Despite

---

4   Unauthorized translations of Western texts nonetheless circulated in the Soviet Union, sometimes
        widely, as part of larger samizdat (self-publishing) networks.

the conviction, the Bulgarian case became a symbol in enforcement circles of the futility of investigating and prosecuting copyright infringement under current Russian law. In the words of one of the investigators, "one piracy case equals three unsolved murders" (Vitaliev 1996).

From 1997 on, the IIPA treated these issues as failures of Russian policy, political will, and training—views visible in its annual Special 301 criticisms of Russian officials for failure to prioritize anti-piracy efforts. But the ineffectiveness of copyright enforcement is difficult to disentangle from the broader problems of institutional development and state power in Russia in the 1990s. New state institutions were fragile and slow to assimilate the vast array of new laws, norms, and procedures created only a few years before. Ongoing fiscal crises—the most severe in 1998—limited the capacity of the Russian state to effectively perform many of its regulatory and law enforcement roles. And the rapid, disorganized privatization of state enterprises—largely abetted by US interests—created massive economic dislocation in which lines between legal and illegal business were often impossible to draw. In this context, the introduction of fully formed international intellectual property norms into Russian society was a predictable failure.

The Bulgarian case remained a point of reference for industry demands for expedited legal procedures, stronger customs controls, and other measures designed to strengthen enforcement (many of which would be implemented in later revisions to the criminal code). In practice, however, the case also closed the door on the period in which the cross-border smuggling of optical discs was the major vector for piracy. As CD/DVD burners became less costly and more portable, pirate production became predominantly domestic and more diverse in location and scale. Small and medium-sized production lines mushroomed in Russian cities, supplementing the large-scale state-licensed factories that pumped out a mix of licensed goods and above-quota pirated copies. The retail side also grew and diversified as small and medium-sized stores sold both licensed and unlicensed goods and as specialty stores emerged to address the chronic deficiencies of the legal market. By the early 2000s, transnational networks had ceased to play an important role in supplying Russian pirate markets. The supply chain for pirated discs had become mostly national.

Russian government interest in accession to the WTO combined with growing pressure from domestic and international copyright interests began to alter this landscape as the decade advanced. Executive and legislative action on several levels introduced changes to Russian IP law and enforcement practices, with enforcement-friendly revisions to the criminal code passed in 2004 and 2007 and a major overhaul of the civil code passed in 2006. As in other countries, efforts were made to streamline enforcement authority and strengthen coordination among the many government agencies involved in anti-piracy efforts. Responsibility for enforcement was consolidated around a handful of agencies and specialized units, including the Department of Economic Crimes, the Militia (or municipal police), and the Ministry of the Interior's "Department K" computer crime unit. Public-private and nonprofit partnerships quickly became the norm in these contexts and continue to play a large role in directing enforcement efforts.

The outcome of these developments was the enforcement push in 2006–7, in which police eliminated many of the small and medium-sized producers, distributors, and vendors of optical discs, but which—according to our interviews with enforcement agents—also reconsolidated production around large, politically protected plants. The enforcement push raised the cost of buying off the police and other law enforcement agencies, strongly favoring the largest-scale and best-connected players. The cartel-like behavior of these enterprises kept optical disc prices in Russia unusually high, with prices of $5 for a high-quality DVD still common. Although prices at the low end of the market have fallen (for example, for homemade compilations), there has been no general collapse of pricing to near the marginal cost of the media, as we have seen in other countries when copying technologies flooded the marketplace.

This centralized model has come under increasing pressure since 2007, though not primarily from the police. In 2004, 675,000 Russians had a broadband connection. By 2007, that number was 4.8 million. By 2009, there were 10.6 million broadband subscribers, with 14 million projected for 2010 (Dorozhin 2007; Kwon 2010). Russians are rapidly joining the global online community, and our limited evidence suggests that they have followed their Eastern European neighbors in embracing peer-to-peer (P2P) services as their primary broadband applications—representing up to 70% of bandwidth utilization according to recent measurements at several Eastern European ISPs (Schulze and Mochalski 2009). Russian authorities, moreover, have been slow to act against companies that have exploited ambiguities in licensing rules to offer their own low-cost online-distribution services. The well-known case of AllofMP3, a Russian website that sold nominally licensed music at $0.01 per megabyte, was a prominent example. The commercial profile of the site made it a major annoyance to international copyright organizations and a regular subject of IIPA and USTR complaints. Although the website was eventually shut down in 2008 after a lengthy legal battle and political intervention, it spawned various clones that continue to operate.

Many Internet service providers, for their part, offer low-cost or free music- and movie-download services as part of their subscription packages—not all of them legal. Nearly all observers attribute the persistence of these quasi-legal businesses to the legal thicket around licensing in Russia, which has permitted extensive manipulation of the rules by local groups and limited judicial recourse by rights holders. The shift toward online distribution has also introduced a lag in law and policing strategies, marked by a lack of effective criminal procedures for commercial online infringement. Such lags also have an important geographical component, with bandwidth, income, policing, and judicial experience concentrated in the capital cities.

Disaggregating Piracy by Sector

In Russia, as in the other countries documented in this report, changes in the organization of piracy and enforcement are part of a wider evolution of media markets and patterns of media access. Overall, legal media markets in Russia have expanded dramatically in the last decade, fueled by rising middle-class incomes and the growth (and global integration) of the Russian film and software industries. The market for recorded music has been in decline since 2004 but largely in sync with the global fall of the CD format.

The rapid growth of the last decade, however, occurred against a backdrop of economic crisis in the 1990s. Present-day media markets remain very small for a country of 145 million—some $220 million for recorded music (IFPI 2009) and $830 million at the box office in 2008 (Berezin and Leontieva 2009)—and highly concentrated in the capital cities of Moscow and St. Petersburg.[5] Per capita spending on music, film, and software is still a fraction that of Western markets, and the prices of a number of signature media goods have dramatically risen—notably movie tickets, which doubled in price between 2004 and 2008 (Berezin and Leontieva 2009).

These market shifts also have important social dimensions. The rebirth of the film market, in particular, also represented a transformation of that market from a massively popular and accessible form of entertainment in the Soviet era to a luxury good largely confined to the urban middle and upper classes. In the late Soviet period, Russians averaged sixteen theatrical visits per year (Padunov 2010), more than triple the US average. By the mid-1990s, that number had fallen to 0.25 visits per capita. In 2008, after a decade of growth, it was 0.83 per capita.

Movies

Between 1991 and 2008, the Russian film industry underwent a near-total collapse, restructuring, and revitalization. Prior to perestroika, Soviet censorship kept Western film and television largely out of reach of Soviet citizens, with the predictable effect of vastly increasing the status of such goods. Economic liberalization released this pent-up demand but provided it few licit outlets. Informal, privately run movie theatres multiplied rapidly, often consisting of little more than conference rooms with video projectors. *Rocky, The Terminator, 9½ Weeks*, and other iconic Western movies were publicly shown for the first time in such theatres. Nearly all this exhibition was based on pirated video cassettes. As home players became more common, rental businesses based on pirated stock also emerged.

Broadcasters also engaged widely in pirate exhibition. Following the privatization of the state TV channels, new station owners routinely broadcasted foreign movies without permission. This practice angered the MPAA, but it had little direct recourse: although foreign rights were legally recognized, there was no infrastructure for enforcing them. US movie studios responded with a boycott of Russia in 1992–93, which ended with the passage of a new copyright law.

---

5   Moscow collects 35% of box office receipts, and St. Petersburg 8%–12% (Anufrieva 2008).

The transition also inaugurated a period of rapid decline for the domestic movie market as public financing for Russian producers and exhibitors disappeared. Annual feature film production dropped from roughly three hundred in the early 1990s to only fifty by 1995. Signs of renewed investment in production and exhibition began to appear in the capitals in the second half of the 1990s. The first Western-style multiplex opened in Moscow in 1996. By 1998, film distribution companies had taken the lead in renovating dilapidated movie theatres.

By the early 2000s, Russians had returned in significant numbers to movie theatres, and theatrical exhibition has continued to grow. The Russian hit *Night Watch* topped the box office charts in 2006, with $30 million in revenues. *Pirates of the Caribbean: At World's End* earned $31 million in 2007. 2008 set new records in (post-Soviet) attendance, with over 120 million tickets sold and $830 million in revenues. *Avatar* passed the $100-million mark five weeks after its release at the end of 2009. Despite the renewal of the domestic industry, Hollywood dominates the Russian box office, accounting for over 80% of theatrical revenues in the last decade.

Music

Throughout the 1990s and even into the early 2000s, the vast majority of Russians used record and cassette players to play recorded music, with piracy largely confined to the latter market. Although the IFPI and the IIPA repeatedly raised concerns about cassette piracy in Russia in the mid-1990s—citing what were almost certainly little more than rough guesses of losses—the market was inconsequential in size and, by Western standards, technologically obsolete. In 1997, only 2% of Russians owned CD players, and the major labels were in no rush to see this change. The international labels working in Russia—at the time, EMI, Sony, BMG, Polygram, and Warner—licensed only cassette rights, not CD rights, to their Russian partners and generally viewed the Russian market as unprofitable.

The music market nonetheless grew rapidly in the early 2000s, fueled by rising urban middle-class incomes and the widespread adoption of the CD. According to the IFPI, the wholesale market peaked at $342 million in 2004 (the high point for CD sales in most countries). Since then, sales have hovered at lower annual levels of around $220 million—still primarily based on CD sales but complemented by an emerging (but for now, tiny) legal digital sector geared toward the Russian cell-phone market. Although the Russian music market is miniscule compared to the United States, Japan, and the United Kingdom, it is still the twelfth largest national market.

Most of the music purchased in Russia is local repertoire—over 70% by most estimates, and higher in the provinces. Unlike most countries, where the four global majors (EMI, Sony Music Entertainment, the Universal Music Group, and Warner Music Group) typically control 80%–85% of the market, the Russian market is dominated by forty to fifty independent local labels. As a relatively formal concert market emerged from the black market in recent years, local labels also took on primary roles as promoters (Alekseeva 2008).

By all accounts, the licit market for music is smaller than the pirate market—and very likely much smaller. In 2006, the IFPI estimated the rate of physical piracy at 67% of the total Russian market (IFPI 2006). Its conventional estimate of digital piracy, also from 2006, is 95%. Because the IFPI shares no details about its research methods, we place no particular confidence in these numbers. But we do think them plausible, and indeed the intervening years have likely shifted the balance further toward the high end as digital technologies have become more widespread.

Software

The skilled and highly educated Russian IT community emerged from the Soviet period with great expectations for the transition to capitalism. These hopes were widely shared: one of the earliest acts of the Russian Duma was the passage of a law granting IP protection to software products and databases (1992).

In the early 1990s, several different operating systems competed on the Russian software market, but as elsewhere, pirated copies of MS-DOS and—soon—Microsoft Windows quickly won out. As new versions of Windows were released, new pirated versions entered into circulation, leaving Microsoft with the dominant position in the Russian operating-system market. The adoption of business tools followed a similar pattern, with Microsoft, Adobe, Corel, Autodesk, and other companies holding commanding positions by the mid-1990s in a thoroughly pirated software market. In 1995, the BSA estimated that 94% of business software in Russia was pirated.

Throughout the 1990s, however, the software market—both legal and illegal—was tiny. Computer adoption in Russian businesses and households was still negligible. By 2000, only 6% of households had personal computers, with the vast majority concentrated in a few large cities (Abraham and Vershinskaya 2001). But economic stability and falling computer prices after 1999 combined to produce a very rapid transition. By 2004, 20% of households had computers (Tapalina 2006); by 2009, 49% had them (Ministry of Communications 2009).[6]

Despite the availability of pirated foreign software, software sales also grew dramatically, climbing to $2.6 billion in 2003 and an estimated $10 billion in 2007. Russian software companies benefited greatly from this expansion. The sector recorded 30%–40% annual growth and emerged as the third-largest destination country for "offshore" programming services after China and India. The current leaders in the market include Russian firms such as 1C Company, Kaspersky Lab, and Center of Financial Technologies—all of which specialize in tools for Russian businesses. In the boom year of 2007, 1C's sales increased by over 90%, driven by its popular accounting suite, a global-hit World War II flight simulator, and other foreign-licensed games

---

6   We have seen widely varying estimates of computer adoption in Russia, leading us to approach this subject with caution. The commonly used replacement rate of one-third of systems per year almost certainly doesn't describe the situation in Russia. Many computers purchased for business purposes end up in Russian homes, either directly or after they have been retired from office use. Boston Consulting Group recently put the total number of PCs in Russia at 45 million, representing an overall 32% penetration rate (Boston Consulting Group 2010).

distributed in the Russian market. Overall, however, the relative position of Russian vendors has eroded as transnationals like Microsoft and Adobe increase their presence in the Russian market (RosBusinessConsulting 2008).

This growth has given Russian software companies a voice in enforcement policy and led to some local research interventions that push beyond and, in some respects, challenge the BSA narrative about marginal improvements in Russian business software compliance. The RosBusinessConsulting review suggests that software piracy was only 25%–30% in the corporate sector in 2006—a far cry from the BSA figure of 80% for the market overall. We have no opinion on the accuracy of this number and note the skepticism of at least one consulted expert. But a significantly lower figure for the corporate sector is reconcilable with the BSA findings. Large businesses are usually the most compliant organizations due to their pricing leverage with vendors, generally sophisticated IT-management practices, and vulnerability to enforcement if piracy becomes too flagrant. Such factors contribute to the differential treatment of big and small business discussed later in this chapter.

Enter the Internet

Inevitably, pirate media markets are shaped by the available consumer infrastructure for audio and video consumption. Today, DVD players are the primary playback devices in Russia, offering backward compatibility with CDs and—increasingly—forward compatibility with MP3 and MP4 files. As in other middle-income countries, this is a very recent development: in 2004, only 6% of Russian households owned a DVD player. By 2007, that number had risen to 51%. According to a 2008 Screen Digest report, twenty-eight million Russian households had a DVD player in 2008, giving Russia the largest installed base in Europe. The growth of DVD piracy, in the past half decade, is both a response to and a driver of this process of adoption.

Music has weaker ties to optical disc media than film, due to the wide range of different storage and playback devices. Sales of digital audio players, for example, doubled annually in the latter half of the decade. The small size of digital audio files facilitates downloading, sharing in bulk, and the amassing of large music collections at moderate cost. Our survey work on media habits[7] indicated that few younger Russian listeners treat CDs as the elements of a personal music collection. Instead, individual listening and collecting is satisfied mostly through digital files that are downloaded or shared among friends. The CD retains a role, however, as a status object in some contexts, notably for gifts.

The geographic distribution of wealth and services also shapes patterns of use. Access to broadband remains very uneven in Russia, with Moscow and St. Petersburg well ahead of other Russian cities. Uncapped, relatively affordable broadband services became available in St. Petersburg only in 2007. White-collar families, especially those with older children in the

---

7   "Consumers of DVDs" survey conducted for this report by the Evolution Marketing Center in Irkutsk in November and December 2008. The project coordinator was K. Titaev.

household, are the primary early adopters, replicating adoption trends for other consumer technologies. In our interviews, members of this group clearly indicated a shift away from DVDs—pirated or otherwise—as the medium of choice for film/video consumption. It was also clear from interviews that access to pirated media is not just a consequence of broadband adoption but an important driver of it. Given the high costs of media, low local incomes, and underdevelopment of other digital services in Russia, P2P is a strong value leader among broadband applications.

As in other countries with rapidly growing broadband infrastructures, Russian P2P activity is directed primarily at top-tier international sites—a list that in 2009 included The Pirate Bay, Demonoid.com, and Mininova. But a large local P2P community has also emerged in the past several years, consisting of around fifty BitTorrent trackers that specialize in Russian-language content and a wide variety of niche genres. The eight largest BitTorrent trackers in Russia counted almost eight million total registered users in late 2009 (not counting overlap between them). These larger sites typically index a wide range of materials, from movies to local TV, games, music, books, educational materials, pornography, and subtitled versions of foreign media in all those categories.

The largest site, Torrents.ru, specializes in film and TV. Our analysis of the geographic distribution of users of Torrents.ru suggests that the site serves a broad regional and diasporic Russian-speaking population, with almost half the users with resolvable IP (Internet protocol) addresses located outside Russia.[8] In addition to providing access to a much wider range of media for Russian broadband users, sites like Torrents.ru also clearly provide local Russian-language media for those living abroad. Torrents.ru's domain name was suspended in early 2010 by authorities, allegedly at the behest of software companies Autodesk and 1C. As with many enforcement actions, the attempt to shut the site down resulted in its relocation outside Russia—in this case to an ISP registered in the Bahamas (enigmax 2010a).

How Piracy Works

Media piracy in Russia has many determinants, ranging from the high price of licit goods relative to local incomes, to police crackdowns on retail vendors, to the failure of licensing regimes to provide much variety in Russian music, film, and software markets. Among these factors, price and income are fundamental. Although Russia is usually described as a middle-income country, GDP (gross domestic product) per capita is around $9,000 and median annual income remains under $5,000.[9] Full-price licit CDs and DVDs, especially for foreign music and film, cost between

---

8   On the basis of a data crawl of Torrents.ru's user index, we identified 156,487 registered users residing in Russia and 70,087, in descending order, in the Ukraine, Germany, Latvia, Moldova, Lithuania, Estonia, Israel, the United States, Kazakhstan, and Belarus. The IP addresses of 98,168 registered users were not resolvable at this level.

9   These are nominal GDP per capita figures appropriate for the comparison of fixed-price goods like DVDs. GDP is often reported in terms of "purchasing power parity," or PPP, reflecting the relatively

$10 and $25 and consequently have a very small market share (according to IFPI sources, full-price CDs account for only 10%–12% of the market). The large pirated optical disc market and, increasingly, the large-scale culture of online and digital piracy cannot be understood outside this price-income mismatch.

The limited selection of media goods offered by the legal market is another crucial determinant. Most Western film and music, for example, is simply unavailable through legal retail channels, with the range of goods falling off still further in the provinces. This situation is by no means unique to Russia but rather reflects global business models in which incentives to compete on price and services in emerging economies stay low. In Russia, this problem is exacerbated by two additional factors: (1) the unparalleled complexity of the licensing environment and (2) the assumption of risk by retailers in the media-distribution chain.

In Russia, the costs associated with unsold stock are borne by the retailer, not the distributor. This assumption of risk has consequences for the availability of media at the retail level: it pushes retailers toward low-risk, well-established, well-marketed products that are less likely to leave them with unsold stock and correspondingly away from more specialized or lesser-known music and film; and it creates incentives to stock much cheaper unlicensed goods, which can often be purchased wholesale at prices of from $0.30 to $1.

The police crackdown in 2006 changed these calculations. First and foremost, it became more dangerous to stock pirated goods, leading many vendors to exit the business. Distribution shifted toward less vulnerable retail networks, including anonymous chains and mobile street vendors. Higher up the distribution chain, complex warehousing networks emerged that separated pirate production and distribution, minimizing risk to distributors.

In turn, this less formalized trade has come under pressure from Internet-based distribution, in the form of both file sharing networks and well-known, nominally "licensed" Russian download sites like AllofMP3 and its successors. Although relatively few Russians possess the combination of broadband service, a modern computer, and digital playback and storage devices that enables full participation in the digital media economy, such infrastructure is growing rapidly and represents a clear challenge to the current organization of both legal and pirate media markets.

Licenses Everywhere

Most debates about piracy and enforcement presuppose the existence of clear distinctions between licit and illicit goods. The IIPA's assertion of $2–3 billion or more in annual losses in Russia since 2003 draws a bright line on this issue. But such distinctions can be complicated on the ground and are uniquely complex in Russia, where the copyright economy has been—and to a considerable extent, still is—mediated by overlapping licensing regimes that govern relationships among creators, publishers, distributors, vendors of media goods, and makers of audiovisual equipment.

---

lower prices of goods and services in many countries. Russian GDP per capita in PPP terms has fluctuated between $15,000 and $16,000.

There have been several efforts to consolidate licensing procedures in the past decade, but the results have been either modest or counterproductive. The situation remains, by most accounts, a mess.

As elsewhere, responsibility for licensing individual works for production and distribution resides first with rights holders and then, under certain circumstances, with collective rights management (CRM) organizations. These collect royalties for airplay, performance, and other use of works; distribute the money to rights holders; and otherwise act to protect artists' interests. In Russia, this scenario was complicated by three provisions of the Copyright Statute of 1993: (1) the law allowed an unlimited number of collective rights management societies; (2) these societies were allowed to represent authors in absentia, without specific contracts to do so; and (3) they were allowed to manage a wide and underspecified range of neighboring rights.

The situation allowed for extensive gaming and abuse. In several cases, publishers and distributors registered as CRM societies and began publishing and distributing work—often without the consent of the rights holders. Nonpayment of fees and royalties was a recurring problem in this context and became the basis of mobilization and lobbying by the IFPI and the RIAA.

AllofMP3 was the best-known exploiter of these loopholes. The Russian web portal sold music online to international audiences at prices far below international norms, realizing a modest profit of $11–14 million annually (Golovanov 2008:2). Mediaservice, the parent company, obtained its licenses from two legally licensed Russian CRM organizations. The legitimacy of these licenses was challenged in 2004 by the IFPI and the RIAA under Article 146 of the Criminal Code. The owner of AllofMP3, Denis Kvasov, was charged with criminal infringement but was later acquitted in 2007 for lack of evidence of actual illegal activity. Under continuing pressure from the IFPI and US groups, the Russian government closed AllofMP3 in 2007, but clones of the site soon opened and continue to operate (although on a much smaller scale).

Reform in 2008 introduced a process of state accreditation of CRM groups, with the aim of consolidating authority around a single society in each of the main domains of culture and entertainment. Henceforth, only the accredited societies would be able to represent authors and rights holders without formal contracts. The law was not retroactive, however, and several of the CRM societies in existence before 2008 continued to operate.

In the copyright area, the Russian Author's Society won accreditation and became the de facto government-backed monopoly. Arguably, this consolidation traded one set of problems for another. The RAO has been repeatedly criticized for a lack of transparency and for failure to deliver collected funds to musicians—all charges vehemently denied by the RAO's deputy director, Oleg Patrin. The organization keeps 30% of its gross licensing revenues[10] and has grown rapidly since 2006, increasing its proceeds from 1.5 billion rubles (approximately $50 million) in 2007 to 2.2 billion ($70 million) in 2008. The RAO charges concert organizers 5% of their proceeds and

---

10   See RAO's official website at http://www.rp-union.ru/en/docs/.

3% of the box office take at movie theatres for "the public performance of music used in films" (Goncharova and Pushkarskaya 2009).

The RAO has made a particular habit of targeting concert promoters found skirting this tax (as in many other countries, live performance is the only high-growth sector of the music business in Russia). In 2008, the RAO attracted attention by suing Yug-Art, a concert organizer, for the "unauthorized public performance" of Deep Purple songs by Deep Purple during its Russian tour. The RAO won an award of 450,000 rubles ($15,000, or $1,000 per song), affirming the principle that all performance revenues must pass through the RAO. In March 2010, the RAO sued a World War II veterans' choir for performing patriotic Soviet songs at a free concert in Samara without signing a licensing agreement (enigmax 2010b). This event provoked a minor uproar in the Russian Parliament and may signal more organized pushback against the RAO's maximalist stance on performance rights.

With such a record, most CRM organizations enjoy low levels of participation and high levels of distrust from rights holders. In 2008, the RAO's coverage of public performance spaces was estimated at only 10%–12% of the total market.[11] Such numbers reflect, to be sure, the difficulty of establishing a consistent and credible framework for performance rights in a country that had historically ignored them but it also reveals clear frustration with the RAO's maximalist view of IP rights and hyperactive, indiscriminate practices of litigation. Many popular musicians now waive their performance rights altogether in order to avoid the RAO.

The RAO's efforts to expand have also created problems. In 2008, RAO associates launched the Russian Organization for Intellectual Property (VOIS) in a bid to become the accredited organization for "neighboring rights," such as those granted to broadcasters or producers. Concerns about the VOIS's lack of transparency regarding royalties and governance, however, led many producers to back a separate group in the accreditation process, the Equal Rights Phonographic Alliance (RFA). By all accounts, the political jockeying for accreditation was intense. The RFA's general director, Vadim Botnaruk, was assassinated during this period, although clear motives for the crime were never established. Ultimately, the VOIS won accreditation in 2009. The RFA continues to operate, however, grandfathered under the 2008 law, and is still the preferred organization of many foreign CRM societies.

The rights management situation remains unresolved in key respects. Other organizations, such as the Russian Association of Allied Rights (ROSP), are vying for accreditation to collect royalties in still other areas. Control over licenses for manufacturers and importers of audiovisual equipment and blank media is one of the prizes, estimated to be worth $50–100 million annually.

Other types of licenses add to the confusion. In addition to the CRM societies, numerous anti-piracy organizations conduct their own "licensing" of the products of their members, often in the form of stamps of approval or authenticity placed on the goods. Such forms of authentication have no legal power but are intended to help signal legitimacy to retailers and consumers.

---

11   Interview with IFPI Russia staff.

Regional and local authorities also issue an array of licenses to commercial vendors, from street vendors to large national chains. These allow trade in CDs and DVDs but in practice have no bearing on whether the products sold are legal. One interesting and controversial variation is the "regional license" (*regionalka*), which authorizes the distribution and sale of media goods at reduced prices (and often reduced quality) within a particular geographical region. Regional licenses have become a common strategy used by Russian distributors to lower prices outside the core Moscow and St. Petersburg markets.

Legal, Grey, and Illegal

Inevitably, the overlapping licensing regimes introduce a wide variety of opportunities for abuse. Different kinds of licenses define different types of illegality beyond the simple infringement of copyright, including violations of the authorized format (CD, DVD, streaming audio or video), the number of authorized copies, the permitted geographical boundaries of distribution, and so on. From the perspective of producers, these diverse violations are part of the larger repertoire of piracy in Russia. They are all sources of rights-holder losses and differ mostly in terms of their legal remedies.

From the perspective of consumers and—arguably—retailers, however, the same range of practices describes a spectrum of white, grey, and black goods—not a dichotomy. From this perspective, not all violations are equal, nor is the legality or illegality of most goods clearly marked or unambiguous. Rather, consumers make efforts to relate differences in perceived legality to differences in perceived quality, with the highest-quality, fully legitimate goods at the top of the hierarchy and the lowest-quality, informally produced goods at the bottom. The language of white, black, and variations on grey circulates explicitly in this context, though by no means consistently with regard to set practices. We provisionally distinguish five "shades" at work in this consumer logic:

1. *White* goods exhibit all the attributes of legal production, including high-quality packaging and printing and, above all, high prices, which can range from 350 to 800 rubles ($14–$32[12]). These are typically sold in large, specialized music-video stores as well as major department stores and supermarkets.

2. *Light-grey* goods are legally produced but involve other practices whose legality or fairness may be in dispute—notably in the case of the parallel importation of CDs or DVDs into Russia, which can contravene trade laws or geographical licensing restrictions.[13]

---

12   Prices are cited at the summer 2008 exchange rate of roughly 25 rubles to the US dollar, when this first phase of work was conducted. Changes in exchange rates can have a dramatic impact on these cross-currency comparisons (by October 2010, the rate stood at around 31 rubles to the dollar), though less on local affordability.

13   The status of such imports is a matter of some legal dispute in Russia. The RAPO and other anti-piracy

3. *Grey* goods, such as above-quota CDs or DVDs, are identical to the legal versions but of dubious origin, signaled by their lower prices (150–250 rubles; $6–$10). Their paths lead primarily to specialized small and medium-sized stores, lower-end supermarket chains, and kiosks.

4. *Dark-grey* goods are factory produced but generally lower quality and visibly unlicensed or under-licensed. Unauthorized song compilations and film collections are prime examples. Such discs are commonly sold at kiosks, by street vendors, and in open-air markets, at 100–120 rubles ($4–$5).

5. *Black* goods are burnt on home computers or produced by small-scale cottage operations. In many cases, these are copies of other pirated copies or burned to disc from downloads. Such pirated media is found at open-air markets and costs from 10 to 100 rubles ($0.40–$4).

Many other genres also circulate in the lower tiers of the retail ladder, including concert bootlegs, self-help videos, evangelical sermons, and pornography. The legal status of these goods is harder to assign as they include a great deal of amateur and non-commercial productions.

Stamps and Stickers

On the street, determining the degree of legality of media goods is often impossible, even for experts. Our interviews suggested that neither shop assistants nor law enforcement officers involved in inspecting retail outlets for counterfeit goods can reliably distinguish legal and illegal copies. Nor, for that matter, can the growing strata of customers willing to pay a premium for legal goods. In practice, people rely first on price as a signal of legal status and second on their tacit knowledge of what constitutes a legal (or almost-legal) copy. A regime of authenticating stickers and stamps placed on the goods themselves has emerged to guide this process but has itself become so byzantine and extensively counterfeited that it only adds to the confusion.

Many rights-holding or rights-issuing authorities attempt to validate products with specially designed stamps or stickers. These serve as signals to consumers of (ostensible) authenticity, but they have no legal authority and are not required by law. Because there are many rights-holder groups and rights-granting authorities, discs can be marked with a wide variety of stamps and stickers—and even multiple stickers. Some carry as many as five. As the chief of the St. Petersburg branch of the RAPO put it:

> RAPO CHIEF: Each copyright holder puts his sticker however he wants to: "I want to protect my property in this way, and so I better put a sticker on."

organizations have challenged its legality, but a number of companies operate openly as importers. Here the legal question is not copyright infringement but, rather, the extent to which the imported goods comply with trade agreements and whether customs are paid. The discounted price of many imported goods often permits vendors to undercut locally produced and licensed goods. See also Olimpieva, Pachenkov, and Gordy (2007).

INTERVIEWER: So the sticker, in general, means nothing?

RAPO CHIEF: In general, it means absolutely nothing.

Representatives of the IFPI-backed recording industry association NFPP (National Federation of Phonogram Producers) reported better luck, but only when the sticker program was actively supported by law enforcement.

Naturally, stamps and stickers are also widely copied and fraudulently applied to pirated goods. In 2003, sixteen of the largest Russian music distributors created a private-public partnership called the Nonprofit Partnership of Distributors (NPD), which promptly issued its own sticker of authenticity.[14] The executive director of the partnership explained to us that, today, the stamp itself needs protection. Even the complex holographic design does not deter counterfeiters:

> We have a stamp and it has many protection features, but it is now also forged. It has the serial number. It has a hologram. It has every possible watermark. So it has a lot of protection. But we are pirated. Our stamp itself is pirated!

Interviews with consumers and music store personnel made it clear that no one can explain the meaning of the different stamps and stickers in any detail. Recognition has not been aided by changes in appearance: the NPD's sticker, for example, "has changed many times" and has existed, in its current form, for "maybe the last four years."[15]

Such confusion in the marketplace leads to indifference to the licensing system. Many stamps are assumed to be fraudulent. Worse, some stamps signal side-deals between the state, private enforcers, and commercial interests that are little more than protection rackets. Several retail-shop owners described these as "pseudo-licenses." Most were aware of a case in St. Petersburg from mid-2005 in which a so-called association of retailers used such licenses to expand their control of the music market. The association struck a deal with the local police to raid only those shops whose products lacked its stamps.

Nonetheless, the NPD and other rights groups have launched several efforts to educate consumers about the differences between legal and illegal optical discs. The key attributes of pirated discs, the NPD suggested on its website in 2005, are:

1. There is no NPD sticker on the cover, or the sticker has been forged.

---

14 In its early days, the NPD was itself dogged by accusations of involvement in the distribution of pirated CDs.

15 NPD representative to the regions, in a round-table discussion, Moscow, June 2008.

2. There is a fake holographic stamp on the cover, or an imitation of a stamp has been made of a light-reflecting material.

3. The cover has apparent signs of scanning (copying) from the original. E.g., on close scrutiny the disc reveals that: the picture on the cover consists of separate points having regular geometrical shapes (square, rectangle, rhomb, and hexagon) or horizontal and/or vertical lines; and/or the contrast and color spectrum do not correspond to the original colors (for instance, the prevalence of one color in the picture, i.e., blue)

. . .

5. The data appear in small print or are blurred or unreadable.

6. There are no indicators of author's rights or copyright—i.e., the Latin letters ©.

. . .

9. There is no logo of the issuing company on the cover.

10. There is no information in Russian on the cover.

11. More than one audio collection or several albums by the same performer are present   on a CD.

12. The disc case indicates that the disc is a CD-R or DVD-R format.

(NPD 2005)

Despite obvious problems related to the use of stickers, many pirated goods are, of course, easily identifiable. Most of the attributes listed by the NPD are clear giveaways: faded colors, blurred pictures, and grammatical mistakes mark the pirate origins of a disc. Curved, uneven, or scratched discs are more likely to be pirated. Street markets and kiosks, especially, are full of pirated discs that make no serious effort to hide their origins. The high end, in contrast, presents serious problems of identification. Although it can be easy to spot an undisguised pirated disc, it has become extremely difficult—even for enforcement officers—to verify a legal copy. In the higher tiers of the Russian media market, licit and pirated discs come from the same assembly lines and are identical.

The Social Organization of Production

According to estimates by the RAPO, the majority of pirated DVDs are produced above-quota at licensed factories.[16] As of 2009, the RAPO believed that there were at least fifty such factories running extra shifts or, in some cases, additional production lines. There are obvious advantages to this method: the oversupply is impossible to distinguish from the authorized production run, greatly complicating enforcement. As the executive director of the NPD put it:

> Before lunch they produce according to the terms of their license, and after lunch they do the pirated run [*piratka*]. And who can tell which they are printing at this very moment—legal or pirated?

Capital-investment and technological requirements at this level are high—as are the quality standards. An industrial replication line can cost $2 million. Production also involves the printing of high-quality jackets and other insert materials, as well as packaging. Output can be massive—as many as 450,000 discs per month per production line. Large plants may have as many as twenty to thirty such lines. When a factory is operating illegally or above quota, it also installs crushers— special machines for destroying the pirated discs in the event of a police raid.

Large-scale production is often broken into stages, with each stage conducted in a separate location. A typical division of labor separates the replication of discs, the printing of accompanying materials, box assembly, packaging, and transportation. This division of functions often results in the delivery of disassembled products to retailers, allowing producers and distributors to shift costs further toward the retail end of the commodity chain.

Smaller production lines played a major role in Russian optical disc piracy earlier in the decade. Built around multipurpose "recording machines," often compact enough to fit in single offices, such lines usually combine the different stages of the production process, from disc burning to the printing of covers and inserts. The quality of output from these operations is typically lower than from the major plants—especially in regard to printing and packaging—but so, too, is the capital investment: dedicated burners on this scale cost, on average, $40,000, allowing pirates to quickly recover their initial investment. The lower cost also means that abandoning the machine in the event of a police raid is less likely to represent an irrecoverable loss. Unlike the major factory production lines, such machines are generally clandestine and unregistered. Production is flexible, fast, and consumer oriented.

Discs burned in homes or small shops on personal computers are also widely available, especially in open-air markets. These generally cater to the poorer strata of the urban population and show little concern for packaging. Vendor stock, in such contexts, is often supplemented by

---

16   According to IFPI sources, above-quota production has never been much of an issue for CDs, which are typically produced on dedicated pirate production lines.

Flooding the Market

According to sources in enforcement, above-quota production is routinely several times larger than the licensed production run, ensuring that the vast majority of copies in circulation are pirated. RAPO estimates regarding several popular films from 2001 to 2004 (well before the DVD boom in Russia) suggest the scale of the practice:

> *Harry Potter and the Philosopher's Stone* (2001). Legal issue: 120,000. Illegal issue: 350,000.
>
> *The Matrix Reloaded* (2003). Legal issue: 200,000. Illegal issue: 500,000.
>
> *Shrek 2* (2004). Legal issue: 220,000. Illegal issue: 450,000. (Vershinen 2008)

Such estimates were generally based on extrapolations of estimates of volume from key points of retail sale—methods that have proved highly approximate. The IIPA in 2010 reported that some seventy million pirated DVDs were produced in Russia in the previous year—apparently applying the older MPAA technique for modeling the pirate DVD market based on the difference between the size of the licit market and estimates of the total production capacity of Russian DVD factories. According to IFPI sources, most current industry estimates are based on the assumption that factories run at 60%–70% capacity. As the chief of the St. Petersburg branch of the RAPO observed, the central fact is that:

> There are too many plants, too many production lines, while the real demand for legal DVDs is not as large. Legitimate orders don't pay for this many production lines.

back catalogs of film and music titles that can be either ordered or burned on demand. Homemade discs can also be found in specialty stores, such as those specializing in rare film or music. Sales personnel in such stores often have networks of trusted clients whom they assist in finding and burning particular albums, compilations, or movies. Such services are inexpensive—usually the cost of the blank disc plus a markup of 20–40 rubles ($0.80–$1.60).

Consolidated Production after the Crackdown

A different pirate economy emerged from the crackdown in 2006 and 2007. Where the older model was characterized by relatively decentralized production and a wide range of retail types, the new model consists primarily of centralized, politically protected manufacturers and, at the local levels, an increasingly informal retail sector marked by shifting legal ownership, greater anonymity of distribution and retail outlets, and growing reliance on street vending by illegal migrants and the working poor.

The crackdown took the sharpest toll, by most accounts, on the middle and lower tiers of producers and retailers.[17] Among the large production facilities, the crackdown produced consolidation and restructuring, with protection from the police and other enforcement authorities becoming the critical differentiator. As large manufacturing facilities have become more dependent on state protection, the mix of protection strategies has shifted. Manufacturing has moved away from open facilities into "closed" sites with private or sometimes public security. The most overt forms of protection involve placement in military facilities beyond the authority of the conventional police:

> INTERVIEWER: I was told that almost every pirate factory now sits on the premises
> of another factory, where there are multiple security checkpoints . . .

> ENFORCEMENT AGENT: In "PO boxes"—that's what we call them. . . . An ordinary
> plant can be inspected by city police, and regional police can raid it too. Now, a PO
> box is classified and has its own security service. No one goes there, not even the FSB
> [Federal Security Bureau], without a permit.

PO Boxes are production lines located on military bases or other premises run by state security services (also known as Restricted Access Regime Enterprise, or RARE, sites). Estimates of the number of these plants have varied over the years, and they are not, in and of themselves, illegal. In 2005, the Russian government put the number at eighteen of the forty-seven registered optical disc plants in Russia. These were allegedly responsible for a large portion of the above-quota production that saturates Russian markets.

Foreign complaints about the RARE sites were a contributing factor to the 2006–7 crackdown, and by most measures the government responded. By 2008, the number of RARE sites used for optical disc production had reportedly fallen to four (IIPA 2010, 2009). A Ministry of the Interior (MVD) police unit called the Eighth Directorate was given the responsibility of policing the RARE sites, but its effectiveness remains a subject of dispute. Confidence in the Eighth Directorate among some industry sources is low, fed by rumors of corruption of the directorate's staff. Criminal investigations have only occasionally resulted in the revocation of licenses, and only a handful of low-level employees of such facilities have been prosecuted.

Less substantiated rumors about state involvement also circulate widely. An owner of a record label in St. Petersburg told us about a plant that used illegal prison labor:

> Some years ago a production facility was raided inside [Prison X]. It shows what a
> remarkable system we have here. . . . Someone in Prison X, who is very powerful,

---

17   Small-scale production now plays a relatively small role in the Russian pirate economy, though the
      situation is dynamic and the IIPA, for its part, has recently begun to signal the reemergence of small
      production lines as a result of government pressure against some of the larger factories (IIPA 2010).

Defending the Release Window

In the film business, the primary goal of enforcement is to delay pirate access to a high-quality copy during the initial release or exhibition window—the period in which a film makes most of its profits. Rapid, widespread distribution of films is both the key to capitalizing on this window and the challenge to maintaining control of copies. In Russia, the transfer from distributors to movie theatres is a particularly vulnerable step in the distribution chain. Typically, 35mm film reels leave Moscow-based distributors two to three days in advance of a movie release. Reaching more remote locations can take as long as ten days. This is often more than enough time for a detour to a specially equipped studio where pirates can produce a high-quality DVD master.

Providing early access to pirate copies of such films is lucrative, fetching between $10,000 and $40,000 according to our sources. Such copies also quickly appear online—although at this stage it is in the pirate's interest to limit competing channels of distribution. For obvious reasons, studios are most upset by pirate releases that precede the official release (although evidence for a strong substitution effect vis-à-vis box office receipts, even for pre-release films, is weak). The next generation of digital film projectors is designed to address this vulnerability in the distribution chain by downloading encrypted satellite feeds of movies directly to the theatre (Vershinen 2008). Among the advantages from an enforcement perspective is that such downloads can be watermarked, giving police a means to trace digital copies back to their source. In 2008, there were 91 digital screens in Russia, out of a total of 1,800 (Berezin and Leontieva 2009).

---

well connected, and clever about networking, bought an optical disc production line. It isn't a cheap line—it costs, probably, one hundred thousand dollars. And decided to install everything in the prison, where the people doing the work don't cost much. In the raid, some millions of discs were confiscated, including one and a half million copies of my discs . . . There was a trial, but nobody was convicted, everybody got away in some way.

The Russian media publish similar stories. One account described the use of psychiatric patients for assembling boxes for pirated video:

[Assembling boxes] proved to be such mindless work that vile entrepreneurs organized a packaging line in a psychiatric clinic. The final products were distributed from there to warehouses across the country. (Vershinen 2008)

Not all our sources credited these examples, and it is quite possible that some of them are urban legends. But sources did consistently corroborate reports about the wider role of state security agencies and the military in large-scale pirate disc production. Such matters are, for obvious reasons, extremely difficult to investigate, and we did not substantiate them independently.

The crackdown also provides a context for one of the unique features of the Russian pirate market: its high prices. In Russia, $5 for a high-quality DVD is typical, and the price of high-end pirated discs has actually slightly increased in recent years. Pirate prices can even equal or exceed the price of certain categories of licensed DVDs, such as region-specific regionalka, which often involve compromises in quality or features. No other country documented in this report comes close to maintaining this degree of price stability. Elsewhere, the proliferation of low-cost burners and the growth of Internet distribution has radically reduced pirate disc prices—generally to between $1 and $2 at retail for a high-quality copy, and often much less at wholesale. At the high end in other countries examined here, such as in the tourist-priced flea markets of Johannesburg or Rio de Janeiro, prices rarely exceed $3.50.

Our conclusion is that relatively successful Russian enforcement in the lower tiers of the supply chain has afforded the large producers a degree of control over supply and, consequently, pricing. Such market power is very likely temporary. Industrial-scale disc production plays a diminishing role in the larger, digital pirate economy, and it is hard to see how high prices will survive the spread of broadband and digital storage, which are already showing signs of circumventing the optical disc supply chain. In this respect, the digital transition in Russia may have an unusual upside in putting the large, state-protected pirates out of business.

Geography

As we have noted throughout this report, the globalization of media industries has not produced a unified global market for copyrighted goods. Although copyright is an international system in which rights established in one country must be honored in another, the licensing system for copyrighted goods is nation-based and requires producers and distributors in each country to separately license foreign goods for distribution. The vast majority of media goods for the Russian market—including Hollywood movies and international pop albums—are consequently produced under license in Russia. There is very little direct importation of media goods from foreign producers, although movie and music industry groups have complained vociferously about the legal loopholes that enable a small, legal import trade in discounted foreign CDs and DVDs.

The relevant geography of both licit and illicit production is thus a Russian one. Among illicit producers, reports suggest that small- to medium-scale producers are present in most cities (the Ministry of the Interior singles out Kazan, Rostov-on-Don, Samara, and Novosibirsk for attention). But most of our interviews described a concentration of industrial-scale production around Moscow and St. Petersburg, tracking broader patterns of growth in high-tech industry, media markets, and income. Several informants indicated further that, in the post-crackdown era, Moscow has become dominant in terms of both the volume and variety of its production and the extent of its regional networks.

Pirate production is also geographically marked within cities. St. Petersburg, like many Russian municipalities, is not only a post-communist but also a post-industrial city. The former Soviet

Garbage Firms

A variety of business ownership patterns have emerged to meet the two requirements of the pirate optical disc market: maintaining economies of scale while minimizing exposure to raids. In one common strategy, legal, registered firms act as commercial landlords for smaller illegal businesses (generally known as "garbage firms" in Russian). The garbage firms are registered to fictitious owners but, in practice, belong to the landlord. In the event of police raids, the garbage firm can be thrown off without jeopardizing the parent firm—whence its name.

Syndicates are another organizational response in which a group of firms operate as a single enterprise but without legal contractual agreements. Capital and other material resources are shared, but liability is not. In the event of a raid, police can generally only charge the targeted firm, minimizing losses to the larger operation.

heavy and high-tech manufacturing sector is both highly developed and very underutilized. Such declining enterprises are the principle sites of large-scale illegal production. In St. Petersburg, they are typically located at the periphery of the city, following the Soviet pattern of industrial development.[18] Smaller production lines and the "warehouse printing plants" described in the next section, in contrast, do not require large premises and are, consequently, more geographically dispersed throughout the city.

The Social Organization of Distribution

The ease of manipulating licensing regimes, notably via above-quota production, is one of the chief reasons why piracy is so difficult to prosecute in Russia: it is often impossible to distinguish legal from illegal copies. The organization of the distribution chain also presents challenges, however, and more so in the wake of the 2006–7 crackdown as pirate intermediaries change their business practices to minimize risk.

Warehouses

Throughout the late 1990s and early 2000s, large pirate manufacturers usually found it advantageous to concentrate production services within single sites, ranging from the illegal acquisition of originals to production, printing, packaging, and storage. From these production centers, the product would be distributed to smaller warehouses, sometimes called "studios" in reference to their size. Increased police action beginning in 2006 resulted in the closure of several major manufacturing lines and increased risk for producers who stored above-license goods in

---

18   A notable exception is the Kirov Factory, one of the oldest and largest industrial plants in St. Peters-
        burg, which is situated in the city center. This plant was raided by police in 2008.

their facilities. Pirate manufacturers responded by creating networks of specialized warehouses that separated production from storage and distribution. According to the RAPO and other sources, one-stop shops are now rare.

A RAPO expert interviewed in our study estimated that there are roughly a thousand warehouses in Russia that distribute pirated products to retail (we found a comparable number listed on Russian-language online forums discussing piracy). Based on assessments by both retailers and the RAPO, St. Petersburg, in 2008, had three or four large optical disc plants making both pirated and legal discs, five to ten large warehouses, and about seventy small warehouses through which different white- and grey-market goods were processed.

The multiplication of warehouses permits a high degree of compartmentalization for firms engaged simultaneously in licit and illicit activity. A single owner may control multiple warehouses, with some dealing in licensed goods while others distribute pirated materials, often in close spatial proximity. This separation provides some protection against losses in the event of a raid.

Personal referrals and long-term partnerships between distributors and retailers also play important roles in the distribution chain. Most suppliers, we were told, cultivate relationships with small networks of retailers and protect them from competitors—indeed suppliers compete for retailers rather than consumers. These relationships reinforce security, help suppliers gauge demand and, above all, lower the likelihood of overproduction, which can ruin profit margins.

Different types of vendors have different structural positions in these distribution networks. Vendors who trade from portable stalls acquire stock from the nearest small warehouse (it may be a rented apartment or a commercial structure) and generally carry it with them. Fixed-location retail shops, as a rule, employ delivery services and have designated personnel for managing stock. Large chains often have their own warehouses to which they transport goods by train or truck. Because of the growing concentration of production, more and more product is now ordered through catalogs sourced to suppliers in Moscow or St. Petersburg.

Working outside this mixed licit-illicit economy is difficult and requires substantial investment in managing both the production and the distribution chain. Leading Russian software companies such as Soyuz and 1C, which do have sufficient scale and resources, control their inventory by purchasing directly from licit producers and by operating their own warehouses. Few smaller businesses have this capacity—or incentive structure.

The Deformalization of Retail

The organization of retail in Russian media markets has changed dramatically over the past decade as piracy and, later, stepped-up local enforcement altered the profitability of retail sales and the cost of entry into recorded media markets. In the formative 1995–2000 period of the St. Petersburg media market, music and video businesses invested heavily in retail store branding. Businesses openly sold an array of pirated and licensed goods and competed for consumer attention. The size of the legal market was small, and enforcement was infrequent—or easily bought off. As the pirate media market grew and raids by different enforcement agencies became more common, market strategies changed. Today, firms trading in music and video are structured to meet the pressure of law enforcement, rather than consumer demand. Stores with pirated goods generally prefer anonymity and generic names, exemplified by the common "CD/DVD" signs that hang from storefronts.

Several informants put the total music/video retail market in St. Petersburg at seven hundred to a thousand "stationary selling points," including established music/video chains, independent music/video stores and kiosks, the day stalls of street vendors, and stalls at weekend open-air markets. The 2009 St. Petersburg Yellow Pages listed sixty "branded" CD/DVD media chains in the city—that is, stores that have a distinctive name under which they formally register their activities, such as Titanik or Nastroyenie. In addition, there are some thirty generic CD/DVD chains. Although indistinguishable from one another by name, their ownership structure varies, with some registered to individuals and others grouped into larger corporate chains.

The shift toward anonymity, mobility, and flexibility in the retail sale of pirated goods is strongest at the level of kiosks and day stalls, which are generally found near train and subway stations and in other high-traffic locations. Foot vendors have also become more common, operating mainly in subway trains. Such vendors are almost always part of larger networks with centralized suppliers and are mainly preoccupied with avoiding the police. The riskiness of the business favors rapid turnover and low prices, which allows for smaller losses in the event of an arrest or raid. The cycle of investment is short, and profit margins are modest. The cheaper categories of pirated music and video are essential to business operations at this level.

Municipal efforts to regulate and occasionally ban these forms of informal trade have proved successful in temporarily disrupting and—in a few cases—destroying street markets, but the broader effect, by most accounts, has simply been the further deformalization of the street trade. As a sales assistant at a specialized music and movie store in St. Petersburg explained:

> Organized pirate markets, large networks, were destroyed [by the police] and
> became disorganized. Now there are Uzbeks with tables, Tadjiks with tables. Earlier
> these were stalls with cash registers. Now they don't have them. Everything can be
> purchased. Everything can be found in the city.

Rather than wiping the pirate trade out, this source argued further, prohibition has stripped it of its last vestiges of formality and transparency, pushing it completely underground and opening it to more harmful and illegal trade and labor practices.

The crackdown of 2006–7 and changes in policy at the local level have also altered this dynamic by substantially increasing the cost of protection. In 2005, according to informants, the cost of maintaining a busy trading spot near a metro station in Moscow was $5,000–$7,000 per month, mostly in rent and bribes to the police and other controlling agencies. Such businesses could generate $4,500–$6,500 a month in profit. Today the margins are much smaller. Maintenance costs—including bribes—are now closer to $10,000 per month, making profitability uncertain for all but the highest-volume vendors. By most accounts, these changes have been less dramatic in provincial cities and towns, where the enforcement push was less intense and sustained and where paying off the police remains cheaper.

The retail market for optical discs continues to bifurcate. Informants generally agreed that the ratio of licensed to unlicensed goods on sale correlates with the size and "formality" of the business. Large retail chains sell a higher proportion of licensed/legal products, while informal businesses tend toward 100% pirated goods. This market structure presents challenges for vendors who want to "go legal." None of the vendors we interviewed believed that licit sales alone provide a viable model for small and medium-sized businesses. Our interviews found a strong desire to legalize among such vendors but also sharp constraints from competition with the lowest-priced street vendors. As one owner of a middle-sized, medium-priced music shop in St. Petersburg argued:

> SHOP OWNER: Say a store sells licensed products, but as soon as you step out you see twenty tents that sell pirated stuff, at half the price. Well, try to be competitive there!
>
> INTERVIEWER: Selling licensed products is more profitable for you?
>
> SHOP OWNER: For us, having fewer problems is more profitable. And as a way to avoid problems, licensed product is of course better. But how can you switch to licensed stuff if there are small stores, tents, kiosks, in the subway and around it, who beat you down with the price? Thank God, all trade was banned in the subway two years ago. I mean, not only from the subway but from everywhere, and it has affected the license situation in a good way. Because why would someone need licensed discs if they have already bought unlicensed ones, and for half-price?

Piracy is less a choice, in this context, than an economic survival strategy for both parties, and it creates predictable tensions between shop owners and street vendors.

The presence of larger chains selling (by all appearances) exclusively legal goods suggests that the important differentiator is not a conventional notion of formal versus informal organization but rather the scale of operations. As one informant suggested, fully legal media retailing in the Russian market is a "luxury business" because of low rates of return and long investment cycles. Only well-capitalized companies with profit centers in other areas can afford to "play by the rules." The recently established music chain Nastroyeniye (Russian for "ambiance") is an example. Nastroyeniye is known for legal, high-quality media goods.It is also a product of corporate diversification strategies,[19] rather than growth from within the vendor market. This was

---

19   In this case, of a large St. Petersburg-based (but with branches nationwide) gambling business called

an important distinction in the eyes of several informants, who viewed the combination of high enforcement and high media prices as a structural advantage for big business and a guarantee of a persistent—if increasingly impoverished—low-end pirate sector.

Vendors and distributors of pirate media are acutely aware of the changes that have come with deformalization. Several characterized them as "hurting the consumer." Some talk about the diminishing quality and variety of pirated goods in the marketplace. Many of what were perceived to be the "best" warehouses in St. Petersburg, in the sense of providing quick access to a wide variety of music and film, were closed in the 2006–7 crackdown.

Consumption

In interviews, enforcement officers often blame the consumer for piracy, citing lack of respect for intellectual property laws, lack of knowledge or confusion about what constitutes piracy, and a general indifference to the moral dimension of piracy in favor of the obvious economic advantages. Our work—involving a survey of DVD consumers in Irkutsk, a focus group in St. Petersburg, and analysis of online discussion groups—suggests that Russian consumers bring a great deal of discrimination to the purchasing of pirated goods and that their moral calculus is more complex than a simple account of ignorance, greed, and theft accommodates. Moral discourse is hardly absent, but it often targets corporate dominance of the media markets and US pressure on Russia in making judgments about fairness and legitimacy in the actions of both pirate and licit producers.[20]

"Piracy," in this context, is a broad concept covering a shifting array of activities, which ordinary Russian consumers often only label on the spot in their interactions with media markets. "Media pirates" are favorably distinguished, notably, from "counterfeiters" who pass off fakes at original prices. Other forms of illegal copying, even when of poor quality, are accepted, and some are routinely praised. Neutral and positive attitudes toward illicit media goods, their trade, and the groups involved in it are strong among Russian consumers. Rather than morally indifferent, Russians are highly sensitive to the political issues surrounding piracy—just not the ones industry stakeholders would like.

THE SOCIOECONOMICS OF CONSUMPTION

Our work indicates that the vast majority of Russian consumers are active participants in the pirate economy, whether through the purchase of unlicensed CDs or DVDs or—increasingly—through the downloading and sharing of digital files.[21] Inevitably, differences in these practices

---

Volcano, which sought out new investments after gambling restrictions came into effect in the city.

20   Russia may be unique in having such positions become the basis of advertising campaigns themselves, as one informant described in the case of AllofMP3.

21   A recent survey commissioned by the International Chamber of Commerce put this number at 89% of the population, with software only slightly behind at 80% (BASCAP/StrategyOne 2009).

are shaped by age, socioeconomic status, and associated variations in purchasing power, access to technology, and cultural capital.

In our survey of three hundred DVD consumers conducted in Irkutsk in December 2008 with the Evolution Marketing Center, college students were the most active viewers of pirated movies and—as elsewhere—lead the way in the shift from optical disc purchases to downloaded or otherwise shared digital files. Personal computers, university-based Internet connections, and—increasingly—home broadband connections are relatively common in this group. Although our survey recruitment method likely oversampled those with high Internet skills, the frequency of reported downloading still surprised us: some 50% of those surveyed reported downloading at least three films or videos per week. Among the larger pool with access to broadband connections, respondents reported downloading, on average, ten albums a month and five to ten films. Despite this shift, the overwhelming majority of consumers in our sample still buy discs from pirate vendors—and among lower-income consumers, almost exclusively so.

Opinions about the importance of purchasing a licensed disc or paying for a download varied. Students led the way in general disregard for licensing: only 17% described it as "very important" or "important" (the other options were "not very important" and "not at all important"). Blue-collar and white-collar workers accorded licensing more importance: 45% and 50%, respectively, for the two categories combined. When asked how much they would pay for a legal DVD, blue-collar workers indicated a maximum average price point of around 140 rubles (about $5.60), and white-collar workers 165 rubles (about $6.60), with students falling in the middle. All these averages are well below the price of licensed DVDs ($14–$20), providing an indicator of the pricing mismatch in Russian media markets.

Lower-income families are, predictably, the most sensitive to price and gravitate toward the low end of the price/quality spectrum of pirated goods. White-collar consumers are at least potentially "swing consumers" motivated by the implied quality of the licensed disc or legal download. And although they still describe legal CDs and DVDs as overpriced, they do occasionally buy them. More generally, they furnish the market for perfect above-quota copies, which sell at a higher price than the low-end pirated goods. Our research strongly suggests that the struggle between pirates and legal distributors is primarily a struggle for this new middle class, which can be "tipped" into the licensed market under the right circumstances.

PATTERNS OF CONSUMPTION

In nearly all cases, respondents drew attention to the complex set of decision points that could override the price barrier. Most often, these were contexts in which the quality or the social function of the purchase was at a premium. Several respondents observed that it was poor form to give pirated copies as gifts, making gift-giving a significant motivator for purchasing licensed CDs.

Our focus group also revealed strong "sampling effects" from downloading in some contexts, in which the discovery of music or movies through pirate sources led to the purchase of licensed

CDs or DVDs.[22] Such purchases generally occurred in the context of collecting. High-quality CDs remain the gold standard in many music collections, and the supplemental materials packaged with a licensed CD provide added value for those with broader interests in music culture. Some respondents also reported purchasing DVDs after sampling downloaded versions—again in contexts where collecting was a primary interest.

Sampling via pirated goods mitigates the problem of poor signaling in cultural markets. Because consumers generally buy cultural goods with only limited information about their likely satisfaction, many of their choices turn out to be wrong. In a context of high prices relative to local incomes, bad choices are especially costly, and the consumption of licensed music and movies remains an expensive hobby. The licit media market in Russia offers few ways to lower these costs of casual cultural consumption. Notably, piracy occupies the place of low-cost movie rental services, which are virtually absent from the Russian market.

Implicitly or explicitly, every person interviewed expected the quality of a pirated CD or DVD to be lower than that of an original, and occasionally so poor that discs may need to be returned. But quality, many respondents made clear, refers to more than the fidelity of the recording. For the majority of Russian families who don't own sophisticated TV and sound systems, properties such as screen resolution, surround sound, and other high-end differentiators in the US and European markets do not greatly matter. The first measure of quality is basic playability. Next come obvious distinctions between poor camcordered copies of new films and DVD-quality prints. Quality also frequently refers to the print materials that accompany the disc, including its inserts and cover. For regular consumers, low-end and low-priced pirated discs are generally purchased in locations near one's home or workplace, where they can be returned if they prove unsatisfactory.

COMPARATIVE PURCHASING POWER

Our findings suggest that piracy in Russia is—first and foremost—a failure of the legal market to price goods at affordable levels. The comparison of GDP per capita in Russia to the prices for legal and pirated CDs and DVDs provides ample evidence of this disconnect. The problem is sharpest with respect to the international goods that dominate Russian markets—notably film and software. As in other countries, international licensing models allow for only modest retail-price discrimination. International CDs, DVDs, and nearly all software at retail are sold—with occasional exceptions—at Western prices, leading to huge differentials in the cost of goods relative to local incomes.

Like a number of other high-piracy countries, Russia has seen a handful of experiments with lower pricing of media goods, including efforts by Columbia Tristar and Warner Brothers to create a "mass model" in 2003–4 by cutting DVD prices to $10 and $15 respectively (Arvedlund 2004); a more significant effort by Russian film distributor ORT-Video, which briefly released its

---

22   The group was too small to derive quantitative estimates of this effect.

How to Watch *Night Watch*

"From my point of view, a competitive-price policy can be quite effective. For example, ORT-Video, an official distributor of the latest Russian box office hit *Nochnoi Dozor* [Night Watch], carried out an experiment by reducing the price of licensed DVD copies of the movie. The price was set at almost the same level as that for illegally duplicated copies. As a result, the official sales of the movie soared, while the number of bootleg DVDs dropped to a trifling 2%. Around three hundred licensed movies are being distributed in DVD format on the basis of the same pricing. A few of them are foreign-made. It is mostly the movies that were already shown in many theatres across Russia; therefore the market demand is steady.

Some of those DVDs are on sale in my stores. I pay a reasonable price when I buy them wholesale, and they are officially licensed products so no brushes with the law are expected. Moreover, lately the legal distributors have started releasing movies on DVD much faster than before. In the past a movie would be available on DVD only two or three months after it was officially released in Russia. Nowadays the delay lasts a month at the longest. These latest developments significantly hamper the illegal trade of the pirates. My message to those who are trying to fight piracy is the following: we are not die-hard perpetrators; we are businessmen trying to succeed in the competitive business environment. Therefore, the market mechanisms should be used for putting an end to piracy."

–A retailer in Veliky Novgorod (Pravda 2005)

catalog at lower price levels; and the low-priced regionalka licensing of DVDs in the provinces. Software companies like Microsoft have experimented with lower-cost, stripped-down versions of their products, such as Windows XP Starter Edition, which limited multitasking and networking features. But these efforts have been controversial, short lived, and largely unsuccessful at making a dent in the pirate market. Modest price drops and lower-quality options may boost sales at the margin but are clearly not compelling alternatives to the pirate market for most Russians.

More radical measures by Russian distributors to compete at pirate prices, for their part, have not been sustained. Reduced-functionality software has fared similarly—pushed into use in some institutional-licensing contexts but not viable in the retail market, where full-featured pirate versions provide better value. As a result, international prices have remained the norm for DVDs, international music, and retail software. The notable exception to this rule is the local music business, which is independent of the majors and consequently less bound by international licensing regimes and pricing norms. In contrast to the movie and software businesses, Russian music labels often compete with pirate pricing—generally in the hope of building audiences for more lucrative live performances.

There is no single price for pirated goods, but rather a price range reflecting differences in the point of sale, perceived quality, and popularity, among other factors. The price differential between licensed and pirated goods also widely varies but is especially stark in the DVD and retail software markets. A licensed international hit movie or album is commonly priced at between

350 and 450 rubles ($14–$18 in 2008–9), with certain titles priced as high as 800 Rubles ($32). Albums published by local labels—representing an unusually large 80% of the total Russian market according to the IFPI (2009)—are almost always significantly lower in price, ranging from 150 to 200 Rubles ($6–$8). Business software at retail, such as Microsoft's Office suite or Adobe's Creative Suite, is priced at Western levels.[23]

High-quality pirated CDs and DVDs, in contrast, generally range between 100 and 150 rubles ($4–$6). Compilation discs—whether the popular collections of ten to twelve films or software packages combining twenty to thirty commercial programs—undercut even this price level and often run as little as 50–100 Rubles ($2–$4).

What do these prices mean to most Russians? Russian GDP per capita is roughly $9,000 in nominal US dollar terms (IMF 2009), representing a little under one-fifth the US GDP per capita. Median Russian income is under $5,000, with higher earners disproportionately concentrated in Moscow and St. Petersburg.

Even in a context of high pirate prices (in comparison to other countries), the market-creating role of piracy is obvious. The price of a high-quality pirated CD or DVD in Russia is comparable to the price of a legal disc in the United States, relative to per capita GDP. The pirate market, in effect, is the only mass market in Russia for audiovisual goods. For software, the discrepancy is much larger. In our street survey of pirate prices in St. Petersburg, we found no stand-alone copies of major productivity software or games but many large pirated software compilations, which retail for a couple dollars depending on the contents. According to Microsoft, software piracy is still prevalent in established retail—present in 25% of software-selling stores overall and up to 70% in more remote regions of Russia (Microsoft Russia 2010). Much of this sector, our sources indicate, has simply moved to non-commercial distribution online.

This general pattern of pricing for legal media (see table 4.1) is consistent with our findings in other countries. The film market is highly integrated at the global level and maintains strongly uniform pricing for DVDs. $14–15 is the floor for DVD pricing in most countries, regardless of the origins of the film. Russian hits like *The Inhabited Island* (2008) retail at the same price as

---

23   Our interviews—though not our fieldwork—revealed noise in the price data for some of these catego-
      ries of goods that we were unable to fully resolve. Microsoft representatives, for example, criticized our
      use of retail prices for small businesses, on the grounds that Microsoft offers small-business discounts
      (analogous to the lower-priced volume licenses that Microsoft negotiates with many large institutions).
      Although our sample is by no means representative, none of the business owners interviewed for this
      report were aware of or took advantage of such programs. From their perspective, the market is split
      between the high-price retail sector and the low-price pirate sector. There are a number of possible
      reasons for this, ranging from inadequate investment in licensing programs by vendors (completing
      markets under such circumstances requires significant local investment) to reluctance on the part of
      small-business owners to formalize their businesses except when absolutely necessary. These explana-
      tions are not mutually exclusive. Our description, in any event, is consistent with the 68%–80% rate
      of piracy in the software market reported by the BSA over the past five years. Similar issues complicate
      music pricing at the periphery of our account, including the release of lower-priced cover albums of
      international hits and short-lived experiments by the major labels with lower domestic pricing.

Table 4.1 Legal and Pirate Prices, 2008-9

Movies

	Legal Price ($)	CPP Price	Pirate Price	Pirate CPP Price
The Dark Knight (2008)	$15	$75	$5	$25
The Inhabited Island (Обитаемый остров, 2008)	$15	$75	$5	$25
Compilations (10–12 films)	—	—	$4	$20

Music

	Legal Price ($)	CPP Price	Pirate Price	Pirate CPP Price
Coldplay: Viva la Vida (2008)	$11	$55	$5	$25
Dima Bilan: Against the Rules (2008)	$8.50	$42	$5	$25
Krematorium: Amsterdam (2008)	$6.50	$32.50	$5.75	$ 28.75

Business Software

	Legal Price ($)	CPP Price	Pirate Price	Pirate CPP Price
Microsoft Windows Vista Home Premium	$260	$1300	–	–
Microsoft Office Small Business 2007 (for Windows)	$500	$2500	–	–
Adobe Photoshop CS4 Extended	$999	$5000	–	–

Entertainment Software

	Legal Price ($)	CPP Price	Pirate Price	Pirate CPP Price
Grand Theft Auto IV, PC (2008)	$20	$100	–	–
Mario Kart, Wii (2008)	$50	$250	–	–

The "legal price" is a widely available retail price for the good in Russia. The CPP, or comparative purchasing power, price is the hypothetical price of the good in the United States if it represented the same percentage of US per capita GDP. "Pirate price" and "pirate CPP price" apply the same principles to the pirate market.

Source: Authors.

Hollywood hits like *The Dark Knight* (2008). The music market is more complex and shows prices ranging from international hits like Coldplay's *Viva la Vida* at the high end, to regional favorites like Dima Bilan's *Against the Rules* in the middle, to local hits like Krematorium's *Amsterdam*, which is sold slightly above the price of high-end pirated goods. The dominance of Russian music labels in the market and the strong promotional function of CD sales clearly impact pricing. Business and entertainment software, in contrast, show no price discrimination to speak of at the retail level (although as in other countries, institutional markets are generally served through lower-priced volume licensing).

Author and Copyright

Author's rights and the commercial features of copyright are distinct issues in Russian copyright law and—our work suggests—in the minds of many Russian consumers when invited to explain how media commodities and media markets work. In our interviews, sympathy for artists and authors was usually strong. Sympathy for the business culture responsible for the commercialization of cultural work was usually nonexistent. Justifications of piracy generally occupied the space between the two.

> They usually buy the right for distributing a film, and it has no relationship to the author himself. It has no impact on the author that I buy an unlicensed disc. We are talking about author's rights and about licensed discs as if they're all in the same terms. These are two different things! (23-year-old male, St. Petersburg focus group)

Several focus group members offered accurate descriptions of copyright, author's rights, and the transfer of rights associated with commercial production—though knowledge of the details of copyright law was infrequent overall. Much more common was the general belief that culture is, fundamentally, a common heritage, a "public domain" that should be accessible to all. When pushed to clarify these positions and reconcile them with existing copyright law, many respondents argued for more limited rights of commercial exploitation. The duration of copyright protection seventy years beyond the death of the author (in Russian law) proved particularly unpopular: "Why should [an artist's] successors own what he created? In my view, after the author's death, his work should belong to everyone, to the society" (41-year-old male, St. Petersburg focus group— higher education, high income, employed in marketing).

Respondent attitudes toward piracy generally combined this bias toward access with broader cynicism about business practices and business culture—both licit and illicit. Although all participants were aware that the sale of pirated optical discs was illegal, few framed this as an important moral dilemma, and none condemned the practice. Some expressed approval of pirate vendors. Several described piracy as occupying an empty niche in Russian cultural markets and pirates as business people who had moved in to fill it. Several reserved their disapproval for the "counterfeiters" who pass off fakes at the full licit price, in distinction from "media pirates" who deal openly in pirated material or who seed content on peer-to-peer websites.

> My attitude toward pirates is neutral. When I find a pirated copy of something I have been looking for for a long time, I feel happy that there are people in this business who help me to fulfill my needs. I feel rather positively toward them. (24-year-old male, St. Petersburg focus group—student)

Any qualms about dealing with the black market disappeared when the subject turned to Internet access. Free downloads on the Internet, and P2P file sharing networks in particular, garnered unambiguous praise from the majority of respondents. Central to this approval was the absence of a profit motive and the resulting lack of consumer implication in the corrupt—and corrupting—wider business culture. In our study, P2P proved to be a very powerful focal point for Russian attitudes toward piracy, combining a self-interested economic rationale with a moral framework that nearly always trumped the claims of rights holders.

Counterfeiting

In Russia, the terms for piracy (*piratka*) and counterfeiting (*kontrafakt*) are used interchangeably in industry and media representations of copyright infringement (a term that, in contrast, is almost never used). Traditionally, however, piracy and counterfeiting refer to distinct phenomena: piracy to the unauthorized copying of the content of expressive works; counterfeiting to the unauthorized application of brand names to cheap copies of goods—often with the implication of consumer fraud. The former violates copyright and the latter trademark.

Where consumers generally treat piracy as a neutral phenomenon, subject to only limited moral censure, the term *kontrafakt* has stronger negative associations with health and safety risks—especially in connection with fraudulent pharmaceutical and alcohol products. As we discuss in chapter 1 of this report, our collective work finds piracy and counterfeiting to be largely disconnected phenomena at the global level. Efforts to precisely duplicate the packaging and presentation of licit discs are rare and, on the Internet, irrelevant. Cross-border smuggling—the basis of the counterfeit-goods trade—has been largely replaced by a mix of local production and Internet distribution.

In Russia, however, the central role of above-quota production in licensed factories creates a zone of overlap between the two phenomena—albeit in ways that erase notions of substandard quality. The place accorded to this type of counterfeiting in Russia is unique in this report. And yet we are hard pressed to find evidence of widespread fraud. In our findings, Russian consumers show remarkably little confusion about what they're paying for—indeed the complex negotiation of price versus perceived quality is popularly understood as a Russian specialty, grounded in the long-standing role of black markets in the distribution of luxury goods (Dolgin 2006).

In cases where the copy is above-quota or otherwise exact, it makes little sense to speak of fraud: the consumer gets exactly what he or she expects. More often, the choice to purchase unlicensed goods reflects a deliberate compromise. Consumers weigh price against perceived quality, not against perceived origins. Discontent with pirated goods, in this context, arises when the paid price is seen as higher than the received quality. Such negotiations shape the spectrum of pirated goods available in Russian markets. There is, for example, a large market for low-quality pirated products offered at low prices, with the corresponding risk assumed by the buyer.

Intentional deceit of consumers is, of course, far from unknown. There are a variety of ways in which packaging can mislead with regard to quality or content, and it isn't difficult to find such products in the retail markets of St. Petersburg or Moscow. But deceptive practices in the pirate market are mitigated by the important role of return customers in many vending contexts—and the resulting importance of trust. Pirate vendors often have a strong customer-service ethos that includes the exchange or return of defective or deceptive discs. Such services often go well beyond what is available in the licit market. As one of St. Petersburg's major music stores makes clear: "Licensed discs are not subject to exchange or return."

It seems likely to us that the deformalization of pirate retail in Russia will put pressure on this service-oriented model. As pirate distribution leaves the sphere of the (however informally) regulated market, connections between vendors and customers become weaker, and more opportunities for outright fraud arise.

LOCAL EFFECTS

The moral calculus around piracy often has strong patriotic overtones. US pressure on the Russian government related to IP and trade issues is widely viewed by consumers as commercial imperialism—a form of aggression rather than an assertion of universally shared rights. The intellectual property of foreign corporations, in this context, circulates at a moral discount and raises fewer concerns about the impact of piracy.

Many of our informants believed, in this context, that domestically produced Russian music and film are pirated less voraciously than foreign material. Although empirical evidence is thin, we see both social and economic factors supporting such a claim. First and foremost is the issue of cost: the economics of licensing in Russia create a strongly bifurcated market for Russian and foreign music. The costs of licensing and producing Russian music are significantly lower than for foreign acts, and the resulting discs are sold at prices much closer to those of their pirated competition. It is far easier for consumers to make the "right" choice when legal CDs cost $6–$8, while pirated copies cost $4–$5. The price of a foreign-licensed CD, in comparison, averages $12 and can go as high as $30. The market for such goods is accordingly tiny, and the range of goods available is very limited.

Our findings also point toward a more tentative distinction between local and non-local artists within Russia. Music retailers in St. Petersburg described a loose system of obligations in which social ties among musicians, producers, record companies, and retailers anchor norms of respect for local commercial products. "No one wants to hurt one's own," said one retailer. Informants described several cases in which representatives of St. Petersburg-based record labels personally reproached music store owners for pirating local bands, and others in which music producers and bandleaders made the rounds before album releases to urge music store owners to refuse pirated products.

The view that local musicians are sheltered from illegal copying was not universally shared, however. As one representative of an anti-piracy organization put it: "They write on the Internet that 'we do not pirate Russians.' Come on, everybody is pirated!"

The Economic Function of Piracy

For obvious reasons, industry research and lobbying paint piracy as a drain on the Russian economy and emphasize losses to copyright holders. The BSA's estimate of total software losses in Russia topped $4.2 billion in 2008 (BSA/IDC 2009). Creative reframings of piracy's effects have also become common in this context. The BSA now produces annual estimates of the number of jobs that would be created in return for small reductions in software piracy. But this account of piracy as a pure loss to the economy is misleading. Piracy is not just a drain on the media economy in Russia—it is a fundamental part of the media economy, deeply woven into a wide range of licit practices and forms of enterprise. The direction of claimed losses also matters greatly. From the perspective of the Russian economy, losses to international rights holders are, strictly speaking, gains for Russian businesses and consumers. Losses to Russian copyright holders, in contrast, represent a more complex reallocation within the Russian economy, in which money is not "lost" but spent in other ways. For countries that import more IP than they export, like Russia, this balance of exchange may be strongly positive.[24]

Even within law enforcement agencies and anti-piracy organizations, views about the impact of media piracy on Russian society vary. Strong negative judgments equating piracy with theft are, of course, commonplace. But so too are alternative accounts, offered less in support than in explanation of piracy. The roles of the informal economy in Russia as a source of additional jobs and second incomes and—in cultural terms—as a support infrastructure for cultural diversity figure prominently in these accounts. These roles are especially strong outside Moscow and St. Petersburg, where the commercial media infrastructure remains very underdeveloped.

SURVIVAL IN THE INFORMAL ECONOMY

Media piracy is attractive to workers in the low-end retail economy because it affords relatively high profit margins compared to other types of trade. According to our interviews with vendors and store owners, the retail markup for licensed media goods runs approximately 80%–150%

---

24   See the analysis of economic gains and losses, including BSA/IDC job-creation estimates, in chapter 1.

for CDs, 80%–200% for DVDs, and 40%–60% for the fading LP market. Unlicensed goods, in contrast, generally carry a 200%–300% markup (over significantly lower wholesale prices). For street vendors further down the economic ladder, other incentives come into play: the informal sector provides jobs to social groups with little access to legal employment—especially the urban poor, migrant laborers, and students. In most other regards, the street vending of CDs or DVDs is comparable to selling other commodities, such as food and clothes, but yields higher incomes.

The informal economy in St. Petersburg is organized around several different types of commerce, ranging from transient vending in subway stations, to more organized sales in open-air flea markets, to established retail shops. By our estimate, there were between 150 and 200 subway vendors in St. Petersburg in 2008–9. Most work from foldable card tables, which can be set up and packed away quickly. Vendors at different stations are almost always part of a larger distribution chain run by a single owner, who takes most of the profits. A vendor typically receives 10% of sales in salary. Our research indicated that a hired vendor typically earns between 500 and 1000 rubles per day ($20–$40), or approximately $450–$900 per month. This is between 78% and 157% of the median income in St. Petersburg (Regnum News 2008). It is also significantly higher than other segments of the street economy. Footwear vendors working at marketplaces around subway stations, by comparison, earn an average of 500 rubles ($20) during the same nine- to ten-hour workday. With an estimated profit of $10–15 million generated annually in the city by pirate street vending, some $1–1.5 million ends up as household income for those employed in the sector.

The sale of pirated goods is thus a relative privilege for people who would otherwise be—at best—part of St. Petersburg's working poor. Roughly 10% of the city's population qualifies as poor by Russian standards. The median monthly income in 2008 was 19,000 rubles, or $770, while the estimated minimum subsistence level is 4,900 rubles, or almost $200 (City Statistics Bureau 2008). For the many families in this category, the additional income from piracy is often the difference between simple poverty and destitution, as one working family member usually supports anywhere between two and four non-working members. The trade in pirated goods has one important further advantage: it is relatively socially acceptable and does not carry the kind of stigma associated with the drug trade, prostitution, or other forms of criminality (Ovcharova and Popova 2005:15, 28, 35).

Well-known open-air markets like Gorbushka in Moscow and Yunona in St. Petersburg play prominent roles in this wider economy of secondhand, grey-market, counterfeit, and pirated goods. But they are not unique. Almost every open-air market in Russia includes trade in software, music and video CDs, and DVDs—all of which are likely to be pirated. Most trade in the cheapest variety of pirated goods—the products of "cottage-industry" production. Used CDs, DVDs, and video cassettes are also common. Some vendors carry more conventional high-end pirated stock purchased from larger suppliers. The prices of CDs and DVDs in these markets are significantly lower than in subway stalls: 50 rubles ($2), as opposed to 100–150 rubles ($4–$6), for a DVD. In these circumstances, a single vendor generally sells between one hundred and two hundred discs over the weekends in which the markets operate, making a 5- to 10-ruble profit ($0.20–$0.40) on each disc. Total profits may range between 1,000 and 2,000 rubles ($40–$80) per weekend. The less-organized, part-time structure of the work in such marketplaces means that many of the vendors are industrial and service workers trying to supplement their incomes. One of our informants was a travel agent who organized trips abroad for Russian tourists.

THE SMALL-BUSINESS DILEMMA

Software is critical infrastructure for most businesses, from basic productivity tools to the more specialized software packages needed as businesses move up the value chain. As in the retail-sales sector, however, operating with fully licensed software in Russia is often a privilege of size. Because software is generally priced at or near Western levels, it represents a disproportionate and often prohibitive investment for small businesses and start-up companies. Although open-source software tools can provide much of the same functionality for free, low rates of open-source adoption and—in some areas—inferior open-source alternatives impose costs on use, especially when commercial software operates as a de facto standard.

Given the choice between free open-source options and free pirated commercial software, Russian businesses almost always opt for the latter. The downstream consequences for the Russian economy—or for any other economy, to the best of our knowledge—have never been adequately calculated.[25] But there is little doubt that pirated software is both an enabler of economic activity and an obstacle to wider open-source adoption.

The role of pirated software as infrastructure for emerging software companies is an open secret in the globalized software industry—sufficiently so that the Romanian president, Traian Basescu, publicly thanked Bill Gates in 2007 for the role that pirated Microsoft software played in IT development in Romania (Reuters 2007). Although Russian firms like 1C and Kaspersky Lab are now large enough and integrated enough into the global software economy to make them domestic advocates of IP enforcement, the economics of small software firms still virtually dictate the use of pirated software.

The price of licensed software for a firm working with professional multimedia tools, such as Flash, can start at roughly $2,000 per employee and rise quickly from there.[26] Software licensing costs can easily represent the largest portion of investment in a new web or design business. Few start-up firms have sufficient capital or prospects for return business to warrant such investment, especially in a context where "free" pirated alternatives are widely available. The owner of a start-up media-production firm in St. Petersburg put it this way:

---

25 See chapter 1 for a discussion of why BSA claims on this subject should be discounted.

26 A typical professional software setup could include the Adobe Creative Suite 5 Master Collection ($2,500) and more basic Microsoft Windows, Microsoft Office, and productivity tools ranging up to $1,000.

We can cover all the office costs for four months for this amount—rent, phone, bank service, taxes . . . What part of turnover is this? Well, it's hard to say—the turnover is not stable at all, as is typical for any small firm. There's just no free money we can take out of the business and invest in software.

The logic is much the same in other software-dependent businesses, such as printing. Margins on print jobs are invariably thin, and such businesses rarely begin on sound financial footing. Software costs per employee or seat can easily run $3,000,[27] and a business with any significant volume would need multiple seats.

While incentives run strongly toward pirated software for small businesses, such choices involve a measure of risk. Small businesses are often visible and potentially vulnerable to police and private investigators. The ability to effectively manage software licenses is itself a luxury of size. Maintaining accurate licensing records for multiple versions of different software tools, over years and across machines, is challenging for firms that lack well-run, professional IT departments and puts even well-intentioned software purchasers at risk. Large corporations, in contrast, are better able to absorb software-licensing costs and, importantly, cut deals with software vendors and defend themselves against charges of infringement—factors that, in turn, make them less likely to be raided. Enforcement tends to fall hardest, consequently, on small businesses, where a raid or a fine can be fatal.

As in the retail sector, consequently, the price differential between licit and pirate media shapes the larger opportunity structures of business in Russia. Because only large, well-capitalized companies can afford to operate consistently within the licit economy, enforcement campaigns confer de facto commercial advantages on those firms by making competition and innovation from below more costly and precarious. Although large businesses would describe the pressure to remain licit as an additional cost, the history of the last fifteen years in Russia is unambiguously one of oligopoly—of self-perpetuating commercial advantage for the largest players. IP enforcement, perhaps unavoidably, has become part of that dynamic.

The Cultural Function of Piracy

The poverty of legal supply is a key to understanding the cultural function of pirate markets. With the exception of international hit albums, most foreign music is not legally available in Russia—even in St. Petersburg, Russia's second-largest media market. Provincial markets are even more poorly served. Legal distributors generally do not stock niche genres or artists that fall below a perceived threshold of international popularity. Manufacturers make similar choices due to the high costs of licensing foreign work.

---

27   A printer would need to buy three to four programs per seat: commonly a lower-end Adobe Creative Suite 5 bundle ($1,800) and the usual Microsoft packages and productivity tools.

The growth of chain stores, centralized production and distribution, and police pressure on smaller retailers are both contributing causes and symptoms of this problem. All work to diminish the role of customer relationships in the retail business. Local feedback loops from customer to retailer to distributor to manufacturer are much less frequent in the new era of large diversified chains—and indeed have become a signature of "unlicensed" retail and production. The market advantage of pirate vendors is thus one not only of price but also of greater proximity to consumer demand and greater freedom to innovate with compilations, mash-ups, and other alternative formats to meet that demand. The greater diversity of content within the pirate market is particularly important to groups who feel strongly connected to international cultural conversations—especially the educated, mostly urban Russian intelligentsia, for whom access to a wide variety of cultural goods is a condition of cultural participation.

Interviews with older St. Petersburg residents reveal a consistent set of observations about piracy's place in the transition from communism. The social consciousness of many educated Russians in the communist era was formed through the consumption of censored Western cultural goods—typically available through the black market and other clandestine networks (Baker 1999). The end of communist censorship unleashed enormous pent-up demand for such works. A wide variety of books, films, and music came to market, but at prices that were prohibitive for all but a few Russians. Piracy—first of books and cassettes, and more recently of discs and digital files— became the main remedy for this problem.

SPECIALTY STORES

To a significant extent, this cultural diversity role in Russian cities has been filled by specialty music/video stores, which often deal in niche film genres and music subcultures. Because legal manufacturers and distributors in Russia cannot efficiently license and produce for small markets, specialty stores deal mostly in pirated goods. Often they are the only suppliers for whole categories of music and film.

The management of such shops often involves a mix of business and vocation. In our research, shop owners recurrently emphasized their sense of contributing to the education of the Russian audience. Such considerations did not obviate moral concerns about piracy but did generally overcome them.

> As for the moral side of things, yes, there is nothing good in the violation of
> somebody's rights. There is nothing to be proud about. But I am doing this not
> in order to make mega-profits; I am doing this because I think that people in this
> country, in this city, should have access to good music and good films. I will not
> become rich doing this, that's for sure. But people will get access to these things.
> (Owner of a well-known St. Petersburg specialty music/video store)

In 2009, there were a handful of specialty shops in St. Petersburg that sold music and film and two or three more that specialized in video only. The social roles these shops play go well beyond the traditional functions of the media retail business. They anchor local communities of music and film collectors and serve a broader enthusiast culture, stocking vinyl records and secondary literature, organizing film showings, and—perhaps most important—promoting local and lesser-known artists. Several of the stores have libraries of MP3 and MP4 files, which can be made into compilations and burned on demand. The dialogue between customers and retailers shapes both the stock of goods and the services offered.

New market niches can be—and often are—created through these customer relationships, and store owners and staff sometimes play an important part in building audiences:

> Yes. I understand that they are pirates. But I also understand that if I don't take a step toward these pirates myself, the listener who is interested in electronica will never learn about the Frans label or the existence of the Violet project, which combines electronica with psychedelia. No Russian magazine will write about this project. And if one day someone wants to organize a concert by Violet, no one would come—or only those few people who either downloaded it from the Internet or bought it from me. (Manager of a St. Petersburg specialty music store)

This manager also views her role as one of integrating Russia into the wider global market for live performance:

> I am thinking about the day when the group Sunn comes to Russia. In order for them to come, I need to prepare the ground, the audience. So that when I come to a promoter and say, "Let's bring Greg Anderson to Russia," I have an answer when he asks me, "Will it pay?" Every promoter has to know whether it will pay. And for that I prepare my audiences.

Because major legal distributors rarely go beyond a narrow stock of international hits, international music scenes in Russia have been built largely around pirate networks.

For specialty store staff, maintaining this loop between taste-building, promotion, and consumer demand in Russia is not easy. One popular independent music store in St. Petersburg works on a regular basis with three warehouses, all of which furnish a mix of licensed and unlicensed CDs and DVDs. The important differentiator, for store owners, is not the license but rather the quality of the service: the range of titles available, the frequency with which they are updated, and the speed with which stock is delivered to the store. Store staff work, especially, to ensure that they stock titles reviewed in international music publications like the *Wire*, which are increasingly accessible online. Suppliers, for their part, routinely frustrate retailers by not updating their offerings with enough frequency. As the same specialty store manager emphasized, "I tell them, if you guys are pirates, at least make sure you bring in new music!"

In our interviews, music producers and musicians expressed ambivalence toward this promotional function. Although many viewed piracy as "theft," a majority also understood that piracy helps build audiences for live shows and can provide an indicator of popularity. As one record promoter argued:

> Nowadays pirates don't pirate anything and everything, thank God. They need to make money. So when I see one of my projects pirated, it means it's good. I have become popular. My promotion has worked well in this project. Pogudin, The King and the Clown, The Pilot, and Night Snipers became popular, yes. Pirates pirated them, and I am very proud of it.

EXPANDING THE MARKET

The market for cultural goods extends far beyond the available licensed material. Pirate production—and sometimes non-commercial production—fills the gap. Documentaries and educational programming, to take two popular genres, almost never appear in licensed form, yet pirated British and American World War II documentaries and military history videos are commonplace in the pirate market. Meditation and yoga videos, self-help audiobooks, language courses, and religious sermons ranging from Orthodox Christianity to Buddhism to Pentecostalism also circulate widely. Although the commercial media in these contexts is clearly pirated, other material—especially religious media—is often produced without express commercial purposes. Circulation, rather than sales, is the primary goal, with the informal marketplace providing the infrastructure.

The pirate market's close connection to consumers sometimes results in product innovation, occasionally to a degree that produces original and valuable new work. Unauthorized subtitling and even dubbing of foreign material into Russian is a prominent example. A trader at St. Petersburg's open-air market Udel'naya described the phenomenon:

> Well, they shoot it on video in a movie theatre and then translate the dialogue. There is a special site, Kvadrat Malevicha, where you can find all kinds of Western TV series, like *Lost* and so on. They offer all the popular Western serials that are not aired in our country at all or only with a big delay. The site translates these series [into Russian] at their own audio studio—yes, they do have their own studio for translation.

The intersection of Russian-language film-dubbing traditions and cheap sound-mixing technologies has also given rise to a popular genre of alternative voiceovers, pioneered by Dmitri

Puchkov (a.k.a. "Goblin"). Many of these alternative dubs are serious efforts to improve the translation of colloquial language. Others are "funny translations" that make satire out of the original material. Goblin's satirical edits of the *Lord of the Rings* films, *The Matrix*, and *Star Wars: Episode I* have been enormously popular and led to the inevitable dual-track DVDs. The same trader noted:

> Recently pirates began to issue films which can switch from "normal" translation to "funny Goblin" and back. The movies have both soundtracks. This is a very popular format in the market now.

Film compilations and "greatest-hits" music compilations populate much of the middle and lower tiers of the pirate market. Although some of these are pirated versions of authorized greatest-hits albums, many are original compilations released by pirate producers. Many have additional tracks or favor songs that became popular in the Russian market. The general decline of the album format also plays a role, as pirates respond to consumer preferences for cherry-picking hits and avoiding filler. Film compilation DVDs—often grouped by genre, actor, or theme—are almost always original pirate productions.

REPRODUCING THE INTELLIGENTSIA

Ten years after the transition from communism, German Gref, the minister of economic development, offered a bleak assessment of Russian cultural institutions: the physical plants of theatres, museums, and other infrastructure for culture were in severe decline; the cultural and informational isolation of many regions of Russia was on the rise; and wages among employees in the cultural sector were falling. As a result, he argued, Russia was playing a diminishing role in global cultural production (Gref 2000). The groups hardest hit by this decline were public employees in schools, universities, orchestras, museums, and other cultural and educational institutions—the core of the Russian intelligentsia. Plummeting incomes and the rising costs of cultural and educational goods made this class of educated urbanites a natural consumer of pirated media. The superior diversity of the pirate marketplace, especially for non-mainstream and foreign imports, solidified this relationship.

In interviews, members of this loose class displayed strikingly favorable attitudes toward pirated goods. Several called piracy their "rescue" or "salvation." What they typically meant was that piracy provided their only access to the world of non-blockbuster media goods—independent music, art-house films, and much Western media. Such access is not a luxury for members of this group, but in many cases the basis of their professional activities as musicians, writers, editors, and producers. Piracy—not the licit market—enables them to participate in the international cultural arena. Consequently, it is also the condition of their survival and renewal as a professional class. As one St. Petersburg film critic and college teacher explained:

> Libraries are so impoverished. I cannot find anything reasonably recent in them—
> anything that came out after 1985. And if I only watched films released in our movie
> theatres, I would have to quit my job. I would not know what to write about! Thanks
> to pirates, I can download books I need. And what would magazines cover if people
> like me couldn't see movies that I know get released over there?

Such views are commonplace in Russia and reinforce a basic finding of this report: pirate distribution plays an important role not only in relation to access and consumption in the media economy, but also production. Below the top income strata in Russia and other emerging markets, the globalization of media culture passes largely through pirated goods.

Law and Enforcement

For several years after the demise of the Soviet Union in 1991, authorities worked to build the legal institutions considered necessary for a modern market economy, including the first Western-style intellectual property laws. The enactment of the laws On the Legal Protection of Computer Software and Information (1992) and On Author's Right and Neighboring Rights (1993) culminated several years of work initiated prior to the transition on adapting European IP norms to the Russian context. Consistent with most interpretations of the Berne Convention, copyright in the new law was described as a private right, with provisions for criminal penalties limited to cases of commercial infringement. Criminal provisions were also written into the Statute on Programs Protection, covering the illegal commercial copying of software. Through 1997, the older Criminal Code provided for up to two years of forced labor and/or a fine in such cases, but the statute was very rarely applied (Golavonov 2008).

After 1992, responsibility for IP enforcement was assigned to the new Russian police agencies, including the MVD (the Ministry of the Interior) and the FSB (the Federal Security Bureau). A variety of non-governmental anti-piracy organizations also soon emerged to ensure that foreign (and over time, domestic) rights-holder interests played a role in guiding enforcement efforts.

The new legal institutions—and law enforcement in general—were nonetheless very weak in this period. Diminished financing and dwindling ranks left the police internally disorganized and short of basic resources. The social and economic turmoil occasioned by the rapid, disruptive transition from communism sent rates of violent crime skyrocketing and allowed for the emergence of powerful organized-crime organizations that co-opted law enforcement and operated with general impunity, especially in regard to non-violent "property crimes." Although the IIPA and other industry groups complained loudly about piracy throughout the 1990s, and particularly about the ineffectiveness of Russian law enforcement, these complaints found little political traction in a context of wider problems of public order and weakness of the rule of law.[28]

The consolidation of a new legal, civil, and economic order did not happen overnight and suffered notable setbacks, including a major economic crisis in 1998. The process of revision of the legal foundations of the new Russian state continued through the early 2000s. With respect to IP policy, an important new factor was the emergence of a domestic IP lobby, built on the recovering film industry and the emerging software industry and closely integrated with and advised by Western companies. This lobby began to push for stronger enforcement efforts and an end to the impunity of pirate activity.

The new enforcement advocates pushed for changes to law, law enforcement, and judicial process. Lawyers who represented plaintiffs in piracy suits in the 1995–2000 period, in particular, had had overwhelmingly negative experiences with the courts and strongly favored efforts to streamline police and court procedures. The lack of coordination among law enforcement agencies was also a major hindrance. Cases were often split among several agencies and frequently slowed or ground to a halt when those agencies could not effectively coordinate the different stages of investigations. The penalties at the end of the process were rarely dissuasive and generally thought to be futile. A common saying among lawyers and prosecutors at the time was that "one piracy case equals three unsolved murders"—a reference to the waste of material and administrative resources associated with copyright infringement cases. For much of the legal community, strengthening the laws against piracy was a pragmatic solution to the wider set of inefficiencies produced by IP laws. For law enforcement, which rarely viewed IP cases as a high priority, the streamlining of case procedures promised to free them to do more of the other things they considered important.

High-level meetings between government and industry were held in 2001 and 2002 and led to the creation of an interagency commission devoted to coordinating IP policy and enforcement efforts.[29] The first major legislative reform took place in 2003, with the amendment of Article 146 of the Criminal Code. The revised Article 146 granted police ex officio powers to launch investigations and make arrests, sped up criminal prosecution, and increased the penalties and other remedies available under the law. This effort, in turn, paved the way for the development of sweeping new IP legislation, known as Part IV of the Civil Code.

Part IV was intended to unify Russian IP law and bring it into full compliance with international

---

28   Court statistics indicate that charges against pirate distributors and vendors were nonetheless filed with some frequency in the period, often for violations of fair competition rather than piracy. Few of these cases went to trial, however, and industry actors, in general, were slow to make use of the civil remedies available in Russian law. The first civil complaint for software piracy, filed by the software company 1C against another company, Nais, went to trial in 1995 and led to the first successful sentences against software "pirates." The Bulgarian case, described earlier in this chapter, went to trial in 1997 (Vitaliev 1996).

29   The commission was chaired initially by Prime Ministers Kasyanov and Fradkov and later by future president Dmitry Medvedev, whose main interest was the development of Russia's IT sector.

IP agreements, including the Berne Convention (in anticipation of WTO accession) and the World Intellectual Property Organization's Internet Treaties. It was also intended to solve a number of long-standing problems in the IP rights arena, such as the chaotic situation in collective rights management. Although the end result achieved many of these goals, not everything went as planned. When a draft bill emerged in 2006, it was widely criticized on both form and substance by industry stakeholders, local IP practitioners, and foreign government officials. The omnibus approach was controversial—viewed as both hard to implement and hard to change. Industry stakeholders also resisted what they viewed as steps back from TRIPS (Agreement on Trade-Related Aspects of Intellectual Property Rights) and TRIPS+ standards, particularly in regard to private copying, secondary liability, and technical protection measures in the digital environment, on which the bill was vague or silent.[30]

When it became clear that rights-holder groups could not kill the bill, they mobilized to change it. Several rounds of debate and revision followed, with the final compromise mostly ratifying the strong interpretations of TRIPS promoted by rights-holder groups. Part IV entered into force on January 1, 2008, superseding all previous law governing IP protection. Elements of the legislation are still being implemented, including the 2010 passage of a 1% levy on blank media and audiovisual equipment as a means of counterbalancing the law's relatively broad allowance for private copies.[31]

Russian efforts to join the World Trade Organization played an important role in the evolution of these IP reforms. WTO rules require that prospective members secure the agreement of major trading partners prior to entry—a requirement that gave the United States a de facto veto over Russian admission.[32] As Russian economic and political stability appeared to bring prospects for admission closer, pressure grew for stronger cooperation on enforcement.

In 2006, following several rounds of talks about trade and IP protection, the United States and Russia signed a Bilateral Market Access Agreement that focused on expanding Russian enforcement efforts. In the wake of the agreement, the government launched the major crackdown on piracy described earlier in this chapter. The Duma also raised penalties for piracy, making damages greater than 250,000 rubles ($10,000) punishable by up to six years in prison. Although Part IV says nothing about how to establish damages, Russian courts have followed the industry practice of using retail prices to value pirated goods, thereby expanding the range of offenses subject to the highest penalties.

---

30   On these changes and the broader arc of IP reform in Russia, see Mamlyuk (2010), Budylin and Osipova (2007), and Golavonov (2008).

31   Among other things, the measure will pour an estimated $100 million per year into yet another opaque collective rights management organization, called the Russian Union of Rightsholders, led by filmmaker Nikita Mikhalkov (Russian Law Online 2010).

32   It was widely expected that WTO member-state Georgia would be the stalking horse for the actual veto, with support from the United States and probably the European Union.

But by most accounts, this momentum was not sustained. Since 2008, Russian enthusiasm for WTO accession has faded—and with it much of the incentive to implement additional US requirements on enforcement. From the perspective of the Russian government, dialogue with the United States involved a continuously shifting set of goalposts, with agreement to one set of demands only producing more stringent demands. Lack of progress on issues unrelated to copyright, such as Russian requests to the United States to repeal the Jackson-Vanick Amendment (a Cold War provision linking tariffs on Soviet-produced goods to emigration policy), created additional sources of tension.

Plans for further changes to Part IV of the Civil Code slowed dramatically in this context, with little movement on the TRIPS+ enforcement standards demanded by the USTR and international rights-holder groups. The absence of clear provisions for secondary liability for Internet service providers (and other web-based services) in cases of online infringement has been one of the touchstones of this debate. Although Russia officially acceded to the WIPO Internet Treaties in 2009, the emergence of a strong Russian ISP lobby in the past few years and the apparent reluctance of the administration to undertake a separate law on Internet regulation has kept this issue off the legislative agenda. The adoption of US-style "notice and takedown" procedures for infringing online content, in particular, remains unlikely for the foreseeable future.

Governmental Actors

Within the executive branch, four agencies have primary responsibility for IP enforcement: the Ministry of the Interior (MVD), the Prosecutor's Office (Prokuratura),[33] the Federal Security Bureau (FSB), and the Federal Customs Service (FTS). Each has its own investigative functions and the right to initiate criminal proceedings for copyright infringement, and each now conducts investigations ex officio—without the traditional need for a complaint filed by the rights holder(s). Prior to 2006, the MVD played a relatively minor role in the investigation of copyright infringement. It was considered a weak and easily corrupted organization, with limited experience in investigating IP crimes. With the amendment of Article 151 of the Criminal Code in 2006, both the MVD and the Prosecutor's Office became more central and active in enforcement efforts. By most accounts, the MVD's weaknesses in enforcement have been addressed or at least minimized through collaboration with anti-piracy organizations, which now guide and advise its operations.

The MVD is itself composed of many individual departments, each specializing in a different domain of criminal activity and type of enforcement. The OBEP (the Department of Economic Crimes) targets pirate production and distribution networks. Department K, the MVD's computer crime unit, specializes in Internet crime and, in theory, bears responsibility for prosecuting copyright violations on the Internet—though in practice such action has been infrequent. The

---

33   Roughly speaking, the Prokuratura in Russia combines functions that in the United States are assigned to the Office of the Attorney General, Congressional investigating committees, grand juries, and public prosecutors.

Department of Public Safety (a.k.a. the Militia) has jurisdiction over street vendors, including the trade in pirated discs.

The Federal Security Bureau has plays a variety of roles in IP protection efforts, initially due to its jurisdiction in contraband cases, but more recently under the pretext that the unregulated financial flows associated with piracy represent a threat to national security (a standard that would include most Russian businesses if applied consistently). In practice, it also acts as an occasional check on the other agencies and has prosecuted cases of corruption in the MVD.

The role of Customs in copyright enforcement has declined significantly since the 1990s, when smuggling of pirated optical discs into Russia was widespread. Pirated discs, in this earlier period, came principally from former members of the Soviet bloc—especially Bulgaria and the Ukraine. Once optical disc manufacturing technology became less capital intensive and more mobile, a local pirate manufacturing base developed that obviated the need for most high-risk smuggling operations. With the decline in this form of physical distribution, Customs is now preoccupied with other types of illegal and counterfeit smuggled goods, including pharmaceuticals, textiles, electronics, and industrial products.

Other agencies play more specialized or occasional roles. The Ministry of Culture and the Russian Cultural Protection Agency together license and administer the activities of the collective rights groups and have been involved in the controversies surrounding AllofMP3 and now the RAO. The Federal Anti-Monopoly Bureau plays a periodic enforcement role when a case turns on "unfair competition" or related anti-competitive business practices. The Ministry of Mass Communication has responsibility for Internet regulation, including laws governing e-commerce and the liability of ISPs, hosting services, and other services for infringement. It also licenses optical disc plants. Other ministries and organizations participate in more specific contexts, such as the Ministry of Economic Development, which is negotiating Russia's entry to the WTO and bears responsibility for Russian compliance with international IP norms.

Non-governmental Actors

Numerous non-governmental organizations also take part in enforcement efforts, but six are preeminent in this area: the Russian Anti-Piracy Organization (RAPO), the Nonprofit Partnership of Distributors (NPD), the Nonprofit Partnership of Software Suppliers (NP PPP), the National Federation of Phonogram Producers (NFPP), the local affiliate of the BSA, and the local branch of the IFPI. These organizations specialize in different areas of enforcement: the RAPO oversees film and video compliance; the NFPP, the NPD, and the IFPI handle sound recordings and music; and the NP PPP and the BSA monitor the distribution and use of computer software.

The RAPO was created in 1997 by US studios and film distributors seeking better representation of their interests in Russia. American studios continue to provide the RAPO's core funding, although Russian films co-produced with US studios are also covered through the organization. In addition to Moscow and St. Petersburg, the RAPO has offices and representatives in a number of provincial cities, as well as its own investigative staff, which works closely with the state police. Until 2004, these investigative networks provided the basis for the MPAA's "supply-side" estimates of pirated disc sales in Russia, which hovered between 80% and 90% of the total market throughout the late 1990s and early 2000s.[34] In recent years, the RAPO has focused on investigations targeting large warehouses and optical disc manufacturing, but several informants described a new shift toward disc sales and—inevitably—Internet monitoring.

The most influential industry group is the NP PPP, which represents major domestic and international software companies, including Microsoft, Adobe, Borland, Symantec, and Autodesk. This international membership substantially overlaps with that of the BSA, leading to many shared interests. Unlike the BSA, the organization also represents virtually all the domestic software suppliers—a constituency that gives NP PPP considerable access to government officials. As a result, the NP PPP is not perceived only as a representative of foreign technology interests but also of the growing domestic software industry and, consequently, as a credible voice for domestic business interests.

The NPD is similar in structure to the NP PPP but focused on the music market. Founded in 2003, it coordinates lobbying and enforcement efforts for the eight largest music distributors in Russia. The NFPP is an IFPI-sponsored competitor and maintains close relationships with international organizations and labels.

Despite efforts in the past decade to routinize coordination and cooperation, relations between law enforcement and industry are complicated. Interests in the enforcement area have become too complex and diverse for simple implementation of industry directives of the kind favored by the IIPA and the BSA. In our research, representatives on both sides of these partnerships characterized the relationships as "cautious." Industry groups are routinely dissatisfied with the scale and effectiveness of police efforts. The RAPO has issued numerous complaints about the unwillingness of local police to raid businesses identified as pirates. NPD representatives described how local precincts offer protection—or "roofs" in Russian argot—to markets or retailers for a modest fee ($300–$500) and how regional representatives to the Duma have blocked NPD efforts to mobilize police to raid markets where they have discovered illegal sales. By the same token, when local businesses feel harassed by aggressive enforcement tactics, they have become much quicker to involve local political authorities and to pursue appeals within the industry groups.

The Effectiveness of Enforcement

There is considerable debate about the effectiveness of IP enforcement efforts in Russia, especially in the wake of the 2006–7 crackdown. Without question, the crackdown hurt the more exposed categories of manufacturers, distributors, and retailers and produced a reconfiguration

---

34   In 2005, the MPAA switched to a consumer-survey method, which estimated the rate of piracy at 81% (MPAA 2005). The MPAA has not conducted a follow-up study.

of production and distribution—centralizing the former and breaking up the latter into less formalized channels. Industry reported measurable drop-offs in piracy in the software and music categories in 2007–8. The RAPO, in particular, claimed a 40% fall in DVD piracy in the major St. Petersburg and Moscow street markets, where enforcement actions were concentrated.[35]

Figure 4.1 Estimated Share of Pirated Products in Russian Markets, 2000–9

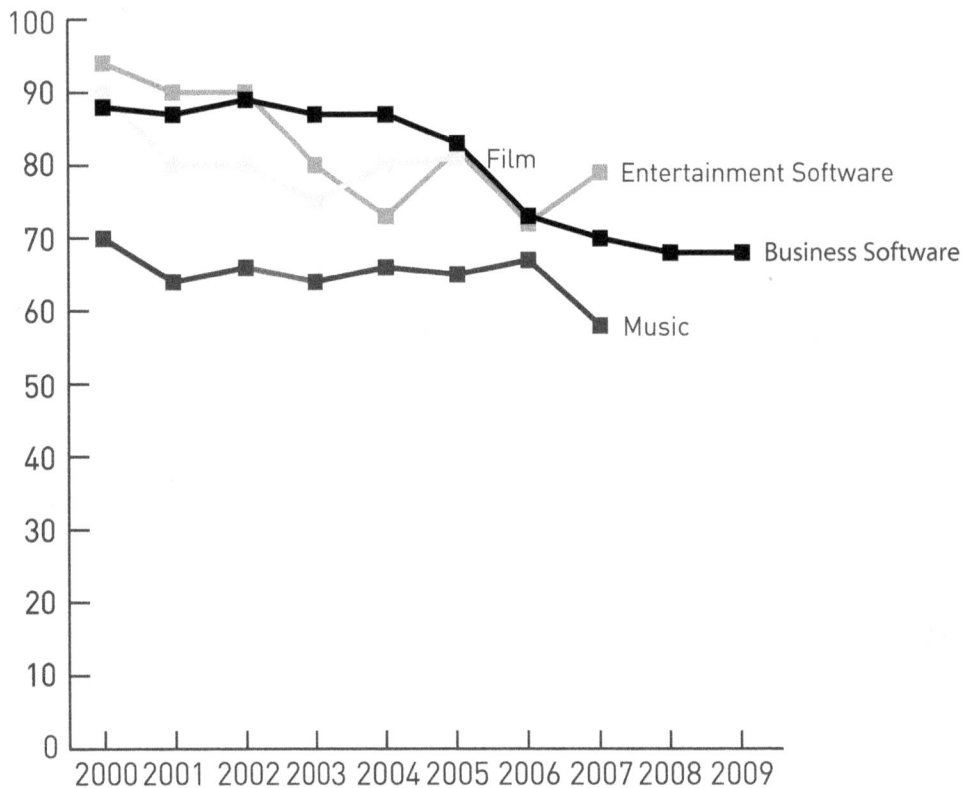

Source: Authors based on IIPA data, 2001-2010.

These numbers fit within the broader industry account of modest decline in the rates of piracy in the Russian market (see figure 4.1). Policy change and stepped-up police efforts are central to this account, but organizational changes within law enforcement are also commonly credited. Despite ongoing tensions, many of the enforcement personnel we interviewed testified to the improvement of cooperation between industry and law enforcement and to improvement, in particular, in the skills of investigators and prosecutors. As an NP PPP representative observed: "We learned how to successfully combat piracy in its traditional form," where traditional refers to the optical disc retail trade.

---

35   Interview with RAPO staff.

In Russia, as in other countries, the effectiveness of enforcement tends to be measured in terms of the scale of street operations, seizures, and resulting penalties or convictions. As in other countries, this practice has been lopsided, with very large numbers of raids and seizures producing a much smaller collection of suspended sentences, fines, and occasional prison terms. Although the IIPA and other rights-holder groups have complained vociferously about the poor record of the courts in Russia, we argue throughout this report that this is not a defect of the national enforcement regime, strictly speaking, but rather the global norm, in which raids scale much more easily than due process and in which courts have remained relatively indifferent to the view that street piracy, especially, constitutes a serious economic crime.

Although this was clearly the case in Russia through 2005, the crackdown in 2006–7 inaugurated a higher level of enforcement and judicial activism. Raids, arrests, and criminal charges increased dramatically, drawing on the streamlined evidentiary and court procedures implemented in 2003. The Russian Supreme Court, which tracks convictions under different articles, reported a sharp increase in criminal copyright convictions under Article 146, from 429 in 2004 to 2,740 by 2007, with numbers holding roughly steady since then. The IIPA, it is worth noting, misreports this data, appearing to conflate different data sets for reported crimes and individuals charged rather than actual convictions (see figure 4.2). This would be a minor point except that the IIPA uses it to overemphasize the magnitude both of the crackdown and of the alleged drop-off in Russian enforcement after 2007—warning repeatedly against the "recent trend of diminished enforcement activity" (IIPA 2010).

Figure 4.2 Criminal Copyright Convictions in Russia, 2004–9

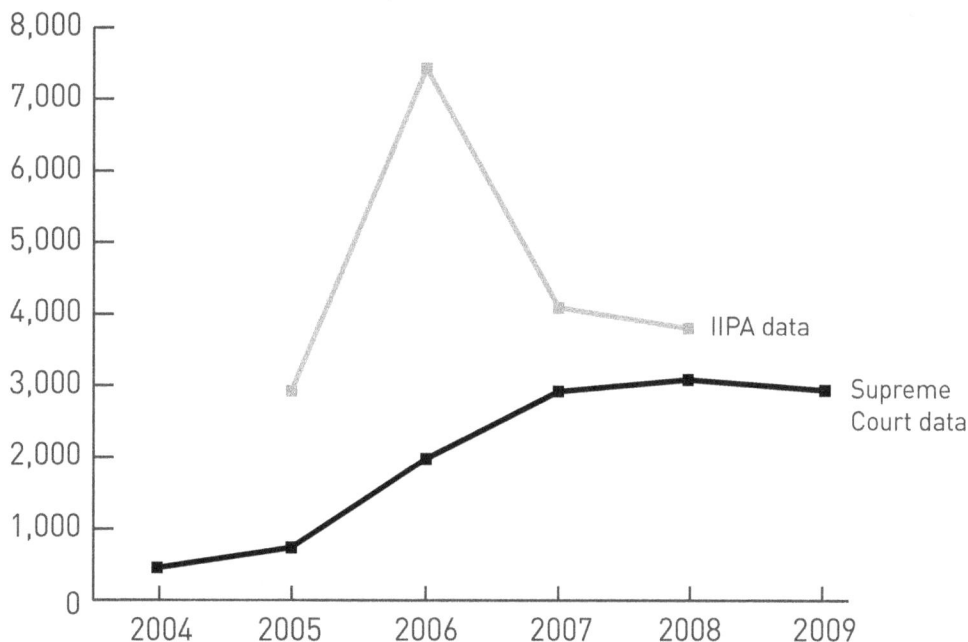

Source: Authors based on IIPA (2006–2010) and Russian Supreme Court data.

Nonetheless, the crackdown was intense. A seven-month campaign in 2007 netted some 4,300 copyright violations discovered through raids, with reportedly 2,000 persons charged and 2 billion rubles ($80 million) recovered through fines and other forms of compensation (Levashov 2007). In the course of a single week-long national police sweep during the campaign, the Ministry of the Interior reported 29,670 "actions"—a fairly astonishing number that produced only 73 criminal cases. Despite the new status of most acts of street and retail piracy as "serious crimes," nearly all these convictions resulted in suspended sentences or small fines (see figure 4.3).

Figure 4.3 Russia: Number and Types of Sentences, 2004–8

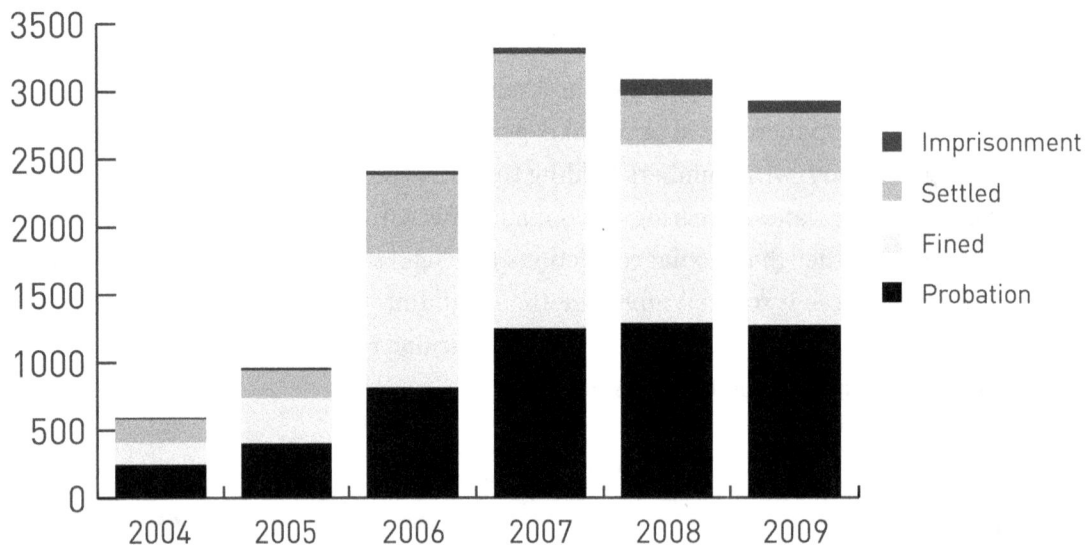

Source: Authors based on Russian Supreme Court data.

What the crackdown meant for the wider prevalence of pirated goods is less clear. Prior to 2007, estimated losses in film and entertainment software showed remarkable—and even implausible— stability, given the dramatic increases in the installed base of DVD players and computers. MPAA losses were pegged at $250 million between 1999 and 2002 and increased only slightly to $266 million by 2005. The ESA reported $240 million in entertainment software losses in 1998 and $282 million in 2006.[36]

Only the RIAA and the BSA reported sharp upward trends in losses in the period. According to the RIAA, music and sound-recording losses to US rights holders tripled, from $170 million in 1998 to a peak of $475 million in 2005. The BSA is a special case: BSA estimates for business software losses fell throughout the late 1990s and early 2000s, dropping from $196 million in 1998 to $93 million in 2002. In 2003, however, the BSA expanded the range of software it

---

36   For a detailed account of industry reporting, see chapter 1.

includes in its studies to cover Microsoft Windows and a variety of consumer products and games. This change reset the baseline for losses, producing a claim of $1.1 billion lost in 2004. Estimated losses then climbed steadily, reaching $4.2 billion by 2008.

By the time of the crackdown in 2006–7, only the BSA and the RIAA were compiling new annual piracy numbers in Russia.[37] In 2007, both reported dips in absolute losses and estimated rates of piracy. The RIAA's findings took place within a largely static market for CDs in Russia, which has averaged $200–230 million per year at wholesale since 2003 (IFPI 2009).[38] The BSA's estimates were made against the backdrop of rapid 30%–40% annual growth in the Russian software market.

Given the pressure on retailers and the focus on institutional compliance, evidence for declining piracy in some sectors of the software market is relatively strong. The BSA's reports are loosely corroborated, for example, by Microsoft's survey of software retailers, which found in 2010 that only 25% sold pirated software—a rate that Microsoft represented as a clear sign of progress (Microsoft Russia 2010). When the software association NP PPP surveyed its 281 institutional members in 2008, 40% perceived a decline in the illicit retail sector represented by small businesses, street vendors, and subway vendors. A decrease in "pre-installment" of pirated operating systems was observed by 45%, with most of the rest citing "no change." When asked about Internet piracy, 44% of members signaled "no change"—a surprising number given the frequency of reports of the growth of Internet piracy.[39] In our view, the declining BSA numbers describe a bifurcated market, in which negotiated volume licenses produce higher rates of compliance among large institutional actors, while consumers and small businesses continue to pirate at very high levels due to the lack of low-cost retail alternatives.

Evidence for change in other markets is more equivocal. The Russian Guild for Development of the Audio and Video Trade—an organization that includes street vendors and almost certainly pirate vendors—estimated that the rate of pirated goods in Russian media markets was stable between 2006 and 2009. In spite of the recent economic crisis, the average retail price for non-licensed CDs and DVDs has remained roughly the same—between 100 and 150 rubles—suggesting no significant change in supply.

Some sectors also have financial incentives to overestimate piracy, complicating evaluations of enforcement. According to one informant, the padding of both box office numbers and piracy estimates is common in the Russian film industry as a strategy for boosting the perceived popularity of films. Such estimates, which the informant described as routinely inflated by $1–1.5 million at the box office, create leverage for studios as they negotiate over subsequent DVD and home video

---

37   The MPAA conducted its most recent survey in 2005, the ESA in 2006.

38   With the exception of a banner year in 2004, according to the IFPI. Russian reporting of record sales has always been considered unreliable.

39   The remainder were roughly equally divided between "increase," "decrease," and "no basis for making the judgment." From an unpublished NP PPP study shared with the researchers.

release rights, as well as for further rounds of investment. When film executives talk about the impact of pirate distribution on new films, the rule of thumb is to cite a 20%–25% estimated loss in box office receipts. Thus the director general of the RAPO, Konstantin Zamchenkov, estimated that the Russian blockbuster *The Irony of Fate 2* lost $10 million when high-quality copies became available shortly after its release (overall, the film made $50 million). Twentieth Century Fox estimated losses to piracy for the same film at $12 million (Vershinen 2008). The basis for these estimates is unclear. In a business with a 10:1 failure ratio, no one—including the studios—understands the alchemy of hits or can accurately forecast results. For nearly a decade, all new films have been pirated within a few days of the initial theatrical release, if not earlier. Whatever the impact on specific films, this phenomenon has not prevented Russian box office receipts from growing over 300% between 2004 and 2008.

For our part, the focus on indicators of success in and around the enforcement effort risks missing the larger point: with a partial exception for the institutional software market, piracy remains ubiquitous in Russia. We see no evidence that this situation has been altered in any significant way by either the changes in law or the crackdown on retail. None of our focus group respondents expressed any difficulty acquiring pirated goods, and our street surveys, conducted in 2008 and 2009—revealed ample and generally undisguised opportunities to purchase discs. Moreover, we see little evidence that industry methods can reliably track small year-to-year changes in the wider prevalence of pirated goods—especially at the consumer level. Industry research methods—including the comparatively solid BSA rates model—introduce too many points of uncertainty, from an inability to measure the scale of digital distribution, to conflicting estimates of the size of licit retail markets. Consumer surveys allow measurement across the different forms and modes of piracy but introduce survey and self-reporting biases when asking consumers about illegal behavior. Because these surveys and underlying data are not made public, their results cannot be evaluated or, in our view, trusted.

Most enforcement personnel, in our experience, are aware of these limitations and do not greatly trouble themselves about such margins of error. Several indicated that slight reported increases or decreases were shaped as much by perceptions of the enforcement effort and the politics around it as by the situation on the ground. Given industry and Russian government interest in managing perceptions of piracy, such skepticism seems warranted.

Selective Enforcement

Debates about effectiveness also tend to obscure the other side of enforcement in Russia: its capture by politically connected actors. Rather than serving (or failing) all parts of the copyright sector equally, enforcement is a scarce resource that confers competitive advantages in the marketplace. Some of these advantages are relatively subtle, as when large firms enjoy more influence with police or prosecutors than small firms. One company's protection, in such contexts, is often another's exposure. But others are cruder and run the gamut from commissioned prosecutions and harassment of competitors to more elaborate forms of extortion and corporate raiding (*reiderstvo*). Such problems are by no means limited to IP rights in Russia, but the mix of corruptible institutions, near-universal legal jeopardy, and scarce actual enforcement creates fertile ground for them.[40]

Much of the practice of capture is public and even celebrated in the form of partnerships between rights-holder groups and law enforcement. Here, domestic interests often trump international ones. Our limited data suggests that national actors generally have greater capacity to mobilize the state than the multinationals. Among the 207 indictments for software piracy initiated by the NP PPP between 2002 and 2008, for example, some 126 were on behalf of the Moscow-based software company and retailer 1C, predominantly in relation to infringement of 1C's widely used accounting suite. Microsoft, in the same period, was the plaintiff in 21 cases. The Russian software companies Konsul'tant and Garant-Service (equivalent to the American legal-research platforms LexisNexis and Westlaw) appear third and fourth on the list, with 9 cases each.

Several domestic film companies have also been successful in mobilizing the police to suppress the pirate DVD distribution of particular films prior to their theatrical release. Such was the case, notably, for the Russian hits *Night Watch* (2004), *Day Watch* (2006), and *The Irony of Fate 2* (2007). Not every film benefits from special police protection. A representative of Channel One, the Russian TV broadcasting corporation that distributed *The Irony of Fate 2,* said in regard to street piracy:

> We simply scared them off. We asked the OBEP to pass the word that our reaction
> [to pirated copies] will be harsh . . . Our access to "administrative resources"
> undoubtedly helped. They would be unlikely to listen to anyone smaller than us.
> (Vershinin 2008)

"Administrative resources," in Russian business parlance, means connections to municipal, regional, or in this case, federal officials. Predictably, such resources are not evenly distributed but track with—and reinforce—influence and size. Large companies, such as Mosfilm or The First, have much more leverage with officials than smaller companies, which translates into better protection for their films.

There are also more aggressive uses of enforcement. Among the St. Petersburg retailers interviewed for this report, the OKO-505 case was a well-known example of anti-piracy enforcement as a form of corporate raiding. OKO ("the eye") was a St. Petersburg-based collective rights management organization established by Dmitry Mikhalchenko, a powerful local businessman involved in the privatization and redevelopment of municipal property. Structured as a public-

---

40 Firestone (2010) describes the wide variety of abuses that have become common in such contexts, including the practice of "intellectual property squatting," or fraudulent registering of trademarks and patents in order to set up a civil suit or criminal charges against the rights holder. Astafiev (2009) notes that the MVD initiated 350 investigations of corporate raiding in 2005. See also Rigi (2010).

The Total Checkup

The director of the St. Petersburg branch of a Moscow-based insurance company described an MVD-directed police raid in summer 2008, triggered by a complaint by a frustrated client who had had an insurance claim denied. The complaint led to what in Russia is called a "total checkup," in which police investigate all aspects of a business's activity. When a review of the company's accounting practices turned up no inconsistencies, police asked company representatives to produce evidence that software on the office computers was licensed. Because the office's equipment was purchased and serviced from Moscow, no such documentation was available onsite. The police confiscated the computers until proof of the licensing status of their copies of Microsoft Windows could be acquired from their Moscow headquarters. No charges were filed during the investigation; nor—according to the office director—was the seizure of the equipment accompanied by proper paperwork. These procedural problems eventually led to the closure of the investigation without charges. But for two weeks the company was paralyzed and could not provide services to its clients. Beyond the initial disruption, the experience has also affected the firm's business practices: the company is now much more careful to avoid conflicts with clients.

private partnership, OKO enjoyed the support of the governor and had close ties to the regional police. In 2006, OKO initiated raids against 505, one of the most popular local music/video retail chains. All fifteen of 505's shops were closed for three days, and most of its stock was confiscated. The public-relations director of 505, Alena Kondrikova, acknowledged that a substantial portion of the media sold at 505 was unlicensed, but it was widely understood that the same was true of other large retail chains such as Titanik, Desyatka, and Aisberg, which remained in business.

The logic of OKO's selective targeting began to emerge in press reports and online forums. OKO pressured music and movie retailers to become paid members of the organization, with the strong implication that this would exempt their operations from police harassment. Some retailers consented; others, like 505, did not. On this basis, 505 filed a complaint with the Antitrust Service (FAS), but as with many raiding investigations the case was not pursued and 505 eventually went out of business (Russian Antitrust Service 2006).

Business Community Pushback

The ramp up of investigations and raids during 2006–7 amplified these problems and produced a powerful reaction in the Russian business community. As a growing number of businesses were disrupted by police raids, software-licensing investigations, and other forms of pressure, IP enforcement began to be identified with police corruption and business takeovers rather than the protection of rights. Shakedowns were common in these contexts, and even under the best of circumstances, police raids could paralyze businesses for days before a case was cleared up or dropped. Pressure to turn in "adequate" statistics on enforcement led to spikes of activity during enforcement reporting periods, with licensed goods—according to enforcement sources—representing up to 30% of seizures.

By 2007, police overreach had become a regular focus of local and federal-level economic summits and soon produced a revised approach to enforcement. Now president, Dmitri Medvedev took the side of harassed business owners, stating, in December 2008, that businesses should not be "terrorized" by the police. In 2009, Vladimir Putin, now prime minister, claimed that most police inspections of businesses were commercially motivated and had "no obvious justification" (Firestone 2010; Beroev 2009). Pushback had become the government position: in early 2009, the Duma passed a new law limiting police inspections of businesses to once every three years.

Not surprisingly, rights-holder groups reacted very unfavorably to these developments, with the IIPA in particular describing the new restrictions as a retreat from the 2006 bilateral agreement with the United States. Recent government investigations of Microsoft and other leading international companies may be further evidence of a divergence of foreign IP interests from the perceived interests of the Russian business community. Although it is hard to attribute coordinated intent to the range of actors operating within this space, the Russian government appears increasingly comfortable with a strong and self-interested position on behalf of its business community. As in other contexts, enforcement policy needs to be seen as a product of this balance of forces.

THE PONOSOV CASE

The politics of selective enforcement, software piracy, and Russian technology policy came together unexpectedly in May 2006, when a police raid on a primary school in rural Perm found that twelve of the school's twenty new computers were operating with unlicensed copies of Microsoft Windows. In November 2006, these findings were transferred to the local court, and the principal of the school, Aleksandr Ponosov, was arrested on suspicion of piracy. Initially, Ponosov was charged with approximately $10,000 in damages to Microsoft—a sum that also exposed him to up to five years in prison. Ponosov pleaded not guilty.

Between 2006 and 2008, the case underwent seven hearings and appeals, demonstrating in the process the difficulty of reconciling the harsh terms of the law with the utterly commonplace use of pirated software. Initially, the presiding judge found Ponosov guilty of infringement but rejected the damage claims. Although Ponosov was set free without penalty, he appealed the ruling. A subsequent hearing imposed a fine of $380 on the vendor who sold the computers—though the question of where the software had been installed remained a point of contention. At a further hearing in May 2007, Ponosov was pronounced guilty and fined 5,000 rubles ($190). Again, Ponosov appealed—both to the regional court and to the Russian Supreme Court. The regional court turned down the appeal, but the Supreme Court recognized it as valid and sent the case back for retrial. This step initiated a series of legal victories for Ponosov. In December 2008, he was pronounced not guilty and awarded legal costs. In July 2009, he was awarded compensation for defamation.

In the course of the prosecution, Ponosov attracted some remarkable public support. In February 2007, former president Mikhail Gorbachev and Duma deputy Aleksandr Lebed sent an open letter to Bill Gates calling on him to intervene to get the charges dropped. Microsoft, for its part, saw a public-relations disaster in the making, and its Russia office publicly disavowed the actions of the regional Prosecutor's Office. Microsoft's intervention, however, was both late and ineffective. Because the case was being prosecuted as a criminal matter, not a civil one, the injured party could not withdraw the charges. The prosecutors dug in.

The number of Ponosov supporters continued to grow. The Federal Agency on Press and Mass Communications of Russia (Rospechat) offered to pay the claimed damages, asserting that "it is implausible to consider a teacher in a rural middle school a major media pirate in our country" (NTV 2007). The Minister of Education of the Perm region also weighed in on Ponosov's side, as did Posonov's students, who picketed the court building during the hearings. Support also came from then president Vladimir Putin, who, when asked about the case at a February 2007 press conference, offered that prosecuting all pirates "under the same blanket" was wrong: Ponosov may have been guilty of buying computers with unlicensed software, but "threatening him with prison is complete nonsense, simply ridiculous" (Putin 2007).

Ponosov himself reacted to the experience by becoming a vocal advocate for free software. Together with former Duma deputy Viktor Alksnis, he founded the Center for Free Technology in 2008, which promotes the development and distribution of free software in Russia. Posonov's own school (from which he has retired) now runs Linux.

Ponosov's troubled prosecution brought the ubiquity of unlicensed software in Russian schools (and by extension, in state institutions) into the open and recast it as a policy, rather than an enforcement, issue. Given the level of political involvement in the case, something now had to be done. The Ponosov case had become a catalyst for technology-policy change at the federal level.

The case highlighted not only the problem of an unlicensed governmental sector but also the difficulty of resolving it at the lower levels of the state administration. A national policy for transitioning state offices to licensed software was proposed as a way forward.

Mass retroactive licensing was discussed but quickly dismissed as prohibitively expensive due to the quantity of unlicensed software in use. The government then tried to cut a deal with commercial vendors. Intense lobbying by domestic and international software interests resulted in the creation of the First Aid program, designed to fund software licensing in Russian schools. Initially, the NP PPP, which represented all the major suppliers of software to the Russian market, was to receive a $200 million contract to purchase licensed commercial software for Russian schools. (The budget for open-source solutions under First Aid, announced later, was around $20 million.) Senior Russian officials, including Dmitri Medvedev, began negotiations with major international software companies, including Microsoft and IBM, for discounts on their products. The stated goal was "100% license purity in Russian schools."

But First Aid quickly ran into trouble with the Ministries of Economic Development and Finance over its "high cost." The Russian leadership pivoted again and convened a meeting with leaders in the Russian IT community—including those working with open-source software—to chart a more economical solution. The new strategy involved a mix of commercial and open-source solutions deployed in three-year pilot programs in three regions of Russia, with plans for expansion into the wider national education system and possibly into other governmental sectors. Pushed by Ponosov and Alksnis, Linux played a role in these programs but not an exclusive one. A wide range of commercial software, including products by Microsoft, Adobe, Corel, and other US firms, were included in the discounted package of software deployed in the pilot programs. Microsoft agreed to include Windows Vista in this package at a 95% discount off the retail price, suggesting the degree of its commitment to the Russian market when open-source alternatives are seriously in play.

The war of maneuver with Microsoft and other vendors continues. In June 2009, the FAS opened an antitrust investigation of Microsoft for its withdrawal of Windows XP from the Russian market—and closed it three months later without filing charges. In March of 2010, Microsoft agreed to provide free copies of Windows 7 to 54,000 Russian schools, with licenses set to expire at the end of 2010. This date also marks the scheduled end of the First Aid program, when license negotiation and the possibility of more widespread open-source adoption will presumably be back on the table.

State Capture

In September 2010, the New York Times published a story on the Russian government's use of software piracy investigations as a means of harassing political activists and journalists (Levy 2010b). The piece establishes a pattern of government raids and criminal charges against opposition actors, going back several years. It also establishes a pattern of complicity by Microsoft's local representatives in these efforts.

As in other cases, raids lead to the seizure of computers, the disruption of work, and a wide array of follow-up criminal charges if unlicensed software is found, including fines and the possibility of long prison terms. Because Microsoft and other software vendors have insisted on using the retail value of software when pursuing criminal charges, a handful of pirated copies can quickly push users into felony territory.

With BSA estimates of piracy in Russia at around 68% of the market—and with the actual rate almost certainly higher in the small-business and nonprofit sector—exposure to felony prosecution is thus the norm in Russia. Complex licensing and arbitrary compliance standards make claims of innocence—such as those advanced by the environmental activist group Baikal in the New York Times story—difficult to establish and dependent on the integrity of the police. In an effort to head off a public-relations fiasco, Microsoft announced a blanket license to activist groups and media outlets in Russia—shielding them from this type of harassment (Levy 2010a).

At one level, of course, these cases have little to do with piracy. Software enforcement is a convenient tool in wider campaigns of political harassment. But the larger matrix of enforcement in Russia—the sharp criminalization and highly selective enforcement of commonplace behavior—makes abuse inevitable. The blanket-licensing solution sidesteps this wider problem but also sets up Microsoft as an arbiter of civil liberties in Russia. When the blanket license expires in 2012, Microsoft will determine whether the political climate warrants its renewal.

Conclusion

Like other countries documented in this report, Russia is in the midst of a transition from optical disc piracy to digital file piracy, conducted largely but by no means exclusively via the Internet. The consumer infrastructure around optical discs, however, is still more developed than the Internet counterpart, and it will be several years before broadband connections, digital playback and storage devices, and recent-vintage computers supplant the optical disc channel for most Russians. Russia will, in the meantime, continue to face the problems chronic to small media markets dominated by multinational companies: licit market growth will be tied to slowly rising local incomes rather than to rapidly dropping technology prices. As we argue throughout this report, this is a recipe for continued high rates of consumer piracy.

In all the countries examined in this report, price competition and service innovation come primarily from competition among domestically owned media industries. The multinationals, our work suggests, simply do not have the incentives to offer significant price cuts in low- and middle-income markets, for fear that these will impact pricing in their larger, more profitable markets. In the software sector especially, piracy assists this policy by providing the vendors a form of de facto price discrimination that generates positive network effects for commercial products, while locking out "free" open-source alternatives. The Ponosov case suggests the complexity behind this balancing act—as well as the pragmatism of the Russian government in angling for advantageous deals with multinationals. The government's strong stated commitment to open source appears to be just one part of this larger strategy of hedging and dealmaking.

In this context, it seems entirely possible that both licit and illicit media markets will continue to grow in the next years. The software business is still riding the wave of Russian computer adoption and optimizing the tradeoffs between piracy and enforcement. The record business is already heavily promotional in orientation, rather than vested in retail disc sales. And movie exhibition continues to set records, coming off the near total destruction of the industry in the 1990s.

In Russia, these developments raise familiar questions about the future of media business models. As elsewhere, we would expect this future to involve lower-priced, more convenient forms of legal media access that compete with the pirate market. In our view, the ramp up in enforcement does little to encourage this transition and quite a bit, in contrast, to reinforce the high-price, high-piracy status quo. Related problems with the criminalization of infringement and the arbitrary application of the law are not unique to Russia but are magnified by the weaknesses of Russian regulation and due process protections. In this context, it is hard not to welcome one short-term outcome of the (still largely pirate-driven) digital transition: the crowding out of state-protected piracy. With that problem headed toward obsolescence, Russians can have a more candid conversation about the costs and benefits of piracy and enforcement and the policies needed to achieve wider access to media.

About the Study

This chapter on Russia combines the efforts of two teams of researchers specialized, respectively, in economic-legal issues and the informal economy. The primary research for the report was conducted by Olga Sezneva, Oleg Pachenkov, Irina Olympieva, Anatoly Kozyrev, and Joe Karaganis. Numerous experts and researchers made additional contributions or provided valuable feedback, including Bodó Balázs, Dmitri Pigorev, Igor Pozhitkov, Maria Haigh, Boris Mamyluk, Kathryn Hendley, and William Pomeranz.

Much of the analysis of the street economy, including enforcement, pricing, availability, and consumer practices, is based on fieldwork in St. Petersburg conducted by Sezneva, Pachenkov, and Olympieva in 2008 and 2009. This work was complemented by some twenty interviews of industry lawyers, judges, and legal scholars and representatives of non-governmental and non-commercial organizations involved in enforcement.

Our broad inquiry into consumer attitudes and values drew on this fieldwork and was complemented by three additional components: a March 2009 focus group with heavy users of unlicensed content in St. Petersburg, a late-2008 survey of three hundred DVD consumers in the city of Irkutsk (conducted by colleagues at the Evolution Marketing Center in Irkutsk), and content-analysis of Russian-language media and online forums.

The analysis of industry and organizational structure and the costs and benefits of copyright enforcement drew on primary interviews, secondary literature, and contributions from our economic and industry research partners, especially Anatoly Kozyrev. A wide array of secondary sources contributed short accounts or expert advice on narrower topics, including pricing, selective enforcement, and other issues.

In order to better understand the shift toward online distribution and its particular Russian inflections, we conducted a data crawl of the Russian BitTorrent site Torrents.ru in March 2009, with the assistance of Bodó Balázs and Dmitri Pigorev.

Key institutional partners in this process included the Social Science Research Council in New York and the Center for International Social Research in St. Petersburg.

Access to sources and confidentiality presented challenges throughout our research. Much of our information about pirate networks, their organizational structure, and above all state involvement came from interviews or media reports. Many statements were off the record.

References

Abraham, Elena, and Olga Vershinskaya. 2001. Информационное общество, вып. 5, с. 44–49. http://emag.iis.ru/arc/infosoc/emag.nsf/BPA/9bf1f1225b9df535c32575b6002b963c.

Alekseeva, Anastasia. 2008. "Недозрелая музыка." *Эксперт*, March 14. http://www.expert.ru/printissues/expert/2008/15/nedozrelaya_muzyka/.

Anufrieva, Anna. 2008. "Кинотеатры остались с кассой." *Коммерсантъ*, April 6. http://www.kommersant.ru/doc.aspx?DocsID=897632.

Arvedlund, Erin E. 2004. "Hollywood Competes With the Street in Russia; To Combat Rampant DVD Piracy, U.S. Film Companies Cut Prices." *New York Times*, April 7. Accessed May 4, 2010. http://www.nytimes.com/2004/04/07/movies/hollywood-competes-with-street-russia-combat-rampant-dvd-piracy-us-film.html?pagewanted=1.

Astafiev, A. D. 2009. *The Role of Russian Internal Affairs Agencies Countering Corporate Raiding*. Primorsky Krai: State Protection Center.

Baker, Adele. 1999. *Consuming Russia: Popular Culture, Sex and Society Since Gorbachev*. Durham, NC: Duke University Press.

BASCAP (Business Action to Stop Counterfeiting and Piracy)/StrategyOne. 2009. *Research Report on Consumer Attitudes and Perceptions on Counterfeiting and Piracy*. Paris: International Chamber of Commerce. http://www.internationalcourtofarbitration.biz/uploadedFiles/BASCAP/Pages/BASCAP-Consumer%20Research%20Report_Final.pdf.

Berezin, Oleg, and Ksenia Leontieva. 2009. *Russian Cinema Market: Results of 2008*. St. Petersburg: Nevafilm Research.

Beroev, Nigina. 2009. "Владимир Путин: «Большинство проверок бизнеса - «заказные» или недобросовестные»." *Komsomol'skaya* Pravda, November 29. http://pskov.kp.ru/print/article/24400/576309/.

Boston Consulting Group. 2010. "The Internet's New Billion: Digital Consumers in Brazil, Russia, India, China, and Indonesia." http://www.bcg.com/documents/file58645.pdf.

BSA/IDC (Business Software Alliance and International Data Corporation). 2009. *Sixth Annual BSA-IDC Global Software Piracy Study*. Washington, DC: BSA. http://global.bsa.org/globalpiracy2008/studies/globalpiracy2008.pdf.

Budylin, Sergey, and Yulia Osipova. 2007. "Total Upgrade: Intellectual Property Law Reform in Russia. *Columbia Journal of East European Law* 1 (1). http://www.roche-duffay.ru/articles/pdf/IP%20Reform.pdf.

City Statistics Bureau. 2008. *Statistic and Analysis Archive*: St. Petersburg. http://gov.spb.ru/day/statistika/stat/.

Dolgin, Aleksandr. 2006. *The Economy of Symbolic Exchange*. Moscow: Infra-M.

Dorozhin, Alex. 2007. "Особенности рынка MP3-плееров в России." *Mobile Review*, September 12. http://www.mobile-review.com/mp3/articles/rus-mp3-features.shtml.

enigmax. 2010a. "Torrents.ru Fights Back After Domain Seizure." TorrentFreak (blog), February 28. http://torrentfreak.com/torrentsru-fights-back-after-domain-seizure-100228/.

———. 2010b. "World War II Veterans Must Pay To Sing War Songs." *Torrent Freak* (blog), March 28. http://torrentfreak.com/world-war-ii-veterans-must-pay-to-sing-war-songs-100328/.

Firestone, Thomas. 2010. "Armed Injustice: Abuse of the Law and Complex Crime in Post-Soviet Russia." *Denver Journal of International Law & Policy 38* (4). http://law.du.edu/documents/djilp/38No4/Firestone.pdf.

Golavonov, Dmitri. 2008. "Transformation of Authors' Rights and Neighbouring Rights in Russia." *Iris Plus: Legal Observations of the European Audiovisual Observatory,* 2008-2.

Goncharova, Olga, and Anna Pushkarskaya. 2009. "Кинотеатрам рекомендовали союз с композиторами." *Коммерсантъ*, September 23.

Gref, German. 2000. "Основные направления социально-экономической политики Правительства Российской Федерации на долгосрочную перспективу." http://www.budgetrf.ru/Publications/ Programs/Government/Gref2000/Gref2000040.htm.

IFPI (International Federation of the Phonographic Industry). 2006. *The Recording Industry 2006 Piracy Report: Protecting Creativity in Music.* London: IFPI.

———. 2009. *The Record Industry in Numbers.* London: IFPI.

IIPA (International Intellectual Property Alliance). 2009. *Russian Federation: 2009 Report on Copyright Protection and Enforcement.* Washington, DC: IIPA.

———. 2010. Russian Federation: *2010 Report on Copyright Protection and Enforcement.* Washington, DC: IIPA.

Kwon, Paul. 2010. *Russia: Telecoms, Mobile, Broadband and Forecasts.* Sydney: BuddeComm.

Levashov, Alexander. 2007. "Милиция отловила 2 тысячи пиратов." *Новости структурного подразделения*, October 10. http://www.adm.yar.ru/uits/print_news.aspx?news_id=504.

Levy, Clifford J. 2010a. "Microsoft Changes Policy Over Russian Crackdown." *New York Times,* September 13. Accessed September 20, 2010. http://www.nytimes.com/2010/09/14/world/ europe/14raid.html?ref=world.

———. 2010b. "Russia Uses Microsoft to Suppress Dissent." *New York Times*, September 11. Accessed September 20, 2010. http://www.nytimes.com/2010/09/12/world/europe/12raids.html?_r=2&p agewanted=1&sq=microsoft&st=cse&scp=4.

Mamlyuk, Boris N. 2010. "Russia & Legal Harmonization: An Historical Inquiry Into IP Reform as Global Convergence and Resistance." Cornell Law Faculty Working Papers, Cornell University, Ithaca, NY, March 4. http://scholarship.law.cornell.edu/cgi/viewcontent.cgi?article=1073&context=clsops_ papers.

Microsoft Russia. 2010. "The Mysterious Customer." Microsoft survey.

Ministry of Communications. 2009. ICT Infrastructure Database. http://www.inforegion.ru/ru/main/ infrastructure/.

MPAA (Motion Picture Association of America). 2005. *The Cost of Movie Piracy.* Washington, DC: MPAA.

NPD (Nonprofit Partnership of Distributors). 2005. "Внимание: пиратство!" http://www.netpiratam. ru/index.php?m=achtung.

NTV. 2007. "Роспечать заступилась за обидчика Билла Гейтса." January 23.

Olimpieva, Irina, Oleg Pachenkov, and Eric Gordy. 2007. *Informal Economies in St. Petersburg: Ethnographic Findings on the Cross-Border Trade.* Belgrade and Washington, DC: Jefferson Institute.

Ovcharova, L., and D. Popova. 2005. *Детская бедность в России. Тревожные тенденции и выбор стратегических действий.* Moscow: UNICEF.

Padunov, Vladimir. 2010. "From Art House to Cine-Plex." http://www.rusfilm.pitt.edu/2010/.

Pravda. 2005. "Fighting Video Piracy in Russia Still Leaves Much to be Desired." March 21.

Putin, Vladimir. 2007. Press conference, February 1. http://www.admhmao.ru/narod_vl/presid/01_02_07.htm

Regnum News. 2008. "Валентина Матвиенко: Средняя зарплата в Санкт-Петербурге - 20 тысяч рублей." http://www.regnum.ru/news/1035139.html.

Reuters. 2007. "Piracy Worked for Us, Romania President Tells Gates." *Washington Post*, February 1.

Rigi, Jakob. 2010. "The Coercive State and the Spectacle of Law in Russia: The Use of Legal Schemes for Expropriation of Middle and Small Businesses." Manuscript.

RosBusinessConsulting. 2008. *A Review of the Russian Software Market and Development.* http://www.marketcenter.ru/content/doc-2-12286.html.

Russian Antitrust Service. 2006. "Сеть '505' подала сигнал SOS." Newsletter, September 18. http://fas.gov.ru/article/a_8484.shtml.

Russian Law Online. 2010. "His Name is Nikita." October 28. http://www.russianlawonline.com/content/his-name-nikita.

Schulze, Hendrik, and Klaus Mochalski. 2009. *Internet Study 2008/2009.* Leipzig: ipoque.

Screen Digest. 2008. *Video Market Monitor*: Russia. London: Screen Digest. http://www.reportlinker.com/p0109234/Video-Market-Monitor-Russia.html.

Tapalina, V. S. 2006. "Экономический потенциал населения России начала XXI века." http://econom.nsc.ru/ECO/arhiv/ReadStatiy/2008_02/Tapilina.htm.

Vershinen, Alexander. 2008. "Vzyali na ispoug." *Smart Money/Vedomasti*, February 18. http://www.vedomosti.ru/smartmoney/article/2008/02/18/4937.

Vitaliev, G. 1996. "Какие законы и как работают в государстве российском." *Софт маркет* 3.

World Bank. 2010. Databank: GDP per capita (current US$). http://data.worldbank.org/indicator/NY.GDP.PCAP.CD.

Chapter 5: Brazil

Pedro N. Mizukami, Oona Castro, Luiz F. Moncau, and Ronaldo Lemos

Contributors: Susana Abrantes, Olívia Bandeira, Thiago Camelo, Alex Dent, Joe Karaganis, Eduardo Magrani, Sabrina Pato, Elizete Ignácio dos Santos, Marcelo Simas, and Pedro Souza

> We have some problems in Brazil at this time.
>
> —*Jack Valenti, president of the Motion Picture Association of America*[1]

Introduction

As in many other developing countries after World War II, Brazilian approaches to intellectual property were shaped by import substitution strategies designed to foster the growth of local industry. High tariffs on imported goods and a narrow scope for patentable technologies were important elements of these strategies. In the case of pharmaceutical patents—which Brazil abolished in 1969—health policy also played a large role: for many categories of medicine, Brazil had sufficient capacity to meet its own needs at low cost.

As the United States led the push for stronger global IP norms in the late 1970s and 1980s, most IP-exporting countries revised their laws to extend protection to emerging fields of technical innovation, including pharmaceuticals and software. Most developing economies were reluctant to follow. Brazil, India, and South Korea, in particular, maintained lower IP protection for such goods, resulting in sharp disputes with the United States in the 1980s.[2]

Tensions over intellectual property protection dominated the Brazil-US relationship during the period, first in the context of US efforts to roll back Brazilian protection of its nascent computer industry (1985) and later in relation to US attempts to force Brazilian adoption of pharmaceutical patents (1987). Brazil acceded quickly to US demands in the first case, establishing copyright for software and removing import restrictions on computer equipment. But it held its ground on pharmaceuticals, leading to US-imposed sanctions under Section 301 of the US Trade Act (Sell 2003:90; Bayard and Elliott 1994:187–208). With negotiations over the new World Trade Organization (WTO) drawing to a close, however, and broader IP obligations for pharmaceuticals imminent, Brazil gave up this position in 1990. Consistent with its obligations under the WTO's TRIPS (Trade-Related Aspects of Intellectual Property Rights) agreement, Brazil established pharmaceutical patents in 1996.

---

1   See Valenti (2003).

2   For a thorough overview, see Bayard and Elliott (1994).

Chapter Contents

219   Introduction

226   The Legal Framework for Copyright Enforcement

228   Criminal Enforcement

230   Civil and Administrative Enforcement

231   The Legislative Agenda

232   Internet Legislation: The State of Play for 2011

233   Graduated Response

235   Copyright Reform

236   The IP Enforcement Network

238   The Enforcement Decade

240   The National Council on Combating Piracy

241   The National Plan

243   The GIPI and Brazilian IP Policy

245   Know Your Enforcement Authorities

246   The Police and the Municipal Guards

247   Revenue, Customs, and the Patent Office

247   The Prosecution Services and the Judiciary

248   Copyright Industry Associations

249   Private Funding, Public Enforcement

251   Cross-Industry Coordinators

253   How Piracy Works

254   The Street Trade

255   The Tri-Border Area and China

256   Organized Crime, Terrorism, and Piracy

257   Proving the Connection

258   Organized Crime and Brazilian Law

261   Law Kim (Kin) Chong

262   Internet Piracy

264   Discografias and Orkut

265   Fan Communities and Subtitling

266   Book Piracy

269   Research, Training, and Education

270   Anti-Piracy and Poetic License

272   To Repress and Educate

274   Mixed Signals

275   Research

276   The Magic Numbers

278   Sector-Specific Research

283   Cross-Sectoral Research

287   Training, Awareness, and Education

289   "Projeto Escola Legal"

292   Conclusion

295   About the Study

296   References

Acronyms and Abbreviations

ABCF	Associação Brasiliera de Combate à Falsificação (Brazilian Association for Combating Counterfeiting)
ABDR	Associação Brasileira de Direitos Reprográficos (Brazilian Association of Reprographic Rights)
ABES	Associação Brasileira das Empresas de Software (Brazilian Association of Software Companies)
ABPD	Associação Brasileira de Produtores de Discos (Brazilian Record Producers Association)
ABPI	Associação Brasileira da Propriedade Intelectual (Brazilian Intellectual Property Association)
ABRELIVROS	Associação Brasileira de Editores de Livros Escolares (Brazilian Association of Textbook Publishers)
ACTA	Anti-Counterfeiting Trade Agreement
AmCham	American Chamber of Commerce
ANGARDI	Associação Nacional para Garantia dos Direitos Intelectuais (National Association for Safeguarding Intellectual Rights)
APCM	Associação Antipirataria Cinema e Música (Film and Music Anti-Piracy Association)
BPG	Brand Protection Group
BSA	Business Software Alliance
CBL	Câmara Brasileira do Livro (Brazilian Book Chamber)
CNC	Confederação Nacional do Comércio de Bens, Serviços e Turismo (National Confederation of the Commerce of Goods, Services, and Tourism)
CNCP	Conselho Nacional de Combate à Pirataria e Delitos contra a Propriedade Intelectual (National Council on Combating Piracy and Intellectual Property Crimes)
CNI	Confederação Nacional da Indústria (National Confederation of Industry)
CPI da Pirataria	Comissão Parlamentar de Inquérito da Pirataria (Parliamentary Commission of Inquiry on Piracy)
DEIC	Departamento de Investigações sobre o Crime Organizado (Department of Investigations of Organized Crime, São Paulo Civil Police)
DRCPIM	Delegacia de Repressão aos Crimes contra a Propriedade Imaterial (Police Unit for the Repression of Crimes against Immaterial Property, Rio de Janeiro Civil Police)
ECAD	Escritório Central de Arrecadação (Central Collecting Office)
ESA	Entertainment Software Association
ESAF	Escola de Administração Fazendária (Superior School of Public Revenue Administration)
ETCO	Instituto Brasileiro de Ética Concorrencial (Brazilian Institute for Ethics in Competition)
Fecomércio-RJ	Federação do Comércio do Estado do Rio de Janeiro (Rio de Janeiro Federation of Commerce)
FGV	Fundação Getulio Vargas (Getulio Vargas Foundation)
FIESP	Federação das Indústrias do Estado de São Paulo (São Paulo Federation of Industries)
FNCP	Fórum Nacional contra a Pirataria e a Ilegalidade (National Forum against Piracy and Illegality)
GDP	gross domestic product
GIPI	Grupo Interministerial de Propriedade Intelectual (Inter-Ministerial Intellectual Property Group)
GNCOC	Grupo Nacional de Combate às Organizações Criminosas (National Group for Combating Criminal Organizations)
GSP	Generalized System of Preferences
HADOPI	Haute Autorité pour la Diffusion des Œuvres et la Protection des Droits sur Internet (High Authority for the Diffusion of Works and the Protection of Rights on the Internet)
IDC	International Data Corporation
IDEC	Instituto Brasileiro de Defesa do Consumidor (Brazilzian Institute for Consumer Defense)
IEL	Instituto Euvaldo Lodi (Euvaldo Lodi Institute)
IFPI	International Federation of the Phonographic Industry
IIPA	International Intellectual Property Alliance
IMC	Inter-Ministerial Committee on Combating Piracy (Comitê Interministerial de Combate à Pirataria)
INPI	Instituto Nacional da Propriedade Industrial (National Industrial Property Institute)
IP	intellectual property
IP address	Internet protocol address
ISP	Internet service provider
LAN	local area network
MPA	Motion Picture Association
MPAA	Motion Picture Association of America
PEL	"Projeto Escola Legal" (Legal School Project)
P2P	peer-to-peer
RIAA	Recording Industry Association of America
SENAI	Serviço Nacional de Aprendizagem Industrial (National Learning Service of Industry)

Acronyms and Abbreviations

SINDIRE-CEITA	Sindicato Nacional dos Analistas Tributários da Receita Federal do Brasil (National Syndicate of the Tax Analysts of the Brazilian Federal Revenue Service)
SNEL	Sindicato Nacional dos Editores de Livros (National Union of Book Publishers)
TBA	Tri-Border Area (between Brazil, Paraguay, and Argentina)
TPM	technical protection measures
TRIPS	Agreement on Trade-Related Aspects of Intellectual Property Rights
UBV	União Brasileira de Vídeo (Brazilian Video Union)
UFRJ	Universidade Federal do Rio de Janeiro (Federal University of Rio de Janeiro)
Unafisco	União dos Auditores Fiscais da Receita Federal do Brasil (Union of the Fiscal Auditors of the Brazilian Federal Revenue Service)
Unicamp	Universidade de Campinas (University of Campinas)
USP	Universidade de São Paulo (University of São Paulo)
USTR	Office of the United States Trade Representative
WIPO	World Intellectual Property Organization
WTO	World Trade Organization

The respite from US pressure was short-lived, however. By the end of the 1990s, Brazil was again in the crosshairs of the Office of the United States Trade Representative (USTR)—this time for alleged failures of copyright enforcement. In 2000, the International Intellectual Property Alliance (IIPA) filed a petition requesting a review of Brazilian trade privileges under the US's Generalized System of Preferences (GSP) program—a request the USTR granted in 2001. In 2002, the USTR placed Brazil back on its "Priority Watch List," where it remained through 2006. Since the creation of the Special 301 process in 1989, Brazil has appeared on the "Watch List" nine times, the "Priority Watch List" ten times, and the "Priority Foreign Country" list—the highest level and the prelude to trade sanctions—once (see table 5.1).

Table 5.1 Brazil's Special 301 Status, 1989–2010

1989	1990	1991	1992	1993	1994	1995	1996	1997	1998	1999
PWL	PWL	PWL	PWL	PFC	SM	PWL	WL	WL	U	WL

2000	2001	2002	2003	2004	2005	2006	2007	2008	2009	2010
WL	WL	PWL	PWL	PWL	PWL	PWL	WL	WL	WL	WL

The Brazilian government responded to this new round of pressure by revamping its approach to enforcement. Special 301 warnings clearly played a role, as did fear of a wider deterioration of trade relations.[3] As one private-sector consultant recollected in an interview:

> It was a big scandal because . . . what happened? Shoe exporters, for example, panicked. People in Rio Grande do Sul went nuts: "I, a shoemaker, who makes shoes for the US . . . I'm going to lose my benefits because there's DVD piracy in Brazil? What do I have to do with that?" Then they lobbied through the CNI [National Confederation of Industry], in Brasilia, to improve protection.

Industry groups had also become much better organized and better able to coordinate pressure on governments, both domestically and internationally. By the turn of the millennium, all of them were pushing to expand enforcement in major emerging markets.

As in many other countries, IP enforcement in Brazil was restructured through policy changes at the legislative and executive levels. And as elsewhere, the scaling up of enforcement was challenging for both state and private actors, resulting in several major reorganizations in less than a decade. The first such effort in Brazil was the creation, in 2001, of an Inter-Ministerial Committee on Combating Piracy (IMC) tasked with coordinating enforcement efforts across the different

---

3   Throughout the 1990s, the United States was Brazil's largest trading partner by a significant margin, with roughly US$20 billion in exports and US$26 billion in imports in 2009. The United States was surpassed for the first time by China in April 2009 (Moore 2009).

Anti-Piracy Spectacle

In Brazil, as elsewhere, the government's commitment to enforcement gives rise to public spectacles meant both to educate the public and to signal cooperation with industry. In 2005, a federal law established December 3 as the National Day of Combating Piracy and Biopiracy. Since then, every December 3 has been marked by the Federal Revenue Service's public destruction of thousands of pirated and blank CDs and DVDs and large quantities of counterfeit goods. In 2009, this spectacle involved the destruction of three tons of seized merchandise.

Globally, such events have given rise to one of the few iconic images of the enforcement wars: the destruction of large piles of pirated disks (crushed by steamrollers, smashed with sledgehammers, or trampled by schoolchildren or, in India, elephants).

---

agencies and ministries with a stake in the issue. Industry concerns about the responsiveness of the committee manifested themselves quickly.[4] Under pressure from the National Congress, the federal executive established a Parliamentary Commission of Inquiry on Piracy (CPI da Pirataria) in 2003. The recommendations of this commission, in turn, resulted in the 2004 creation of a new anti-piracy organization within the Ministry of Justice, called the National Council on Combating Piracy and Intellectual Property Crimes (CNCP). In response to industry demands, both government ministries and private-industry groups were seated on the council, providing industry groups a level of direct access to government that the USTR has described as a model of public-private partnership (USTR 2007:30).

The CNCP soon became the main forum for anti-piracy efforts in Brazil and—through regular media coverage—the main public face of Brazilian enforcement. It also became the principal forum for developing IP enforcement policy at the federal level, culminating in the release of a National Plan on Combating Piracy in 2005 (and a substantially revised version in 2009).

Government attentiveness to the enforcement agenda was soon rewarded: Brazil was demoted from the Priority Watch List to the Watch List in 2007. Industry complaints have not stopped, though. Although the USTR (2010:29) expresses satisfaction with the government's efforts, noting that "Brazil continued to show a commitment to fighting counterfeiting and piracy and to strengthening its enforcement efforts," it still complains about "significant levels of piracy and counterfeiting." Concerns about patent law, book and Internet piracy, and Brazil's refusal to sign the WIPO (World Intellectual Property Organization) Internet Treaties are also cited as reasons

---

4   Less than a year after its creation, the IIPA complained that "the IMC has not produced any document, has not taken any action, nor has it manifested any indication that it intends to take any action. Indeed, the only thing that we have heard from the commission is that it needs considerably more time to develop its ideas. This lackadaisical attitude in the face of debilitating piracy is simply not tolerable, and should not be countenanced. The private sector has plenty of ideas about actions that the government could take that would begin to address the piracy situation. The IMC cannot be permitted to ruminate indefinitely" (IIPA 2002:73).

for keeping Brazil on the "Watch List." The IIPA, which in 2006 noted Brazil's "definite shift in political willingness" to combat piracy, has also kept up the pressure. In its 2010 recommendations to the USTR, it continues to complain about the lack of deterrent penalties for piracy, the low number of convictions, and the spread of online copyright infringement, among other issues, leading to an extensive list of demands for further legislative action and government engagement.

Given its extensive cooperation with both the US government and IP stakeholders, the Brazilian government now routinely contests the portrayal of Brazil as a pirate nation. The self-promotional activities of the CNCP, which celebrate cooperation with the private sector and CNCP victories against piracy, are an important part of this pushback (Ministério da Justiça 2005a, 2005b, 2006, 2009).

Major changes in Brazilian policy regarding the Internet, intellectual property, privacy, education, and law enforcement hinge on the stakeholder politics and evidentiary practices underlying these two narratives: Brazil, the pirate nation, and Brazil, winning the fight against piracy. Although Brazilian legal literature on copyright law has grown and, on balance, improved in the past decade, the wider interaction of law, policymaking, enforcement practices, and consumer behavior remains very poorly documented and, perhaps above all, very poorly integrated into broader synthetic accounts that can provide perspective on these issues.

This chapter is an attempt to tie together several of these lines of inquiry, enabling what should become much more routine scrutiny of public debate and policymaking on copyright, piracy, and enforcement. In contrast to the other country chapters in this report, the Brazil chapter focuses primarily on the domestic politics of enforcement policymaking, on the evidentiary discourses that frame policy debates, and on the wider efforts to build an "intellectual property culture" in which piracy withers away.

After nearly a decade of expansion of enforcement activities, and as Brazil faces decisions about extending enforcement practices to the Internet, we also think that some effort to evaluate the country's overall enforcement campaign is timely. From our perspective, the choice between Brazil as a pirate haven and Brazil as a stalwart in the war against piracy is inapt. Although the Brazilian government has done a great deal in the past decade to comply with US demands, we have seen no evidence—in Brazil or elsewhere for that matter—that suggests that piracy is on the decline.[5] On the contrary, it shows every sign of growing as technologies for copying and sharing media become cheaper, more widespread, and more varied.

What we see, instead, is an enforcement debate in which the cooperation of the Brazilian state, not the impact of its initiatives, has become the main measure of success. We see a debate in which the common front between state and industry actors against hard-goods counterfeiting hides considerable disagreements over how to move forward in the emerging "culture of the copy" (Sundaram 2007), with little government or public enthusiasm for the expansion of "repressive

---

5   In absolute terms, at least. Piracy may, on the other hand, make up a shrinking share of some markets as they grow—as the Business Software Alliance argues with respect to software piracy in Brazil.

measures" and little industry interest in conversations about business models or access to media. "Educational measures" have become the preferred way out of this stalemate, in the hope that "respect for intellectual property" can be built over time. But a close look at the content of these measures reveals a degree of disconnection from consumer experience that makes such cultural change extremely unlikely. What we can hope for, instead, is a more honest, transparent, and accountable politics of intellectual property in Brazil, in which policymaking is calibrated to the needs—and realities—of contemporary Brazilian life.

The Legal Framework for Copyright Enforcement

> Brazil's law enforcement agencies, various municipal authorities, and prosecutors all have authority to enforce copyright infringement. More resources should be provided to law enforcement.
> - IIPA 2009 Special 301 Report: Brazil

During the first years of the Special 301 process, Brazil was subject to constant criticism of its allegedly inadequate standards for protecting intellectual property. Brazil's problems, the USTR and American industry groups argued, included both weak laws and weak enforcement of those laws.

When Brazil introduced new, TRIPS-compliant IP legislation in the mid-1990s, the USTR responded approvingly. As a reward for the 1998 enactment of "modern laws to protect computer software and copyright" (USTR 1998), Brazil was delisted from Special 301 for the first (and so far, only) time in the history of the program.

Brazilian copyright law now exceeds TRIPS requirements in key respects. When the current copyright statute came into force in 1998, the term of protection was increased from the life of the author plus sixty years to life plus seventy years—both in excess of the Berne Convention standard adopted in TRIPS. The circumvention of technical protection measures (TPM), such as encryption on DVDs, was made a civil offense, and the list of exceptions and limitations to copyright was significantly narrowed. In practice, some of the most obvious exceptions and limitations in Brazil fall well short of international standards, with the rules governing the reproduction of "small excerpts" of larger works (Law 9.610/98, Article 46, II), for example, ambiguous to the point of providing no meaningful guidance (Mizukami et al. 2010; Souza 2009; Branco 2007). Consumers International's *IP Watchlist Report* rates Brazilian legislation as the seventh worst in the world due to the obstacles it creates to access to knowledge (2010:2).

Further strengthening of IP protections, especially with regard to the Internet, remains a major industry concern. Brazil has notably refused to adopt some of the post-TRIPS standards that have emerged in the past fifteen years, such as the 1996 WIPO Internet Treaties.[6] But since the late 1990s, enforcement, not stronger laws, has been the central preoccupation of industry stakeholders, the IIPA, and the USTR. Much of their attention has fallen on two issues: (1) the expansion of police investigations and arrests (ex officio) and (2) faster, more reliable punishment through the criminal justice system.

For the most part, industry groups and the USTR have continued to praise the institutional side of Brazil's efforts in these areas—notably the role of the CNCP and corresponding improvements in coordination between law enforcement agencies at the federal level. But the courts are a different story: there have been very few convictions for piracy and fewer serious fines or prison sentences. The IIPA (2010:146) blames "a long litany of systemic problems and bottlenecks" in both civil and criminal cases for this situation, as well as a lack of "prosecutorial attention" to copyright infringement.

As the IIPA itself notes, however, these problems are not specific to IP enforcement but relate to Brazilian law enforcement in general. A crowded prison system,[7] complex procedural law,[8] and a court system operating well above capacity[9] are simply unable to manage the flow of cases that would follow from mass enforcement of copyright laws, particularly when there are so many more serious offenses to be prosecuted. In the words of a software industry representative:

---

6   According to an informant within government, there is strong sentiment that these treaties are not in Brazil's best interests and that current international IP law provides an adequate level of protection for rights holders. The main reason for Brazil's refusal to sign the WIPO Internet Treaties, according to this informant, is concern that requirements for the legal protection of technical protection measures used for digital media could override the already scant limitations and exceptions included in domestic copyright law, coupled with skepticism about their efficacy. This stance is largely political: Brazilian copyright law already provides protection for TPM and rights-management information in Law 6.610/98, Article 107, at a level that some authors view as compliant with the treaties, even if the IIPA claims otherwise (Ascensão 2002). Brazil has also objected to other treaties that legitimize the WIPO Internet Treaties as "soft law" standards, such as the Budapest Convention on Cybercrime, which relies in part on the WIPO Internet Treaties.

7   According to official data, in December 2009, Brazil had a total prison population of 473,626 prisoners (all imprisonment regimes included) within a system designed for 294,684. Ministry of Justice, Prison System Database, http://portal.mj.gov.br/etica/data/Pages/MJD574E9CEIT-EMIDC37B2AE94C6840068B1624D28407509CPTBRNN.htm.

8   Procedural law is mainly found in the lengthy Civil Procedure Code and Penal Procedure Code. Together with the copyright and software laws, these form the core of copyright and enforcement legislation. Border regulations are also relevant (Decree 6.759/09), as are a few municipal ordinances, mainly in the context of street-level piracy.

9   Brazilian state courts of appeal had an average workload of 2,180 cases per judge in 2009 (CNJ 2010:133). At the lower level, state courts had an average workload of 2,931 cases per judge (ibid.:228).

It's not that few people get arrested; they do get arrested. It's just that few people remain in prison. Why? Because of the order of priority between murder, robbery, rape, and the crime of piracy. So I think what frustrates authorities the most is not just arresting criminals but keeping them under arrest. Why? Because the Brazilian prison system is in need of restructuring, and [that] frustrates some of these professionals, who could be fighting this crime in a more effective way.

Industry requests for special IP courts and dedicated police units are easy to understand in this context but also hard to justify when weighed against other social needs. For the most part, the Brazilian government has ignored these requests, shielded by Article 41.5 of TRIPS, which states that members have no obligation to create a judicial system for IP enforcement distinct from the enforcement of law in general. Nor are members required to dedicate more public resources for IP enforcement than for law enforcement in general.[10] A major restructuring of the Brazilian criminal justice system of the kind desired by the IIPA (and by many other critics) will not happen anytime soon, nor will resources for law enforcement dramatically increase in the near future. The future of enforcement in Brazil, in this environment, looks much like the present.

Criminal Enforcement

Under Brazilian law, all infringement is subject to criminal prosecution. Two laws are paramount here: Article 12 of the Software Law and Article 184 of the Penal Code (which applies to all copyrightable works other than software).[11] Infringement can be pursued as either a criminal or civil matter, but in practice industry stakeholders place most of the burden on criminal law, reflecting assumptions about the deterrent effects of criminal penalties and—more important—industry preferences for shifting enforcement costs to the public sector. Criminal prosecution is usually carried out by public prosecutors and other law enforcement authorities. The costs of civil litigation, in contrast, fall more heavily on plaintiffs.

---

10 TRIPS Article 41.5 was based on a proposal by India and is one of the few provisions in the enforcement section of the agreement where "developing countries' views made a difference" (UNCTAD/ICTSD 2005:585). It effectively serves as a safeguard against industry demands for preferential judicial systems.

11 This dual track has led to differences in the penalties and types of prosecution applicable to otherwise very similar types of infringement. When the last major copyright-related legislation was passed in Brazil in 2003—Law 10.695, which reformed parts of the Penal Code and the Penal Procedure Code—the Software Law (Law 9.609/98) was not updated to match its counterparts.

Figure 5.1 Penalties for Copyright Infringement under Brazilian Law

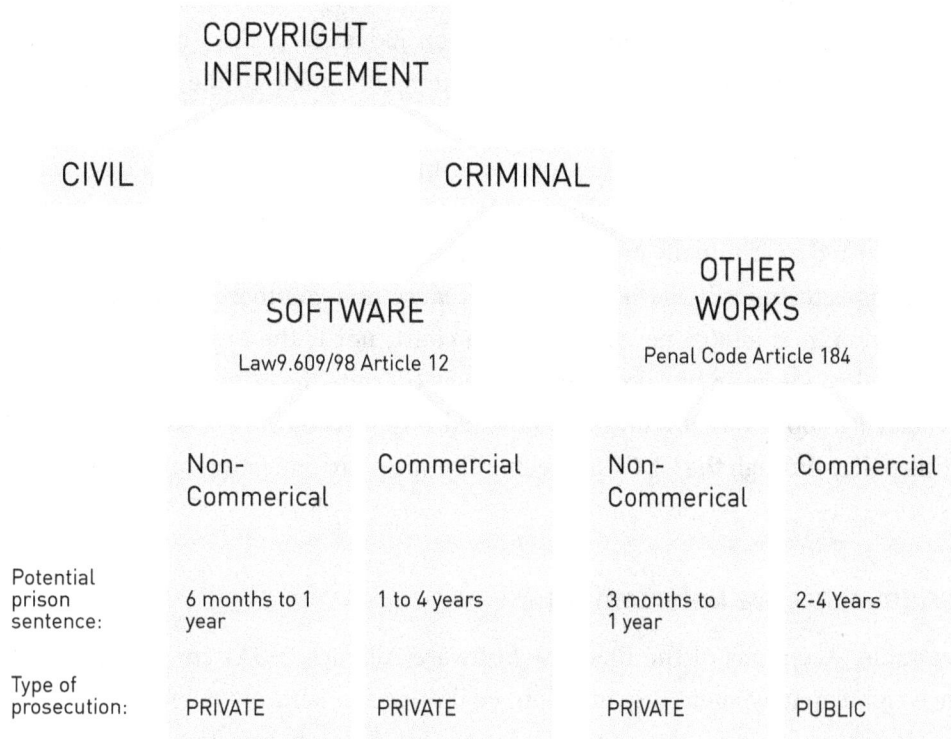

COPYRIGHT
INFRINGEMENT

CIVIL   CRIMINAL

	SOFTWARE Law9.609/98 Article 12		OTHER WORKS Penal Code Article 184	
	Non-Commerical	Commercial	Non-Commerical	Commercial
Potential prison sentence:	6 months to 1 year	1 to 4 years	3 months to 1 year	2-4 Years
Type of prosecution:	PRIVATE	PRIVATE	PRIVATE	PUBLIC

The high demands on police and judicial resources mean that criminal copyright enforcement in Brazil is highly selective. For the same reason, Brazilian judges are generally reluctant to apply the full penalties available under the law. The informally observed minimum-penalty policy ensures that penalties throughout the system are usually applied at the minimum required level (Nucci 2009). When a case of commercial copyright infringement is successfully prosecuted, consequently, the initial result may be a two-year sentence, but this is nearly always commuted to a less severe penalty, such as community service. Where crimes have a minimum penalty of one year or less of imprisonment—as with non-commercial software-related infringement— prosecution can be suspended at the discretion of the prosecutor and judge.[12] This has nothing to do with commitments to IP enforcement per se, but rather with the triaging of cases in an overcrowded criminal justice system.

Although the general rule in Brazil is that crimes are publicly prosecuted, Brazilian criminal law also has provisions for private prosecution before a criminal court. In the copyright field, this applies in cases of software infringement and non-commercial infringement, which must be privately prosecuted.[13] In such cases, police have a duty to investigate if the victim requests it,

---

12   Law 9.099/95, Article 89.

13   This is one of the major differences between enforcement under the Software Law and the Penal Code— and a source of irritation for the software industry, which would like to shift prosecution costs to the

but prosecution is conducted entirely by the victim, rather than by a public prosecutor. Copyright infringement for commercial purposes, in contrast, is always investigated by the police and publicly prosecuted—though prosecutors can petition judges to archive cases based on lack of evidence and are not required to follow up on indictments made in the course of preliminary investigations.[14]

Despite or, arguably, because of the scope of criminal copyright liability, large categories of infractions fall below the enforcement threshold—especially those committed by consumers. No one has been arrested or criminally prosecuted for taping a TV show, for example, despite the fact that time-shifting is technically illegal in Brazil. Because non-commercial infringement requires private prosecution, it is almost never pursued in court, nor is the buying or receipt of pirated and counterfeited goods, even in commercial settings (despite the criminalization of *receptação*, the act of receiving goods that are illegally obtained or produced).[15] The Ministry of Justice has endorsed this policy through the CNCP and generally speaks of educating rather than prosecuting consumers.

Civil and Administrative Enforcement

With the notable exceptions of the Business Software Alliance (BSA) and ECAD, the Brazilian performance rights management organization, civil litigation is rarely pursued by rights holders in Brazil.[16] The non-commercial infringement subject to such litigation rarely warrants the time and expense. BSA use of the civil system is different, first because its primary targets are businesses, and second because it has succeeded in characterizing infringement within businesses as the publication of "fraudulent editions" of the work, which are subject to very high penalties.[17]

Following in the footsteps of the Recording Industry Association of America (RIAA) in the United States, the IFPI (International Federation of the Phonographic Industry) and its Brazilian

---

public sector. There is a basis for this, according to the IIPA (2010:153), in a provision of the 1998 Software Law for any crime of software copyright infringement that also involves tax evasion to be publicly prosecuted. However, fiscal offenses in Brazil only result in criminal prosecution once administrative procedures have come to a close (see Súmula Vinculante 24, issued by the Supreme Federal Court).

14 Private prosecution is still the rule for the crimes of trademark and patent infringement, making this a major item on the legislative agenda for IP industry lobbyists, who would like to see prosecution costs shifted to the public sector. Penalties related to patent and trademark infringement include fines or imprisonment, ranging from three months to a year. Some specific cases, such as the sale of fake medicines, are more severely punished, with penalties of between ten and fifteen years in prison (Penal Code, Article 273).

15 Penal Code, Article 180.

16 Criminal prosecution is left in the hands of the local software association, the ABES (Brazilian Association of Software Companies).

17 Law 9.610/98, Article 103. The law provides for damages of up to three thousand times the value of the infringing work in such cases, allowing for very high awards even where routine office software is concerned. For further background on these cases, see Souza (2009:297–305).

affiliate the ABPD (Brazilian Record Producers Association) have explored civil litigation strategies for consumer infringement. In 2006, suits were initiated against twenty users of P2P (peer-to-peer) networks as a test of the receptiveness of Brazilian courts to mass litigation against file sharers. This effort has been unsuccessful so far, primarily due to concerns that requests for identification of the users would violate privacy protections. When a judge ordered ISPs (Internet service providers) to comply with IFPI requests for information, the process also ran up against a lack of data-retention requirements in Brazilian law, which made the identification of users impossible (IIPA 2009:161–62).

To the annoyance of the copyright industries, Luiz Paulo Barreto, then president of the CNCP (and, from February 2010 to January 2, 2011, Brazil's Minister of Justice), came out against the lawsuits. The IIPA, in particular, was not pleased: "The head of the CNCP has expressed publicly his disagreement with the campaign, which has had a detrimental effect on judges evaluating the cases and diluted the needed determent" (IIPA 2007:215).

Administrative law offers an additional set of enforcement strategies for rights holders. These are often faster, less rigorous, and most important, cheaper for rights holders than judicial procedures—not least because the burden of storing or destroying seized goods falls entirely on the public sector.[18] Border-control, zoning, commercial-licensing regulations, and customs and sales taxes all provide common triggers for administrative action. At Brazil's borders, for instance, suspect cargo can be seized by customs agents either following a complaint from rights holders or ex officio (without a complaint) if copyright or trademark infringement is suspected. In cases where goods have already entered the country, administrative procedures can be invoked if infringement can be associated with tax evasion or the misuse of public space. Zoning ordinances, for example, have been used repeatedly to shut down pirate markets (usually only temporarily), including one of the most famous in São Paulo, the Galeria Pagé (G1 2009). Police investigations and criminal charges are generally unnecessary in these contexts.

The Legislative Agenda

Industry representatives interviewed for this project unanimously shared the view that stronger enforcement is necessary but expressed different degrees of satisfaction with the current framework. Improved enforcement of existing law was, overall, a more frequent concern than demands for substantive legislative change. Nonetheless, industry associations have devoted considerable time and energy to building a legislative agenda on enforcement involving modifications to procedural law, higher penalties for copyright infringement, and stronger measures against online infringement.

Legislative proposals incorporating many of these recommendations have been taken up by a congressional commission of the Chamber of Deputies, the CEPIRATA (Special Commission

---

18   In a criminal proceeding, industry bears the costs of the storage of goods seized in raids until the case is finally adjudicated. Every seized item, moreover, must be examined by a publicly appointed expert.

to Analyze Legislative Proposals that Aim to Combat Piracy). Between May 2008 and August 2009, the CEPIRATA held a number of hearings and seminars to evaluate legislative strategies for strengthening enforcement. These included diminished requirements to retain and store seized goods for evidence; the right to base charges on a sample of infringing goods, rather than a complete itemization of them (a relevant issue when dealing with large quantities of pirated discs, for example); and the harmonization of penalties for software and other types of infringement (Câmara dos Deputados 2009b). The first two demands are now part of Bill 8052/2011, presented to the National Congress by President da Silva on December 31, 2010 and drafted based on CNCP recommendations.

Internet Legislation: The State of Play for 2011

There is no specific legislation regarding data retention or ISP liability in Brazil. Current general-liability rules offer no strong guidance on the issue, and judges have tended to apply strict liability based either on the Civil Code or the Consumer Rights Code, though divergent opinions do occur (Lemos et al. 2009). The application of copyright law to online intermediaries and their users, consequently, remains unsettled in Brazil.

Data-retention requirements and a number of other measures for facilitating the policing of online activity came to the foreground in a debate catalyzed by the 2005 "Azeredo Bill" (Bill 84/99), so named after its main proponent in Congress, Senator Eduardo Azeredo. The Azeredo Bill is an attempt to strengthen the legal infrastructure for investigating and prosecuting Internet crime. It is partly inspired by the Council of Europe's 2001 Budapest Convention on Cybercrime but has origins in older Brazilian proposals as well. The bill is backed by a number of powerful public and private organizations, including the FEBRABAN (Brazilian Bank Federation), the Brazilian Federal Police, and the IIPA. The IIPA supports the bill in the hope that it will facilitate the prosecution of individuals and non-commercial intermediaries involved in file sharing and other forms of online infringement—in the first instance, by requiring longer retention of data on user behavior by service providers. In interviews, a private-sector informant involved in enforcement described the bill specifically as an instrument for suing users of file sharing networks.

Public reaction to the Azeredo Bill, however, was strongly negative and well organized. Concerns about user privacy and anonymity became the basis for a large-scale campaign called the Mega Não movement (literally, "mega no," as in a big "no" to Bill 84/99), which staged a number of public demonstrations and media debates. Perceptions of the bill as a stalking horse for stronger copyright enforcement also featured prominently in this opposition.

In 2009, the intensity of resistance led to calls for a wider public consultation on potential changes to Internet law. In response, the Ministry of Justice initiated an ongoing process, called the "Marco Civil,"[19] that aims beyond policing and enforcement toward a broader spectrum of

---

19   Roughly translated, the Marco Civil means "a civil framework" for Internet law, as opposed to the criminal framework proposed by the Azeredo Bill.

Internet regulation issues, including net neutrality, Internet access, and users' rights. Public consultation concluded in May 2010, and a draft bill will be sent to Congress in early 2011.

The Marco Civil does not deal with IP law directly but rather with securing the openness of the Internet, with fundamental user rights and principles, and with ISP liability for infringement conducted over their networks. These boundaries also reflect a decision to separate consideration of copyright issues proper into a second draft bill presented for public consultation in mid-2010 by the Ministry of Culture.

Predictably, the Marco Civil consultation revealed a wide range of tensions between different rights and interests. Freedom of speech, anonymity, privacy, and access rights were topics of heated debate. Around two thousand contributions were received, including some thirty by organizations such as the IIPA, the IFPI, and the Brazilian consumer rights association the IDEC (Brazilian Institute for Consumer Defense).

Meanwhile, the Azeredo Bill has stalled. According to an informant within Congress, "projects that are controversial with the social networks [that is, civil society groups] are being avoided by government." An attempt in 2009 by Deputy Bispo Gê to introduce "graduated-response" legislation that would require ISPs to disconnect users accused by rights holders of multiple infringements resulted in so strong a backlash that he had to withdraw the proposal (Pavarin 2009b).

GRADUATED RESPONSE

Despite the failure of the Bispo Gê bill, graduated response is still a central part of the industry agenda for Brazil. Some of the contributions to the Marco Civil, such as that of the ABPD, explicitly asked for a regime based on either the French HADOPI (High Authority for the Diffusion of Works and the Protection of Rights on the Internet) model or the British Digital Economy Act—both of which include procedures for terminating the Internet access of repeat infringers. The Marco Civil's net neutrality provisions were also criticized for failing to differentiate legal from illegal content and for creating obstacles to technological solutions to curb P2P file sharing.

Changes to statutory law, however, are only one of the ways forward for graduated response in Brazil. Responding to recording industry requests, the Ministry of Culture has been presiding over an ISP working group that facilitates meetings between ISPs, telecommunications companies, and the recording, film, and software industries. The working group is focused on achieving a consensus between ISPs and copyright industries on the issue of P2P file sharing. In the words of a private-sector member:

> We can find out who is [infringing on copyrights], via the IP [Internet protocol] address of that user. The biggest discussion is around who owns the information regarding this IP address . . . And then we discuss to what extent the ISP is responsible for the crime, or not. Does the ISP invade [users'] privacy or not? Does the ISP have that right or not? That's what we discuss concerning Brazil and Brazilian legislation . . .

> As IP owners, we believe that yes, the ISP is co-responsible and that it should act at least in educating users that they cannot commit this crime. A comparison that I like to make is the following: I'm the landlord of a house. Do I lack responsibility for what goes on inside? Depending on the actions of my tenant, yes I'll be held responsible. So why not take this same concept to ISPs? [ISPs] are the landlords [of the Internet], but what about what happens there? We have great support from ISPs, not in relation to peer-to-peer, but for example, in [certain cases of] piracy. Pirate software being sold through the Internet, [fake] certificates of authenticity being sold through the Internet. When we discover that, we ask providers to take down that information, and they effectively take them down. What we're discussing along with the Ministry [of Culture] is the question of peer-to-peer specifically.

Unlike the Marco Civil consultation, this discussion has been carried out entirely behind closed doors. News of the working group's existence emerged only in 2009. The IIPA's 2009 country report for Brazil mentions it, as do the Ministry of Justice and the ABPD in their sections of the annual report from the CNCP (Ministério da Justiça 2009). Media coverage of these meetings has so far been nonexistent.

In our interviews, members of the ISP working group expressed fear that the negative publicity and controversy surrounding similar laws in other countries (such as the French HADOPI law) could hurt attempts to implement graduated-response measures in Brazil. Informants mentioned that the preferred strategy was to associate the Brazilian approach with the "British model," which at the time (before the passage of the Digital Economy Act and the June 2009 publication of the Digital Britain report) envisioned only reducing the bandwidth of so-called heavy uploaders rather than cutting off their Internet access altogether.

Although this work was far from reaching its conclusion when our interviews took place, there seemed to be consensus among industry representatives that bandwidth reduction was the appropriate penalty for P2P infringement, exercisable through the guilty party's ISP contract. As one film industry representative observed:

> I'm not in favor of the invasion of anyone's private life, or for the surveillance of whatever someone's doing . . . No one is. I just think that there should be controls, in the sense that the contracts that people sign with provider companies be enforced. Every contract between an individual and an ISP presupposes that the individual will not engage in illegal conduct, will not make [illegal] downloads. That's in the contract. So I'm not in favor of knowing the sites you browsed last night, but if there is [an illegal] download, there should be a manner for us to alert this person and to make him or her stop doing that.

This means of implementing graduated response has the notable advantage, for rights holders, of diminishing or obviating the role for judicial or administrative oversight: rights holders would simply notify ISPs that infringing content was shared by a particular IP address, and the ISP would relay the notification to the user. After repeat violations (the so-called three strikes), the ISP would impose the penalty. If subscribers were to challenge this action, they would do so in the context of a contractual dispute, not as a deprivation of a basic right. This strategy also sidesteps the need for new law.

The compatibility of such penalties with existing Brazilian law, especially consumer law, remains an open question, even in the context of voluntary contracts for services. Variations on this model in which the courts play a role have also been discussed, including versions based on France's HADOPI law, which incorporates a brief judicial review of disconnections. None of the industry representatives interviewed for this report, however, described a need for checks and balances, or the involvement of public authorities, or a procedural framework that would allow users to respond when charged with infringement. Such issues, in our view, are likely to create serious legal hurdles for the implementation of a graduated-response model in Brazil.

COPYRIGHT REFORM

Copyright reform began to be discussed seriously at the Ministry of Culture in 2005, leading to a series of conferences collectively called the National Forum on Authors' Rights. The forum was established with the goal of bringing attention to the shortcomings of Brazilian copyright law and to announce the Ministry of Culture's intention to work on a copyright reform bill. Beginning in December 2007, eight multi-stakeholder conferences were held, focused on specific aspects of copyright law. One of the main topics was the role of the executive branch in copyright matters, including its proposed return to active supervision of societies that collect royalties for music rights holders. Parallel to the forum, the Ministry of Culture has consulted with representatives of all the copyright industries and with the ECAD (Central Collecting Office), the association of collecting societies for recorded music, as well as with a few non-industry NGOs (non-governmental organizations). In June 2010, the Ministry of Culture published its draft copyright reform bill and placed it under public consultation in a form similar to that of the Marco Civil.

Among the affected industries, the publishing sector has emerged as the most adamant opponent of the reform process, holding the position that current legislation is adequate. Together with the ABPD and the collecting societies, the publishers have formed a coalition called the National Committee on Culture and Authors' Rights (CNCDA) to campaign against the public supervision of collecting societies, which they present as the "statization of authors' rights."[20] On the opposite side, a network of individuals and institutions favorable to the general contents of the draft bill has formed to defend the need for copyright reform.[21]

---

20   See the CNCDA website at http://www.cncda.com.br/.

21   A list of the parties involved can be found on the Reforma da Lei de Direito Autoral (Reform the Copyright Law) website at http://reformadireitoautoral.org.br/lda/?page_id=317.

From a consumer perspective, the Ministry of Culture's draft bill is a clear improvement over current Brazilian law. The current draft would significantly expand the list of exceptions and limitations to copyright, greatly facilitating access to educational materials and bringing the law into closer alignment with actual practice in most educational institutions.

The draft bill also takes a rational approach to technical protection measures for digital goods, authorizing the circumvention of TPMs and rights-management information for works in the public domain or in other contexts where limitations and exceptions to copyright apply. It also treats the creation of obstacles to the legal use of copyrighted works, or to the free use of works in the public domain, as analogous to improper circumvention and subject to the same penalties. This position is already drawing fire from industry groups, which view current Brazilian law as "weak on technological protection" (IIPA 2010:152).

The IP Enforcement Network

3.1 (3) Each party shall, as appropriate, promote internal coordination among, and facilitate joint actions by, its competent authorities responsible for enforcement of intellectual property rights.

3.1 (4) Each party shall endeavor to promote, where appropriate, the establishment of formal or informal mechanisms, such as advisory groups, whereby its competent authorities may hear the views of right holders and other relevant stakeholders.
- *Anti-Counterfeiting Trade Agreement (ACTA), final text, November 15, 2010*

Organizations specializing in copyright enforcement are not a recent phenomenon, with some dating to the early twentieth century (Johns 2009:327–55). In Brazil, the video distributors association the UBV (Brazilian Video Union) is one of the older living examples, with a history of anti-piracy activism going back to the early days of the home-video market in the 1980s (Bueno 2009). Two features of the current enforcement environment, however, set it apart from earlier efforts.

The first is the appearance of strong coordinating bodies designed to bring together public and private actors who would not otherwise cooperate easily. Since 2004, this role at the federal level in Brazil has belonged to the CNCP. Other organizations from the private sector, such as the FNCP (National Forum against Piracy and Illegality), the FIESP (São Paulo Federation of Industries), and the American Chamber of Commerce (AmCham) act as inter-industry coordinators at the federal and state levels.

The second and related feature is the emergence of a complicated ecology of enforcement functions and roles, with a loose division of labor across institutions. Where before it was possible to look at enforcement as the product of a handful of specific state and private actors, today we need to speak more properly of the consolidation of an enforcement industry, whose products include legal, training, and advocacy services operating across the public and private sectors. The Brazilian version of this industry, in turn, needs to be understood as part of the broader international IP advocacy and policymaking network—the local branch of a larger enterprise.

The rapid proliferation of such organizations in the past decade makes any comprehensive account challenging and somewhat tedious for the non-aficionado. We have accordingly focused only on the most active organizations in the Brazilian enforcement landscape. All such organizations, in any event, engage in some combination of the following activities:

1. Enforcement support, involving the provision of direct assistance to authorities in the investigation and prosecution of IP infringement and related crimes. Assistance can be either material (including funding provided by rights holders) or logistic, with the provision of training and operational support for enforcement measures, such as raids.

2. General advocacy, in the form of public relations, lectures, education, the production of research, and anti-piracy/pro-IP marketing.

3. Lobbying, or advocacy targeted directly at lawmakers and focused on specific legislative change.

4. Coordination of the above activities between different sets of actors. Increasingly, such activity drives the enforcement ecosystem.

With the formation of the CNCP, the National Council on Combating Piracy and Intellectual Property Crimes, industry succeeded in creating a powerful forum for coordination at the federal level. The CNCP has emerged as a platform for both interagency and public-private coordination and has furnished a model for similar efforts at the state and municipal levels.[22] One sign of the CNCP's effectiveness is the adoption of strong anti-piracy discourse by the government, with the National Plan on Combating Piracy as its centerpiece. Piracy is now very much a part of the public agenda in Brazil.

To a large extent, the CNCP and the National Plan are the products of international pressure brought to bear in Brazil over the past decade—primarily by the US government but also by the multinational copyright industries. But it would be wrong to see these developments simply as impositions. Informants from the public sector who are involved in IP policymaking in Brazil were all supportive of the institutional outcomes that followed US pressure in the early 2000s.

---

22   The IIPA claims that industry relationships with federal authorities are now good and that the "bottleneck as far as physical piracy of music and movies is concerned lies in the federation state and municipal levels" (IIPA 2010:148). Accordingly, the IIPA now focuses on law enforcement coordination across levels of the Brazilian federation and has pushed for the creation of joint state and municipal task forces (IIPA 2010:140).

Nevertheless, there are considerable differences in the agendas of the public and private sectors (and even within those sectors) that result in important differences in what stakeholders mean by "combating piracy."

The Enforcement Decade

After bruising struggles with the United States over IP policy in the 1980s and early 1990s, the renewed USTR and industry push on enforcement in 1999 and 2000 met with considerable cooperation on the Brazilian side. The IIPA's 2000 petition to the USTR set out the terms of this dialogue. The petition cited Brazil for "unacceptably high levels" of piracy in all industry sectors, non-deterring criminal penalties, a low number of convictions, and delays in civil and criminal cases. Between 2001 and 2005, facing the threat of the removal of trade privileges, Brazil developed institutional arrangements for addressing private-sector concerns on enforcement.

Table 5.2 Brazilian IP Enforcement Timeline

2000	The IIPA petitions the USTR to put Brazil under GSP review (asking for the removal of Brazil's trade privileges).
2001	The IIPA's request is granted by the USTR. The Brazilian government establishes the IMC.
2002	The USTR promotes Brazil from the Special 301 "Watch List" to the "Priority Watch List."
2003	The CPI da Pirataria begins its proceedings. In parallel, a Congressional Anti-Piracy Caucus is created, as well as the private, cross-industry association that would become the FNCP.
2004	The CPI da Pirataria publishes its final report, recommending the creation of a new public-private entity to replace the IMC. The CNCP is created.
2005	The CNCP publishes its first National Plan on Combating Piracy, listing ninety-nine anti-piracy and anti-counterfeiting initiatives.
2006	The USTR ends its GSP review of Brazil. The IIPA recommends Brazil's demotion to the "Watch List."
2007	The USTR demotes Brazil from the Special 301 "Priority Watch List" to the "Watch List."

In some respects, little changed on these fronts in the ensuing years. By 2010, complaints regarding online infringement had eclipsed those regarding optical disc piracy, but the underlying industry concerns with high rates of piracy and weak enforcement remained largely the same.

With respect to the politics of cooperation, however, much had changed. Recent IIPA reports show government and industry on much better terms, visible in the greater "political willingness" of the Brazilian government to combat piracy (IIPA 2006:199).

The IMC, the Inter-Ministerial Committee on Combating Piracy, was created in 2001 as a direct response to the threat of exclusion from US trade privileges under the Generalized System of Preferences. The IMC was composed entirely of representatives of government ministries, including the Ministry of Justice, the Ministry of Science and Technology, the Ministry of Culture, the Ministry of the Treasury, the Ministry of Foreign Affairs, and the Ministry of Development, Industry, and Foreign Commerce. The private sector contributed to IMC activities on an invitation-only basis.

The major industry associations were dissatisfied with this relationship[23] and continued to press for changes to the institutional framework for enforcement policy. Industry dissatisfaction also meant USTR dissatisfaction. Trade pressures on Brazil were maintained throughout the early 2000s and led to the Parliamentary Commission of Inquiry on Piracy, established by the Chamber of Deputies in 2003.

The commission held a full year of hearings and investigations, involving representatives from all the IP industry associations, as well as law enforcement authorities and US government emissaries. Legislative ties with the United States were established, and a group of Brazilian deputies traveled to Washington, DC, to meet with members of the US Congress's International Anti-Piracy Caucus. Among its several recommendations, the commission proposed the creation of a public-private forum for coordinating and developing anti-piracy initiatives. This proposal became the CNCP.

Much of the commission's inquiry was focused on counterfeiting and contraband rather than piracy. But the copyright industries were the most vocal stakeholders in the process and were heavily represented in the initial formation of the CNCP. The recording, software, publishing, and film industry associations all had individual seats on the new council, with only a single seat provided for the industrial-property sector, given to ETCO (Brazilian Institute for Ethics in Competition), representing fuel, beverage, medicine, tobacco, and software companies.

---

23   "Several IIPA members have met individually and in small groups with the IMC chairman, as well as other senior Brazilian officials, including the Minister of Justice. A list of suggested actions was presented to the IMC chairman; however, the IMC never implemented these suggestions. The industry has never met with all members of the IMC. The industry has never received any official communication from the IMC regarding any of its decisions or actions, although informally, the copyright industries were advised that no decisions were made and nothing was planned. In sum, the IMC has not shown any willingness to work with the private sector or the U.S. government. Furthermore, the IMC chairman promised enforcement actions in October and November 2001, but nothing was done. The IMC has no agenda for 2002, as far as the industries are aware" (IIPA 2002:73).

The National Council on Combating Piracy

The CNCP is essentially the older IMC restructured to meet industry demands. Membership was expanded to include seven organizations representing the private sector. A panel of external partners was also created, which participates in the council's sessions on an ad hoc basis. To date, this panel has included almost exclusively individuals and organizations with ties to the IP industries.

There have been a number of changes in the composition of the CNCP over the six years in which it has been active. The publishing industry representative, the ABDR (Brazilian Association of Reprographic Rights), was one of the initial members but did not have its seat renewed. The IP lawyer association the ABPI (Brazilian Intellectual Property Association) also left, after two two-year terms. The CNC (National Confederation of the Commerce of Goods, Services, and Tourism), the CNI (National Confederation of the Commerce of Goods, Services, and Tourism), and the BPG (Brand Protection Group), all industrial-property groups, were included to create a more balanced distribution across the different IP industries. In its current incarnation, the copyright industries have three seats on the CNCP, reduced from an initial four.

Figure 5.2 CNCP Composition, December 2009 to Present

Public Sector, Legislative	Public Sector, Ministries	Federal Law Enforcement	Private Sector, Copyright	Private Sector, Industrial Property
Chamber of Deputies	Justice*	• Federal Highway Police	ABPD (Recording Industry)	Brand Protection Group
Senate	Culture	• Federal Police	ABES (Software Industry)	National Confederation of Industries (CNI)
	Development, Industry, and Foreign Commerce	• National Public Security Office	MPA (Film Industry)	
	Foreign Affairs	Federal Revenue Service**		National Confederation of Commerce (CNC)
	Labor			ETCO (fuel, beverages, cigarettes, medicine, software)
	National Treasury			
	Science and Technology			

Presides over the CNCP and directs the Federal Highway Police, Federal Police, and NPSO.

** organized under the National Treasury*

Members from the public sector are split into three groups: (1) seven participants from different ministries, all with responsibilities that intersect piracy and counterfeiting; (2) two members from the technical staff of the houses of the National Congress; (3) and four representatives from federal law enforcement agencies, under the authority of either the Ministry of Justice or the Ministry of the Treasury (figure 5.2). The president and executive-director of the CNCP are appointed by the Ministry of Justice.[24]

As a consultative body, the CNCP has several functions ranging from studying and proposing anti-piracy measures to supporting the training of law enforcement agents. Among these, the most important has been the drafting and execution of the National Plan on Combating Piracy.

The National Plan

Prior to the creation of the CNCP, one of the IIPA's chief complaints about Brazil was the absence of a national plan for combating piracy that incorporated the demands of the private sector. The CNCP addressed this deficiency shortly after its creation with the release of the first National Plan in 2005.

According to a public-sector CNCP participant interviewed for this report, the 2005 plan was crucial from a political standpoint: Brazil was still under GSP review and "there was an atmosphere of reciprocal accusations" between the public and private sectors. Writing the report brought both groups to the table to discuss conflicts that had been simmering for a long time. For this informant, the creation of the CNCP and the publication of the National Plan represented important steps both internally, for stakeholders, and externally, in gaining US government recognition that Brazil was engaged in the fight against piracy. The desire for a truce between public and private sectors drove the drafting of the plan, trumping practical considerations in many respects. The result, in the words of another informant, was "not very manageable." The report announced some ninety-nine activities to be undertaken. Many of these were not under the authority of the CNCP or the federal government but belonged to the judiciary or state and municipal administrations.

The weaknesses of the plan were not lost on the CNCP. In 2009, a second plan was released, which pared back the ninety-nine action items to a somewhat more manageable twenty-three. This narrowing of the focus was not welcomed by all participants due to concerns that it turned government approval for anti-piracy efforts into a comparatively scarce resource. According to

---

24   Changes in leadership have partly set the tone for the CNCP's activities. The first executive-director, Márcio Costa de Menezes e Gonçalves, was an IP lawyer appointed by industry, who left to found the anti-piracy organization the ICI (Intellectual Capital Institute) at the end of his mandate in 2008. The second executive-director, André Barcellos, was a career public servant from the Ministry of Planning. Some industry informants expressed clear nostalgia for the Gonçalves era, when they believe there was greater industry control of CNCP activities. Luiz Paulo Barreto served as president of the organization through early 2010, and subsequently replaced Tarso Genro as Minister of Justice. He left that office in early 2011.

one private-sector informant, the most valuable assets of the CNCP are its brand and stamp of approval on awareness campaigns, and the new plan made the latter more difficult to obtain.

Table 5.3 The CNCP's Priority Projects

Project	Objective	Coordinator
Piracy-Free Cities	Create public-private anti-piracy councils at the municipal level.	ETCO
Legal Fair	Reduce the supply of illegal goods in street markets and popular fairs.	ETCO
Commerce against Piracy	Unite shop owners in a national awareness campaign against piracy.	CNC
Anti-Piracy Portal	Build an interactive web portal to provide a better communication channel between the CNCP and its members and the public and to disseminate awareness campaigns.	ABES
Partnerships with ISPs	Find solutions to the problem of P2P file sharing and online copyright infringement based on "partnerships" between industry and ISPs.	Ministry of Culture

Among the five priority projects selected from the twenty-three (table 5.3), the furthest along, as of late 2010, is the Piracy-Free Cities project, which involves the creation of regional anti-piracy councils to coordinate efforts at the municipal level.[25] This project was first implemented in the city of Blumenau, in the state of Santa Catarina, in 2007. The model was subsequently extended to the cities of Curitiba, Rio de Janeiro, São Paulo, Brasília, and Ribeirão Preto, with more to come according to the project sponsor, ETCO (2009:30–33).[26]

---

25   A complete list of National Plan projects (some of which are not really "projects" at all) can be found on our project website at http://piracy.ssrc.org/resources.

26   There have been other attempts to replicate the public-private structure of the CNCP at the state level, but not systematically. São Paulo's Inter-Secretarial Committee on Combating Piracy (Comitê Inter-secretarial de Combate à Pirataria) was formed in 2006 (a private-sector informant who used to be a member of the committee described it as "not very active" as of late 2009).

So far, these state and local councils have not departed much from existing policy and are limited to dealing with street markets for infringing goods. They act as fora for pressuring municipal authorities to use their powers of regulating and policing public space, and they participate in awareness campaigns. They have no special powers over online infringement, with the exception of local ordinances affecting cybercafes, or LAN (local area network) houses, as businesses offering Internet access in Brazil are known, and other physical sites of Internet access.

The GIPI and Brazilian IP Policy

In theory, the CNCP is a strictly consultative body, with no clear authority to deliberate on legislation. Despite the participation of congressional staff members from the Senate and the Chamber of Deputies, the inclusion of a "legislative fine-tuning" project in the current National Plan (which so far has contributed to the pending anti-piracy legislation Bill 8052/2011), the CNCP's main function is to coordinate enforcement actions under the existing authority of its participating public agencies. Similarly, private-sector lobbying is mostly done outside the CNCP, directly between private-sector groups and members of Congress.

Brazil maintains an important—if not always clear-cut—distinction between enforcement policy and general IP policy, with the final word on Brazilian domestic and international IP policy set not by the CNCP but by the higher-level Grupo Interministerial de Propriedade Intelectual (Inter-Ministerial Intellectual Property Group), or GIPI. Although representation within the GIPI significantly overlaps that of the CNCP (see figure 5.3), the GIPI is a purely public body, with an explicit mandate to balance the interests of rights holders and the public in setting IP policy. The GIPI's role in enforcement is accordingly much different than that of the CNCP and is framed in terms that require the consideration of enforcement's "broader meaning": its "social accordance with intellectual property legislation in its ensemble, acknowledging both the rights granted to rights holders and the limitations and exceptions that are present and necessary in every legislation."[27]

According to a public-sector participant in both the CNCP and the GIPI, discussions about enforcement policy that shade into general IP policy are systematically moved to the GIPI, as are sensitive issues more generally. The move to the GIPI ensures, in particular, that there will be "no pressure from the private sector." The government's position on any bill involving intellectual property rights is formed within the GIPI, including, for example, the Ministry of Culture's draft copyright reform bill. This does not mean that the GIPI ignores input from the private sector, but the existence of different levels of policy fora ensures that public-sector opinion on IP policy has some autonomy from private-sector influence. In our interviews,

---

27   According to the "Lines of Action" for the GIPI as listed by the Ministry of Development, Industry, and Foreign Commerce on its website, http://www.mdic.gov.br/sitio/interna/interna.php?area=3&menu=1783.

opinions about this autonomy tended to split along predictable public-private lines, with private-sector representatives regretting the independence of the GIPI (one noted in particular the greater influence of the private sector on the GIPI prior to the formation of the CNCP) and public-sector informants describing that independence as "a very positive thing within the state."

Figure 5.3 GIPI and CNCP Overlap

GIPI CNCP

●   Chambers of Foreign Commerce (CAMEX)
●   Civil House
●   Ministry of Health
●   Office of Strategic Affairs (SAE)
●   Ministry of Agriculture
●   Ministry of Environment

●   ●   Ministry of Justice
●   ●   Ministry of Culture
●   ●   Ministry of Science and Technology
●   ●   Ministry of Foreign Affairs
●   ●   Ministry of Development, Industry, and Foreign Commerce
●   ●   Ministry of the National Treasury

     ●   Ministry of Labor and Employment

     ●   Federal Police Department
     ●   Federal Highway Police Department
     ●   National Public Security Office
     ●   Federal Revenue Service

     ●   Federal Senate
     ●   House of Deputies

     ●   National Confederation of Industry
     ●   National Confederation of Commerce

     ●   Brazilian Record Producers Association (ABPD)
     ●   Brazilian Association of Software COmpanies (ABES)
     ●   Brand Protection Group (BPG)
     ●   ETCO

In practice, the GIPI's independence allows it to act as a de facto graveyard for the more extreme enforcement proposals. When the Motion Picture Association of America (MPAA) tried to gather the CNCP's support for an anti-camcording bill that would have made carrying a camera into a movie theatre a crime,[28] debate was moved to the GIPI, where the proposal was rejected.

The participating ministries in the GIPI specialize in different fields of IP law, sometimes with further in-house divisions. The Ministry of Culture bears overall responsibility for copyright policy, managed through the Directorship of Intellectual Rights (the DDI), as well as issues related to traditional knowledge. Software copyright policy (including software policy more generally) is an exception to this rule and belongs under the Ministry of Science and Technology. Patent and trademark policy is concentrated in the Ministry of Development, Industry, and Foreign Commerce and the INPI (National Institute of Industrial Property), a.k.a. the Brazilian patent office, which, although not a permanent member of the GIPI, sits at every meeting involving industrial IP. Although it is linked to the Ministry of Development, the INPI is an autonomous body and tied to the global patent-office network described by Peter Drahos (2010). Traditional knowledge is handled by the Ministry of Agriculture and the Ministry of Environment, which houses the CGEN (Genetic Resources Management Council). The Ministry of Foreign Affairs, for its part, has the DIPI (the Division of Intellectual Property), which relays Brazil's internal policy positions to international IP forums like WIPO and the WTO.

Know Your Enforcement Authorities

Copyright enforcement, and particularly criminal enforcement, involves coordination between authorities at three levels—federal, state, and municipal—as well as efforts by rights holders to train and support enforcement agents (see figure 5.4).

Figure 5.4 Law Enforcement Authorities

Federal	State	Municipal
Customs/Federal Revenue Service	Civil Police	Municipal Guard
Federal Highway Police	Military Police	Municipality
Federal Judiciary	State Revenue Service/Patent Office	
Federal Police	State Judiciary	
Federal Prosecution Service	State Prosecution Service	

---

28   Stronger anti-camcording provisions are a universal feature of MPAA and IIPA lobbying. In Brazil, camcording movies is already a crime, and by IIPA accounts an infrequent one: in 2010, the IIPA noted only twenty-three cases.

THE POLICE AND THE MUNICIPAL GUARDS

Brazilian police are divided into federal and state forces and further into forces that specialize in either the prevention or the investigation of criminal offenses. Each Brazilian state has a Civil Police force, tasked with the investigation of criminal offenses and the gathering of evidence, and a Military Police force, in charge of crime prevention and immediate response. At the federal level, the Federal Highway Police acts as the preventative police force, complementing the work of investigative Federal Police. In piracy-related matters, the Federal Police is highly constrained and can only act if the criminal activity involves contraband or the irregular entry of goods into Brazilian territory.[29] Otherwise, state Civil and Military Police are the competent authorities.

A few states have created specialized Civil Police units for IP enforcement. The most significant is Rio de Janeiro's DRCPIM (Police Unit for the Repression of Crimes against Immaterial Property), created in 2003 for "political reasons" related to industry pressure and "furnished and equipped mostly by private funds," according to a law enforcement expert interviewed for this report. The DRCPIM is often referred to as a success story and a model to be followed by other states.[30] In the early 2000s, the state of São Paulo also established a specialized anti-piracy unit, under its DEIC (Department of Investigations of Organized Crime). Bahia, for its part, has the GEPPI (Special Intellectual Property Protection Group) within its Civil Police, established in 2007. The creation of additional special units is part of the current National Plan, but the actual decisionmaking resides with state government authorities.

Cybercrime units are being created throughout the state Civil Police departments, with an initial focus on bank fraud, child pornography, and hate speech—and a mandate that could plausibly be extended to copyright infringement. The flagship cybercrime division is the Federal Police's URCC (Cybercrime Repression Unit). The Federal Police is also one of the strongest supporters of the Azeredo Bill, which would amplify police powers in the investigation of online crimes, including file sharing.

Municipalities do not have police forces but are authorized by the Brazilian Constitution to maintain a Municipal Guard for the purposes of protecting local "goods, services and facilities" (Article 144, § 8). Not every municipality has such a force, and they sometimes coexist with city-hall staff in charge of fiscal or zoning matters. The tension between these authorities and street vendors is one of the main local factors shaping street piracy in Brazil.

---

29   Or, occasionally, in cases authorized by the Minister of Justice where there are "interstate or international repercussions" and a need for "uniform repression." Law 10.446/02.

30   The CNCP's fourth report, *Brasil Original*, describes the DRCPIM's activities in glowing terms (Ministério da Justiça 2009:71–79).

REVENUE, CUSTOMS, AND THE PATENT OFFICE

There are independent revenue services at all three levels of the federation, each with authority over different types of taxation. Taxes on imports and exports, for example, are the domain of the Federal Union, and consequently, of the Federal Revenue Service, which also manages Customs. These authorities play an important role in combating street-level piracy. Better coordination between the Federal Revenue Service and the Federal Police is one of the most visible results of the CNCP's activities.

An employee union at the Federal Revenue Service, the SINDIRECEITA (National Syndicate of the Tax Analysts of the Brazilian Federal Revenue Service), which represents fiscal analysts, has also become involved in anti-piracy education via its sponsorship of the "Pirata: Tô Fora" (Piracy: I'm Out) public awareness campaign. The campaign is an outreach initiative tied to the notion of "fiscal education"—of educating citizens to pay their taxes—that informs a number of SINDIRECEITA activities. Similar concerns inform anti-piracy education efforts at the ESAF (Superior School of Public Revenue Administration) under the Ministry of the Treasury.

The INPI, the Brazilian patent office, plays a key role in training, education, and IP advocacy, broadly in the service of "promoting a culture of respect for intellectual property rights" (INPI n.d.:21). The INPI has been heavily invested in education since 2005, sending staff to anti-piracy seminars and workshops, offering courses through its Intellectual Property Academy, and collaborating with the industry association FIESP in publishing intellectual property primers.

THE PROSECUTION SERVICES AND THE JUDICIARY

As of late 2010, Rio de Janeiro is the only state that has a special public prosecution office for IP crime.[31] Piracy and contraband do get the attention of other public prosecutors, and particularly in São Paulo there has been increased activity following the creation of a special Integrated Action Program (PAI) on piracy. A national prosecutors group, the GNCOC (National Group for Combating Criminal Organizations) had also become known in recent years for encouraging federal and state prosecutors to ramp up activity against IP crimes, but sources say its activity on this front has dropped off.

Judges and prosecutors, for their part, have considerable autonomy and are free, in particular, of the strong hierarchical constraints that shape police and executive-branch agendas. A few judges and prosecutors have taken on piracy as a personal cause and publicly speak on the subject; prosecutor Lilian Moreira Pinho and judge Gilson Dipp are the most visible examples. But there has been very little broader institutional capture: both the judiciary and the prosecution services are comparatively insulated from industry influence.

---

31   The Sixteenth PIP (Criminal Investigation Prosecution Office). This is the office that receives the cases investigated by the DRCPIM special police unit.

Copyright Industry Associations

The ABPD is the Brazilian branch of the global recording industry and is affiliated with the IFPI.[32] The recording industry also has strong political influence via the ECAD, the umbrella organization for collecting societies in Brazil. Although it is not involved in anti-piracy enforcement, the ECAD is a major actor in IP lobbying and advocacy and is one of the main opponents of the copyright reform draft bill presented by the Ministry of Culture in 2010.

The film industry is represented in Brazil by the MPA (Motion Picture Association), the international arm of the MPAA. Video distributor association the UBV is also very vocal in lobbying and general IP advocacy.

The ABES represents Brazilian and foreign software companies and provides enforcement support for the business and entertainment software sectors. It includes the US-based BSA and the ESA (Entertainment Software Association) as special members and has often acted in concert with them. The BSA, for its part, also operates independently in Brazil and carries out a wide range of civil litigation against companies suspected of copyright infringement.

The publishing sector is represented by several organizations. The book trade, including distributors and retailers, is represented nationally by the CBL (Brazilian Book Chamber) as well as the ABDR. The ABDR was founded in 1992 to act as a reproduction rights organization and is currently the main representative of publishers in the enforcement network. It focuses on enforcement support, especially in regard to copy shops and universities, but it is also the main IP advocacy and lobbying organization for publishers. Also important in the publishing sector are the SNEL (National Syndicate of Book Editors), the ABRELIVROS (Brazilian Association of Textbook Editors), and the ABEU (Brazilian Association of University Presses).

Enforcement Support Organizations

Industry associations have created a variety of units (or, occasionally, separate organizations) to support enforcement. This work ranges from interventions with public authorities, such as making complaints to the police, assisting with investigations and prosecutions, and participating in the training of public agents, to broader surveillance and direct action, such as monitoring the Internet and sending takedown notices when infringing content is found. Several of these groups also provide material and financial support directly to law enforcement.

Arguably the most active of these groups is the APCM (Film and Music Anti-Piracy Association), an anti-piracy organization created in 2007 through a merger of the main film and recording industry enforcement groups. The software industry has a working group within the ABES that acts as its enforcement support unit. The ABDR plays a similar role within the larger matrix of organizations representing the publishing industry.

---

32   Its current members are EMI Music, Sony Music Entertainment, Walt Disney Records, Universal Music, Warner Music, the Argentinian company Music Brokers, and the Brazilian companies MK Music, Paulinas, Record Produções e Gravações, and Som Livre.

The APCM is involved in a wide range of activities, including the creation of awareness campaigns and more general copyright advocacy and lobbying. But it is best known as an enforcement organization. To date, its work has primarily targeted street vendors and online communities engaged in infringement rather than individuals. In the online environment, the APCM has pursued the service providers and administrators of file sharing sites and similar platforms, issuing cease-and-desist letters or takedown notices when infringing content can be identified. Police became involved for the first time in such an action in 2010, following the APCM's complaint that infringement conducted within the online community Brasil Séries constituted a commercial service subject to public prosecution (Zmoginski 2010). When users of these services protest, the APCM—not the recording and film industries—receives most of the backlash. It is probably best to understand this buffering function, too, as part of the APCM's role.

Other groups operate more narrowly in the anti-counterfeiting arena but partly overlap anti-piracy efforts. The BPG (Brands Protection Group) is one such group, founded in 2002 and recently admitted as a CNCP member. The BPG supports investigations, raids, and prosecution on behalf of its associates: Nike, BIC, Swedish Match, Louis Vuitton, Chanel, Henkel, Souza Cruz, and Philip Morris.[33] The ABCF (Brazilian Association for Combating Counterfeiting) is another, representing Souza Cruz, Xerox, Abbot, Mahle, Technos, Philips, Motorola, and Johnson & Johnson, among others. On its website, the ABCF claims that one of its chief contributions to enforcement is the direct financial support of police stations.[34]

PRIVATE FUNDING, PUBLIC ENFORCEMENT

Although by law, policing is a strictly public function in Brazil, in practice, organizations such as the APCM are deeply involved in most aspects of law enforcement, including investigations, raids, and prosecutions. Boundaries between public and private law enforcement have been blurred to the point of irrelevance in this area. As Alex Dent has argued, it is often more accurate to characterize the relationship as an official stamp on private enforcement rather than as private support for a public function. This is how a typical APCM operation unfolds:

> The process begins with their hotlines: the public calls in to report instances of
> piracy, which the APCM then organizes into files. As soon as these files are coherent,
> they are sent to the mayor's office and to the Civil Police. If the police are short of
> officers to go on a raid, the APCM can send along extra people. Similarly, if the police
> lack the necessary transportation for people and confiscated product, the APCM
> rents a van. The APCM is often called upon to provide the garbage bags into which

---

33   In 2009, the BPG struck a partnership with the São Paulo Public Prosecution Service to provide public prosecutors with "technical and operational support" as well as "human and material resources" (Ministério Público do Estado de São Paulo 2009).

34   See "Doações," under "Realizações," on the ABCF website, http://www.abcf.org.br/.

the offending products are placed. It takes the confiscated product, catalogues and then destroys it. The APCM may then need to call a locksmith to repair a door that was broken in the course of a raid. And finally, the association often buys printer cartridges so that the police officers can print their reports on police-station printers. The agency views all this as crucial support, since the red tape and hierarchy of Brazil's police forces cause requests for action of any kind to move very slowly. The APCM, the staff pointed out, have a very quick response time and need not go through elaborate procedures just to get ink, rent a truck, buy garbage bags, or hire a locksmith.

The point is that organizations such as the APCM are on the front lines of the actual policing, providing both the impetus and the logistical support.[35]

Dent describes the situation in São Paulo, but our team witnessed similar arrangements in Rio de Janeiro's special enforcement police unit, the DRCPIM. Due to the finite resources of the private organizations, however, such partnerships have been limited to a handful of states and cities in the Brazilian federation.[36]

In our interviews, law enforcement agents generally expressed appreciation for the financial and logistical support. As one informant noted, "They have resources we don't have, and we can do work they can't." Private involvement with the public functions of the law does, nonetheless, raise concerns about the independence of police functions. In 2009, the São Paulo Public Prosecution Service launched an investigation of the APCM's donations to the DEIC, the São Paulo Civil Police unit in charge of intellectual property crimes. According to the prosecutor, these donations—which include a refrigerator and a new floor for the police unit—could lead to charges of "administrative improbity" (Tavares and Zanchetta 2009), possibly resulting in serious civil and administrative penalties.[37] Such arrangements also make it impossible to determine the actual size and budget of enforcement efforts—a basic and, to date, missing datapoint in a debate about the expansion of public responsibilities.

---

35   Alex Dent, unpublished working paper, 2009.

36   The APCM's workforce of thirty is concentrated mostly in the state of São Paulo. Four employees in São Paulo are "intelligence agents," responsible for monitoring street markets and gathering information on cases of copyright infringement. The states of Minas Gerais, Rio de Janeiro, Rio Grande do Sul, Pernambuco, and Santa Catarina have one intelligence agent apiece. When necessary, the APCM hires freelance workers (to help, for instance, with collecting pirated media in raids). Six employees work exclusively on monitoring of the Internet and online communities (Muniz 2009b).

37   The APCM's director, a former member of the Federal Police, has claimed that all its donations are legal. As of late 2010, the matter remains unresolved.

CROSS-INDUSTRY COORDINATORS

Over the past decade, the copyright industries have become much more adept at making common cause with other industry sectors—including sectors with business models that do not rely heavily on IP. A number of cross-industry associations have emerged to ensure that common elements in the agendas of different industries become opportunities for collaboration and resource sharing. There are many such groups, but much of the action revolves around the FNCP and ETCO, which represent the industrial and manufacturing sector; the CNI and the CNC—the Brazilian confederations of industry and commerce, respectively; the American Chamber of Commerce and the US Chamber of Commerce; and the IP lawyer association the ABPI.

The FNCP (the National Forum against Piracy and Illegality) was founded by lawyer and economist Alexandre Cruz during the Parliamentary Commission of Inquiry on Piracy to assist manufacturers in combating piracy, counterfeiting, and related crimes such as smuggling and tax evasion. The organization was formally established in 2004, and Cruz served as its president through 2009. Its members include 3M, HP, Xerox, Adidas, and Philip Morris, but at present no major media companies or copyright industry associations.[38] The FNCP's activities, according to a private-sector informant, have become "too broad" to justify membership for the copyright sector, but it seems more likely that the opposite is true: despite areas of overlap between the enforcement agendas of the copyright industry and other IP industries, the FNCP focuses only on hard-goods piracy and related legislative lobbying.

The CNI and the CNC, the Confederation of Industry and the Confederation of Commerce, represent employers within the "syndicate system" that organizes much of Brazilian economic life. Both are national bodies that confederate state-level associations, which in turn aggregate sector-specific syndicates ranging from clothing to steelworks to filmmaking. Individual businesses occupy the lowest level. Through this organizational structure, the CNI is said to represent over 350,000 businesses. The CNC claims to represent more than five million. Because the Brazilian Constitution forbids the creation of more than a single syndicate covering the same "economic or professional category" in the same territorial area (Article 8, II), connections to the CNI or the CNC are ubiquitous in the business sector.

Both organizations are highly influential and experienced participants in the Brazilian legislative process. Both are members of the CNCP, and both manage a variety of affiliates that play more specialized roles in the copyright and enforcement arena, including an alphabet soup of educational and cultural organizations, such as the SESI (Industry Social Service), the SENAI (National Learning Service of Industry), and the IEL (Euvaldo Lodi Institute) under the CNI and the SESC (Commerce Social Service) and the SENAC (National Learning Service of Commerce) under the CNC. The IEL, for instance, in partnership with the SENAI and the INPI (the Brazilian patent office), maintains the Intellectual Property Program for Industry, which hosts IP training

---

38   The ABDR used to be a member and still claims to be on its website, http://www.abdr.org.br/site/

seminars and publishes primers for journalists and educators as part of its effort to "disseminate intellectual property culture" (IEL 2009:44). Other affiliates have also acquired important roles in the anti-piracy ecosystem, sometimes overshadowing the role of the larger associations. This is the case with the FIESP, the industry federation of São Paulo, and with the Fecomércio-RJ, the commerce federation of Rio de Janeiro, both of which are active in IP advocacy and anti-piracy efforts.

The US Chamber of Commerce is one of the main actors in international IP lobbying and has been an active force in Brazilian intellectual property advocacy since 2004.[39] As we discuss later in this chapter, it is the primary sponsor of one of the main longitudinal domestic surveys on the consumption of pirated and counterfeited goods.

The American Chamber of Commerce, or AmCham as it is usually called, is easy to confuse with the US Chamber of Commerce and has many of the same interests, but is in fact a different organization. There are 115 AmChams around the world, all affiliated with the US Chamber but governed independently. Among AmCham Brazil's five thousand members, roughly 85% are Brazilian and 10% American.[40] AmCham's claim to fame in enforcement networks derives mostly from its "Projeto Escola Legal" (Legal School Project) campaign in Brazilian schools, which is sponsored by several industries and enjoys the support of the CNCP. AmCham also maintains an anti-piracy taskforce, which contributes to the training of public agents and maintains relations with the USTR and the US Trade and Development Agency.

ETCO (the Brazilian Institute for Ethics in Competition) is one of several organizations that lobby on behalf of manufacturers in Brazil and help promote anti-piracy discourse in the private and public sectors.[41] ETCO was founded in 2003 and has been a CNCP member since the beginning. Its membership draws primarily on the beverage, pharmaceutical, fuel, and tobacco industries but also includes information technology companies such as Microsoft. Judging by ETCO's publications, ethics in competition means campaigning for reductions in business taxes, cutting regulation, making labor legislation more flexible, and fighting to eradicate the informal economy and piracy (both depicted as "scourges"). ETCO literature is highly moralizing[42] but also tends to be more grounded in academic work than that of other organizations.

Last but not least in our overview, the ABPI (the Brazilian Intellectual Property Association) is an association of IP lawyers that has become active in IP advocacy and lobbying (though its stated mission relates more to studying intellectual property and related fields). Because most ABPI

---

39   Interview with US Chamber representative Solange Mata Machado, published in the SINDIRECEITA's (2006) online anti-piracy bulletin.

40   AmCham Brasil, "Quem Somos," http://www.amcham.com.br/quem-somos.

41   Others—all significantly smaller than ETCO—include the ANGARDI (National Association for Safeguarding Intellectual Rights) and the IBL (Legal Brazil Institute, or Brazilian Institute for the Defense of Competition), which represents electronics manufacturers.

42   ETCO's slogan is "now illegality will have to face ethics." ETCO, "Quem Somos," http://www.etco.org.br/texto.php?SiglaMenu=QSM.

members provide legal services for industry, its work has a clear industry tilt. The ABPI publishes a journal on IP law, organizes conferences, and has a number of working groups dedicated to a wide range of IP topics, including enforcement. It was a member of the CNCP for most of the council's history, before losing its seat in 2010 to the BPG. It also serves as the Brazilian chapter of the AIPPI (International Association for the Protection of Intellectual Property), self-described as "the World's leading non-government organization for research into, and formulation of policy for, the law relating to the protection of intellectual property."[43]

How Piracy Works

> Every time I go out on the streets and see a street vendor, I feel like kicking his stall.
> *–Tânia Lima, executive director of the Brazilian Video Union*[44]

Despite the speed of the digital transition, physical piracy still matters in Brazil. Broadband quality is generally poor (IPEA 2010), making large downloads problematic even for privileged users. Only 42% of the population of Internet users in Brazil have access through home connections, with LAN houses the primary means of access for 29% (CETIC.br. 2009).

Accordingly, IP enforcement in Brazil still focuses on the traffic in hard goods, such as optical discs, and on the role of the large informal sector in production, distribution, and retail. Because much of the trade in counterfeit and contraband goods implicates transnational smuggling networks—especially across the porous "tri-border" zone with Argentina and Paraguay—government and private-industry associations have generally found common ground on enforcement. Arguably the most effective measures taken by the CNCP during its first years involved bringing the Federal Police, the Federal Revenue Service, and affected industries into closer coordination on border control. Other widely supported initiatives have focused on the regulation and policing of Brazil's informal-retail sector, including kiosk malls and street markets.

A convergence of industry and government perspectives may be harder to achieve around Internet and book piracy. These two types of piracy have figured in the past three USTR reports as reasons for keeping Brazil on the "Watch List." Book piracy—mostly involving the photocopying of educational materials—involves a very different field of actors than those associated with street vending and the optical disc trade. Much the same is true of Internet piracy, leading to different forums and policy coalitions, with outcomes that are harder to predict. The CNCP, in particular, is a less central protagonist in these debates.

---

43   AIPPI website, http://www.aippi.org.

44   NBO Editora (2009).

The Street Trade

Since the early 1990s, many of Brazil's largest municipalities have tried to formalize their local informal sectors, both as a matter of improving public order and in an effort to bring street vendors into the sphere of regulated and taxed business activity. The relocation of street vendors to centralized, better-policed markets—the so-called camelódromos—has been a central feature of these efforts. Because space in these locations has proved too scarce to accommodate all interested workers, these measures have also produced new forms of informality along this border of relative privilege, notably in the form of illegal markets for permits (Itikawa 2006; Mafra 2005). In addition to the publicly regulated malls, privately owned kiosk malls have also emerged as places where legal and illegal goods are traded.

Vendors, street markets, and malls trade in more than just counterfeit goods. Industrialized junk food, for instance, is one of the main products sold by street vendors (Gomes 2006:220). Regular business occurs alongside illegal activity. Accordingly, rights holders often pursue pirate vendors via regulation outside the framework of intellectual property or criminal law, including municipal ordinances related to zoning and construction. Such actions can disrupt or force the relocation of trade in pirated and counterfeited goods. Some of the more notorious malls, such as the Stand Center and the Promocenter in São Paulo, were closed through the enforcement of such ordinances (Bertolino 2007).[45] But in general, the informal sector remains highly fluid, and commerce can relocate quickly to other areas. In the past few years, Internet auction sites and online communities have also become a channel through which pirated and counterfeit merchandise is sold.

Despite the informality of the sector, street vendors are usually represented by associations through which they act politically and interact with government (Ribeiro 2006; Itikawa 2006; Braz 2002). Such politics is almost always local, and tensions between vendor associations and law enforcement authorities, politicians, and city hall are relatively common. Two CNCP projects focus on these local-level interactions: a program to coordinate law enforcement agents and rights holders at the municipal level (the Piracy-Free Cities project) and a series of efforts to more fully formalize vendor networks (the Legal Fair project). Thus far, these initiatives have had only limited impact. Ultimately, piracy and counterfeiting are only part of the larger set of commodity flows and social interactions that shape the informal economy. Ownership patterns and forms of regulation are quite diverse in these settings, as are the social relations that structure the operation of markets and malls. Markets with a permanent address, for example, have very different dynamics than the more precarious vendor networks that move from one location to the next—often in function of police or other municipal pressure. Even within superficially similar markets, such as the Uruguaiana market in Rio de Janeiro and the Campinas Camelódromo or

---

45   In 2008, the Stand Center's administrators and sixteen storeowners were sentenced to pay R$ 7 billion to the ABES for software piracy (CBN/O Globo Online 2008).

Galeria Pagé in São Paulo, relationships with police, municipal authorities, and supplier networks can vary widely.

Vendors often specialize in specific categories or genres of goods and sometimes in quite narrow categories such as the sale of video games for a particular platform or local music subgenres. The distribution chains in these settings also show considerable variation. Some vendors operate on a consignment basis and split profits with their suppliers at the end of each day. Other businesses are structured around wholesaler/retailer networks, while still others employ vendors for a fixed daily rate. Vendors can also operate their own production lines, acquiring blank discs, sleeves, and other materials from networks of suppliers but doing all the burning, printing, and assembly themselves (Pinheiro-Machado 2004:112; Mafra 2005:50–51, 92; Rodrigues 2008:87).

By most accounts, optical disc production is primarily domestic, small-scale, and decentralized in Brazil. As a vendor in the Uruguaiana market in Rio told us, "There's no Tony Montana here. It's not like in the drug trade." Rather than a handful of large, central producers of pirate discs, there are many smaller producers using equipment that is readily available on the consumer market. (In November 2010, duplicators for the simultaneous burning of eleven DVDs could be bought for R$ 1,390 ($806) through the auction site Mercado Livre.) Content is usually obtained from online sources and burned to low-priced blank media imported from overseas. In an attempt to break this supply chain, industry groups have campaigned in the past for price floors for blank media (IIPA 2005:67–68). But such proposals are controversial and face major challenges of implementation. Moreover, the window in which the pirate economy might have been vulnerable to such tax strategies is rapidly closing as distribution and consumption shift to all-digital channels.

THE TRI-BORDER AREA AND CHINA

Although there are nine points along the Brazilian border where three countries meet, industry, government, and media attention focuses overwhelmingly on the "Tri-Border Area" (TBA) with Paraguay and Argentina—the entry point, by most accounts, for a large share of the merchandise distributed in Brazilian street markets. Although clandestine networks and shipping undoubtedly play a role in this traffic, the border zone is also a common destination for organized shopping tours from all over Brazil. Brazilian citizens have a monthly tax exemption of R$300 ($180) to bring goods into the country for non-commercial purposes, and many act as laranjas (literally "oranges," meaning proxies) for sacoleiros ("baggers")—informal importers who supply camelôs (street vendors) with varied goods. The term commonly used for this activity is *contrabando formiga* (ant contraband), reflecting the small quantities of goods brought by a large number of border crossers.

Life and commerce in the TBA have been well documented in ethnographic work by Rabossi (2004), Pinheiro-Machado (2009, 2004), and others.[46] These accounts tend to emphasize the

---

46   Also of note is work by Davi (2008), Rodrigues (2008), Goularte (2008), and Martins (2004). A good source for academic information on the TBA is the Observatorio de la Triple Frontera website: http://www.observatoriotf.com/.

enormous social and political complexity of the TBA as a migration hub for diverse national and ethnic groups and as a recurring site of social and political conflict on many levels: between Brazil and other states; between laranjas and sacoleiros; between federal, state, and local authorities; between formal and informal workers; and among the culturally and ethnically diverse populations that live there. Descriptions of the TBA as a lawless haven for organized crime, such as those found in copyright industry reports, are not helpful in understanding the area or its many problems—piracy included.

A variety of supply chains serve the informal market in Brazil. Much of the traffic is sourced to China, which has become a major presence in the region's licit and illicit economies.[47] The China-Paraguay route is, by most accounts, the most important of these.[48] But not all counterfeited goods enter in this fashion. Brazilian borders are long enough to provide numerous points of entry for clandestine goods and persons. For some articles, such as counterfeit leather goods and clothing, production also takes place within Brazilian territory.[49]

ORGANIZED CRIME, TERRORISM, AND PIRACY

As in other countries, Brazilian industry groups make claims of connections between piracy, international organized crime, and occasionally terrorism. As elsewhere, such charges tend to trickle down from international reports into the discourse of local enforcement agents. Claims linking piracy and organized crime emerged in the late 1990s, driven initially by music-industry-group reporting on the illicit global CD trade (IFPI 2001). By the early 2000s, the IIPA was inveighing against the role of criminal organizations in the Brazilian trade, arguing that "organized crime elements, from within and outside Brazil, exercise control over the production and distribution of infringing copyrighted products" (IIPA 2001a:52). In 2010, the argument was much the same: "Organized crime is deeply involved in piracy in Brazil. Not only are Chinese and Middle East groups operating in the border with Paraguay, but they also control the distribution of pirate DVDs in the black markets at the end of a complex chain of command" (IIPA 2010:144).

Statements by public officials regarding organized-crime linkages have become more common since the Parliamentary Commission of Inquiry on Piracy (2003–4), which began the process of public adoption of industry discourse. Key Brazilian authorities have provided support for these accusations. Luiz Paulo Barreto, the first president of the CNCP and, more recently, the Minister of Justice, has argued that the main target of the government's anti-piracy efforts are the "big

47   According to the Brazilian Ministry of Development, Industry, and Foreign Commerce, China became Brazil's largest trading partner in 2009 (Ministério do Desenvolvimento, Indústria e Comércio Exterior 2010).

48   It is also a convoluted route: landlocked Paraguay relies on the Brazilian ports of Santos (São Paulo) and Paranaguá (Paraná) for overseas trade, guaranteed through agreements dating back to 1941 and 1957.

49   Including Minas Gerais and Paraná (O Estado de São Paulo 2009a).

mafias that have been established in Brazil, China and Korea" (Agência Brasil 2009)—not the street vendors who are, in actuality, the most common targets of police action. The current CNCP president, Rafael Thomaz Favetti, is even more forceful in stressing connections between piracy and organized crime.[50]

The TBA figures centrally in such accounts and especially in recent industry efforts to link piracy and terrorism. Alleged connections between Arab immigrants in the TBA and terrorist organizations have appeared periodically in the news since 1992, when unknown actors bombed the Israeli embassy in Buenos Aires. The bombing of the Argentine-Israeli Mutual Association in 1994 triggered another round of free association on the subject, even though evidence for viewing the TBA as a terrorist haven, as Costa and Schulmeister (2007:28) put it, is "meagre and imperfect." In contrast to claims of organized-criminal involvement, piracy-terrorism linkages in the TBA have been strongly contested by Brazilian authorities. No direct links between terrorist groups and criminal activity in the region have been established. Instead, accusations rely on purported remittances by Arab immigrants to groups such as Hezbollah (Amaral 2010).

PROVING THE CONNECTION

In interviews, informants in both the public and private sectors generally argued that there is some sort of connection between piracy and organized crime. But the lack of evidence offered for such assertions was consistently striking. In many cases, informants simply repeated charges found in industry advocacy materials, to an extent that made it clear that industry literature provides the main source of information on the subject. In these cases, links were presented as self-evident, often following "tip of the iceberg" reasoning in which street vendors are cast as the endpoints of vast distribution networks controlled by international criminal organizations. Some informants did try to build stronger cases, generally by describing piracy as part of the wider global traffic in illegal goods, from counterfeit clothing to cocaine, tobacco, and firearms.

The latter argument fits well within the "dark side of globalization" narrative advanced by books such as Moisés Naím's *Illicit* (2005). Naím, a prominent journalist and the editor of the American journal *Foreign Policy*, was mentioned explicitly by two informants with ties to the Federal Revenue Service, whose views had clearly been influenced by the book. Naím has been embraced by IP industry groups and has testified on their behalf before the US Senate.[51] In 2008,

---

50   Favetti has argued that piracy has "no social causes" and that "this idea that people who work in piracy are unemployed or doing odd jobs [to make ends meet] is no longer true. The confidential data we have from the police of Brazil [confirm] that piracy is controlled by organized crime" (Agência Brasil 2010).

51   Pirating the American Dream: Intellectual Property Theft's Impact on America's Place in the Global Economy and Strategies for Improving Enforcement: Hearing Before the US Senate Committee on Banking, Housing, and Urban Affairs, Subcommittee on Security and International Trade and Finance, 110th Congress, April 12, 2007 (testimony of Moises Naím, Editor in Chief, *Foreign Policy*), http://banking.senate.gov/public/index.cfm?Fuseaction=Hearings.Hearing&Hearing_ID=cd1f3746-1926-4da0-a8b7-6bfd6435e583.

the US Chamber of Commerce funded a National Geographic documentary based on *Illicit*. A wide variety of other industry-produced reporting also circulates in this space, including the Alliance Against IP Theft's Proving the Connection (n.d.), the IFPI's *Music Piracy: Serious, Violent and Organized Crime* (2003), and the voluminous MPAA-sponsored RAND report on *Film Piracy, Organized Crime, and Terrorism* (Treverton et al. 2009).

Much is at stake in "proving the connection." Domestically, linking IP enforcement to organized crime is a powerful way to elevate the issue in the eyes of government and public opinion. It is also a strategy for drawing the attention of law enforcement authorities to offenses that police and prosecutors normally consider less serious. Alleged links between piracy and organized crime were crucial for the inclusion of the IP enforcement agenda in certain law enforcement circles. The Civil Police unit in charge of investigating piracy in São Paulo, for instance, is hosted by a department specialized in organized crime, the DEIC; the national public-prosecutor workgroup the GNCOC was created to act against criminal organizations but eventually included piracy in its list of concerns.

Internationally, organized-crime and—still more significantly—terrorist linkages introduce piracy into the circuit of policy communities and fora dealing with bilateral and multilateral security. Pressure for stronger enforcement agreements at the World Customs Organization, and now in the recent Anti-Counterfeiting Trade Agreement (ACTA), reflect this new fusion of security and IP discourse. Tying online piracy to digital threats—identity theft, child pornography, bullying, cyberwarfare, and so on—is an application of the same strategy to fight file sharing. Industry's involvement in the Azeredo Bill debate is a clear example of this shift.

ORGANIZED CRIME AND BRAZILIAN LAW

There is little consensus about the definition of organized crime—"an ever-changing, contradictory and diffuse construct," in the words of one of its scholars (von Lampe 2008:7). But in law, the concept exists primarily to set boundaries for the use of exceptional legal regimes to target criminal activities that the regular instruments of law enforcement have difficulty addressing. The application of organized-crime statutes generally requires less concern for the rights of the accused than in less serious offenses and allows for more invasive procedures, such as wiretapping.

Copyright infringement has not traditionally fallen under the umbrella of organized-criminal activity, either in terms of the application of the law or—equally important—in terms of the administrative organization of policing and enforcement. As a consequence, industry groups have worked to expand definitions of organized crime to encompass copyright infringement and to produce accounts of piracy that emphasize alleged connections with more conventional forms of criminal activity. The two processes are linked: the broader the definition, the easier it is to prove the connection.

The *Proving the Connection* report, for example, concludes that "the imperative is not to over-elaborate the term, rather to emphasize that it describes 'a group or network focused on illegally

Views on Piracy and Organized Crime

"It's a mob thing. The business of piracy [is] to sell corruption logistics: 'I can guarantee that your container will arrive there.' And then I sell [container space] to the toys guy, to the weapons guy, to the drugs guy, to the CD guy, to the DVD guy, to the software guy, to the clothing guy, to everyone. Because even [São Paulo luxury boutique] Daslu had to find a way to get its underpriced clothes to Brazil. How does one do that? How can we get [clothes] through ports, airports, escape routes, and not be searched? To profit as much as they did, what did they do? They brought stuff from abroad. Everybody pays 'X'; they pay 'X divided by 10' and get taxed over that value. [Goods] arrive here cheap as hell. While if I am to manufacture [the goods] at the Manaus Free Zone, I need to hire staff, pay labor dues, pay taxes, all that. Or import legally." (Private sector, consultant)

"You have large-scale, international piracy; big criminal organizations, like in China. Lots of pirate goods come from China, as they manufacture them on a large scale. . . . And then you have domestic piracy, don't you, with small salesmen who practice this kind [of piracy] both in manufacture and in commerce." (Public sector, enforcement)

"Well . . . people insist on that a lot, don't they? They say there's a link between piracy and organized crime. I think the link exists, but it's not the rule. I've done fieldwork in piracy. There are lots of small fish doing counterfeiting. Guy has his computer, records a CD, a DVD, and then goes peddling on the streets, without the backing of any criminal organization. Of course, there are [also] powerful people behind [piracy]. The relation occurs more frequently in frontier regions. We often seize cargo and find cigarettes, drugs, pirated goods, CDs; so that shows that the same people doing [drug] traffic are bringing guns to the country, bringing pirated goods. This is the relationship I see with organized criminality." (Public sector, enforcement)

"It's not like what many people say, 'You guys are stretching it, there's no connection.' If we follow the reasoning that piracy is an informal activity, drug trafficking is [also] an informal activity. All the money, the resources that are moved informally, the government has no knowledge, no idea of the money that runs through these channels. We can't, initially, discard the possibility that this connection exists. Both drug trafficking and piracy are activities that run through informal channels. . . . So there is a connection. Resources run through channels that government has no knowledge of. We can't affirm that all piracy is connected with the drug trade, but a big part of it is, and we've been noticing that everyday." (Private sector, enforcement support)

Views on Piracy and Organized Crime, Continued

"Every crime of piracy is, at the end of the chain, linked to more active, broader criminal organizations with all sorts of businesses, involving, among others, drug dealing, arms contraband, and organized crime, terrorism. It's something that's very serious. So piracy is a very powerful business all over the world; it moves five hundred billion dollars all over the world, more than drug traffic. Therefore, piracy is absolutely connected to other types of crimes and is a branch of broader criminal organizations." (Private sector, film industry)

"This connection is not well proven. Since both [drug] traffic and black market deal with illegal goods, it's natural for piracy to fit within this activity. However, this connection is mostly stressed by media and rights holders in order to capture public attention, [to imply] that if you buy a pirate good, you're contributing to your son's access to drugs. Actually, this connection is very, very weak. It's, as I say, sensationalism, a sort of publicity to call attention to piracy's negative effects, when in reality [links to the drug trade] are not well established." (Private sector, consultant).

obtaining profits in a systematic way, involving serious crimes with societal consequences'" (Alliance Against IP Theft n.d.:4). The RAND study, for its part, recommends "expanding the definition of organized-crime statutes to include commercial-scale piracy and counterfeiting tied with other criminal activity" (Treverton et al. 2009:145). The IIPA demanded, in its April 6, 2001, post-GSP-hearing brief on Brazil, that the Inter-Ministerial Committee on Combating Piracy draft and propose legislation supporting "the principle that medium- and large-scale piracy falls within the definition of an organized crime scheme" (IIPA 2001b).

Brazilian law has no definition of "organized crime" or "criminal organization" (see, generally, Pitombo 2009), despite the existence of legislation authorizing special means of evidence gathering and investigation for acts practiced by "quadrilha or bando" or "organizations and criminal associations of any kind" (Law 9.034/95, Article 1). There is no definition of what count as "organizations and criminal associations of any kind." A "quadrilha or bando" is any lasting association between three or more people with the intent of committing crimes—a criterion broad enough to encompass virtually all aspects of the pirate economy, from large-scale smuggling operations to small-scale vending. Membership in such groups is a crime in and of itself (Penal Code, Article 288).[52]

---

52   Beyond the primary criminal statute, Law 9.034/95, the Brazilian law on drug trafficking (Law 11.343/06) refers to the concept of "organized crime" (Article 33, §4) for the purpose of authorizing judges to reduce the penalties of informers who testify against former associates. The law establishes stronger penalties and a looser threshold for criminal association, now defined as involving two or more people. No more substantive definition is provided. Brazil is also a signatory to the United Nations Convention against Transnational Organized Crime (the Palermo Convention), but the convention's definition of "organized criminal group" applies only to criminal activity that occurs in a transnational context and focuses on the practice of crimes defined as "serious" (meaning, according to Article 2(b),

We are more sympathetic to narrower definitions that emphasize provable links to larger criminal organizations, such as the Camorra, the Yakuza, local or international drug cartels, Brazil's Comando Vermelho, and so on. We see little systematic evidence of these connections to date. Advocacy pieces, for the most part, rely on cherry-picked examples to make the broader case and offer grossly simplified accounts of the dynamics of street markets, street vendors' relations with local authorities, and other features of the informal economy. Industry public awareness campaigns are often the most flagrant in this regard. A recent UBV anti-piracy video, for example, portrays a closed economy tying together drug dealers, street vendors, and consumers of pirated media. Such allegations are important because they associate copyright infringement with "crimes that the public are really scared of" (Drahos and Braithwaite 2002:27), as distinct from the mundane acts of copying and informal commerce in which they routinely engage.

Street vendors involved in media piracy, for their part, tend to view organized crime through a similar lens and often take offense when accused of such associations. None, in our interviews, viewed the simple fact of the organization of the supply chain as significant, and several characterized piracy as an alternative to activities such as drug trafficking. Our interviews in the Uruguaiana market in Rio strongly confirmed earlier ethnographic work on this front (Mafra 2005:94; Gomes 2006:229; Braz 2002) and echo the views of Mexican vendors presented in the next chapter in this report.

LAW KIM (KIN) CHONG

The arrest of smuggler/businessman Law Kim (or Kin) Chong during the initial parliamentary inquiry in 2003–4 became a touchstone in the Brazilian conversation about piracy and organized crime. A naturalized Brazilian citizen of Chinese origin and the owner of several stores in popular shopping centers in São Paulo (O Estado de Sao Paulo 2009b), Chong came to public attention in 2004 when he was arrested while attempting to bribe Deputy Medeiros, the president of the Parliamentary Commission of Inquiry on Piracy, in an effort to buy protection for his businesses (Rizek and Gaspar 2004). Since 2004, he has had continuous problems with law enforcement and São Paulo municipal authorities. He was arrested again in 2007 and 2008 under accusations of smuggling, tax evasion, and money laundering (G1 2008a, 2008b). Chong served a brief jail term in 2004; the 2008 case was dropped by prosecutors for lack of evidence.

Called a "notorious piracy kingpin" by the IIPA (2006:208), Chong has become the emblematic figure of the Brazilian anti-piracy effort. He has been frequently used as an example of the connections between piracy, counterfeiting, and organized crime and is featured regularly in news articles on these issues.[53] Constant reliance on the Chong case, however, also tends to highlight the

---

"an offence punishable by a maximum deprivation of liberty of at least four years or a more serious penalty"). It is thus significantly narrower than the crime of criminal association contained in Article 288 of the Brazilian Penal Code.

53   Chong has also served as a convenient personification of Brazilian anxiety toward Chinese immigration and China's rise as an economic power. Usually implicit in the coverage of the case, such anxiety

lack of a wider case for linkages between organized crime and piracy.

There are many reasons to doubt the broader account of connections, chief among them the rapidly diminishing role of transnational smuggling in the piracy of media goods—in distinction from other kinds of hard goods. The IIPA began to describe this shift as early as 2001, observing that "there are also growing numbers of small duplication facilities which assemble CD burners" in the country and that the "MPA has noted the beginnings of optical disc piracy, previously not present in Brazil" (IIPA 2001a:55–56). The following year's report observed that "piracy has changed from an international industrial profile to a domestic semiprofessional effort." While looking for other ways to invoke organized-crime discourse, they noted that "the distribution of product, however, remains highly organized" (IIPA 2002: 76). The conflation of piracy and counterfeiting in most anti-piracy discourse is also very unhelpful in this context. Of the major figures raised into the media spotlight by enforcement efforts in the past decade, including Roberto Eleutério da Silva and, more recently, Paulo Li, only Chong has been accused of media piracy. Da Silva is best known as a cigarette smuggler (Castanheira 2003). Li is associated with cell phones and electronics (Rangel 2010). In our view, the most compelling case for connections between street piracy and transnational groups involves the trade in blank discs, which most accounts describe as originating largely in China. But we see little evidence of more systematic connections beyond the blank disc trade, and the Chong case alone does not provide evidence of wider linkages between the pirate economy and organized crime.

Internet Piracy

Despite constant complaints about Internet piracy in IIPA reports going back at least to 2001, Brazilian enforcement policy has strongly emphasized hard-goods piracy and commercial-scale infringement over the past decade, consistent with long-standing interpretations of the threshold for criminal liability under TRIPS. This has included a number of actions directed at Internet-based commercial infringers, notably through big operations such as the I-Commerce I and II operations of the Federal Police, which targeted commercial infringement through online channels.[54] But until very recently, it has not involved the expansion of enforcement activity to the wide variety of sites that enable individual sharing of files, from BitTorrent trackers to file locker

---

sometimes shades into xenophobic stereotypes, as in Deputy Medeiros's book about the Parliamentary Commission of Inquiry on Piracy, *A CPI da Pirataria*. In the book, Chong is described as a "cold little Chinese man" and a "moral monster" with "painfully slanted eyes," but also a "predictable" man due to his "millennial obedience" (Medeiros 2005:96–97).

54   In 2006, Operation I-Commerce I, targeting the sale of pirated goods over sites such as Mercado Livre and Orkut, had the Federal Police executing seventy-nine court orders for raids in thirteen states and in the Federal District of the Brazilian federation. Thirteen individuals were arrested in flagrante delicto, and fifty-seven others were formally accused (Tourinho 2006). 2008's Operation I-Commerce II had the same objective, with forty-nine court orders across nine states and the Federal District, mobilizing two hundred police officers (IDG Now! 2008).

services. File sharing became an explicit government concern only in 2008, when, in response to recording industry demands for more government involvement, the Ministry of Culture convened their ISP working group to develop an agreement on the implementation of a graduated-response system in Brazil.

At present, there are no specific liability rules for ISPs. In practice, however, rights holders, ISPs, and other Internet services have adopted a set of informal norms around notice and takedown that has proved very compliant with industry demands. ISPs have typically acted quickly in response to takedown requests (IIPA 2010:148–49). Major auction sites like Mercado Livre regularly comply with industry notices when infringing goods or advertisements for those goods are found on their sites (Nintendo 2010). Industry groups have generally expressed satisfaction with the level of cooperation from Brazilian ISPs and other services in such cases.

Peer-to-peer file sharing poses a very different problem. With P2P technologies, infringing content is hosted on users' hard drives, not on central servers. Most such systems require only minimal intermediation by torrent-tracker sites or their equivalents (such as ed2k links or Direct Connect servers, both of which are popular in Brazil), raising questions about the liability threshold for P2P services.

The recording industry is the leading actor in the campaign against file sharing in Brazil. The film industry is definitely on board but still sees hard-goods piracy as its main target (IIPA 2010), in large part due to the higher bandwidth and technology requirements of video piracy over the Internet. In 2006, Brazilian record industry group the ABPD (with the support of the IFPI) tested the waters of personal liability by initiating some twenty civil suits against individual file sharers, based mostly in São Paulo. The cases ran into difficulty almost immediately when the ABPD could not obtain personally identifying information from the relevant ISPs based on IP addresses collected through surveillance of P2P sites (IIPA 2008). In one case, a judge refused to grant the orders to force the ISPs to release the data. In another, the orders were granted, but the ISPs had already purged the data, as there was no legal obligation to retain it. Under such circumstances, efficient pursuit of individual file sharers was (and remains) effectively impossible. In 2008, industry support consolidated around the Azeredo Bill, which mandates three-year data retention and ISP cooperation in releasing personal information. Shorter data-retention requirements are likely to become law through the Marco Civil process. To date, government commitment on privacy issues, however, remains strong, and more sweeping measures, such as the "three-strikes" law proposed in 2009, seem unlikely.

As elsewhere, Brazilian enforcement faces growing difficulties as the use of file sharing technologies proliferates. LAN houses, as mentioned, have been an important means of both Internet access and Internet piracy for lower-income Brazilians and enable both local network sharing and an extensive culture of face-to-face transfers of physical media—the so-called sneaker-nets (Biddle et al. 2002).[55] And although we have seen no conclusive studies of the subject, the

---

55   The IIPA (2008) cites record industry studies attributing 20% of online piracy to cybercafes.

use of P2P systems is clearly complemented in Brazil by the widespread use of other types of services, including file locker sites such as Megaupload, 4shared, and RapidShare. A variety of community sites and online forums index and link to material stored on these services, creating large communities sharing music, film, books, and software. Industry groups now issue a constant stream of takedown notices to the owners and administrators of blogs and community sites.

DISCOGRAFIAS AND ORKUT

One of the biggest file sharing skirmishes, thus far, has taken place around Discografias—a large community site devoted to music sharing. Discografias is not a P2P site but rather a community built on Orkut, a Google-owned Facebook competitor and the most popular social networking service in Brazil.[56] Within Orkut, any user can set up a public or private "community," consisting of a list of members and related communities, a message board, and a simple tool for polling. Although limited, these features have proved rich enough to enable large communities to consolidate around shared interests.

The Discografias community specializes in the sharing of music through links to external storage services (for example, file locker sites like RapidShare, Megaupload, and 4shared). Strictly speaking, Discografias is a hub for other communities with more specialized roles in the music-sharing process. One community, for example, is used for the posting of requests for particular songs or albums, which are then answered in the main community. Another is used as a general index.

Founded in 2005, Discografias had, by most estimates, a community of 921,000 registered users by early 2009 and almost certainly a much larger casual-user community: membership is not required for reading the boards and accessing the download links (Muniz 2009a). In March 2009, however, its moderators received a cease-and-desist notification from the APCM, requesting the removal of links to infringing content. The moderators decided to comply with the APCM's request (Pavarin 2009a), and virtually all the content on the site was deleted. Paulo Rosa, director of the ABPD, celebrated with the following statement: "The closure of the illegal file sharing network is a significant step in curbing online piracy in Brazil. The communities affected represented the largest user group in a social network dedicated to exchanging links for the purpose of illegal copying" (IFPI 2009b).

The APCM action succeeded in fragmenting the Orkut music-sharing community but not in destroying or even significantly dissuading it. Several similar communities were created immediately to reconstitute the sharing network. Members of the original Discografias community also quickly rebuilt the database of links and soon launched Discografias-A Original (Discographies-The Original) (Folha Online 2009). By April 2010, membership in the new Discografias had

---

56   According to an IBOPE Nielsen estimate, Brazil had an astonishing 39.9 million active Orkut users in August 2010 (Aguiari 2010).

grown to some 760,000 members.[57] Another result of these skirmishes is more clearly articulated resentment of the APCM. In November 2010, the I Hate APCM community had almost 12,000 members.

Fan Communities and Subtitling

Content-distribution networks for film and video in Brazil are strongly grounded in fan culture. Fans have structured themselves into networks of communities engaged in the translation and distribution of foreign movies and, especially, TV shows. These translator (or "legender") communities have adopted many of the practices of "fansubber" and "scanlator" communities for Asian anime and manga and now often compete to adapt specific series or types of content into Portuguese.[58] The translating work in such communities is resolutely non-commercial: the main incentive is prestige.

Fan-based subtitling tends to be extremely efficient. As soon as a new episode of a popular TV show hits the Internet, typically immediately following its initial broadcast, teams start working on subtitles. A completed set of titles for an episode can be done within four hours of the original broadcast (Agência Estado 2008)—with revisions and refinements processed over the following days. There are at least thirty teams of legenders actively working in Brazil and many independent translators working on their own. Most of those involved meet to publish, discuss, and polish their work on a website called Legendas.tv. Legendas.tv neither hosts nor links to infringing video. It only distributes subtitle files in formats, such as .srt, which contain the text and timing for the translations and which are easily integrated into common video files like .mkv and .avi.

The two pieces—the copied videos and their subtitles—come together on other sites, such as those hosting Orkut communities, online forums, and other file sharing communities. These often link to both the subtitle files and the content they refer to, sometimes conveniently packaged. All these communities overlap considerably: it is difficult to separate the activity of subtitling from the ecosystem of file sharing and fan culture. Regular participants in private or public file sharing communities may be, at the same time, respected members of subtitling groups, active posters on Web bulletin boards, avid purchasers of original DVDs, and file hoarders with extensive collections

---

57   In November 2010, the community was deleted once again and recommenced the process of rebuilding. Searching for "discografias" among Orkut communities reveals many interrelated communities, some of them very specific, such as Justin Bieber-Discografia (9,000 members as of November 2010).

58   Anime-translator communities based in the United States have engaged in heated debates about the relation of their work to copyright, and some of the larger ones limit their activities to work that has not been licensed for official distribution. Implicitly, and sometimes explicitly, these communities envision themselves as building a global market for anime—with some justification. The extensive pirate fan base led to wider licensing practices beginning in the early 2000s, though these are still only infrequently extended beyond the United States and other major markets. Brazilian fan communities generally translate foreign content regardless of such considerations. When justifications are called for, they usually involve the weakness of Brazilian distribution, which includes broadcasting delays and the limited availability of many categories of legal media goods.

of pirated films, songs, TV shows, and software. These communities are an example of the wider shift from consumers to users that Yochai Benkler (2000, 2006), among others, has written about.

Conflict between industry and subtitlers in Brazil dates back to 2006, when the ADEPI (Association for the Defense of Intellectual Property), a predecessor of the APCM, sent cease-and-desist notifications to Lost Brasil, the largest community site for fans of the ABC TV show *Lost* (Mizukami et al. 2010). Because dialogue is covered under Brazilian copyright, any unauthorized transcription is an infringement. The APCM has continued to target subtitler communities, such as Legendas.tv and subtitling team InSUBs. Because these sites were originally hosted in the United States, the APCM was able to serve their Internet hosts with takedown notices under the US Digital Millenium Copyright Act. Both sites were closed and quickly re-established on servers outside US territory. Both actions produced a significant backlash among fans, culminating in an attack on the APCM's website.

As of late 2010, both Legendas.tv and InSUBs continue to operate, along with several other related websites. There have been no further legal actions announced against subtitlers.

Book Piracy

After several years off the USTR radar, concerns about Brazilian book piracy reappeared in the Special 301 reports for 2008, 2009, and 2010. The issue is not a new one, of course. The photocopying of books and articles has been common in Brazilian schools and universities since copy machines became widely available in the 1980s. By nearly all accounts, the phenomenon is highly concentrated in and around higher education, fueled by high book prices, inadequate library collections, and the narrow range of in-print works in Portuguese (Craveiro, Machado, and Ortellado 2008; IDEC 2008).

Brazil is hardly unique in this regard. Academic book copying has been a basic feature of higher education in most developing countries (see, for example, the South Africa chapter in this report) and a basis for recurring international conflicts over copyrights and enforcement since the seventeenth century (see chapter 9). But where most university and publisher groups have sought to compromise—rhetorically, if not always in practice—the Brazilian situation is distinctive for having devolved into a serious and ongoing conflict (Mizukami et al. 2010).

Poor copyright legislation set the stage for this conflict. Article 46, II, of the Copyright Act of 1998 (Lei 9.610/98) allows for a single copy of "small excerpts" of works in the case of personal copying—in other words, when the beneficiary of the copy is also the party doing the copying. There are, however, multiple interpretations of what constitutes a small excerpt, ranging from 10% at some universities (Craveiro, Machado, and Ortellado 2008) to 49%, according to one legal scholar (Pimenta 2009:80).

The ABDR, a Brazilian publishing industry group, has taken an almost uniformly hard line on licensing and fair use. The ABDR revoked all its licenses to university copy shops in 2004 and ramped up its enforcement, copyright advocacy, and lobbying efforts. Searches and seizures against university copy shops were particularly aggressive in 2004 and 2005 and continue to this day.[59] ABDR members also print their own warnings and legal interpretations in books, typically claiming that even the reproduction of small excerpts is illegal, often with misleading references to the article in the law that explicitly does authorize excerpting (Mizukami et al. 2010; Souza 2009; Mizukami 2007). Within the larger enforcement network, this idiosyncratic enforcement agenda has produced few opportunities for collaboration with other industry groups and resulted in the ABDR's loss of its seat on the CNCP when its term expired.

With the exception of two problematic proposals we describe later in this section, the ABDR has resisted discussing new business or licensing models that might alleviate the problem of access to materials in Brazilian universities. It is also strongly opposed to the copyright law reform bill proposed by the Ministry of Culture (ABDR n.d.b), which offers a broader and, above all, clearer list of exceptions and limitations. Since negotiations with the ABDR have reached a stalemate, copy shops and universities have generally gone their own way, either adopting a zero-tolerance policy against copying books and articles, even when permitted by law, or creating their own parallel interpretations of copyright law.

In 2005, three universities in São Paulo stepped into this debate by issuing policies that clarified and expanded the scope of the private-copy exception, effectively establishing their own de facto copyright regime (they were joined by a fourth university, in Rio de Janeiro, in 2010). The new university policies stick close to Brazilian law but offer much broader interpretations of what constitutes a small excerpt. The University of São Paulo (USP), for example, interprets it to include full book chapters or journal articles. Additionally, it has authorized users to fully copy works that have been out of print for more than ten years, foreign works that are not available in the Brazilian market, works in the public domain, and works whose authors have expressly granted authorization to copy. The last two are unquestionably legal to copy; the first two are not. Pontifical Catholic University of São Paulo (PUC-SP) and the Getulio Vargas Foundation-São Paulo (FGV-SP) adopted resolutions similar to USP's (Jornal PUC Viva 2005; PublishNews 2005).

In 2005, students from the three institutions formed a short-lived movement to support the university policies: To Copy a Book Is a Right (Copiar Livro é Direito) (Magrani 2006). In September 2010, a raid by the DRCPIM on the copy center of the School of Social Services at the Federal University of Rio de Janeiro (UFRJ) provoked a similar resolution by that university, authorizing the full copying of book chapters and articles (Boghossian 2010).

These university interventions in the copyright arena have been controversial, to say the least. Predictably, they have provoked the anger of the ABDR and, through the ABDR, come to the attention of the IIPA and the USTR. Pressure from the IIPA, visible in its reports since 2005, has especially targeted the University of São Paulo, which, as a state institution, is under the

---

59   These shops are typically independent, formally constituted businesses but are sometimes run by student organizations or by the universities themselves, through their libraries.

jurisdiction of the Ministry of Education. Attempts to pressure the Ministry of Education into taking action against photocopying date back at least to the Parliamentary Commission of Inquiry on Piracy, which published a formal recommendation on the subject (Câmara dos Deputados 2004: 276–77). But the ministry has resisted acting and has stayed out of the campus policy battle. It shows no signs of changing its position. Interviews with officials from several ministries revealed considerable resentment of industry's framing of infringement in educational contexts as piracy. The deliberate blurring of terms, including the conflation of copyright infringement with other forms of hard-goods trafficking and public-safety hazards, appears to have backfired at this level.

Stepping back from the details of the fight, the Ministry of Education has ample basis for circumspection and an understandable reluctance to endorse the idea that educational copying is a form of theft from the private sector. The government is the single biggest buyer of textbooks in Brazil through five ministry programs (Cassiano 2007). And according to one recent study, roughly 86% of the books written in Brazil and used in the country's universities have benefited from investments made by the public sector (taking into account the variety of ways in which the university subsidizes research and writing) (Craveiro, Machado, and Ortellado 2008:28).

Although the cited dollar amounts are trivial, with the IIPA's estimated losses implausibly stuck at $18 million for much of the past decade, the university breakaway represents a challenge to an industry enforcement campaign built on unanimity of rhetoric across a wide public front, if not actual uniformity of practice.

The ABDR and the publishers see other threats as well. The illegal sale of teachers' versions of textbooks is one area of contention and forms the basis of an anti-piracy campaign by textbook-publisher association ABRELIVROS. Digital piracy, for obvious reasons, is also a growing concern. The ABDR recently created a unit specialized in identifying infringing files and sending takedown notices, following the example set by the APCM.[60]

---

60   Generally unmentioned in these complaints are the numerous charges against Brazilian publishers for publishing plagiarized translations of public-domain literature. Translator Denise Bottman's blog extensively documents the subject: http://naogostodeplagio.blogspot.com/.

Research, Training, and Education

"Many pirate products cause serious harms to health. Are you aware of this information?"
— *A 2008 survey question*[61]

It is impossible to analyze industry research, training, and education programs in Brazil outside the context of IP advocacy. Despite pretensions to objectivity, nearly all these programs are efforts to convince authorities and consumers of the harms produced by piracy and counterfeiting and, conversely, of the benefits of strong IP protection and enforcement. As an ensemble, they are designed to produce a stronger "culture of respect" for intellectual property and a collective hardening of attitudes toward piracy. This is, above all, a campaign of knowledge and ideas, built on efforts to define the terms of the piracy debate. It is also a multimodal campaign that comprises research, outreach in schools and among professional groups, media campaigns, and the very effective capture of print and broadcast journalism, which has made press releases, photo ops, and industry-generated stories into staples of Brazilian news coverage.[62]

There is relatively little about the content of these initiatives that is uniquely Brazilian. Nearly all borrow heavily from international templates, marking another side of the international coordination among industry groups. The strong moralization of anti-piracy discourse is present throughout, whether directed at children or filtered through nationalistic accounts of economic development. The strategic conflation of terms is there, too, notably in the effort to boost the harms attributed to piracy through association with the more dangerous forms of counterfeiting and criminal activity. And the endless gaming of numbers and statistics is there, with a range of local actors producing a circular and opaque Brazilian discourse on piracy losses. The progressive (and increasingly official) undermining of these claims in international contexts and the gradual pullback of the industry from new research has done little so far to stem their use in official Brazilian circles.[63]

61　From *Pirataria: Radiografia do Consumo* (The Consumption of Pirate Products), commissioned by Fecomércio-RJ (the Rio de Janeiro Federation of Commerce) and conducted by Ipsos in 2008.

62　This section is informed by the gathering of over five hundred news articles focused on the following themes: (1) arrests of street vendors and individuals engaging in the mass duplication of copyrighted content, (2) alleged connections between piracy and organized crime, (3) training and education of law enforcement agents and the public, (4) legislative proposals to strengthen the IP enforcement legal framework, (5) copyright reform, (6) opinions from content producers and researchers on piracy, (7) industry losses, and (8) new business models conceived to deal with the problem of piracy. An unpublished FGV Opinion report analyzing news collected over the period of May through September 2008 also served as a source. No effort to quantify the occurrences of these topics was made; news articles were collected solely for qualitative analysis.

63　This report echoes the growing official skepticism of industry research found in recent Organisation for Economic Co-operation and Development and US Government Accountability Office reports.

The results of these programs and the overall campaign are somewhat contradictory and, we would argue, in flux. Domestically, Brazilian government and industry discourse has converged in recent years around "the fight against piracy" at exactly the moment when the evidentiary discourse around piracy has been delegitimized. We saw this repeatedly in interviews with public officials involved in the enforcement effort, who often discounted industry claims about losses while holding on to the purposes of the anti-piracy agenda. We also see it in the apparent heterogeneity of Brazilian positions in different policy contexts—notably in the disconnect between the domestic enforcement debate and Brazil's international policy positions on intellectual property, which have been sharply at odds with industry and US wishes at WIPO, the WTO, and other global forums. This latter subject elicited considerable disagreement in our interviews and is examined in more detail in the following pages.

Anti-Piracy and Poetic License

For a subject that elicits so much public attention, definitions of piracy in Brazilian law are surprisingly scarce. In fact there is only one, in the decree that established the CNCP in 2004.[64] Even here we don't learn much: the CNCP decree simply states that piracy is understood as copyright infringement. For a definition of copyright infringement, the decree points back to Laws 9.609 and 9.610 of 1998—the Brazilian software protection and copyright acts, respectively. There is no definition of counterfeiting in the decree—an odd omission for an institution largely focused on counterfeiting, but a telling one given the CNCP's persistent conflation of the two terms.[65]

Nonetheless, the two terms are clearly distinguished in Article 51 (footnote 14) of the TRIPS agreement—the primary framework for international law on copyright and enforcement. TRIPS ties "counterfeiting" to trademark infringement and "piracy" to copyright infringement and uses that distinction to anchor the different protections and enforcement regimes applicable to different types of goods. Goods can infringe one or the other, or in some cases both, when the good reproduces both the expressive content and the brand of an original.

The conflation of the terms in industry discourse is not accidental. It is used to tie copyright infringement to a wide range of public-safety and health hazards associated with counterfeit medicines, toys, and other substandard goods, and it allows industry research to paper over serious gaps in the evidentiary record around copyright infringement—a subject we discuss at more length later. As we have argued repeatedly in this report, the first but by no means only problem with such conflation is that the practices that define piracy and counterfeiting have largely diverged as the pirate economy moves toward cheap, personal digital-reproduction technologies.

---

64 Decree 5.244/04.

65 Law 9.610/98, the Brazilian authors' rights law, defines contrafação (counterfeiting) as any unauthorized reproduction of protected content. This definition was inherited from older legislation——it was already part of Law 5.988/73——created for a technological context in which the physical good mattered more than the digital. It is not in compliance with the TRIPS definition.

The weak legal and factual basis of this conflation is no secret among public- and private-sector actors in Brazil's enforcement debate. In our interviews, private-sector informants were aware that "piracy" and "counterfeiting" are different in law and on the ground but generally had no qualms about using "piracy" as a catchall term. Speaking about industry awareness campaigns, one informant argued, "If we want to develop anti-piracy values, then a DVD is as important as medicines, as adulterated fuel, or any other [counterfeit] product, whether clearly identified or not."

Even crimes that may be only circumstantially related to IP infringement, such as smuggling and tax evasion, get pulled under the piracy umbrella. As a different private-sector informant described it, "[Piracy] is not a technical term, it's not a legal term. It's a colloquial term that people understand and that the country understands. So we take up [the term] 'piracy' and mention 'illegality' next to it, under poetic license, so we may be understood. When you speak about 'piracy,' everyone understands what that is."

Informants from the public sector directly involved in IP policy debates were more cautious. The TRIPS definitions matter to them, and they are concerned with setting a clear boundary between the two terms. As one official noted, "So there's this confusion. From a legal standpoint—international, even—there is a very explicit definition [of piracy]: infringement in the field of copyright law. Domestically, this term has been used in a much broader fashion, even beyond intellectual property."

In its public communications—and even in its use of its name—the CNCP actively propagates this confusion. Although officially named the National Council on Combating Piracy and Intellectual Property Crimes, the CNCP typically drops the "Intellectual Property Crimes" from its title. Press releases usually avoid the word *contrafação*. The Brazilian press, predictably, has picked up the more colorful terminology and extended it further, applying it to virtually any form of fraud or sale of illegal goods. In a typical example, Folha Online, the news website of media conglomerate Grupo Folha, applies the term *piratas virtuais* (virtual pirates) to con artists responsible for "phishing" schemes involving fake online sales or banking sites (Carpanez 2006).

Creative misuse of piracy terminology extends to much of the industry research conducted in Brazil. Here, the conflation also has practical value: it allows for results to be used by more than one industry sector, creating a simplified, self-reinforcing discourse about various types of losses. Rates of piracy or losses due to piracy, in these contexts, commonly refer to "piracy and counterfeiting." (This is the case, for example, in the Ipsos and IBOPE surveys discussed later in this chapter.) We explore this in some detail in the following pages, in the context of alleged job and tax-loss numbers.

To Repress and Educate

With the creation of the CNCP and the drafting of the National Plan, all underlying questions about the goals of intellectual property protection and enforcement were swept under the rug. Enforcement policymaking became, on the surface at least, a discussion about which anti-piracy measures were most effective. At the CNCP, these measures were divided into three categories: repressive, economic, and educational.

According to a CNCP informant, the initial negotiations over the National Plan were acrimonious, with public- and private-sector actors trading accusations about who was most responsible for Brazil's high piracy rates and ineffective enforcement. Public-sector actors blamed the unwillingness of the private sector to develop lower-cost business models or to bear more of the burden of investigation. Private-sector actors blamed the inefficiency of the courts and the inability of the police to fully enforce the law. As one public-sector official put it, "Intellectual property rights are private rights. So rights holders, when rights are infringed, can resort to the judiciary to enforce those rights. But it's another thing to create legislation obligating the state to permanently monitor if that right is being enforced."

During the drafting of the plan, the public and private sectors agreed that strictly repressive measures—raids, seizures, arrests, and lawsuits—would not be sufficient to deter piracy.[66] Repressive measures would have to be complemented by economic measures—a gesture toward the range of business-model, tax, and licensing issues that shape markets for goods. There would also have to be new educational measures designed to raise consumer respect for intellectual property. Every agent involved in anti-piracy work interviewed for this report referred to these three categories, even when critical of some of the assumptions of the National Plan.

There was much less agreement, however, on the appropriate balance between the three types of activity. Much of this tension remains unresolved, with the larger consensus providing cover for ongoing disputes over the division between public and private responsibilities. As one public-sector official described it:

> For rights holders, the tendency is always to want to strengthen rights and to ensure that those rights are enforced in some way. So it is important to stress that intellectual property rights are essentially private rights. Does the state have an interest [in enforcing those rights]? Yes, of course the state has an interest, but private parties must also assert those interests before the state [by conducting investigations and filing complaints].

---

66   The first CNCP report and the submission of the CNCP's then executive-secretary Márcio Gonçalves to the third meeting of WIPO's Advisory Committee on Enforcement also mention a fourth group of "institutional" measures, which are not defined but are described as legislative reform that would facilitate enforcement (Ministério da Justiça 2005; Gonçalves and Canuto 2006). The use of this fourth category has since been abandoned. It still figured in the second CNCP report, also without definition, but was dropped for the third report (Ministério da Justiça 2005b:62; 2006).

The substance of this disagreement is complicated in Brazil. TRIPS makes it clear that intellectual property rights are fundamentally private rights, to be enforced in most cases by the rights holders themselves through civil action. In most countries, industry associations hire networks of private investigators and lawyers to identify infringing activity and file complaints with the police, and it is the complaint that triggers law enforcement involvement. In Brazil, the criminal status of copyright infringement makes the public/private distinction largely moot with respect to print and audiovisual goods.[67] The state has, in principle, assumed the full burden of enforcement, at least as far as commercial infringement is concerned.

In practice, however, police resources are far too limited to fully enforce the law, and the courts and prisons far too overburdened to ensure meaningful rates of prosecution or harsh penalties. Such constraints lead to industry pressure for greater public investment and for stronger criminal provisions. The public sector, in turn, tries to ensure that the private sector continues to play a role in investigations and complaints—relying on the formally private status of copyright to justify this role. The result is the uneasy balance described earlier in this chapter, with extensive private subsidization and coordination of police action.

Coordination between the Federal Police, the Federal Revenue Service, and the Federal Highway Police has improved since 2004, and the number of seizures, raids, and arrests has risen. But this appears to be as far as the public sector is willing to go or, in fact, is able to go. As the main interface between the public and private sectors, the CNCP has come in for criticism from the private sector in this regard. One informant complained that the CNCP's current activities are just "more of the same." It is hard, nonetheless, to imagine how much more effective the CNCP could be in its current capacity; most of what can be done in coordinating law enforcement at the federal level has been done. Coordination at the state and local levels is still incomplete, but the new National Plan addresses that as well.

The public and private sectors seem to be at an impasse regarding repressive measures. The private sector wants more rigorous enforcement; the public sector either cannot or is not willing to provide it. When the topic turns to economic measures, the situation is reversed. The public-sector view of economic measures generally involves re-engineering business models to address the issues of cost and access that fuel piracy. Private-sector representatives have stonewalled such proposals and responded with requests for tax cuts. Because this is manifestly not a serious response to the problem, the result is another stalemate. Work on economic measures at the CNCP, consequently, has been anemic at best. As one private-sector informant put it: "So the music companies, the recording companies, and the cinema and video companies, the MPA, they can't talk about pricing. Then they say [to the CNCP's former head], 'Luiz Paulo, we're not going to talk about pricing, and you can't talk about pricing.' And that's it."

---

67   As explained earlier, software copyright is covered under a separate statute, with private prosecution the rule even in cases of commercial criminal infringement.

Educational measures, in contrast, provide a middle ground where the two sides can generally reach consensus. The basis of this consensus is that neither the public nor the private sector is to blame for the prevalence of piracy and counterfeiting in Brazil. Rather, the blame falls on consumers, who are ignorant of the law, of the harms caused by piracy, or both, and thus in need of education. This implies a longer-term project—a "gradual change of perceptions in society by understanding the harmful effects of illegal products and their high social costs. The aim is to replace the idea that piracy brings benefits and a cheap . . . way to satisfy consumers' needs" (Barcellos 2009:3).

Educational projects are funded and developed mainly by the private sector, in many cases with the explicit approval of government—indeed one of the CNCP's roles is to give official government sanction to these initiatives. Nearly all are advocacy campaigns in disguise, promoting industry-friendly narratives on piracy that avoid the controversial issues that generate stalemates in the CNCP (or, for that matter, that describe actual consumer experience with pirated goods). As we describe in more detail later in this section, different types of campaigns target different audiences, from an ABES road tour touting the economic costs of piracy to local authorities, to training programs for judges and prosecutors, to the "Projeto Escola Legal" campaign in Brazilian elementary and secondary schools, which runs children through a truly disgraceful set of propaganda exercises. Self-reflection is not on the menu, and to the best of our knowledge, none of these programs have been subject to independent evaluation. Indeed, like so many other aspects of the enforcement agenda, what they signal is not success or even progress in the struggle against piracy but simply cooperation between industry and public authorities.

Mixed Signals

Little of this domestic political tension is visible on the international stage. In fact, Brazil has been one of only a handful of developing countries to publicly articulate a clear international agenda on IP independent of the enforcement conversation with the United States. In particular, Brazil has played a leading role in establishing a new basis for IP policymaking at WIPO: the 2007 Development Agenda, which requires that social and economic development be the primary consideration in the formulation of new IP policy, including less rigid application of "one-size-fits-all" global IP norms.

Although little of this international conversation has touched directly on enforcement, there are signs of change on this front. After a three-year hiatus, WIPO held a meeting of its Advisory Committee on Enforcement in late 2009, during which Brazil proposed a new, independent research initiative on the impact of piracy and enforcement. Negotiations over the proposed Anti-Counterfeiting Trade Agreement—a maximalist agreement designed to take responsibility for enforcement away from representative bodies like WIPO and the WTO—have also pushed enforcement to the fore. Like the other major industrializing countries that chart semi-independent paths on intellectual property, Brazil was left off the list of countries invited to develop the new agreement.

Superficially, Brazil's international actions are at odds with the story of convergence between government and industry interests on enforcement, circulated mostly by the CNCP for domestic audiences. When questioned about this, officials responsible for the government's IP policy are often adamant that there is no contradiction. As one official put it:

> Sometimes the idea that Brazil acts differently domestically and internationally is advanced by external actors. This is not a fact. It's a deliberate fabrication, made to create obstacles to international negotiations. . . . [Many] of the actors who sit on the GIPI also sit on the CNCP, and through this overlap we've been trying to [harmonize] Brazilian positions.

This obvious anxiety about mixed messages is suggestive of the very delicate line that the Brazilian government walks in regard to foreign audiences on these issues. So far, the WIPO Development Agenda conversation has been relatively silent on the subject of enforcement—in our view reflecting the de facto substitution, in developing countries, of low enforcement for low IP protection after the latter option was foreclosed by TRIPS. The CNCP "convergence" thus occupies a different political space than Brazil's public international positions—a luxury that could disappear if, for example, the ACTA agreement becomes an effective new international standard. In the meantime, the harmonious public face of the CNCP has paid political dividends. The IIPA has held the CNCP and Brazil's National Plan up as models for other countries. The stronger street and border policing facilitated by the CNCP, in particular, won Brazil a respite from its annual inclusion on the Special 301 "Priority Watch List"—and in fact, the CNCP's fourth report suggests that this downgrade was one of the most important outcomes of increased enforcement (Ministério da Justiça 2009:89, 135). These two international stances—the CNCP for dialogue with the United States and the Development Agenda for international forums—represent a balancing act whose equilibrium is at risk as new demands come from all sides. As a CNCP councilor from the public sector explained, "Generally, in the field of intellectual property, the government acts as one. Everyone holds the same position. Except when it comes to enforcement, which is concentrated at the CNCP."

Research

The delegitimation of industry research that we have seen in other countries and documented in chapter 1 of this report is readily visible among enforcement experts in Brazil. "I don't think they're reliable at all," a law enforcement official told us when asked about industry numbers. Such views were widespread among the public-sector representatives on the CNCP. A representative holding one of the ministry seats elaborated:

> [What we] defend at the CNCP is the development of independent rates of piracy and counterfeiting. We defend that because we feel this is too serious an issue for

the government not to [use its] technical capacity to produce an official picture of the problem of piracy and counterfeiting in Brazil. And if we work with the private sector, [we need] to say 'OK, you got to this number. What was the methodology?' Let's sit together and go over the methodology. If we feel, in our analysis, that the methodology is adequate, we'll have no problem in supporting the numbers. I just think that the government needs, before supporting the numbers, to know how they were produced. They might be true; I'm not necessarily saying that the methodology is not good. But for the government to stand by these numbers, they need to go through some kind of check.

Lack of transparency was a constant refrain of the public-sector skeptics. Among the major domestically produced studies, only a couple make serious efforts to explain their methods, data sources, or underlying assumptions. None, in our view, provided enough information to independently evaluate the research. In several cases, statistics and reports are cited that have no sources—or erroneous sources—as we document in the following pages. In our view, transparency at this level is a minimum condition for credibility, both for the industry and for the government officials responsible for policymaking and law enforcement.

Nevertheless, these critiques have had no discernable impact on how numbers are used or how studies are produced in Brazil. Government officials—including at the CNCP—have not been held accountable for or otherwise dissuaded from repeating ungrounded claims. Not all the bad data comes from industry. One ministerial informant described how Federal Revenue Service numbers that combine different categories of seized goods (pirated, counterfeit, and contraband) are used by the CNCP to show that Brazil's efforts are having an effect on piracy in particular: "The CNCP has tables on seizures that we always divulge internationally. In every international meeting we show [them], always saying, 'Look, the statistics have certain issues, but notice the evolution.'" Again and again, the narrative drives the data. Consistently, the media has played along, turning headline numbers into boilerplate stories that often repeat industry press releases verbatim.

THE MAGIC NUMBERS

Investigation of three of the most frequently used numbers in the Brazilian enforcement context reveals the contours of the problem. Claims that the global value of the pirate market is $516–600 billion, that two million jobs have been lost to piracy in Brazil, and that R$ 30 billion ($17.6 billion) in tax revenues is lost annually have become touchstones of the Brazilian enforcement debate.[68]

---

68   This problem is in no way unique to Brazil, and our work here contributes to a wider body of efforts to trace the origin of piracy's many "magical numbers," including recent studies by the US Government Accountability Office and Ars Technica (GAO 2010; Sanchez 2008). The GAO, for its part, could find no credible basis for widely circulating estimates of losses to US business ($200–250 billion), lost jobs (750,000), or lost automotive-parts sales ($3 billion).

Sourcing these numbers quickly becomes a challenge. The most common citations attribute the global estimate to Interpol, the jobs estimate to research originating at the Brazilian university Unicamp (Universidade de Campinas), and the tax-loss estimate to the Brazilian union Unafisco (Union of the Fiscal Auditors of the Brazilian Federal Revenue Service). The CNCP is such a frequent user that many news reports simply source the numbers to the Ministry of Justice, the CNCP's parent. In a typical example:

> According to [then executive secretary of the Ministry of Justice and CNCP
> president] Luiz Paulo Barreto, the globalization of the economy has also brought
> the internationalization of piracy, which earns around US$560 billion annually
> worldwide—even larger than the drug trade, whose operations are estimated at
> US$360 billion. Based on these numbers, Interpol has come to consider piracy the
> crime of the century. "The power is too great, and no one can fight that alone. We
> need partnerships. This is the idea behind the new National Plan for Combating
> Piracy, which has three axes: repression, education, and economics," he stressed.
> (Agência Brasil 2009)

Association with the Ministry of Justice effectively launders the numbers and allows them to circulate through other government channels, including judicial decisions. In March 2010, a single São Paulo judge quoted these figures in three separate decisions, drawing from documents provided by the prosecution that in turn quote material pulled from the CNCP's website. With Barreto conferring legitimacy on the numbers, both prosecutor and judge have turned them into boilerplate text for use in cases of criminal copyright infringement:

> The statement made by the President of the "National Council on Combating Piracy,"
> Luiz Paulo Barreto, also Executive Secretary of the Ministry of Justice, added to
> the books through the counter-arguments provided by the prosecution, bears
> transcription here: ". . . piracy provokes a reduction of two million jobs in the formal
> market. Brazil, according to the secretary, loses R$ 30 billion yearly in tax revenue.
> Globally, Interpol (the international police) considers piracy the crime of the century,
> amounting to US$522 billion/year, much more than the drug traffic, US$360
> billion/year" (information obtained through informative pieces published by the
> organization's website).[69]

Our investigation was unable to substantiate any of these estimates. With regard to the Unafisco estimate on tax losses, an informant from the public sector categorically claimed that

---

69   The text is reproduced in three decisions: Vote 20.252, AC 990 09 217763-0 - Bauru, TJSP/1 Câmara
      Criminal; Vote 20.253, AC 990.09.236431-6 - Olímpia, TJSP/1 Câmara Criminal; and Vote 20.254, AC
      990.09.229941-7 - Mirandópolis, TJSP/1 Câmara Criminal.

the number "does not exist" at all. Another indicated that it is at best a guess since it would be extremely difficult to provide a reliable estimate of fiscal losses potentially generated by piracy and counterfeiting. Arguably more conclusive is the fact that Unafisco conducts no research. We consider the number purely fictional.

A search for the basis of the Interpol numbers also leads to a dead end. According to the CNCP, the numbers were first disclosed during the Second Global Congress on Combating Counterfeiting and Piracy (Ministério da Justiça 2005b:7), but no mention could be found in the documents hosted on the congress's website. The First Global Congress (2004) factsheet on the impact of piracy and counterfeiting mentions that "in 2000, trade in counterfeit goods reached an estimated US$450 billion—larger than the GDP [gross domestic product] of all but 11 countries and about the same size as the total GDP of Australia." It also makes reference to an FBI (US Federal Bureau of Investigation) estimate of the economic impact of counterfeiting in the United States that has been debunked by the US Government Accounting Office (GAO 2010).

The estimate of the two million jobs lost to piracy presents a similar puzzle. Most of the time, the number is attributed to Unicamp, a public university located in the city of Campinas, without further detail. In only one instance did we find a more specific credit: to Unicamp economist Marcio Pochmann (Indriunas 2006).[70] When we contacted Pochmann, however, he directed us to a study on informal vendors commissioned by the City of Campinas and stressed that the two million estimate was related to how many jobs could be generated through the formalization of the street trade overall in Brazil. This study was unavailable through public channels, and we only obtained it after one of the members of the research team was kind enough to digitize his printed copy (CESIT/SETEC 2001). No mention of the two million estimate could be found in the report.

Sector-Specific Research

There is little recent sector-specific research on piracy in Brazil, and most of the main industry groups have dropped their annual updates. The last film piracy study by the MPAA was in 2005. The last ESA numbers come from 2006, and the last publishing industry figures from 2007. By 2010, only the BSA and the RIAA were reporting numbers. Of these two, only the BSA publicly releases its reports. This decline in reporting is not unique to Brazil. All the industry groups (except the BSA) have had to reconsider how they measure piracy as the pirate economy shifts from physical to digital distribution. And all have been under sustained criticism for their research assumptions and lack of transparency.

Discussions of film piracy in Brazil still look back to a 2005 study commissioned by the MPAA, in which Brazil was one of twenty-two countries surveyed. Despite constant MPAA criticism of Brazil on this front, the study placed Brazil near the bottom of the list for rates of piracy, at

---

70   We also found a reference crediting the number to the McKinsey Global Institute, without further speci-
        fication (Gazeta Mercantil 2004).

22% (compared, for example, to 29% in India, 62% in Mexico, and 81% in Russia). IIPA reports continue to relay MPAA complaints about "growing" rates of DVD and Internet piracy. If there is any empirical evidence of this growth, the MPAA has not shared it. Requests by members of this project for detailed data from the 2005 study were refused, as were requests made by the US GAO and a team of OECD (Organisation for Economic Co-operation and Development) researchers commissioned by the International Chamber of Commerce to study global counterfeiting and piracy. A Brazilian film industry representative interviewed in 2009 reported that a new and supposedly broader and more rigorous study was underway,[71] but no new study has been released. Other sources indicate that cost-cutting at the MPAA has put a hold on new large-scale research.

In the publishing sector, in 2004, the ABDR replaced the CBL as the main industry source for loss estimates in Brazil—an event followed by the ABDR's removal from the CNCP and the intensification of the organization's conflict with universities. In piracy research circles, the transition was marked mostly by the unexplained rise of US-publisher loss estimates from $14 million, where they had been stuck since 2001, to $18 million, where they remained until 2008.

A publishing industry informant indicated that, in fact, no new research has been conducted since 2002, when the ABDR (with financing from the Spanish reproduction-rights organization CEDRO) commissioned a study by the marketing research firm A. Franceschini Market Analyses, also known as the Franceschini Institute. This study was never published, but its conclusions were widely circulated in industry literature. Notably, it estimated that university students in Brazil had illegally copied some 1.935 billion pages from books and articles (how it reached this estimate is unclear). Based on this number, it derived an estimate of R$ 60 million ($35 million) in industry losses—though here too the method was unclear (ABDR n.d.b:1).[72]

Other numbers also circulate in this space without citation. In a primer on piracy and authors' rights, the ABDR cites a figure of R$ 350 million ($206 million) in annual losses due to book piracy.

---

71   As the representative described the new study (in terms that make it sound very much like the old study): "We get samples in the target countries, then we do interviews with people, involving knowledge about pirated products, consumer habits, how many times you don't buy an original product if you buy a pirated one, [and] if you buy a pirate product, what is the effect the pirate product—in our case a pirate DVD—if you buy a pirate DVD, is it going to substitute the purchase of an original DVD [or] is it being just used to sample the product? You buy the pirate DVD because it's there, then you take a look, see if you like it or if you're not sure, and then of course you buy the original DVD to have the extras, inlays, all the nice stuff. So [both substitution and sampling] are considered in the total volume of losses for the industry. Price levels, situations, the types of products you usually buy from pirates or not—usually the big titles, the most expected ones. We have some definition, let's say, of a target audience that tends to be slightly more male [than female], so there's also a profiling element to this research."

72   The Franceschini Institute was also responsible for the first phase of the *Retratos da Leitura no Brasil* (Portraits of reading in Brazil) research, commissioned by the CBL, the SNEL, and ABRELIVROS in 2000. It is possible that there is overlap between the piracy study mentioned by the ABDR and the first phase of the *Retratos da Leitura no Brasil* study, regarding methodology or data collected. The report encompasses an earlier presentation on the initial phase of work (Amorim 2008). The presentation itself is available for download on the ABRELIVROS website: http://www.abrelivros.org.br/abrelivros/01/images/stories/arquivos/dados_retratos_2001.ppt.

It describes this number as "estimated from data on book sales over 8 years, with a comparison between the current number of book sales and the number of new teaching institutions and new students enrolled yearly" (ABDR n.d.a:4). In other contexts, R$ 400 million is used, representing "over 50%" of the market for academic and technical books (Cafardo 2007). When asked about these figures, a member of one of the most important book-trade associations in Brazil claimed not to know of any particular studies of the subject but pointed to the ABDR as the source. In our view, there does not appear to be any current research here worthy of discussion—or even much pretension to it. The published data is inadequate to understanding the scope and significance of book copying in Brazil and should not be credited in discussions of either enforcement or the publishing industry's numerous business-model problems in Brazil.

With regard to the recording industry, local record producers association the ABPD has commissioned consumer surveys in the past through the consultant group Ipsos Insight. These furnish one of the first links in the great chain of piracy research, passed on to international recording industry group the IFPI for inclusion in its periodic market and piracy reports, across to the US-based group the RIAA, which massages the findings into industry loss estimates, and from there into IIPA reports, where they form the basis of claims that piracy "has decimated the local legitimate music industry" (IIPA 2010:143).[73] How this happens is something of a mystery. Neither the IFPI nor the RIAA would disclose how they aggregate national data into larger international models or how, in the RIAA's case, they transform those findings into loss estimates (the IFPI, it is worth noting, does not estimate "losses" and in the past has only offered estimates of the "street value" of pirated goods).

Moreover, we could not determine how—or even whether—these models are updated. The most recent data on the ABPD's website is from 2006, when Ipsos apparently surveyed 1,200 Brazilians about their music-consumption habits. This study provides the basis of the ABPD's claims of over R$ 2 billion ($1.17 billion) in losses due to illegal music downloads, representing "over three times the revenue in the official market . . . for genuine CDs and DVDs, which at the time was R$ 615 million [$370 million]."[74] There is no account of how ABPD arrived at this estimate.

An analysis by the University of São Paulo's Research Group on Public Policies for Information Access (GPOPAI) details several likely flaws in publishing and recording industry numbers, with

---

73   According to the IFPI (2009a), recorded music sales are off roughly 40% since the peak year of 2004. Whether this is decimation is a matter of perspective. The legal market was always miniscule, with per capita music consumption in Brazil, at its peak, of around one-tenth the value of the US market. Furthermore, there are a number of vibrant, lucrative, mostly performance-based local music scenes that do not figure into recording industry sales statistics (Lemos and Castro 2008; Mizukami and Lemos 2010).

74   ABPD, "Pesquisa de Mercado," Música na Internet, http://www.abpd.org.br/musicaInternet_pesquisa. asp. Recording industry numbers can be confusing because the IFPI measures market size at wholesale, while many local industry groups report only retail sales—often resulting in significant differences in reported size. The ABPD number is the wholesale number reported by the IFPI.

the caveat that the inquiry was limited by the refusal of both the ABPD and the ABDR to provide details on their studies (GPOPAI 2010).

Research coming out of the business software sector is in its own league. Compared to the other industry associations, the Business Software Alliance is a veritable engine of piracy and economic-impact studies. The BSA, via its consultant the IDC (the International Data Corporation), produces and disseminates new Brazil data for at least two annual reports—its global software piracy report, focusing on rates of piracy and estimated losses, and a set of secondary reports on the impact of piracy on jobs, tax revenues, and the information technology sector. These reports use the good-news/bad-news format that virtually defines advocacy research in this field. The good news is that the business software piracy rate has been in slow decline in Brazil, falling from 64% in 2005 to 56% in 2009. The bad news is that because of rapid growth in the Brazilian informatics market, the total value of pirated software has grown from $766 million in 2005 to $1.6 billion in 2007 to $2.2 billion in 2009 (BSA/IDC 2010).

Responding to nearly a decade of criticism of its assumption that a pirated copy equals a lost retail sale, the BSA no longer categorizes these numbers as losses to business but rather as the notional "commercial value" of unlicensed software. Although the BSA has a relatively robust model for estimating rates of piracy, it shares the wider industry aversion to releasing the data it uses, which in our view should disqualify that model from serious consideration in policy contexts. Few other readers have been so demanding, however, and the BSA and its local representative the ABES are arguably the most skilled among the industry groups at managing media attention to research releases. Every new BSA report is greeted with articles in major newspapers, as well as in many local news sources. These articles almost always reproduce the good-news/bad-news format. Criticism, doubts about methodology, and dissenting views are completely absent from the media coverage analyzed for this report.

The BSA/IDC reports have a major impact on public conversations because they are integrated into a very comprehensive communication campaign. S2Publicom (formerly S2 Comunicações), the company that provides press assistance for the ABES, times the release of BSA numbers to coincide with the ABES Road Show, a traveling law enforcement training program co-sponsored by the ABES, the BSA, the ESA, and the APCM. As the Road Show moves from state to state, the ABES releases localized estimates of the jobs, industry profits, and tax revenues that would be generated if the piracy rate were reduced (see table 5.4 for examples of the releases associated with the *Sixth Annual BSA/IDC Global Software Piracy Study*, published in 2009).[75] These ensure a more or less continuous stream of news stories that report the numbers and rehearse the larger claims about business software piracy. The press schedule for the *Fifth Annual BSA/IDC Global Software Piracy Study*, released in 2008, was particularly intense, with more than twenty press releases in different states. The numbers for São Paulo alone were published five times, but with

75   ABES press releases can be reviewed on the S2Publicom website: http://www.s2publicom.com.br/imprensa/ClienteReleasesS2Publicom.aspx?Cliente_id=345.

different values between the initial and subsequent press releases. The BSA attributes these differences to its practice of revising its numbers as more data becomes available—punctiliousness we applaud. But the revisions also include dramatic differences in the numbers of jobs allegedly created in response to a 10% decrease in the piracy rate.

Table 5.4 ABES Press Releases for the Localization of the Sixth Global Software Piracy Study

Date	Region	Losses	Gains (if piracy reduced by 8%)
April 28, 2010	Federal District	R$ 121 million	2,100 local jobs (direct + indirect) R$ 180 million profits for local industry R$ 29 million in local taxes
April 14, 2010	Santa Catarina	R$ 126 million	2,300 local jobs (direct + indirect) R$ 187 million profits for local industry R$ 30 million in local taxes
March 24, 2010	Ceará	R$ 63 million	1,100 local jobs (direct + indirect) R$ 93 million profits for local industry R$ 15 million in local taxes
March 9, 2010	São Paulo	R$ 1.1 billion	19,500 local jobs (direct + indirect) R$ 1.6 billion profits for local industry R$ 261.4 million in local taxes
November 26, 2009	Rio Grande do Sul	R$ 213 million	3,800 local jobs (direct + indirect) R$ 315 million profits for local industry R$ 51 million in local taxes
November 12, 2009	Paraíba	R$ 27 million	500 local jobs (direct + indirect) R$ 40 million profits for local industry R$ 6.5 million in local taxes
October 7, 2009	Mato Grosso do Sul	R$ 33 million	600 local jobs (direct + indirect) R$ 49 million profits for local industry R$ 8 million in local taxes
September 24, 2009	Amazonas	R$ 54 million	965 local jobs (direct + indirect) R$ 80 million profits for local industry R$ 13 million in local taxes
August 18, 2009	São Paulo	R$ 1.1 billion	19,500 local jobs (direct + indirect) R$ 1.6 billion profits for local industry R$ 261.4 million in local taxes

The first press release for the state of São Paulo put this number at 3,700;[76] subsequent releases put it at 19,500.[77] The basis for this shift is unclear since the IDC does not show its work and the

76   ABES, "Pirataria de Software Causa Prejuízo de R$ 737 Milhões para Economia de São Paulo, press release, March 28, 2008, http://www.s2publicom.com.br/imprensa/ReleaseTextoS2Publicom. aspx?press_release_id=21116.

77   For example, ABES, "Pirataria de Software Causa Prejuízo de R$ 898 Milhões para Economia de São

ABES does not disclose the procedure used for the localization of the BSA/IDC study. But such swings appear to be relatively common, and we have documented them in both China and India. In our view, they point to the real function of the studies in supplying numbers to conjure with—magical numbers—rather than robust estimates of economic impact. The problematic reasoning behind these estimates, explored in detail in chapter 1, begins with the basic misreporting of the primary costs and benefits of software piracy. Losses to US software providers are, in a first instance, gains to Brazilian businesses and consumers. Piracy is not merely a loss to legitimate markets but also a vast subsidy for other kinds of economic activity dependent on software infrastructure. An adequate account of the economic impact of software piracy would have to take account of both sides of the equation. The IDC has never done so.

CROSS-SECTORAL RESEARCH

Although the big international industry groups have pared back their research agendas in Brazil and elsewhere, the attention generated by the CNCP and the National Plan since 2004 has produced its own small boom in domestic research. A wide range of research consultants now populate this space, hiring out their services to private-sector clients. The FIEMG (Federação das Indústrias do Estado de Minas Gerais—Federation of Industries for the State of Minas Gerais), the FIRJAN (Federação das Indústrias do Estado do Rio de Janeiro—Federation of Industries for the State of Rio de Janeiro), the newspaper *Jornal de Londrina*, and the marketing firm Instituto Análise have all sponsored studies, as has the Instituto Akatu, a consumer-rights non-governmental organization, with funding from Microsoft.[78]

The most important of these are two recently launched longitudinal surveys of consumer behavior in relation to pirated and counterfeit goods. One, begun in 2006 by Ipsos, was commissioned by the Fecomércio-RJ (the Rio de Janeiro Federation of Commerce). The other, begun in 2005 by the research firm IBOPE, is funded primarily by the US Chamber of Commerce. The resulting reports offer very similar perspectives and share, in particular, an unmistakably accusatory tone toward

---

Paulo," press release, September 25, 2008, http://www.s2publicom.com.br/imprensa/ReleaseTexto-S2Publicom.aspx?press_release_id=21931.

78   The Akatu study, done through the consulting firm Fátima Belo-Consultia and Estratégia, is interesting insofar as it encapsulates a contradiction inherent in much of the consulting work of private firms. Many of these studies are commissioned as strategy pieces for the production of more effective advocacy campaigns but end up simply as minor contributions to those campaigns—launched in the media to rehearse familiar anti-piracy claims. Methodologically, the Akatu study is a typical example of this subgenre. It employs a mix of desk research and unspecified qualitative research. Data from US Chamber of Commerce/IBOPE and Fecomércio-RJ/Ipsos reports are used to substantiate statements such as: "[Consumers] are aware that piracy is associated with organized crime" (Instituto Akatu 2007a). Digging deeper, Akatu activity reports for 2007 and 2008 reference seven focus groups conducted for the report, but this is as far as we could go in determining their methodology (Instituto Akatu 2007b:12; 2008:11–12). Contacting Akatu did not prove helpful.

consumers, who insist on buying pirated products despite their awareness of the numerous harms. Attacks on informality are another undercurrent of these studies, particularly in those sponsored by ETCO, which represents manufacturers. ETCO is an idiosyncratic entity in the anti-piracy network because it is the only one that regularly engages with academic research.

The annual Fecomércio-RJ/Ipsos national consumer survey looks primarily at consumption and consumer motivations in relation to pirated goods. Full reports were published in 2006, 2008, and 2010; partial results were released in 2007 and 2009[79]. The 2010 survey was published in conjunction with a public awareness campaign called "Brasil sem Pirataria" (Brazil without Piracy), built around the menacing slogan: "Those who buy pirated products pay with their lives."

Table 5.5 Percentage of Brazilian Population that Has Purchased Pirated or Counterfeited Goods Within the Previous Year

2006	2007	2008	2009	2010
42%	42%	47%	46%	48%

Source: Authors based on Fecomércio-RJ/Ipsos (2010) survey data.

The reports document a slight rise in the percentage of Brazilians who have consumed pirated and counterfeit goods within a year, just outside the margin of error of 3%, over the five-year period (table 5.5). Although we have a hard time drawing any conclusions from this apparent trend, the report paints a bleak future for commerce in Brazil, claiming that "piracy seduces an increasing portion of average citizens" (Fecomércio-RJ/Ipsos 2008:4).

Some of the survey questions can be extracted from the published results. Questions asked in 2009 include:

> Do you believe that the use of these products can bring you any negative consequences?
> Do you believe that piracy causes unemployment?
> Do you believe that piracy "funds" organized crime?
> Do you believe that piracy harms commerce's profits?
> Do you believe that piracy harms manufacturers or artists?
> Do you believe that piracy "feeds" tax evasion?
> Many pirate products cause serious harms to health. Are you aware of this information?"

---

79   The Fecomércio-RJ/Ipsos report for 2010 includes all previously released results.

Although the survey describes itself as a study of piracy, only three categories (out of the fifteen to nineteen used in different years) refer to copyrighted goods—DVDs, CDs, and computer programs. The rest deal with hard-goods counterfeiting.

On a topic loaded with potential for survey bias, the Fecomércio-RJ does not hide its objectives. Everything leads in the direction of confirming harms. The responses to the question "Do you believe that piracy 'funds' organized crime?" (some 60% answered yes in the 2009 survey, down from 69% in 2008) are usually reported to the media as confirmations of the claim and as indictments of the population's moral shortcomings. The study is rather clear on the motivations for piracy, with 94% of consumers buying pirated and counterfeited products because they are cheaper. Price is followed by access, with 12% responding that pirate goods are "easier to find" and 6% that they are "available before the original product." Based on these results, the Fecomércio-RJ draws the two obvious policy recommendations: (1) more investment in awareness campaigns and (2) lower business taxes. Increased penalties for intellectual property crimes also appear among the recommendations, but with much less apparent urgency.

The other major longitudinal survey is an annual US Chamber of Commerce/ IBOPE study conducted in partnership with several organizations, including the Brazil-US Business Council and the ANGARDI. Together with the BSA reporting, these studies arguably have the most mind share with the press, industry groups, and policymakers.[80]

The stated goals are much the same as those of the Fecomércio-RJ/Ipsos survey. They include measuring the consumption of pirated and counterfeited goods, surveying consumer attitudes, and estimating quantities of illegal traffic in major Brazilian cities, including São Paulo, Rio de Janeiro, and Belo Horizonte. The most recent results, published in 2008, examined the consumption of ten product categories, from fake toys to fake motorcycle parts. Once again, only one of these categories (computer/electronic games) related to copyrighted goods. Again, the study described its topic as piracy.

By the standards of its peers, the US Chamber/IBOPE reports are a model of transparency. They explain in detail how the survey is conducted and how losses to piracy are calculated. But in other respects, they repeat many of the mistakes common to the genre. This is true in the details, as when they reach to establish the role of organized crime in counterfeit street vending through similarities in the prices of goods in Rio and São Paulo.[81] And it is true of the larger model of losses that anchors the study. Its fourth-edition report claims that piracy leads to losses amounting to R\$

---

80   We did not have access to the US Chamber's report for the 2006 edition of the survey research. A presentation produced by IBOPE that contains tables for 2006 can be downloaded from the IBOPE website: http://www.ibope.com.br/opiniao_publica/downloads/opp_pirataria_dez06.pdf.

81   "The similarity of results obtained in Rio de Janeiro and São Paulo in terms of pricing is an indicator for the structured nature and organization of the crime that exists behind this market" (US Chamber of Commerce/IBOPE 2007). This is possible but by no means established here. Street vendors operate in larger, highly interconnected markets that share price information.

18.6 billion ($10.9 billion) in taxes across seven sectors of the economy, with a total market value of R$ 46.5 billion ($27.3 billion) (US Chamber of Commerce/IBOPE 2008).

The report calculates the size of the pirate/counterfeit market by assuming a one-to-one correspondence between pirate/counterfeit copies and lost retail sales of genuine products. This assumption ignores the obvious impact of price on the propensity to purchase—an impact that may be fairly low in the context of essential goods but very high in the context of luxuries or discretionary purchases like CDs and DVDs. Industry groups have been slow to integrate these "substitution effects" into their models because they are very hard to measure and invariably undercut claims of losses. But at this point, all the major international groups, including the BSA, the RIAA, the IFPI, the ESA, and the MPAA have moved away from the one-to-one loss model. Neither the ESA, the BSA, nor the IFPI characterize their findings as loss estimates but only cite the street or commercial value of the pirated goods. In the case of software, the BSA's chief researcher at the IDC is on record as describing the actual substitution rate as "perhaps one in ten in developing countries" (Lohr 2004). Estimates of substitution rates for CDs in high-income countries generally range from 10% to 30%—and are certain to be lower in countries where price-to-income disparities are greater.

A further problem with the US Chamber/IPOBE survey is that there is no way of telling how much of lost sales, even if properly calculated, would actually represent lost tax revenue. Here the study assumes, like all other industry studies, that money spent on counterfeited or pirated goods simply disappears from the national economy. This is manifestly untrue. The revenues from informal trade circulate through the economy—including into the regulated economy in ways that are subject to taxation. The savings to consumers from purchasing cheaper counterfeit or pirated goods do not vanish but enable other purchases, which may be taxed. For our part, we find it entirely plausible that the government suffers a net loss in tax revenue due to the high percentage of informal economic activity in Brazil, and we think that the formalization of the economy and the suppression of dangerous counterfeit products are important development goals.[82] The current industry studies, however, are not designed to measure tax losses but rather to produce maximalist accounts of harms and sell the government on more expensive measures to combat them. As the IBOPE study argued: "It was verified that the piracy of products in only three sectors—clothing, tennis shoes and toys—deprives the country of at least R$ 12 billion (US$7 billion) a year in taxes. This value would be sufficient, for example, to cover 26% of the social security deficit" (US Chamber of Commerce/IBOPE 2005).

---

82   It is a separate and much less clear question, in our view, whether the substitution of cheap goods for expensive ones in the media sector negatively impacts overall social welfare. As is discussed in chapter 1, the primary effect is likely to be strongly positive.

Training, Awareness, and Education

> Combating piracy is more than a necessity; it's an ideology—[the belief] that we can
> take this country to another level. I'm a firm believer in Brazil; I'm a firm believer in
> the potential of the Brazilian economy, of Brazilians as citizens, as human beings. I
> believe in the difference we can make in the world. Now, we'll only be able to take
> this country to another level when we respect intellectual property and, more than
> that, invest in education. Education is, in my point of view, the driving force for the
> development of a nation, and Brazil has [the potential] to be a big, beautiful country.
> —*Software industry representative*

The view that strong intellectual property protection is a requirement of development and economic growth is widespread among industry and opinion leaders in Brazil, reinforced by three decades of global-trade orthodoxy and the association of IP enforcement with law enforcement in general. With additional repressive and economic measures blocked by disagreements at the CNCP, the wider indifference of Brazilians to this perspective has become a predictable focus of attention. Since the middle of the last decade, industry groups have invested heavily in training, education, and awareness-building programs directed at a wide range of Brazilian audiences, including law enforcement, consumers, and children. These programs are presented as part of the long-term battle to establish a stronger "intellectual property culture" in Brazil (INPI 2010).[83]

Training sessions, seminars, and courses for law enforcement professionals occur throughout the year in Brazil. The most frequent training programs are part of the ABES Road Show, the traveling outreach program organized by the business and entertainment software industries, with the collaboration of the film and music industries. In one session of the Road Show we observed in São José dos Campos, São Paulo, the main topic was how to identify pirated games and circumvention devices. Other activities included presentations on intellectual property enforcement made by the APCM and the CNCP. Seminars are usually reserved for judges and public prosecutors, who are less likely to accept being "trained" but will certainly participate in events with a more scholarly flavor.

How much judges and prosecutors learn from these events is hard to gauge. Law enforcement officials at this level have considerable leeway to develop their own understanding of the matters

---

83   In the public sector, the national patent office (the INPI) is the most visible advocate of this perspective and recently cooperated with the FIESP in publishing a series of IP primers targeted at journalists, teachers, and entrepreneurs. The INPI's place in a larger Brazilian strategy of global competitiveness is described by Peter Drahos: "The strategy then appears to be to invest in a patent office that opens the grant gates, let it play a major role in spreading patent culture through a multitude of training courses, and with the assistance of the US build a court system that really understands intellectual property and hope that a sufficient number of Brazilian firms are able to capture economically significant monopolies" (Drahos 2010:255).

they act upon. What they actually do with that understanding, in contrast, is more easily measured. Rates of prosecution, conviction, and incarceration for copyright infringement remain extremely low in Brazil, and complaints about prosecutorial and judicial follow-through remain a regular feature of IIPA reports.

Studying anti-piracy public awareness campaigns in Brazil is a dismal exercise. Demagoguery and scare tactics are the norm, often to a degree that reads as comedy rather than instruction. All are localizations of templates developed at the international level, and all hit the same simple messages: "you wouldn't steal a car"; "kidnapping, guns, drugs . . . the money that circulates in piracy is the same money that circulates in the world of crime"; "tomorrow I will sell drugs in my school because of that DVD"; and "thank you ma'am, for helping us to buy weapons!" are typical. (Three of the four quotes come from recent spots produced by the UBV, the organization of Brazilian film distributors, which has developed a particular specialization in the genre. The spots run on TV, in theaters, and in DVD preview materials.) With a few exceptions, these campaigns are produced by the private sector with private money.[84] Two of the larger campaigns—"Pirata: Tô Fora!," maintained by the SINDIRECEITA, and the educational initiative "Projecto Escola Legal," supported by AmCham—are endorsed by the CNCP.

By all appearances, educational initiatives are a bottomless pit for public and industry resources—capable of demonstrating the cooperativeness of the state but not actual impact on attitudes or practices. To the best of our knowledge, none of them has ever been evaluated.[85] And this is, in our view, the problem with much of the enforcement agenda.

As Alexandre Cruz was stepping down as president of the National Forum against Piracy and Illegality, he explained one of his regrets: "We didn't make good use of the space we have on TV. Unfortunately, we didn't have enough resources to make a campaign that would have an impact, to change behavior" (FNCP 2009). At about the same time, the FNCP announced a partnership with Zazen Produções, producers of the hit films *Tropa de Elite* (2007) and *Tropa de Elite 2* (2010), to make a theatrical feature film about piracy that stressed its links with organized crime.

This is a considerably more ambitious and canny outreach strategy. José Padilha, director of *Tropa de Elite*, has found a fertile niche in making films that take on "Brazilian problems" as their premise (Pennafort 2008). *Tropa de Elite*'s controversial but highly engaging narrative about police action against Rio's gangs is often interpreted as condoning police brutality in the service of a zero-tolerance approach to crime. This depiction clearly resonated with Brazilian audiences—and with Brazilian pirates. An estimated 11.5 million people saw pirated versions of the film after

---

84   The few public-sector efforts that we could find include an anti-piracy manual published by the Legislative Assembly of the State of Rio de Janeiro and an issue of Revista Plenarinho, a comic book published by the Chamber of Deputies, with support from the ESAF, part of the Ministry of the Treasury (Braga 2010; Câmara dos Deputados 2009a).

85   StrategyOne examined two hundred campaigns for the International Chamber of Commerce without identifying a single evaluation component (BASCAP/StrategyOne 2009).

an employee at Drei Marc, the company subtitling the film, put it on the Internet prior to its theatrical release (Martins, Ventura, and Kleinpaul 2007). When Gilberto Gil, then Minister of Culture, admitted to owning a pirated copy, Padilha began a public crusade against piracy, speaking out and demanding stronger action. Due to Padilha's influence, *Tropa de Elite* 2 enjoyed a preemptive anti-piracy campaign by police, which guarded working prints of the film so that it would not leak—a first in Brazil for this form of private capture of enforcement (Araújo 2010).[86]

Public-sector participation in enforcement is most visible in and around educational initiatives. Some simply involves opening the doors of schools and universities to anti-piracy and anti-counterfeiting organizations. Representatives of the ABES regularly speak to university students, for example, and students are invited to attend parts of the ABES Road Show. In other contexts, such as the "Projeto Escola Legal," public-sector participation is much more direct.

THE "PROJETO ESCOLA LEGAL"

> To reject pirate products is, therefore, a small personal investment that each
> Brazilian can make in the development of our country.
> –"Projeto Escola Legal" teachers' manual[87]

From our perspective, the American Chamber of Commerce's "Projeto Escola Legal" (PEL—Legal School Project) campaign exemplifies most of the issues raised by this report, from the blurring of private and public power, to the misuse of terms and numbers, to thinly disguised advocacy masquerading as education. As the flagship educational initiative endorsed by the CNCP, it is also one of the main outcomes of the stalemate between repressive and economic measures in enforcement policymaking.

PEL is described as an ethics and civics program for students aged 7 to 14. Coordinated by AmCham, the project has involved a wide array of partnerships with public authorities and industry sponsors. The primary government support comes from the CNCP, which counts PEL as one of its strategic projects in the National Plan, under the umbrella of a broader "Piracy out of Schools or Education against Piracy" initiative.[88] Other public-sector support comes from the INPI, the offices of education of the states of São Paulo and Goiás, the Public Prosecution Service of Goiás, and the Federal Regional Court of the Fourth Region (which covers the states of Rio

---

86   The outreach to Padilha remains unique. The only other case we found involved ABES sponsorship of a 2006 storyline in a Globo soap opera called *Páginas da Vida* that involved drastic consequences for a network of computers after the use of a pirated program. See the synopsis on Globo.com's website: http://paginasdavida.globo.com/Novela/Paginasdavida/0,,AA1367031-5742,00.html.

87   AmCham-Brasil (2010:14).

88   In December 2010, as this report was going to press, CNCP officers indicated to the authors that CNCP support for PEL had recently been withdrawn. If so, this development has not yet been announced.

Grande do Sul, Santa Catarina, and Paraná). The primary public subsidy is, of course, at the local level, where hundreds of teachers and administrators contribute their time to implementing the PEL agenda.

Industry sponsors vary from year to year. The current list includes the ABES, the BSA, ETCO, Interfarma (Associação da Indústria Farmacêutica de Pesquisa—Association of Pharmaceutical Industry Research), Merck Sharpe & Dohme, Microsoft, the MPA, and Nokia. Local store-owner associations often participate, as does the Bar Association of São Paulo and the ABPI, an association of IP lawyers. We were unable to obtain an overall budget or a breakdown of the public and private costs of the initiative, but it has grown dramatically since it was piloted in 2007.

PEL was launched in 2007 at five São Paulo city schools (four public, one private). In its first year, it involved 93 teachers and 1,433 students. By 2009, AmCham claimed that PEL was present in 117 schools (94 public, 23 private) in four cities, involving 953 schoolteachers and 22,000 students. Two additional cities were slated to be added in 2010 (AmCham-Brasil 2010:5). The PEL campaign engages entire school communities, including parents, but the main focus is the training of teachers. Through PEL, teachers are taught how to integrate themes and activities related to piracy and counterfeiting into the regular curriculum, across the different subjects they teach. Students are seen as replicators of the core content of the program, extending anti-piracy messaging to their friends and families. Media coverage of school activities is part of the program and adds a further dimension to its outreach.

PEL is implemented through a yearly cycle of events and workshops, inaugurated by a one-day seminar for schoolteachers in each of the participating cities. This Educator Awareness on Combating Piracy Forum (Fórum de Conscientização de Educadores no Combate à Pirataria) is followed by workshops at individual schools, usually with direct industry participation. Schoolteachers receive a manual, *ABC do PEL* (PEL's ABCs), containing the core themes, ideas, and arguments of the project, as well as complementary material on piracy offered by AmCham and the campaign's industry sponsors. Teachers are encouraged to work anti-piracy exercises into the curriculum under the general themes of civics and ethics. Ethics is one of the "transversal themes" of Brazil's National Curriculum Parameters, which are meant to guide and connect the content and activities for all basic education subjects. PEL updates its lessons on its website and connects them to local news about piracy and counterfeiting. Months later, schools host anti-piracy assemblies, where students showcase PEL-inspired works created in the classroom.

As elsewhere, "piracy" in PEL literature is used all-inclusively to describe not only copyright infringement but also counterfeiting and contraband. Among the list of fourteen products used to illustrate intellectual property crime, only three can be pirated in the proper sense—CDs, DVDs, and software. Although the *ABC do PEL* glossary includes entries on patents and trademarks, it does not have one on copyright.[89] These definitions, and a few passages on the importance of

---

89   A post on the PEL website titled "What are author's rights and the public domain?" goes as far as failing to differentiate copyrights from patents: "The rights of authorship for creations last for a determined

intellectual property for innovation and development, comprise the entire discussion of intellectual property in the manual. No mention is made of exceptions and limitations, nor of balancing users' rights with the rights of content owners.

A parody of hard-line positions on intellectual property and piracy might look very much like the PEL teachers' manual. The manual presents a radical version of "intellectual property culture" in which the legal and the moral perfectly coincide, with no gray areas. It associates a general notion of civic duty with blind respect for laws. As a counterpoint, the text provides a broad sketch of a lawless society where "guns and bribery" are the norm (AmCham-Brasil 2010:6).

The materials can appear bizarre. "Survival and the physical and moral integrity of individuals are ensured by the existence of laws," the PEL manual tells us, which are created to protect society from harm (AmCham-Brasil 2010:6). Because piracy is illegal, it is harmful to society, and more specifically "harms health," "generates unemployment," "provokes tax evasion," "infringes intellectual property," "harms the economy," "damages equipment," "produces clandestine waste," "practices unfair competition," and "finances organized crime" (AmCham-Brasil 2010:11). Each of these consequences is described in detail and connected to consumer behavior: "It is not an exaggeration," the manual says, "to affirm that by buying a pirated product, an individual is worsening his own chances of getting a job, or even provoking the unemployment of one of his relatives or friends" (AmCham-Brasil 2010:10).

A section of the manual entitled "Dealing with Complicated Questions" contains responses to questions or rationalizations about piracy that students or colleagues may raise. These questions touch on the high price of media goods, the role of piracy in access to culture, and the hypothetical effects on employment if informal workers stopped selling pirated goods. In response to a 7- to 14-year-old who, concerned with business taxes, volunteers, "I buy pirated products because the taxes on the genuine goods are too high!" the teacher should have a ready answer: "First of all, by saying that, you are affirming that you would rather give your money to bandits than the government. You would rather see your money being transformed into guns and drugs for organized crime, instead of more schools, hospitals, and security for the people. This is the choice you make when you consume pirated goods" (AmCham-Brasil 2010:15). To the student who suggests that piracy provides the only affordable access to cultural goods, the model teacher brings logic to bear: "The production of movies, music, books, etc., is vast, and therefore, if we cannot buy a ticket to watch a movie, we can't say that we do not have access to culture, but only to that specific movie, in that specific place, and that specific moment." The manual then offers a list of alternatives to piracy, including the suggestion that students pitch in for a DVD rental to watch at home (AmCham-Brasil 2010:16).

PEL's other problems can probably be inferred by the reader at this point. Most of the well-known "magical numbers" are in play here, without citation. And the teachers' manual adds some

---

time. For example: a pharmaceutical company researched and developed a new medicine. It will require a patent for this creation and will have a copyright for 20 years, which is the period of validity of a patent in Brazil" (Projeto Escola Legal 2010).

of its own: the R$ 30-billion (US$17-billion) figure, supposedly representing tax losses due to piracy, is presented as US$30 billion—boosting the value by about 70%. The manual suggests that Brazil's GDP could be 40% higher each year without piracy, which may or may not be a reference to estimates of the size of the informal economy. Predictably, this chicanery is transformed into a lesson plan for "solving mathematical problems with data from research on piracy and pirated products, with statistics and calculations of the losses that piracy causes to the country's economy" (AmCham-Brasil 2010:19).

There has been no attempt to measure the impact of PEL on schoolteachers, students, and families—though the project is reaching a scale and level of ambition where such evaluation seems necessary. Given the absurdity of much of the discourse, we will guess that the impact is very low. But impact on rates of piracy is not the only goal of the PEL program—and possibly not even its primary goal. PEL is, above all, a marketing campaign—for intellectual property protection in general, of course, but also for the principle of state commitment to anti-piracy and for the specific brands that figure in the PEL curriculum.[90] While investments in the first of these may be a matter of ideological commitment on the part of private-sector supporters, regardless of demonstrable results, the latter two deliver more concrete benefits to both the state and the private sector. Our concern is that such irresponsible interventions in public education will continue to grow, not because of their effectiveness, but because they represent the path of least resistance in an otherwise stalemated enforcement debate. If we take seriously the idea that "education is the driving force for the development of a nation," as our software industry informant put it in his defense of the "ideology" of intellectual property, we should begin by canceling the "Projeto Escola Legal."

Conclusion

> "I'm not going to speak about enforcement because it's a waste of time."
> —*Private-sector consultant*

We are, in many respects, at the end of the piracy decade, in which cheap digital technologies fueled an explosion of unauthorized access to media goods. We are also, in a narrower sense, at the end of the enforcement decade, marked by the growth and spread of multinational enforcement industries in the developing world and by the internalization of their agendas by public authorities. Brazil provides a very clear example of both arcs, with the latter running from

---

90   AmCham seems to be aware of the tension between such brand-driven marketing and the demand that consumers essentially refrain from buying. The "Facing the Problem" section of the PEL teachers' manual tries to argue that marketing-driven consumerism must be tempered by the desire to be "honest and conscious citizens"—especially when consumers can't afford the genuine articles (AmCham-Brasil 2010:18).

copyright industry pressure at the beginning of the millennium, to the creation of the CNCP a few years later, to the wider adoption of the enforcement agenda by public authorities at all levels. As the industry groups repeatedly note, Brazilian government cooperation is itself a major success of the enforcement agenda. The once largely private functions of IP enforcement have been assumed by the public sector. But our work also suggests that such cooperation is relevant only at the margins of the larger digital media economy: its impact on the overall availability of pirated goods has been minimal. This is not a Brazilian problem; it is a global one—a direct consequence of the massive, increasingly democratic digital revolution.

Arguably, this has also been a decade of Brazilian resistance to maximalist IP agendas, marked by Brazil's advocacy of the Development Agenda at WIPO, refusal to support the WIPO Internet Treaties, and opposition to ACTA. Although such independence gives the appearance of tension within Brazilian IP politics, there is also a complementary dynamic at work. Domestic compliance enables Brazil's progressive international role. Brazil can stand up in favor of the Development Agenda, in part, because it has appeased industry demands at home. Domestically, this international agenda is almost invisible: the message for consumers is almost always about government and industry cooperation in the war against piracy.

The confluence of interests between public and private sectors in Brazil has two primary features: a strong collective interest in combating the illicit trade in hard goods and a general consensus around the need to educate consumers. But as disc piracy gives way to online piracy, this consensus shows signs of fraying. The conflation of piracy and counterfeiting, so useful when it comes to physical piracy, has not made an easy transition to the online copyright debate. The CNCP's only project in this area, the ISP working group, is not likely to succeed in implementing graduated response in the face of strong public opposition to such measures. The CNCP is still very much a forum for cooperation against counterfeiting and street piracy, with little power to do much against digital, non-commercial piracy. Industry, accordingly, has taken this debate elsewhere, recasting the enforcement agenda as part of a broader set of measures for online security, child protection, and the fight against cybercrime. As the controversy surrounding the Azeredo Bill illustrates, this conflation makes for volatile politics in Brazil.

As every chapter of this report documents, piracy is first and foremost a response to sharp constraints on media access. The failure to address pricing and distribution issues, our work suggests, ultimately makes the investment in enforcement and awareness campaigns moot. Although pirates are increasingly understood as "underserved customers" in some sectors of the copyright industry,[91] this concept has not penetrated very far into Brazilian IP debates. Although the CNCP mandate includes cooperation on "economic measures," this side of the dialogue has been completely blocked—caught between its framing as a business-model issue by the public

---

91   As game publisher Valve's Jason Holtman suggested (Masnick 2009).

sector and as a tax-relief issue by the private sector.[92] This, in our view, is the elephant in the room at the CNCP.

The first test of the adequacy of business models under local conditions is simply the presence or absence of goods in the market.[93] By this standard, Brazil fares poorly. For physical goods such as music CDs and DVDs, high-cost licensing creates a high-priced and culturally impoverished market of the kind documented throughout this report. With regard to digital platforms, Brazil is far down the list when it comes to industry internationalization strategies. As of late 2010, Brazilians had no access to iTunes, Spotify, Hulu, or the PlayStation Network and were only recently granted access to a (functionally restricted) version of Xbox Live. Some of these issues are clearly amenable to public-private cooperation but remain at the margins of a discourse that emphasizes the moral failings of consumers, questionable links between piracy and organized crime, and inflated loss numbers. The CNCP's recent endorsement of the Fecomércio-RJ campaign message that "those who buy pirated products pay with their lives" is a good example of this pattern of avoidance of serious debate on these issues. Brazil's upcoming hosting of the World Cup in 2014 and the Olympics in 2016 is, unfortunately, likely to strengthen this trend as multinational corporate sponsors bear down and as Brazilian officials seek paths of least resistance through the many challenges associated with these high-profile events.

Still buried beneath the enforcement agenda is the question of access to data and research. The domination of policy debates by opaque industry research not only violates the basic principle that public policy should be made with publicly available data but also represents a tremendous lost opportunity for a collective, cooperative exploration of business models that could expand both the production of new media and access to it for Brazilian consumers. In the absence of a real dialogue on those matters, it is all too likely that the next decade will look much like the last one—not an end to the piracy era but its beginning; not an end to the enforcement ramp up but further costly escalation.

---

92   Or in the form of hopeless token efforts like the ABDR's "Pasta do Professor"—a licensing scheme for educational materials that combines sharply limited content with serious privacy issues and "non-transferable" physical copies.

93   As the IIPA stated in its 2002 Brazil report, "Piracy of products for Sony PlayStation is 100% because Sony is not in the market."

About the Study

This chapter is based on research conducted between 2008 and 2010, coordinated by the Overmundo Institute and conducted by researchers at the Center for Technology and Society and FGV Opinion—both at the Getulio Vargas Foundation in Rio de Janeiro. The overall project was supported by a grant from the International Development Research Centre (IDRC). Much of the work was qualitative, grounded in interviews with some twenty-five actors from IP enforcement networks, as well as in informal talks with many more individuals. Data was also gathered from participant observation of industry seminars and training sessions and from visits to known pirate markets and the main police unit specialized in intellectual property crimes, the DRCPIM, in Rio de Janeiro. Finally, the report relies on a wide range of documents produced by industry and government, both textual and audiovisual. Legal analysis was a factor at all stages of the research, with standard texts in Brazilian legal literature taken as yardsticks for the interpretation of law.

Interview subjects are anonymous in this work. For reference, informants were grouped according to their general institutional affiliation, and care has been taken to strip statements of any identifying information.

Pedro N. Mizukami (CTS-FGV), Oona Castro (Overmundo Institute), Luiz Fernando Moncau (CTS-FGV), and Ronaldo Lemos (CTS-FGV) were the lead researchers and the authors of the report that became the present chapter.

Olívia Bandeira (Overmundo Institute) was responsible for the analysis of news articles and contributed many insights. Thiago Camelo, based at the time at the Overmundo Institute, interviewed street vendors from the Uruguaiana market in Rio de Janeiro, and Eduardo Magrani (CTS-FGV) provided supplementary research.

Elizete Ignácio dos Santos, Marcelo Simas, and Pedro Souza, from FGV Opinion, offered assistance with interviews and analysis of industry research. Susana Abrantes and Sabrina Pato, independent researchers working with FGV Opinion, carried out most of the interviews, with guidance from both the Overmundo Institute and the CTS-FGV. Alex Dent contributed a report based on ethnographic work with the APCM and the FNCP and with market workers at the Camelódromo de Campinas in São Paulo. Joe Karaganis was a constant help with all stages of the research and write-up.

Special thanks are due to Raul Murad, Pablo Ortellado, Rosana Pinheiro-Machado, Fernando Rabossi, Gustavo Lins Ribeiro, Allan Rocha de Souza, Hermano Vianna, and José Marcelo Zacchi, who provided invaluable feedback on an early version of this chapter during a workshop held in Rio de Janeiro in 2010.

References

ABDR (Associação Brasileira de Direitos Reprográficos—Brazilian Association of Reprographic Rights). n.d.a. *O Que É Direito Autoral* (cartilha).http://www.abdr.org.br/cartilha.pdf.

_____. n.d.b. *Revisão da Lei de Direitos Autorais: Uma Ameaça à Educação.* http://www.linhaselaudas.com.br/site/imgmateriasmidia/Cartilha_bx.pdf.

Agência Brasil. 2009. "Ministério da Justiça Busca Parceria de Municípios para Combater Pirataria." *Correio Braziliense.* http://www.correiobraziliense.com.br/app/noticia/brasil/2009/12/01/interna_brasil,158351/index.shtml.

_____. 2010. "Pirataria Não Tem Causa Social, Diz CNCP." *Info Exame*, August 26. http://info.abril.com.br/noticias/internet/pirataria-nao-tem-causa-social-diz-cncp-26082010-2.shl.

Agência Estado. 2008. "Legenda de Lost Fica Pronta 4 Horas Após Exibição nos EUA." *Último Segundo*, February 6. http://ultimosegundo.ig.com.br/mundo_virtual/2008/02/06/legenda_de_lost_fica_pronta_4_horas_apos_exibicao_nos_eua_1180447.html.

Aguiari, Vinicius. 2010. "Orkut É 8 Vezes Maior Que Facebook no Brasil." Exame.com, August 25. http://exame.abril.com.br/tecnologia/noticias/orkut-8-vezes-maior-facebook-brasil-590857.

Alliance Against IP Theft. n.d. *Proving the Connection: Links between Intellectual Property Theft and Organized Crime.* http://www.allianceagainstiptheft.co.uk/downloads/reports/Proving-the-Connection.pdf.

Amaral, Arthur Bernardes. 2010. *A Tríplice Fronteira e a Guerra ao Terror.* Rio de Janeiro: Apicuri.

AmCham-Brasil. 2010. *ABC do PEL* (PEL's ABCs). http://www.projetoescolalegal.org.br/wp-content/uploads/2010/02/ABC-do-PEL-2010.pdf.

Amorim, Galeno, ed. 2008. *Retratos da Leitura no Brasil.* São Paulo: Imprensa Oficial.

Araújo, Vera. 2010. "Tropa de Elite 2 Montou Operação de Segurança para Evitar Pirataria do Filme." *O Globo*, November 6. http://oglobo.globo.comrio/mat/2010/11/06/tropa-de-elite-2-montou-operacao-de-seguranca-para-evitar-pirataria-do-filme-922963533.asp.

Ascensão, José de Oliveira. 2002. *Direito da Internet e da Sociedade da Informação.* Rio de Janeiro: Editora Forense.

Barcellos, André. 2009. *Fight against Piracy and Counterfeiting in Brazil: Progresses and Challenges.* Paper presented to the WIPO Advisory Committee on Enforcement, Fifth Session, September 29. http://www.wipo.int/edocs/mdocs/enforcement/en/wipo_ace_5/wipo_ace_5_8.pdf.

BASCAP (Business Action to Stop Counterfeiting and Piracy)/StrategyOne. 2009. *Research Report on Consumer Attitudes and Perceptions on Counterfeiting and Piracy.* Paris: International Chamber of Commerce. http://www.iccwbo.org/uploadedFiles/BASCAP/Pages/BASCAP-Consumer%20Research%20Report_Final.pdf.

Bayard, Thomas O., and Kimberly Ann Elliott. 1994. *Reciprocity and Retaliation in U.S. Trade Policy.* Washington, DC: Institute for International Economics.

Benkler, Yochai. 2000. "From Consumers to Users: Shifting the Deeper Structures of Regulation Towards Sustainable Commons and User Access." *Federal Communications Law Journal* 52 (3): 561–79.

_____. 2006. *The Wealth of Networks: How Social Production Transforms Markets and Freedom.* New Haven, CT: Yale University Press.

Bertolino, Robson. 2007. "Shopping Popular É Emparedado na Avenida Paulista." G1, December 19. http://g1.globo.com/Noticias/SaoPaulo/0,,MUL233178-5605,00-SHOPPING+POPULAR+E+E MPAREDADO+NA+AVENIDA+PAULISTA.html.

Biddle, Peter, Paul England, Marcus Peinado, and Bryan Williams. 2002. *The Darknet and the Future of Content Distribution.* Microsoft Corporation. http://msl1.mit.edu/ESD10/docs/darknet5.pdf.

Boghossian, Bruno. 2010. "UFRJ Vota Regulamentação do Xerox em Seus Câmpus." *O Estado de São Paulo*, September 23. http://www.estadao.com.br/estadaodehoje/20100923/not_imp613958,0. php.

Braga, Ronaldo. 2010. "Cartilha contra a Pirataria Vai Orientar Consumidor. Serviço Já Recebeu Mais de 200 Notícias." O Globo, July 1. http://oglobo.globo.com/rio/mat/2010/07/01/cartilha-contra-pirataria-vai-orientar-consumidor-servico-ja-recebeu-mais-de-200-denuncias-917035216.asp.

Branco, Sergio Vieira, Jr. 2007. *Direitos Autorais na Internet e o Uso de Obras Alheias.* Rio de Janeiro: Lumen Juris.

Braz, Camilo Albuquerque de. 2002. "Camelôs no Sindicato: Etnografia de um Conflito no Universo do Trabalho." Undergraduate thesis, Department of Anthropology, University of Campinas.

BSA/IDC (Business Software Alliance and International Data Corporation). 2008. *Fifth Annual BSA/IDC Global Software Piracy Study:* 2007. Washington, DC: BSA.

_____. 2009. *Sixth Annual BSA/IDC Global Software Piracy Study*: 2008. Washington, DC: BSA.

_____. 2010. *Seventh Annual BSA/IDC Global Software Piracy Study*: 2009. Washington, DC: BSA.

Bueno, Zuleika de Paula. 2009. "Anotações sobre a consolidação do mercado de videocassete no Brasil." *Revista de Economía Política de las Tecnologías de Información y Comunicación 11* (3): 1–22.

Cafardo, Renata. 2007. "Contra Xerox de Livros, Cópias Legais." *O Estado de São Paulo*, August 25. http://www.estadao.com.br/estadaodehoje/20070825/not_imp40719,0.php.

Câmara dos Deputados. 2004. *Comissão Parlamentar de Inquérito—Finalidade: Investigar Fatos Relacionados à Pirataria de Produtos Industrializados e à Sonegação Fiscal: Relatório.* http://apache.camara.gov.br/portal/arquivos/Camara/internet/comissoes/temporarias/cpi/ encerradas.html/cpipirat/relatoriofinal.pdf.

_____. 2009a. *Revista Plenarinho 4* (7). http://imagem.camara.gov.br/internet/midias/Plen/swf/ revistaAnimada/pirataria/revista.swf.

_____. 2009b. *Comissão Especial Destinada a Analisar Proposições Legislativas que Tenham por Objetivo o Combate à Pirataria: Relatório Final.* http://www.camara.gov.br/internet/sileg/

MontarIntegra.asp?CodTeor=681759.

Carpanez, Juliana. 2006. "Saiba Como Funcionam os Golpes Virtuais." *Folha Online.* http://www1.folha.uol.com.br/folha/informatica/ult124u19456.shtml.

Cassiano, Célia Cristina de Figueiredo. 2007. "O Mercado do Livro Didático no Brasil: da Criação do Programa Nacional do Livro Didático (PNLD) à Entrada do Capital Internacional Espanhol (1985-2007)." PhD diss., Department of Education, Pontifical Catholic University of São Paulo.

Castanheira, Joaquim. 2003. "A Máfia do Cigarro Pirata." *Istoé Dinheiro,* no. 315. http://www.istoedinheiro.com.br/noticias/11363_A+MAFIA+DO+CIGARRO+PIRATA.

CBN/O Globo Online. 2008. "Stand Center É Condenado a Pagar Multa de 7 Bilhões por Pirataria de Software." *O Globo.* http://oglobo.globo.com/sp/mat/2008/08/06/stand_center_condenado_pagar_multa_de_7_bilhoes_por_pirataria_de_software-547608643.asp.

CESIT/SETEC (Centro de Estudos Sindicais e de Economia do Trabalho/Serviços Técnicos Gerais). 2001. *Trabalhadores do Comércio Ambulante de Campinas: Diagnóstico sobre as Condições de Trabalho.* Campinas: SETEC.

CETIC.br. (Centro de Estudos sobre as Tecnologias da Informação e da Comunicação). 2009. *Tic Domicílios e Usuários 2009*: Total Brasil. http://www.cetic.br/usuarios/tic/2009-total-brasil/index.htm.

CNJ (Conselho Nacional de Justiça). 2010. *Justiça em Números 2009: Indicadores do Poder Judiciário—Justiça Estadual.* http://www.cnj.jus.br/images/conteudo2008/pesquisas_judiciarias/jn2009/rel_justica_estadual.pdf.

Consumers International. 2010. *IP Watchlist Report 2010.* London: Consumers International. http://a2knetwork.org/summary-report-2010.

Costa, Thomaz G., and Gastón H. Schulmeister. 2007. "The Puzzle of the Iguazu Tri-Border Area: Many Questions and Few Answers Regarding Organised Crime and Terrorism Links." *Global Crime 8* (1): 26–39.

Craveiro, Gisele, Jorge Machado, and Pablo Ortellado. 2008. *O Mercado de Livros Técnicos e Científicos no Brasil: Subsídio Público e Acesso ao Conhecimento.* São Paulo: Research Group on Public Policies for Information Access, University of São Paulo.

Davi, Elen Patrícia de Jesus Silva. 2008. "Trabalhadores na 'Fronteira': Experiências dos Sacoleiros e Laranjas em Foz do Iguaçu-Ciudad del Este (1990/2006)." Master's thesis, Department of Humanities, Education and Literature, State University of Western Paraná.

Drahos, Peter. 2010. *The Global Governance of Knowledge: Patent Offices and their Clients.* Cambridge, UK: Cambridge University Press.

Drahos, Peter, and John Braithwaite. 2007. *Information Feudalism: Who Owns the Knowledge Economy?* New York: New Press.

ETCO (Instituto Brasileiro de Ética Concorrencial—Brazilian Institute for Ethics in Competition). 2009. "CNCP Lança Projeto Cidade Livre de Pirataria." *Revista ETCO 6* (14).

Fecomércio-RJ (Federação do Comércio do Estado do Rio de Janeiro—Rio de Janeiro Federation of Commerce)/Ipsos. 2006. *Pirataria: Radiografia do Consumo*. Rio de Janeiro: Fecomércio-RJ/ Ipsos.

_____. 2008. *Pirataria: Radiografia do Consumo*. Rio de Janeiro: Fecomércio-RJ/Ipsos.

_____. 2010. *Pirataria no Brasil: Radiografia do Consumo*. Rio de Janeiro: Fecomércio-RJ/Ipsos. http://www.fecomercio-rj.org.br/publique/media/estudo.pdf.

First Global Congress on Combating Counterfeiting and Piracy. 2004. "Factsheet: The Impact and Scale of Counterfeiting." http://www.ccapcongress.net/archives/Brussels/Files/fsheet5.doc.

FNCP (Fórum Nacional contra a Pirataria e a Ilegalidade—National Forum against Piracy and Illegality). 2009. Newsletter, December 10. http://www.forumcontrapirataria.org/v1/abf.asp?idP=565.

Folha Online. 2009. "Internautas Montam Comunidade 'Discografias—O Retorno!'no Orkut." *Folha Online*, March 17. http://www1.folha.uol.com.br/folha/informatica/ult124u535979.shtml.

GAO (US Government Accountability Office). 2010. *Intellectual Property: Observations on Efforts to Quantify the Economic Effects of Counterfeit and Pirated Goods*. GAO-10-423. Washington, DC: GAO. http://www.gao.gov/new.items/d10423.pdf.

Gazeta Mercantil. 2004. "Pirataria Causa Prejuízo de R\$ 84 Bi ao Ano." *Gazeta Mercantil,* December 12.

Gomes, Maria de Fátima Cabral Marques. 2006. "O Trabalho Ambulante na Globalização: Resistência, Lutas e Alternativas para a Transformação das Condições de Vida e Trabalho." In *Cidade, Transformações no Mundo do Trabalho e Políticas Públicas: A Questão do Comércio Ambulante em Tempos de Globalização*, edited by Maria de Fatima Cabral Marques Gomes, 217–31. Rio de Janeiro: DP&A/FAPERJ.

Gonçalves, Márcio Costa de Menezes e, and Alex Canuto. 2006. *Public Policies on Combating Piracy in Brazil*. Paper presented to the WIPO Advisory Committee on Enforcement, Third Session, May 15–17. http://www.wipo.int/edocs/mdocs/enforcement/en/wipo_ace_3/wipo_ace_3_14.pdf.

G1. 2008a. "Law Kin Chong É Preso pela PF Mais Uma Vez." G1, April 26. http://g1.globo.com/ Noticias/SaoPaulo/0,,MUL427345-5605,00-LAW+KIN+CHONG+E+PRESO+PELA+PF+MAIS +UMA+VEZ.html.

_____. 2008b. "Justiça Concede Liberdade a Law Kin Chong." G1, April 28. http://g1.globo.com/ Noticias/SaoPaulo/0,,MUL429525-5605,00-JUSTICA+CONCEDE+LIBERDADE+A+LAW+KIN +CHONG.html.

_____.2009. "Após Reabertura, Galeria Pagé Volta a Funcionar Normalmente Nesta Quarta." G1, June 3. http://g1.globo.com/Noticias/SaoPaulo/0,,MUL1181210-5605,00-APOS+REABERTURA +GALERIA+PAGE+VOLTA+A+FUNCIONAR+NORMALMENTE+NESTA+QUARTA.html.

Goularte, Cláudia Cardoso. 2008. "Cotidiano, Identidade e Memória: Narrativas de Camelôs em Pelotas (RS)." Master's thesis, Department of Sociology and Politics, Universidade Federal de Pelotas.

GPOPAI (Grupo de Políticas Públicas para o Acesso à Informação—Research Group on Public Policies for Information Access). 2010. *Estimativas sobre o Impacto de Cópias não Autorizadas de Livros e Discos na Produção Industrial.* São Paulo: GPOPAI, University of São Paulo.

IDEC (Instituto Brasileiro de Defesa do Consumidor—Brazilian Institute for Consumer Defense). 2008. "Copiar é Preciso." *Revista do IDEC,* no. 121: 20–23.

IDG Now!. 2008. "PF Inicia Operação I-Commerce 2 para Combater Pirataria Online." IDG Now!, July 1. http://idgnow.uol.com.br/internet/2008/07/01/pf-inicia-operacao-i-commerce-2-para-combater-pirataria-online/.

IEL (Instituto Euvaldo Lodi—Euvaldo Lodi Institute). 2010. *Relatório Anual* 2009. Brasília: IEL.

IFPI (International Federation of the Phonographic Industry). 2001. *IFPI Music Piracy Report.* London: IFPI.

————. 2003. *Music Piracy: Serious, Violent, and Organized Crime.* London: IFPI. http://www.ifpi.org/content/library/music-piracy-organised-crime.pdf.

————. 2009a. *The Record Industry in Numbers.* London: IFPI.

————. 2009b. "Largest Music Community Trading Illegal File Links in Brazil Shut Down." News release, March 31. http://www.ifpi.org/content/section_news/20090326a.html.

IIPA (International Intellectual Property Alliance). 2000. "Request for Review of the Intellectual Property Practices of Brazil in the 2000 Annual GSP Country Eligibility Practices Review." Letter to the USTR, August 21. http://www.iipa.com/gsp/2000_Aug21_GSP_Brazil.pdf.

————. 2001a. *IIPA 2001 Special 301 Report:* Brazil. Washington, DC: IIPA.

————. 2001b. *IIPA Post-Hearing Brief for the Country Practices Review of Brazil in the 2000 GSP Annual Review.* Washington, DC: IIPA.

————. 2002. *IIPA 2002 Special 301 Report*: Brazil. Washington, DC: IIPA.

————. 2005. *IIPA 2005 Special 301 Report*: Brazil. Washington, DC: IIPA.

————. 2006. *IIPA 2006 Special 301 Report*: Brazil. Washington, DC: IIPA.

————. 2007. *IIPA 2007 Special 301 Report*: Brazil. Washington, DC: IIPA.

————. 2008. *IIPA 2008 Special 301 Report*: Brazil. Washington, DC: IIPA.

————. 2009. *IIPA 2009 Special 301 Report:* Brazil. Washington, DC: IIPA.

————. 2010. *IIPA 2010 Special 301 Report*: Brazil. Washington, DC: IIPA.

Indriunas, Luís. 2006. "As Armas contra a Pirataria." *Indústria Brasileira* 6 (70): 18–23.

INPI (Instituto Nacional da Propriedade Industrial—National Industrial Property Institute). n.d. 2007-2010: *Planejamento Estratégico.* http://www.inpi.gov.br/menu-esquerdo/instituto/o-instituto_versao-passada/planejamento/resolveUid/cc4c8b3c59708876a74e9ceb6cc36615.

_____. 2010. "Fiesp Quer Ampliar o Uso da Propriedade Intelectual na Indústria." http://www.inpi.gov.br/noticias/fiesp-quer-ampliar-o-uso-da-propriedade-intelectual-na-industria.

Instituto Akatu. 2007a. "Projeto: Pesquisa e Formulação de Argumentos, Pela Perspectiva do Consumo Consciente, para Apoio a Campanhas Anti-Pirataria." Presentation, Instituto Akatu.

_____. 2007b. *Relatório de Atividades 2007.* São Paulo: Instituto Akatu. http://www.akatu.org.br/quem_somos/relatorio_atividade/relatorio-de-atividades-2007.

_____. 2008. *Relatório de Atividades 2008.* São Paulo: Instituto Akatu. http://www.akatu.org.br/quem_somos/relatorio_atividade/relatorio-de-atividades-2008.

IPEA (Instituto de Pesquisa Econômica Aplicada). 2010. "Análise e Recomendações para as Políticas Públicas de Massificação de Acesso à Internet em Banda Larga." *Comunicados do IPEA,* no 46.

Itikawa, Luciana. 2006. "Vulnerabilidade do Trabalho Informal de Rua: Violência, Corrupção e Clientelismo." *São Paulo em Perspectiva 20* (1): 136–47.

Johns, Adrian. 2009. *Piracy: The Intellectual Property Wars from Gutenberg to Gates.* Chicago, IL: University of Chicago Press.

Jornal PUC Viva. 2005. "Consun Aprova Normas para Xerox Dentro da PUC." *Jornal PUC Viva,* no. 543. http://www.apropucsp.org.br/jornal/543_j03.htm.

Lemos, Ronaldo, and Oona Castro. 2008. *Tecnobrega: O Pará Reinventando o Negócio da Música.* Rio de Janeiro: Editora Aeroplano.

Lemos, Ronaldo, Carlos Affonso Pereira de Souza, Sergio Vieira Branco Jr., Pedro Nicoletti Mizukami, Luiz Fernando Moncau, and Bruno Magrani. 2009. *Proposta de Alteração ao PLC 84/99 / PLC 89/03 (Crimes Digitais).* Rio de Janeiro: Center for Technology and Society, Getulio Vargas Foundation. http://virtualbib.fgv.br/dspace/bitstream/handle/10438/2685/Proposta_e_Estudo_CTS-FGV_Cibercrimes_final.pdf?sequence=1.

Lohr, Steve. 2004. "Software Group Enters Fray Over Proposed Piracy Law." *New York Times,* July 19. http://www.nytimes.com/2004/07/19/technology/19piracy.html.

Mafra, Patrícia Delgado. 2005. "A 'Pista' e o 'Camelódromo': Camelôs no Centro do Rio de Janeiro." Master's thesis, Department of Social Anthropology, National Museum, Federal University of Rio de Janeiro.

Magrani, Bruno. 2006. "Copiar Livro é Direito!" *Cultura Livre.* http://www.culturalivre.org.br/index.php?Itemid=48&id=53&option=com_content&task=view.

Martins, Cleber Ori Cuti. 2004. "As Fronteiras da Informalidade: A relação da Prefeitura e da Câmara de Vereadores de Porto Alegre com os vendedores ambulantes." Master's thesis, Department of Political Science, Federal University of Rio Grande do Sul.

Martins, Marco Antônio, Mauro Ventura, and Bianca Kleinpaul. 2007. "Acusado de 'Vazar' Cópia de Tropa de Elite Pode Pegar 4 Anos de Prisão; Ator Será Ouvido." *O Globo*, August 29. http://oglobo. globo.com/cultura/mat/2007/08/29/297500579.asp.

Masnick, Mike. 2009. "Valve Exec: Pirates Are Just Underserved Customers." *Techdirt* (blog), January 19. http://www.gamepolitics.com/2009/01/16/valve-pirates-are-underserved-customers.

Medeiros, Luiz Antonio de. 2005. *A CPI da Pirataria: Os Segredos do Contrabando e da Falsificação no Brasil*. São Paulo: Geração Editorial.

Ministério da Justiça. 2005a. *O Brasil contra a Pirataria*. http://portal.mj.gov.br/combatepirataria/ services/DocumentManagement/FileDownload.EZTSvc.asp?DocumentID={99AE21FA-03BE-4533-880D-1A1DB045BC62}&ServiceInstUID={F8EDD690-0264-44A0-842F-504F8BAF81DC}.

_____. 2005b. *Conselho Nacional de Combate à Pirataria: II Relatório de Atividades*. http:// portal.mj.gov.br/combatepirataria/services/DocumentManagement/FileDownload.EZTSvc. asp?DocumentID={489C7807-1A01-4EC1-AA63-81B157A9380D}&ServiceInstUID={F8E DD690-0264-44A0-842F-504F8BAF81DC}.

_____. 2006. *Conselho Nacional de Combate à Pirataria: III Relatório de Atividades*. http:// portal.mj.gov.br/combatepirataria/services/DocumentManagement/FileDownload.EZTSvc. asp?DocumentID={525B7986-030A-4CCB-B2B5-FE1066D00792}&ServiceInstUID={F8E DD690-0264-44A0-842F-504F8BAF81DC}.

_____. 2009. *Brasil Original: Compre Essa Atitude*. http://portal.mj.gov.br/combatepirataria/ services/DocumentManagement/FileDownload.EZTSvc.asp?DocumentID={BF79FA7E-3B98-4476-A3BA-E93470C0329F}&ServiceInstUID={F8EDD690-0264-44A0-842F-504F8BAF81DC}.

Ministério do Desenvolvimento, Indústria e Comércio Exterior. 2010. "Ajustes na Balança Comercial 2009 colocam China como Principal Parceiro Comercial do Brasil." News release, January 14. http://www.desenvolvimento.gov.br/sitio/interna/noticia.php?area=5&noticia=9560.

Ministério Público do Estado de São Paulo. 2009. "Termo de Cooperação que Entre Si Celebram o Ministério Público do Estado de São Paulo e o Grupo de Proteção à Marca – BPG." http://www. mp.sp.gov.br/portal/page/portal/cao_consumidor/arquivos/Convenio-firmado.pdf.

Mizukami, Pedro Nicoletti. 2007. "Função Social da Propriedade Intelectual: Compartilhamento de Arquivos e Direitos Autorais na CF/88." Master's thesis, Department of Constitutional Law, Pontifical Catholic University of São Paulo.

Mizukami, Pedro Nicoletti, and Ronaldo Lemos. 2010. "From Free Software to Free Culture: The Emergence of Open Business." In *Access to Knowledge in Brazil: New Research on Intellectual Property, Innovation and Development*, edited by Lea Shaver, 13–39. London: Bloomsbury Academic.

Mizukami, Pedro Nicoletti, Ronaldo Lemos, Bruno Magrani, Pereira de Souza, and Carlos Affonso. 2010. "Exceptions and Limitations to Copyright in Brazil: A Call for Reform." In *Access to Knowledge in Brazil: New Research on Intellectual Property, Innovation and Development*, edited by Lea Shaver, 41–78. London: Bloomsbury.

Moore, Malcolm. 2009. "China Overtakes the US as Brazil's Largest Trading Partner." *Telegraph*, May 9. http://www.telegraph.co.uk/finance/economics/5296515/China-overtakes-the-US-as-Brazils-largest-trading-partner.html.

MPAA (Motion Picture Association of America). 2005. *The Cost of Movie Piracy*. Washington, DC: MPAA.

Muniz, Diógenes. 2009a. "Orkut Perde Sua Maior Comunidade para Troca de Música." Folha Online, March 16. http://www1.folha.uol.com.br/folha/informatica/ult124u535222.shtml.

————. 2009b. "Saiba Como Age o Esquadrão Caça-Pirata da Internet Brasileira." *Folha Online*, April 22. http://www1.folha.uol.com.br/folha/informatica/ult124u554387.shtml.

Naím, Moisés. 2005. *Illicit: How Smugglers, Traffickers and Copycats are Hijacking the Global Economy*. New York: Doubleday.

Nintendo. 2010. Special 301 Comments on Piracy of Nintendo Videogame Products. http://ap.nintendo.com/_pdf/news/590705863.pdf.

Nucci, Guilherme de Souza. 2009. *Individualização da Pena*, 3rd ed. São Paulo: Editora Revista dos Tribuinais.

OECD (Organisation for Economic Co-operation and Development). 2008. *The Economic Impact of Counterfeiting and Piracy*. Paris: OECD.

O Estado de São Paulo. 2009a. "Tênis, Bonés e Roupas São Piratas Made in Brazil." *O Estado de São Paulo*, February 7. http://www.estadao.com.br/estadaodehoje/20090208/not_imp320225,0.php.

————. 2009b. "Box no Shopping de Law com Preço de Iguatemi." *O Estado de São Paulo*, February 7. http://www.estadao.com.br/estadaodehoje/20090208/not_imp320215,0.php.

Pavarin, Guilherme. 2009a. "Comunidade Discografias É Fechada no Orkut." *Info Exame*, March 16. http://info.abril.com.br/aberto/infonews/032009/16032009-4.shl.

————. 2009b. "Bispo Gê Pede Retirada do Projeto contra P2P." *Info Exame*, August 19. http://info.abril.com.br/noticias/internet/bispo-ge-pede-retirada-do-projeto-contra-p2p-19082009-35.shl.

Pennafort, Roberta. 2008. "Sem 'Cidade de Deus' não haveria 'Tropa', Diz Padilha no Rio." *O Estado de São Paulo*, February 18. http://www.estadao.com.br/noticias/arteelazer,sem-cidade-de-deus-nao-haveria-tropa-diz-padilha-no-rio,126634,0.htm.

Pimenta, Eduardo Salles. 2009. *A Função Social dos Direitos Autorais da Obra Audiovisual nos Países Ibero-Americanos*. Rio de Janeiro: Lumen Juris.

Pinheiro-Machado, Rosana. 2004. "'A Garantia Soy Yo': Etnografia das Práticas Comerciais entre Camelôs e Sacoleiros nas Cidades de Porto Alegre (Brasil) e Ciudad del Este (Paraguai)." Master's thesis, Department of Social Anthropology, Federal University of Rio Grande do Sul.

————. 2009. "Made in China: Produção e Circulação de Mercadorias no Circuito China-Paraguai-Brasil." PhD diss., Department of Social Anthropology, Federal University of Rio Grande do Sul.

Pitombo, Antonio Sergio A. de Moraes. 2009. *Organização Criminosa: Nova Perspectiva do Tipo Legal*. São Paulo: Revista dos Tribunais.

Projeto Escola Legal. 2010. "O Que São Direitos Autorais e Domínio Público?" (What are author's rights and the public domain?). February 4. http://www.projetoescolalegal.org.br/?p=842.

PublishNews. 2005. "Cópias Piratas: Editoras Obtêm Liminar contra FGV-SP." *PublishNews*, November 23. http://publishnews.com.br/telas/noticias/detalhes.aspx?id=20818.

Rabossi, Fernando. 2004. "Nas Ruas de Ciudad del Este: Vidas e Vendas num Mercado de Fronteira." PhD diss., Department of Social Anthropology, National Museum, Federal University of Rio de Janeiro.

Rangel, Rodrigo. 2010. "Chinês É Apontado Como Chefe do Contrabando de Celulares." *O Estado de São Paulo*, May 5. http://www.estadao.com.br/noticias/nacional,chines-e-apontado-como-chefe-do-contrabando-de-celulares,547005,0.htm.

Ribeiro, Gustavo Lins. 2006. "Economic Globalization from Below." *Etnográfica10* (2): 233–49.

Rizek, André, and Malu Gaspar. 2004. "Corruptor de Policiais." *Veja*, no. 1861. http://veja.abril.com.br/070704/p_092.html.

Rodrigues, Ivanildo Dias. 2008. "A Dinâmica Geográfica da Camelotagem: a Territorialidade do Trabalho Precarizado." Master's thesis, Department of Geography, Universidade Estadual Paulista.

Sanchez, Julian. 2008. "750,000 Lost Jobs? The Dodgy Digits Behind the War on Piracy." *Ars Technica*, October 7. http://arstechnica.com/tech-policy/news/2008/10/dodgy-digits-behind-the-war-on-piracy.ars.

Sell, Susan. 2003. *Private Power, Public Law: The Globalization of Intellectual Property Law*. Cambridge, UK: Cambridge University Press.

SINDIRECEITA (-Sindicato Nacional dos Analistas Tributários da Receita Federal do Brasil —National Syndicate of the Tax Analysts of the Brazilian Federal Revenue Service). 2006. *Receita contra Pirataria*, no. 3 (September). http://www.sindireceita.org.br/index.php?ID_MATERIA=7030.

Souza, Carlos Affonso Pereira de. 2009. "O Abuso do Direito Autoral." PhD diss., Department of Civil Law, State University of Rio de Janeiro.

Sundaram, Ravi. 2007. "Other Networks: Media Urbanism and the Culture of the Copy in South Asia." In *Structures of Participation in Digital Culture*, edited by Joe Karaganis. New York: Social Science Research Council.

Tavares, Bruno, and Diego Zanchetta. 2009. "MP Investiga Doações Feitas ao DEIC. *O Estado de São Paulo*, December 19. http://www.estadao.com.br/estadaodehoje/20091219/not_imp484563,0.php.

Tourinho, Gustavo. 2006. "Operação I-Commerce Já Levou 17 para a Prisão." G1, October 16. http://g1.globo.com/Noticias/Brasil/0,,AA1312578-5598,00.html.

Chapter 6: Mexico

John C. Cross

Introduction

Mexico is usually listed among the largest producers and consumers of pirated goods. Among the countries cited by the International Intellectual Property Alliance (IIPA) in its Special 301 recommendations, it routinely places in the top ten in losses claimed by US companies: seventh or eighth in software, second or third in recorded music, first in film, and fourth or fifth in video games (IIPA 2010; 2008; 2006). In terms of per capita losses, Mexico is generally surpassed only by Russia and Italy. Such numbers have helped assure Mexico a spot on the Special 301 "Watch List" of the Office of the United States Trade Representative (USTR) since 2003 and the "Priority Watch List" since 2009.

The politics and geography of US-Mexico relations make Mexico a particularly difficult case for the US copyright industries. The long, porous border facilitates trafficking of all kinds—people, drugs, counterfeit goods, arms, and inevitably, pirated materials. Mexico is also usually the first and largest Spanish-language market for movies and music produced in the United States and consequently serves as a gateway for the illegal distribution of new releases to the rest of the Spanish-speaking world.

For these reasons, the growth of piracy in Mexico since the 1990s has been an object of persistent attention from the IIPA and the USTR. Substantial pressure has been brought to bear on the Mexican government to crack down on piracy within its borders. However, this pressure rarely dominates other factors in the US-Mexico relationship. High levels of illegal migration and drug trafficking along the border make intergovernmental cooperation the highest priority in bilateral talks and confrontational engagements over intellectual property (IP) policy or enforcement unlikely.

Mexico provides an important context for understanding not only the growth dynamics of piracy but also the factors that make enforcement extremely costly, both economically and politically. The country has suffered through a series of devastating economic crises over the last several decades, including the current one. It has an average per capita income of less than one-third that of its northern neighbor (CIA 2010) and a political system that faces recurring and often serious challenges to its legitimacy, from the Zapatista rebellion, to the corruption of the police and armed forces by drug cartels, to perceptions of perpetual subordination to the United States. Enforcement actions on behalf of US and multinational corporations play into this dynamic, especially when they appear designed to restrict the local availability of cheap goods.

Chapter Contents

305 Introduction

307 Optical Discs and the Informal Economy in Mexico

309 The History of a Pirate Market

310 The Social Organization of Piracy in Tepito

313 Political Incorporation

315 Law and Enforcement

318 Private-Sector Pressure

318 Attitudes toward Piracy

320 Pirate Justifications

321 Pirate Populism

322 Conclusion

324 About the Study

324 References

Mexican officials inevitably weigh the potentially significant costs and the uncertain benefits of such actions. As a US Embassy official in Mexico noted in 2005:

> Some government leaders are reluctant to crack down on piracy out of fear that this could lead to social unrest, and many Mexicans believe cheap knock-offs offer a preferable alternative to what they view as overpriced products sold by greedy American firms. There are also corrupt politicians and law-enforcement officials who protect IPR [intellectual property rights] violators, from the street vendor level up to ringleaders of notorious markets like Mexico City's Tepito. (US Embassy 2005)

The first sentence concisely summarizes the political challenges and risks for Mexican officials. The second, however, does less well in portraying the political, legal, and social practices that structure piracy in Mexico.

Our work suggests that piracy in Mexico needs to be understood within three broad contexts:

- Piracy is not organized to a significant degree by gangs, drug cartels, or other large organizations, even in notorious markets such as Tepito, but instead is carried out primarily by networks of smaller family-based producers and vendors. There are consequently few "ringleaders" whose arrest could have a significant impact on the pirate economy. This is what makes targeted investigations of piracy ineffective and larger, sweeping enforcement actions relatively high risks for social unrest.

- Street vendors in Mexico have a long history of resisting administrative attempts at repression while working with political allies within the government. Consequently, most piracy takes place not at the disorganized margins of the market economy but within a highly organized sector of the informal economy, which has long experience in acquiring and successfully managing political capital. Common notions of corruption are very hard to apply in this context.

- While Mexico has adjusted its legal system repeatedly to accommodate treaty obligations under TRIPS (Agreement on Trade-Related Aspects of Intellectual Property Rights), NAFTA (North American Free Trade Agreement), and other agreements, the country has a distinctive legal culture and a penal code, in particular, that reflects the broader social and political compromises of Mexican history. These differences have kept Mexico at odds with international rights-holder groups in certain respects and with the broader project of "harmonization" around stronger IP norms and enforcement practices.

As in other countries, piracy in Mexico is part of a dynamic informal sector that reacts to changes in enforcement and, above all, changes in technology. The shift from cassette and VHS tapes to CDs and DVDs in the 1990s allowed for much faster, cheaper, and higher-quality copying—factors that produced an explosion in the street sale of pirated goods. The proliferation of very inexpensive disc burners and, increasingly, broadband connections is forcing a further reconfiguration as prices drop and alternative sources become more widely available.

Optical Discs and the Informal Economy in Mexico

In the 1970s, media piracy primarily involved illegally produced vinyl records, taped music, and later, video cassettes. The process of copying these media was slow and generally resulted in a significantly inferior product. Reproduction required expensive equipment and, consequently, was organized on an industrial scale. The high costs of production meant that pirated goods were not significantly cheaper than licensed products. Both remained expensive relative to low Mexican incomes.

By the mid-1990s, this equation had begun to change. Music CDs had become widely available in Mexico. Software CDs and, by the late 1990s, movie DVDs had begun to appear. The consumer infrastructure lagged behind these new arrivals but grew rapidly in the following years. DVD-player penetration soared in the first decade of the new millennium, from 14.7% of households in 2003 to

Acronyms and Abbreviations

ACTA	Anti-Counterfeiting Trade Agreement
AFI	Agencia Federal de Investigación (Federal Investigation Agency)
APDIF	Asociación para la Protección de los Derechos Intelectuales sobre Fonogramas (Association for the Protection of the Intellectual Property Rights of the Phonographic Industry)
BASCAP	Business Action to Stop Counterfeiting and Piracy (an initiative of the International Chamber of Commerce)
BSA	Business Software Alliance
GATT	General Agreement on Tariffs and Trade
IIPA	International Intellectual Property Alliance
IMPI	Instituto Mexicano de la Propiedad Industrial (Mexican Institute of Industrial Property)
IP	intellectual property
IPR	intellectual property rights
ISP	Internet service provider
MPAA	Motion Picture Association of America
NAFTA	North American Free Trade Agreement
PGR	Procuraduría General de la República (Office of the Attorney General)
PRI	Partido Revolucionario Institucional (Institutional Revolutionary Party)
PROFECO	Procuraduría Federal del Consumidor (Attorney General for Consumer Affairs)
TRIPS	Agreement on Trade-Related Aspects of Intellectual Property Rights
USTR	Office of the United States Trade Representative
VCD	video compact disc

47.9% in 2006 (Scott 2008). The number of computers per capita also climbed steadily beginning in the late nineties, from roughly 4 per 100 persons in 1998 to 15 per 100 in 2008 (ITU n.d.)

The combination of expanding consumer infrastructure and new copying technologies proved explosive. As costs of production dropped and quality rose, prices in Mexico City for pirated music and video plummeted, from US$5 per unit in 2000 to $1 or less by 2005. Licit CDs and DVDs in Mexican stores, in contrast, ranged from $20 to $40.[1] As early as the mid-1990s, the informal economy had begun to shift toward optical disc sales to exploit this gap. Street vending and other "microbusinesses" became the primary distribution infrastructure for recorded film and music as demand rose and as more individuals entered the informal economy in order to meet that demand (Ferriss 2003).

The GATT (General Agreement on Tariffs and Trade) provisions and NAFTA also played a role by easing restrictions on the importation of cheap materials from Asia. Inexpensive personal computers allowed many more people to set up their own production facilities. Huge consignments of blank optical discs, mostly from China, fueled the supply side and pushed prices even lower (Brown 2003).

Table 6.1 Internet Access Figures for Mexico (Broadband and Dial-up), 2000–2009

	Users	Population	Penetration
2000	2,712,400	98,991,200	2.7 %
2004	14,901,687	102,797,200	14.5 %
2005	17,100,000	103,872,328	16.5 %
2006	20,200,000	105,149,952	19.2 %
2008	27,400,000	109,955,400	24.9 %
2009	30,600,000	112,468,855	27.2 %

Source: Mexican Association of the Commercial and Advertising Industry on the Internet (Asociación Mexicana de la Industria Publicitaria y Comercial en Internet—AMIPCI) and the International Telecommunications Union.

As Internet access and particularly broadband access have increased in Mexico, street-based optical disc piracy has encountered its first serious competition. In 2004, only 14.5% of the total population had access to the Internet (Miniwatts Marketing Group n.d.), compared to 68% in the United States (NTIA 2009). By 2009, 27.2% of Mexicans had regular Internet access (see table 6.1), and the percentage of broadband users had grown dramatically—now approaching 20% of households. The rapid growth of broadband, in particular, is fueled by basic telephony needs: cable service is often easier to install in middle-class neighborhoods than a landline (Paradis

---

1   Some businesses do sell cheaper legal CDs and DVDs, but these are usually overstocks.

2008). The pressure on street pricing for optical discs is already apparent: pirated CDs and, in particular, DVDs are available for roughly the cost of blank optical discs in a retail store.[2]

The History of a Pirate Market

Discussions of enforcement in Mexico usually revolve around Tepito—an inner-city neighborhood in Mexico City famous for street vending, crime, counterfeit goods, and now, optical disc piracy. As the 2005 IIPA report observed:

> Well known pirate marketplaces remain largely outside the reach of law
> enforcement—most notably the district of Tepito. Without a government-initiated,
> sustained campaign against well known pirate marketplaces like Tepito, the situation
> in Mexico is unlikely to change dramatically, regardless of the otherwise fine
> intentions and work of PGR [the Procuradoría General de la República, equivalent to
> the Office of the US Attorney General]. (IIPA 2005)

Tepito is the center of a region-wide production and distribution network for pirated optical discs. By most accounts, it dominates local production, though not sales. The number of vendors selling pirated CDs in Mexico City is estimated at between 30,000 and 70,000. Tepito has a total of 8–10,000 street stalls, of which roughly one-third are dedicated to the sale of pirated music, movies, or computer software.[3] Most vendors from other areas of the city use Tepito as their wholesale market.

Many of the other goods for sale in Tepito are counterfeited—designer clothes and bags, cosmetics, toys, and accessories of all kinds. "Here, everything is pirated," one vendor asserted, referring to cheap counterfeit imports from China that look like name-brand goods. Nonetheless, vendors rarely try to fool customers into believing that they are purchasing legitimate goods. Pirated DVDs, for example, are often labeled as "clones" or otherwise marked as pirated goods.[4]

Originally a marginal Indian settlement in the marshy swamps of Lake Texcoco, Tepito became a vast slum area during the pre-revolutionary period as impoverished migrant families moved into the neighborhood. Tepito's status as a center of street vending was established in the 1920s

---

2   A music-industry-funded Ipsos Bimsa survey of 400 people from 2009 has tried to quantify this shift in Mexico. Predictably, it finds very rapid growth of online piracy, claiming that 4.7 billion songs were downloaded illegally in 2008 by some 14 million Mexicans, representing roughly 99% of the online market.

3   Interview with Alfonso Hernandez, Director of the Centro de Estudios Tepiteños (CETEPIS), July 12, 2008.

4   In which case, the goods are pirated but not counterfeited: their content is reproduced, but they are not presented as the original goods.

when "El Baratillo," the city's secondhand goods market, was relocated to the neighborhood.[5] Tepito subsequently became known as the "thieves' market" because the ubiquitous secondhand goods on sale were sometimes of dubious origin. But much of the Tepito economy was licit. The area became well known for workshops that recycled used goods for poorer customers and for its ability to produce cheap knockoffs of products available at higher prices in more exclusive neighborhoods. Leather workshops, especially, produced shoes and garments of all kinds. As Mexico embraced import substitution policies in the 1950s and '60s, imposing high tariffs on the importation of select luxury goods, Mexican industry displaced much of the artisanal craftwork of the neighborhood, and many Tepiteños branched out into contraband. Efforts during this period to incorporate the street vendors into regulated, city-built public markets failed, prompting most vendors to return to the streets by the late 1960s (Cross 1998).

By the 1980s, Tepito had become the center of *fayuca* (contraband) products in the city, selling illegally imported shoes, clothes, and especially electronic goods, such as television sets and VCRs. Many of the goods were imported, but fake trademarks would often be applied locally. Because tariffs on goods sold in department stores could reach 100%, Tepito was a bargain for those who weren't intimidated by stories of thieves and bandits. Government officials railed against the "unpatriotic" selfishness of the neighborhood and the criminal gangs that supposedly ran it, but their attempts to curtail the illegal trade—even to the point of cordoning off the entire area with customs officials or police—had no lasting effects.

Repeatedly, Tepito shifted its commercial focus to adapt to new economic conditions and opportunities. When protectionist policies were abandoned upon Mexico's entry into GATT and later NAFTA, the price advantage of *fayuqueros* relative to legal goods fell sharply, driving out much of the contraband business. Today, the sale of electronics, for example, has almost completely disappeared.

The Social Organization of Piracy in Tepito

The production of DVDs and CDs in Tepito is well organized, with a relatively complex division of labor. The largest disc wholesalers work out of the old, centrally located workshop areas, which are less vulnerable to surprise raids. Boxloads of their products, often marked only with a number, are sold by the hundreds on the street. CD and DVD covers, in turn, are duplicated separately by print shops and sold in another area of the market, also in bulk. Individual vendors buy the CDs and covers separately and put them together within plastic "jewel" cases (or sometimes just plastic sleeves), which they purchase in yet another area. Vendors often do their own "final assembly," sometimes at their stalls while they wait for customers. Smaller producers work out of their own homes with a few burners, using friends or family as workers.

---

5   Tepito was at that time considered to be a peripheral location, although today it is virtually in the center of the metropolitan area.

The Pirate's Life

Geraldo (not his real name) is an example of this process of economic dislocation and adaptation. His father used to run a successful leather workshop employing dozens of workers to produce handbags for an upscale Mexico City department store called Paris. But when tariffs on Asian goods were lowered, the store shifted to imported bags, forcing his father out of business. His father's next and last venture was a small taco stand in Tepito, which his family helped run until he died. His widow, unable to work the stand herself, rented out the space to Korean merchants and lent Geraldo and his brother money to purchase a street stall, where they sold imported baby clothes. The brother left, discouraged by low sales, and Geraldo struggled on until he made an arrangement with a friend in 2000 to sell pirated music CDs. Pirated CDs were still relatively expensive and the supply was limited, but over time he was able to establish relations with better suppliers and eventually buy his own CD burners. By this point Geraldo was earning enough to rent an apartment outside Tepito and send his children to private schools. A year before our interview, however, his site in Tepito was raided and his equipment (fifteen burners) and CD materials were confiscated. After that, the family had to give up the rented apartment and the private schools. With a loan from friends, he was able to buy some old burners and start to build up his business again. When we next met, Geraldo had ten working burners and was back in business.

Much of the production—and almost all the sales—is conducted by family-based businesses. "Geraldo," for example, has a small apartment in Tepito with his wife, teenage son, and two daughters. He spends his afternoons burning music CDs from master copies acquired elsewhere in the market—simultaneously running three machines that have three to four burners each. After school, the family watches TV while folding copies of CD covers. Still later, they sit at the dinner table assembling the jewel cases, covers, and CDs. By the end of the evening, they will have two hundred to three hundred pirated CDs, which Geraldo will sell at his stall while the kids are at school. The kids usually help him set up and take down the stall, as well as taking over for brief periods when Geraldo goes to select the CD and DVD "masters" that he will copy in the future.

Not all family arrangements follow this model. In some cases older siblings or cousins divide the labor—one handling production and another sales. There are also family businesses that specialize in one or the other. Such arrangements are common because optical disc piracy is now a very low-cost enterprise. All the necessary elements, moreover—from burners, to blank discs, to jewel cases, to covers—are readily and legally available. Computer towers are sold in Tepito itself or at a nearby computer market. Blank CDs, DVDs, and VCDs (video compact discs) are also sold in bulk in the neighborhood, often delivered straight to the stalls or residences of the producers. Prices for blanks are so competitive that they actually fluctuate during the day by a few pesos per hundred and are lower than in any retail store. Cases are sold by the boxload and are often also delivered to the door—although as prices have dropped, some vendors have economized further by switching to thin plastic sleeves. An entire street is dedicated to the sale of the "covers" copied from original CDs and printed by the hundreds. All these preliminaries are perfectly legal. The only act that violates the law is the physical "burning" of copyrighted material onto a blank disc for the purpose of sale.

Some vendors handle all aspects of the process themselves. Geraldo starts by acquiring covers of the discs he wants to sell and, if necessary, a master—a high-quality copy of the original disc. Blank discs and cases are delivered to his stall or his house by "runners" who move through the stalls of vendors offering their products. The use of family labor in assembling the final product dramatically lowers production costs, allowing producers and vendors to reduce markups to approximately ten cents per piece. The low overhead means that, in Tepito, CDs and DVDs can be sold for less than a dollar. On the high end, good-quality copies of new releases cost between one and two dollars.

Streets in Tepito are generally organized by type of product—music, movies, or computer software—and vendors tend to specialize further by genre, particularly among movie and music vendors, who are the vast majority. As a result, despite the huge size and chaotic appearance of the market, it is usually fairly easy to find what one is looking for by just asking around. Vendors are generally very knowledgeable about their genre and their stock, and the larger ones, generally in back streets, have sizeable "back catalogs" of material stacked up in boxes or bags.

The spatial distribution of small-scale and large-scale vendors also reflects the organization of the wider market. The largest producers, as well as vendors of printed covers and other materials, such as jewel cases, are usually in the interior streets of the neighborhood since they service the wholesalers. Smaller wholesalers and retailers are located at the "entrance" to the neighborhood from downtown, where subway stations and a major traffic artery bring more casual customers. Generally, prices are cheaper closer to the center of the neighborhood. A CD selling for $0.50 in the center may sell for $1 at the periphery and $2 downtown or in suburban markets. This price elasticity reinforces vendor claims that there is no gang or cartel control of the market, which would monopolize aspects of the trade and maintain higher prices.

Even with this high level of competition driving bargain pricing, several sources among the vendors indicated that Tepito's dominance is beginning to wane. Vendors who used to wholesale to other vendors coming from as far away as Puebla or Guadalajara report that their clients are purchasing their own computers or finding other suppliers closer to home. The Internet—while still a relatively small factor in terms of consumer access to pirated goods—plays a huge role in providing pirate producers direct access to the source material itself, further diminishing the need for privileged distributors.[6]

---

6   The IIPA commented on this decentralization in its 2009 report, noting: "Although Tepito and San Juan de Dios remain dominant sources for the manufacture and commercialization for different types of illegal products, Plaza de la Computación and Plaza Meave are increasingly becoming sources of pirated products. There remain at least 80 very large, very well-known, 'black markets' in Mexico, many of which are well organized and continue to be politically protected."

Political Incorporation

Organization at the level seen in Tepito requires a degree of complicity with political authorities, and indeed there is a long history of incorporation of street vendor organizations into Mexican party politics—notably in connection with the dominant Institutional Revolutionary Party (Partido Revolucionario Institucional—PRI), which ruled Mexico continuously (with some name changes) from the 1930s to the 1990s. This symbiotic relationship emerged in the 1950s after the PRI banned street vending. Street markets were slowly allowed to reappear under a system of agreements brokered with PRI politicians. Vendors were encouraged to organize as civil associations affiliated with the PRI, and vendor leaders were made into local caciques in return for support of the PRI at rallies and during elections (see Cross [1998] for a more detailed description of this process). In exchange, the occupation of specific streets by the merchants was acknowledged through a semiofficial system of "tolerances."

Over time, thousands of different street vendor organizations emerged, with over forty in Tepito alone. Today, these organizations protect over 300,000 street vendors throughout Mexico City, including roughly 10,000 in Tepito. Most of these organizations have access to one or more political patrons, to whom they can turn for help should local officials try to remove them. This structure has survived the democratization of Mexican politics over the past two decades, though it has become more politicized by it—notably after the victory of the right-wing Partido de Acción Nacional in 2000, when some organizations broke with the PRI and shifted their allegiance to the left-wing Partido Revolucionario Democratico.

The strength of the street vendor organizations provides cover for the subset of vendors who deal in pirated media. At the same time, however, these associations have no legal authority over their members. A leader can discipline a member for keeping a dirty stall or for exceeding his allotted space but cannot make a legal determination about whether or not the vendor is selling a pirated product (as opposed to a clearly illegal product, such as marijuana). When the conservative government threatened in 2004 to apply a conspiracy statute to leaders who "harbored" pirates, a local PRI official countered that "the leader [of the vendor association] isn't a policeman. He can't denounce his members to the police because his people could denounce him also for defamation. They aren't police officers or legal scholars to know what is legal and what isn't . . . it's like a witch hunt."[7]

Of course, street vendor leaders know that their members are selling pirated goods, and in general they turn a blind eye to it. Piracy allows their members to make a livelihood and thus pay their dues. Furthermore, the history of conflict and accommodation between street vendors and the state means that leaders have a strong tendency to see any punitive policy as an attack on their hard-won de facto right to sell in the streets. When the leader of a large vendor organization

---

7   Specifically, Article 164 and Article 164 bis of the Mexican Penal Code, modeled after racketeering laws in the United States. Interview with Jorge Garcia Rodriguez, president of the Commission of Commerce of the Assembly of Representatives of the Federal District. He was also the leader of a confederation of organizations that included street vendors.

publicly raised the possibility of enforcing an anti-piracy policy in response to the government's 2004 racketeering threat, angry vendors demanded a meeting. With over a hundred vendors packed into his office, the leader found himself under attack by his own affiliates. While a few supported him, the vast majority argued from a position of economic necessity and that "custom makes the law." Ultimately, the leader backed down, signaling that local market delegates could make their own decisions. In the end, the racketeering statute was never applied.

Tepito's history of struggle positioned it well to become the main wholesale market for pirated goods in Mexico. This advantage went beyond the political protection afforded by street vendor organizations. Residents protected each other from the police. Vendors could run into their own buildings or those of friends or relatives in the event of a raid. As the street market grew, police raids into core areas of the neighborhood became more difficult. The thick tangle of customers, residents, vendors, and stalls was hard to navigate and the lack of cooperation slowed down or obstructed coordinated police action. This social solidarity within the neighborhood produced a spatial organization in which the most clearly illegal activities take place deeper in the network of streets, surrounded by buffer zones of retail vendors who can relay information and more easily afford to run and leave their merchandise behind (Cross and Hernandez 2009).

The political power of street vendors remains unsettling to many in the Mexican establishment. Elite Mexican opinion in the press and in official statements often attributes these arrangements to "mafias" and "gangs." Statements by the IIPA and by Mexican branches of the international industry organizations have adopted this line of argument—and indeed have gone further in trying to conflate media piracy with the drug trade and other forms of violent criminal activity. A 2009 report authored by the RAND Corporation and sponsored by the Motion Picture Association of America (MPAA)—ambitiously called "Film Piracy, Organized Crime, and Terrorism"—adopts this template of guilt by (spatial) association:

> Tepito's resistance to law enforcement makes it terrain for fencing and piracy and
> provides a haven for the more dangerous criminal enterprises of narcotics and arms
> trafficking. Drive-by shootings have become commonplace. The Tijuana drug cartel
> once was said to be ensconced in the neighborhood, using local children to distribute
> cocaine throughout the capital. The Federal Investigation Agency (AFI) led an early
> morning anti-piracy raid of warehouses in Tepito in October 2006, confiscating tons
> of discs and 300 burners capable of producing 43,200 pirated DVDs per day. To
> illustrate what a cesspool of crime Tepito became, according to authoritative press
> accounts, six raids were made between April and July 2008, one of which resulted in
> the seizure of 150 tons of counterfeit material. By late 2006, when Mexican President
> Felipe Calderon moved to evict residents and street vendors from Tepito, it had
> become Mexico's premier "narco-neighborhood." (Treverton et al. 2009:108–9)

While it is undeniable that crime in Tepito is common, there is little in this study—or in our own findings—to warrant the conclusion that these activities are organizationally linked. To date, the combination of strong vendor organizations and low profit margins has been a powerful force for ensuring vendor autonomy—from the police on one side and the drug cartels on the other.[8]

Law and Enforcement

The conflation of narcotrafficking and media piracy in the RAND piece underscores a basic constraint on IP enforcement in Mexico: the violent, destabilizing, and corrupting effects of narcotrafficking dwarf the harms attributed to media piracy and so are very unlikely to be comparably viewed by most politicians and law enforcement officials. Estimates of drug cartel profits in Mexico range from $8 billion to $24 billion, derived primarily from marijuana sales in the United States (Cook 2008). Drug violence has claimed more than 28,000 lives in Mexico since 2006 (BBC News 2010). Twenty thousand Mexican troops occupy major drug-transit cities near the US border (Booth 2008). In this context, with law enforcement fighting a battle that threatens the Mexican state, media industry efforts to tie the two "wars" together do a disservice to both countries.

Nonetheless, Mexico has been under significant US pressure to shift policing resources toward anti-piracy efforts. The Mexican government has accommodated these requests in several important respects, including the granting, in April 2010, of ex officio authority to the police to allow them to act against suspected pirates without a prior complaint from rights holders.[9]

8   The RAND study's section on Mexico appears to rely exclusively on newspaper stories and interviews with representatives of the copyright industry. There is no indication of any attempt to speak with the parties involved in any of their primary examples. The weakness of this approach is clear in the discussion of a struggle between two street vendor leaders, which resulted in the death of a family member. This tragic ending to a conflict over street space and membership in the two organizations becomes, in the RAND report, part of a misleading litany of piracy-inspired violence. (The author has been in touch with both groups involved in this incident for twenty years; neither has a significant number of vendors selling pirated products.) The complexity of the optical disc production and distribution chain is taken as prima facie evidence of "organized criminal syndicates." Los Ambulantes/Tepito (Street Vendors/ Tepito) is listed as a criminal organization on par with the Yakuza and Chinese Triad (xiii). In effect, the RAND authors have classified the community as a criminal gang. It is worth noting, too, that studies of narcotrafficking in Mexico fail to mention any connections to media piracy—though connections between narcotrafficking and human and arms trafficking, kidnapping, and other serious crimes are well documented (Cook 2008; UNODC 2007).

9   Previously, ex officio authority was off the table because copyright infringement was characterized as a "private complaint" rather than a public matter, following long-standing Berne Convention and TRIPS traditions on this point. Chiefly, this distinction forbade police from conducting on-the-spot arrests or confiscations of goods where people were producing, selling, or buying pirated products. In practice, it meant that the sale of pirated goods often proceeded unimpeded within sight of police officers. As a private matter, the injured party (usually an agent representing a rights-holder organization) had to file a detailed complaint (querella) in order to trigger an investigation or raid. This information, in turn, had to be investigated before a court order could be obtained that allowed agents to arrest the accused person and search his or her property. Although the IIPA describes this as a major source of inefficiency in Mexican enforcement, the process was simple enough to permit three to four thousand raids per year

Criminal penalties have also been scaled up, with new law specifying up to ten years in prison for anyone involved in the production or wholesale of pirated goods. (Retail sales on the street are subject to a lower penalty of five years.) Other measures remain under discussion, including the loosening of evidentiary requirements for search and seizure, the expansion of agencies that work directly with rights-holder organizations, and the use of the racketeering statute in piracy cases (which carries a sentence of from twenty to forty years for "organized" criminal activity, defined as involving three or more people).

As the IIPA (2010) notes, however, strong penalties on the books and—by their count—some 4,000 thousand raids in 2008 and 3,400 in 2009 translated into few actual arrests and only a handful of convictions. The IIPA claims that only fifty-seven Mexicans were serving jail time for piracy convictions in 2010, attributable to factors ranging from the inadequate prioritization of piracy on the part of judges and federal agencies to the difficulty of prosecuting copyright cases under Mexican intellectual property law.

A central constraint on the enforcement agenda is that, under Mexican law, copyright infringement applies only to acts *con fin de especulación comercial*—conducted for purposes of commercial gain. While industry groups have argued that this applies to any act of copying, on the principle that "profit results from any realized cost savings" (Segovia 2006), most legal authorities in Mexico regard commercial gain as connected to sales. At present, this provision appears to protect both private copying and file sharing. An investigator for the prosecutor's office assured the author that making copies for oneself or for friends is legal under current Mexican law. There have been no prosecutions for file sharing; nor is there law clarifying ISP (Internet service provider) or other third-party liability for exposing or linking to infringing content. Copyright industry groups have lobbied for legal sanctions for both types of activity and are additionally discussing a version of the controversial three-strikes law to empower industry groups to terminate the Internet service of copyright infringers (IIPA 2009). Mexico was also one of only two developing countries to participate in the Anti-Counterfeiting Trade Agreement (ACTA) negotiations, which many observers expect will create pressure for such changes in national law (the other was Morocco).[10]

Mexico's police forces also present a complex picture. Control over the local police is highly decentralized in response to long-standing distrust of police power. IPR enforcement is carried out almost exclusively by federal authorities—specifically the PGR and the Federal Investigation

by industry and police units. Our research in Mexico was completed before the Chamber of Deputies amended the relevant laws on this point, so we are unable to gauge its impact. It may be low. PRO-FECO, the Attorney General for Consumer Affairs, has ex officio authority but has used it sparingly—drawing IIPA criticism on this point in 2008. In other countries (see, for example, the India and Russia chapters), broader ex officio power has not substantially changed the situation on the street, as police resources remain scarce and policing priorities fall elsewhere.

10　A US State Department cable from 2007 reported that Mexican IPR officials were "keen to highlight their increasingly active role in the international arena, stressing their willingness to join the Anti-Counterfeiting Trade Agreement (ACTA) negotiations and push-back against Brazilian efforts to undermine IPR in international health organizations." (Wikileaks cable 07MEXICO6229, December 2007).

Agency (Agencia Federal de Investigación—AFI). Traditionally, neither agency has had ex officio powers, and it remains unclear how their practices will change with the recent expansion of ex officio authority.

When authorized, raids range from small-scale busts to large-scale operations involving hundreds of police. The latter often elicit fierce opposition from vendors. In a fairly typical case from August 2003, the International Federation of the Phonographic Industry reported that "Mexican law enforcement authorities (LEAs) and the anti-piracy group APDIF Mexico conducted two raids on targeted locations in the notorious Tepito district which led to violent clashes with criminal gangs operating in the area" (IFPI 2004). When the author visited the neighborhood shortly after this raid took place, the market was booming as though nothing had happened. As the IIPA observes:

> Raids in Tepito and other large pirate markets are only conducted at night, as it is unsafe for law enforcement to run actions during the day. Such raids are largely ineffective as the same shops reopen and simply continue their business. (IIPA 2009:65–66)

Organizing raids into districts such as Tepito requires intense planning. PGR officials not only have to coordinate with the rights-holder organizations making the complaint but must also rely on local riot police, who are needed to force entry into hostile neighborhoods and conduct crowd control. These layers of coordination make it difficult to maintain secrecy. PGR officials claim that police officers themselves tip off local residents—a situation that has led to considerable PGR distrust of local police. More generally, however, large police contingents moving into dense neighborhoods make at least some advance notice inevitable.

While vendors usually run from the police, they have occasionally reacted with taunts and—in some cases—violence. In 2008, for example, an anti-piracy operation using three hundred riot police was fought off for three hours (Notimex 2008). An operation in 2005 led to a child being shot by a police officer, resulting in a temporary ban on anti-piracy raids in Tepito. The IIPA's observation that Tepito is too dangerous for police in the daytime needs to be understood in this context. Although confrontations clearly put police at risk, the greater danger is that large-scale resistance can lead to bystanders being hurt or killed. It is this risk—and its high political cost—that leads the police to operate primarily at night when the streets are free of stalls and pedestrian traffic.[11]

---

11   This is not to say that Tepito is unpoliced. Small police patrols do circulate in the market—often heavily armed with submachine guns. But these are "preventive" police, whose primary job is to dissuade violent crime, not to bother vendors.

Private-Sector Pressure

Other government agencies than the PGR also operate in the IP enforcement sphere and—though they lack the power to make arrests—can levy fines and impose other non-criminal penalties against infringing vendors and businesses. The Attorney General for Consumer Affairs (Procuradoría Federal del Consumidor—PROFECO), the Copyright Office, and the Mexican Institute of Industrial Property (IMPI) are the most prominent among these. These agencies work closely with industry groups and often contribute to the investigative work that informs complaints. As in other countries, Mexican groups are often affiliates of US-based or multinational industry associations, like the International Federation of the Phonographic Industry, the MPAA, and the Business Software Alliance (BSA), among others. The agendas and lobbying efforts of these different layers of rights-holder groups are usually closely aligned and often combine in broader alliances that can coordinate legislative and enforcement efforts on local, national, and international levels. In 2006, such an alliance was formalized through the creation of Mexico's Institute for the Protection of Intellectual Property and Legitimate Commerce—combining representation from the Association for the Protection of Film and Music (APCM), the Mexican Association of Phonogram and Videogram Producers (AMBROFON), the National Producers of Phonograms (PRONAPHON), the BSA, and the MPAA.

Attitudes toward Piracy

The contradictions of enforcement in Mexico are sharpened by the general indifference of the public—and even, in interviews, of some enforcement officers—to the moral and economic arguments against piracy. There have been two recent consumer surveys on these issues—one carried out in 2006 by PROFECO and another in 2009 by the consultant group Strategy One on behalf of BASCAP (Business Action to Stop Counterfeiting and Piracy), an anti-piracy initiative funded by the International Chamber of Commerce. Both studies asked similar questions and arrived at broadly similar results.

Regarding the scale of piracy in Mexico, large majorities of respondents in both studies reported buying pirated and counterfeit goods.[12] PROFECO put this figure at 75%; BASCAP at 87%. The BASCAP study helpfully differentiated among categories of goods and found that

---

12   The PROFECO survey queried some 1,425 people over the age of 18, at 81 "interception points" in the
      Mexico City metropolitan area. These "points" were primarily located in stores, markets, and govern-
      ment buildings in areas selected to represent different income levels. While obviously not random, this
      study is one of the only large-scale sources on Mexican consumer attitudes toward piracy. If the study
      has a bias, it lies in its overrepresentation of older and more educated people (those likely to do the
      shopping or go to government buildings). Some 30% of the respondents had a college degree, only 23%
      were under the age of 28, and no one under the age of 18 was included—the population most likely to
      purchase pirated goods according to the BASCAP study. As a result, the survey almost certainly under-
      estimates the prevalence of piracy and its wider acceptance among Mexican consumers. The BASCAP
      survey was based on online interviews with 1,000 people, complemented by focus group findings.

71% of respondents had purchased pirated CDs or DVDs, and 55% pirated software—numbers significantly higher than those for traditional counterfeit goods, like clothing or luxury items. The studies also found predictable correlations with age, with over 90% of respondents in their late-teens and twenties reporting purchases of pirated or counterfeit goods, followed by lower percentages in older age brackets.[13] The vast majority of these acquisitions were made on the street: according to PROFECO, some 93% for music CDs, 92% for movies, 84% for video games, and 50% for computer programs.[14]

Only small minorities expressed agreement that piracy imposed social costs: in the PROFECO study, 31% agreed that it hurt producers, 26% that it caused unemployment, and 21% that it hurt the economy. Only 1% expressed concern that piracy led to greater corruption (PROFECO did not bother listing other forms of crime). Ignorance of the law was not a factor. Nearly all respondents—89%—indicated that they knew that selling and purchasing pirated goods was illegal. The BASCAP study, which was concerned primarily with field-testing anti-piracy messages around these issues, found that only 16% agreed with the claim that the proceeds of piracy went to criminals and only 2% with the claim that buyers supported "a business based on stealing others' idea or art."

This indifference to moral and economic arguments against piracy becomes sharper still in the PROFECO study's breakdown of reasons offered by the 25% who did not purchase pirated products. Among this group, only 9% (or 2.4% of the whole sample) cited concern about how piracy "affected the economy of the country"; only 4.7% (or 1.2% of the sample) refused to buy pirated goods because it was illegal. In contrast, 47% of this group cited "low quality" as their primary reason, and 28% took the (overlapping) position that they preferred originals.

Overall, the PROFECO survey shows that most respondents focus on the relationship between quality and price. Price was cited by 71% of respondents as the most important factor driving their purchases of pirated goods. At the same time, complaints about quality were the largest single concern: of the sample, 68% claimed to have had some type of problem with pirated products—most commonly with video or music quality (61%). Only 12% indicated that pirated goods were more readily available than legal versions.

As for their reaction to extensive enforcement efforts and education campaigns, 86% of respondents believed that piracy had increased over the previous two years; 51% agreed with the view that the government was doing "nothing" about it, and 44% indicated "a little."

---

13   In the PROFECO study, age was a decisive factor: only 33% of those over 67 acknowledged buying pirated goods. The BASCAP study showed a narrower but still significant disparity.

14   The survey didn't specify other locations where pirated or counterfeit goods could have been purchased, but small shops and stalls in public-market buildings are also very informal and tend to be sources for pirated and/or counterfeit goods. This is especially true of computer programs, most of which are sold at the "Plaza de la Computación"—a huge public market in Mexico City's downtown area that specializes in computer hardware and software—or in adjacent small stores rather than on the street in markets like Tepito.

Pirate Justifications

The arguments given by consumers for why they purchased pirated media products line up closely with the justifications given by those producing and selling the goods.[15] In interviews with thirteen vendors, all were aware that piracy was illegal and subject to severe punishment under Mexican law.

The most frequent justification for selling pirated goods was the inevitable combination of economic need and lack of other opportunities. "We know it isn't legal, but it leaves us 2 or 3 pesos and that pays our salaries," one vendor said. Often this defense is combined with criticism of the government: "There are no jobs here, and if they do provide jobs, your expenses are 200 pesos but you only earn 100. Can you live on that?" A leader of Tepito vendors put it more eloquently, noting that despite low incomes, the piracy trade supports many families, and "the government doesn't produce jobs, but it does produce poverty."

Most vendors do not believe that their actions constitute a significant harm to society—a point several made in explicit distinction from the sale of drugs. As one argued: "If you didn't have a job, would you rather deal drugs, steal, or sell pirated goods?" Another vendor, asked if he thought piracy was an honorable profession, responded: "No, but what honorable job can there be? To switch from piracy to theft?" When asked whether it wouldn't be better to avoid both, the vendor responded: "That is the proof of our honesty then—for us to die of hunger! They talk of honesty, but they don't know about our needs."

The impact on licit business is, nonetheless, something that several vendors felt quite keenly in relation to their own business trajectories. Several had started by selling legal CDs that they had obtained on sale from distributors. One woman noted: "They didn't sell a lot, but you earned enough." This model worked when vendors were able to buy originals at a discount, usually when particular CDs were unpopular or retail stores were overstocked. One vendor explained that this secondary market dried up when stores began to destroy their unsold stock instead of reselling it. Customers also pushed for access to newer songs and compilations that the discount model couldn't provide.[16] Piracy provided superior customer service in this respect and, as several vendors noted, introduced an impossible competitive dynamic: "If I turn legal, but the guy next to me still sells pirated goods, what do I do?"

Over time, the prevalence of poverty and illegality in the vendor community clearly has a normalizing effect on these choices. As one vendor indicated, "The truth is that one becomes

---

15   This section is based on interviews carried out in 2004 and 2005. See Cross (2007) for a broader description.

16   Or, for that matter, that were simply unavailable. Pirates routinely produce their own "mix tapes" of popular hits that cater to local tastes. One vendor even lamented that when he makes mix tapes, other vendors simply copy them.

accustomed to it—being here you see everything and get used to everything." Younger participants have simply grown up in a pirate economy. When I asked a vendor's daughter if she felt guilty about the source of her family's income, she responded simply, "I never thought about it."

Pirate Populism

After economic need, the most common vendor rationales for piracy were criticisms of the culture industries, often situated within a wider critique of US and international dominance of Mexico's terms of trade. When I asked a middle-aged couple whether piracy was a form of robbery, the man said yes but then added, "Let me explain. Who robs more, them or us? What have the record companies done for the country? What have the movie studios done? What have the presidents done for the country, to make jobs?" His wife added, "They just worry about themselves." The man became so excited that he stood over me to make sure I wrote down every word: "That free trade agreement makes the rich richer and the poor more screwed because to benefit from trade you have to have a lot of money. Now [Mexican companies] are all transnationals, but the poor are worse off."[17]

Many vendors, in this context, saw themselves as providing a public service that the transnationals refuse to deliver. "As someone who sells pirated goods, I screw the industry. But who am I helping?" one asked rhetorically, then answered himself: "The people." Another suggested: "With the minimum [Mexican] salary [about 50 pesos or $5 per day], it isn't possible to buy an original disc for 200 or 300 pesos. They will spend their entire weekly wage. They come here and can find the same quality . . . but we can make it cheaper." Still another added: "The need of popular culture is to have culture that is accessible for the people. But [the industry] just makes money and more money."

The vendors' defense of piracy fuses the two main ideas that shape attitudes toward piracy in Mexico: (1) the paramount question of inequality, with the pirates providing the only low-cost access to many kinds of cultural goods; and (2) a politicized, nationalist reading of piracy that attributes high prices to (mostly US) profiteering and that views domestic anti-piracy efforts as a form of subordination to foreign interests.

Although there is a clear self-justifying motive at work, these vendor interviews reinforce and arguably complete the picture of Mexican indifference to the arguments of government and industry groups visible in the PROFECO and BASCAP surveys. In Mexico, the rationale for piracy is economic and populist. Most of the time, economic justifications rise to the top. IP enforcement,

---

17   This viewpoint was shared to a greater or lesser extent by at least half my informants and is promulgated by a local newsletter put out by an anonymous group that calls itself the "Pirates of Tepito." In one issue they responded to the director of a movie called *Don de Tepito*, who had complained loudly when his "director's cut" was distributed widely in the market a month before the official release date. "Who is he to complain," the article asked, "when his lousy movie makes everyone in Tepito seem like a criminal or a drug dealer?"

on the other hand, is viewed as a foreign (and mostly US-driven) agenda, in which the Mexican state acts illegitimately on behalf of outside interests. Anti-piracy efforts take their place, in this context, in a long history of popular resentment of US dominance of the US-Mexico relationship. Such views are hardly marginal; rather, they were basic features of PRI discourse over decades of one-party rule.

The scope of such views cannot be overestimated, and—in our view—they shape the sometimes schizophrenic approach of the Mexican government to legal change and cooperation with rights-holder groups. In 2005, I interviewed a PGR official charged with IP enforcement in Mexico City. In response to a question about the new criminal penalties for piracy, he said, "I'm convinced that raising the penalties is not the solution. It is a social and economic problem more than a delinquency problem. . . . I would prefer to be grabbing drug traffickers rather than pirates." Like the pirates, he attributed much of the responsibility for piracy to the industry itself: "It is also a problem of the artists. It isn't possible that a disc that costs 200 pesos [$20] has just one good song, while all the rest are garbage!" Why, then, does the Mexican government invest so much in anti-piracy efforts? "It is mostly the internationals—that is, the gringos squeeze us to carry out these operations." Like several of the vendors, the official repeated the comment about gringos to make sure that I wrote it down.

Conclusion

The economic and political factors surrounding media piracy in Mexico almost never figure in industry reporting, but they are the elephant in the room of IP enforcement. The IIPA's Mexico reports—so critical to maintaining US pressure on the Mexican government—touch only obliquely on the mix of indifference and hostility that greets enforcement efforts and indeed rarely mention the Mexican public at all, except as the target of industry-sponsored education campaigns, such as the menacing "Think About It" (*Piénsalo Bien*) campaign initiated by the IMPI and the BSA in 2008. In our view, the PROFECO and BASCAP surveys cast strong doubt on the value of these initiatives. There are very few Mexicans who are uninformed about piracy or confused about its legality. And there are very few for whom this knowledge has any deterrent effect.

For nearly a decade, the copyright industry has waged a campaign to connect piracy to Mexico's flourishing drug trade. The advantages of doing so may seem obvious: Narcotrafficking represents a serious crisis for the Mexican state and a basis for expanded governmental and police powers. Tying piracy to narcotrafficking allows industry groups to capture new public resources for the anti-piracy effort. The use of new organized-crime statutes, the extension of ex officio powers to the PGR and local police, and the formation of specialized IP enforcement units are part of this wider effort to shift enforcement responsibilities and costs onto an expanded security state.

As this chapter has argued, however, the case for substantive connections between street piracy and the drug trade is thin—based largely on guilt by association and reliant on the general disrepute of Tepito and other street vending neighborhoods to cover gaps in the evidentiary chain.

It also runs against much of what we know about the informal economy in Mexico. Street vendors are well organized and politically protected for reasons that have nothing to do with the drug trade and much to do with their history of economic struggle and resulting incorporation into alliances with major political parties.

These disconnects between the official account of piracy in Mexico and the facts on the ground point to a persistent upper boundary in the enforcement agenda. Despite constant pressure from the United States and copyright industry groups, the Mexican government has not fully committed to the material and political costs of pervasive street-level enforcement. The diverse "failures of cooperation" cited in IIPA Mexico reports need to be understood in this context—not simply as products of inefficiency or lack of understanding but also as a dynamic process of balancing the demands of trade partners against the possible domestic costs of such efforts. It is hard to imagine short- or medium-term circumstances in which this balancing act would change. Yet, such decisions are inevitably negotiated outcomes, and the Mexican government is not a unified actor in these discussions. Different agencies have adopted different de facto positions on enforcement. Mexican trade negotiators involved in the ACTA process, for example, endorsed policies that would sharply impact how the PGR and other enforcement agencies prioritize and conduct their anti-piracy efforts—though only time will tell how much such formal agreements are worth on the streets.

As in many other countries, piracy in Mexico is the product of a complex interaction of forces—among them, the widespread availability of digital media technologies; the high cost of licit media goods; severe, persistent economic inequality; and popular indifference or hostility to enforcement efforts. Because the enforcement agenda of the industry groups does not recognize much less address these issues, those groups seem destined to remain on a war footing, struggling to break an economy built on basic economics and ubiquitous consumer behavior. Here, the drug war analogy seems more apropos.

About the Study

This chapter draws on research conducted by Dr. John Cross over some twenty years of work on the informal economy and urban poverty in Mexico City. Many of the interviews with Tepito vendors and other community members were held in 2004 and 2005. Most of the other interviews—including those with enforcement officers—took place in 2008.

References

BBC News. 2010. "Q&A: Mexico's Drug-Related Violence." August 25. http://www.bbc.co.uk/news/world-latin-america-10681249.

Booth, William. 2008. "Mexico Drug Cartels Send A Message of Chaos, Death." *The Washington Post*, December 4. Accessed June 4, 2009. http://www.washingtonpost.com/wp-dyn/content/article/2008/12/03/AR2008120303953.html.

Brown, Jack. 2003. "Mexico's Music Business Meltdown." *Salon.com*. Accessed March 22, 2010. http://www.salon.com/technology/feature/2003/06/09/mexican_piracy/.

CIA (US Central Intelligence Agency). 2010. "The World Factbook: Mexico." https://www.cia.gov/library/publications/the-world-factbook/geos/mx.html.

Cook, Colleen W. 2008. *CRS Report for Congress: Mexico's Drug Cartels*. Washington, DC: Congressional Research Service. Accessed June 4, 2009. http://fpc.state.gov/documents/organization/105184.pdf.

Cross, John. 1998. *Informal Politics: Street Vendors and the State in Mexico City*. Palo Alto, CA: Stanford University Press.

———. 2007. "'¿Somos Piratas y Qué?' Globalization and Local Resistance: The Case of Cultural Piracy in Mexico City." Paper presented at the annual meeting of the American Sociological Association, New York, NY, August 11.

Cross, John, and Alfonso Hernandez. 2009. "Divergent Theory and Identity Construction: The Role of Space and Community in Tepito, Mexico City." Paper presented at the annual meeting of the American Sociological Association, San Francisco, CA, August 8.

Ferriss, Susan. 2003. "Off-the-books Economy in Mexico Booms; Failed NAFTA Promises Spur Microbusinesses." *Atlanta Journal-Constitution*, December 21.

IFPI (International Federation of the Phonographic Industry). 2004. *The Recording Industry Commercial 2004 Piracy Report*. Accessed March 22, 2010. http://www.ifpi.org/content/library/Piracy2004.pdf.

IIPA (International Intellectual Property Alliance). 2005. *International Intellectual Property Alliance 2005 Special 301 Report: Mexico*. Washington, DC: IIPA. Accessed June 4, 2009. http://www.iipa.com/rbc/2005/2005SPEC301MEXICO.pdf.

_____. 2009. *International Intellectual Property Alliance 2009 Special 301 Report*. Washington, DC: IIPA. Accessed June 4, 2009. http://www.iipa.com/2009_SPEC301_TOC.htm.

_____. 2010. *Special 301 Report on Copyright Protection and Enforcement: Mexico*. Washington, DC: IIPA.

Ipsos Bimsa. 2009. "Descargas Ilegales de Música en Internet en México."

ITU (International Telecommunications Union). n.d. "ICT Statistics Database." http://www.itu.int/ITU-D/ict/statistics.

Miniwatts Marketing Group. n.d. "Mexico Internet Statistics and Telecommunications Reports." Internet World Stats. http://www.internetworldstats.com/am/mx.htm.

Notimex. 2008. "Estalla la violencia por operativo en Tepito." Accessed June 5, 2009. http://www.terra.com.mx/articulo.aspx?articuloId=642696.

NTIA (US National Telecommunications and Information Administration). 2010. *Digital Nation: 21st Century America's Progress Toward Universal Broadband Internet Access*. Washington, DC: US Department of Commerce.

Paradis, Isabelle. 2008. "Mexico's Telecom Revenue Soared by 27.9% to US$28.6 billion in 2007." Hot Telecom. Accessed June 4, 2009. http://www.hottelecoms.com/cp-article-september2008.htm.

PROFECO (Procuraduría Federal del Consumidor). 2006. "Resultados de la Encuesta Sobre Piratería." http://www.profeco.gob.mx/encuesta/mirador/pirateria_2006.zip.

Scott, David. 2008. *World Video Spending Stabilises*. London: Screen Digest. http://www.screendigest.com/reports/08_11_yp1/SD_08_11_WorldVideoSpendingStablises/view.html.

Segovia, Amadeo. 2006. "Piratas: Parecen . . . Pero no Son." PROFECO. Accessed March 30, 2009. http://www.profeco.gob.mx/encuesta/brujula/bruj_2006/bol19_piratas.asp.

StrategyOne. 2009. *Research Report on Consumer Attitudes and Perceptions on Counterfeiting and Piracy*. Paris: Business Action to Stop Counterfeiting and Piracy, International Chamber of Commerce. http://www.iccwbo.org/uploadedFiles/BASCAP/Pages/BASCAP-Consumer%20Research%20Report_Final.pdf.

Treverton, Gregory F. et al. 2009. *Film Piracy, Organized Crime, and Terrorism*. Santa Monica, CA: Rand Corporation.

UNODC (United Nations Office on Drugs and Crime). 2007. United Nations Office on Drugs and Crime Annual Report 2007. New York: UN Publications. Accessed June 4, 2009. http://www.unodc.org/documents/about-unodc/AR06_fullreport.pdf.

US Embassy (Mexico). 2005. "Intellectual Property Rights: Overview of Mexico's IPR Environment." Accessed May 26, 2009. http://www.usembassy-mexico.gov/eng/IPRtoolkit_overview.html.

Chapter 7: Bolivia

Henry Stobart

Introduction

Bolivia is one of the poorest and most socially unequal countries in South America, with a GDP (gross domestic product) per capita of US$4200—around one-eleventh that of the United States.[1] Proportionally, it has the largest indigenous population in the region and also one of the largest informal sectors, accounting for roughly two-thirds of all economic activity.[2] Rates of music, video, and software piracy are estimated to be among the highest in Latin America. The most recent International Intellectual Property Alliance report on Bolivia cited rates of 90% for music piracy and 80% for software piracy (IIPA 2006). The current level of music piracy is probably similar to that of neighboring Peru, which the IIPA listed at 98% for 2009 (IIPA 2010).

Bolivia has not traditionally been a large domestic producer of pirated music and film. Until recently, most of the pirated goods sold in Bolivia were imported from Peru, which has a more developed industrial base and a larger media sector. As in the other countries documented in this report, however, the plummeting cost of reproduction equipment has fostered large-scale domestic production of pirated media and much wider distribution—fueling a pirate mass market that has all but destroyed the tiny, vastly more expensive licit market. In the process, that mass market has also provided access to recorded media for the first time to an emerging population of young consumers, built on Bolivia's remarkable demographic wave (the median age in Bolivia is under 22, and almost 40% of Bolivians are under 15 years of age).

The IIPA has long called for the revision of Bolivia's copyright laws and for stronger enforcement. But beyond the creation of a national intellectual property service (SENAPI) in 1999 and unfulfilled plans, in 2001, to overhaul copyright and create a special police unit dedicated to enforcement (in conformance with a US-led regional trade agreement), recent Bolivian governments have shown few signs of interest in policy change (IIPA 2008). As in other countries, this outcome reflects a range of domestic and international pressures, from the higher priority accorded to cooperation

---

[1]   Like the Mexico chapter, this chapter is primarily a work of individual scholarship and therefore provides a narrower account of piracy than the four large country reports (South Africa, Russia, Brazil, and India). Its focus is music piracy, and secondarily, the optical disc market.

[2]   According to Schneider (2002), the informal economy—mostly street vendors and itinerant workers—represents 67% of Bolivia's total economy. Bolivia did not figure in Vuletin's (2008) account of the informal economy in Latin America. The methodological challenges of measuring the informal economy make all such figures approximate and precise comparisons difficult.

Chapter Contents

327   Introduction

329   The Transformation of the Recorded Music Industry

330   The New Wave

331   Formats and Reproduction Equipment

331   A Short History of Bolivian Piracy

334   From Pirates to Legitimate Distributors?

335   Indigenous Politics

337   About the Study

337   References

with US efforts against drug trafficking to the very real prospect of social unrest if police were to disrupt informal markets on behalf of foreign commercial interests—an explosive dynamic in recent Bolivian history.[3] Although there is widespread sympathy among consumers for local artists, global media companies—and especially American companies—are viewed with distrust. Our research showed little or no popular concern for their loss of income, nor is there a compelling account in circulation of how those losses impact the lives of most Bolivians.

Intellectual property (IP) enforcement has also taken a backseat to the drama surrounding Bolivia's role in the international community following the election of president Evo Morales in December 2005. Widely hailed as the first "indigenous" president in the Americas, Morales has been a polarizing figure, generating strong support among poor Bolivians[4] but also strong opposition among Bolivian elites and in the international community. Ties to Iran and to Hugo Chavez's government in Venezuela have complicated external relations, especially with the United States. Following Morales's expulsion of the US ambassador in September 2008 for allegedly "conspiring against democracy" (BBC 2008), the Bush administration expelled Bolivia from the Andean Trade Promotion and Drug Eradication Act (ATPDEA), which provides for duty-free exportation of goods to the United States.

While many had hoped that relations with the United States would improve following the election of Barack Obama, this has not been the case: Obama ratified Bolivia's expulsion from the ATPDEA, which took effect in July 2009 (USTR 2010). Tensions between the two countries have filtered down into other areas of policy cooperation, such as anti-narcotrafficking and IP enforcement. Stalemate on these issues, compounded by the very small size of the Bolivian music, film, and software markets and by uncertainty

---

3   See, for example, the Water Wars in 2000, fought over efforts to privatize Cochabamba's water supply at the insistence of the World Bank, or the subsequent Gas Wars in 2003, fought over control of profits from Bolivia's natural gas reserves.

4   Morales won a national recall referendum in August 2008 with 68% of the vote. He was reelected for a second term in December 2009 with 63% of the vote.

about Morales's commitment to international trade agreements, may explain why the IIPA dropped its Bolivia reporting after 2006. Although there are still domestic forums for copyright industry interests—notably SENAPI—the IIPA's long wish list of legal reforms and stronger enforcement mechanisms appears to be, for the near future at least, a dead letter.

The Transformation of the Recorded Music Industry

Piracy has been widely blamed for the almost complete collapse of Bolivia's "legal" music-recording industry and for the exodus of multinational record companies from the country. In 1995, recording industry profits in Bolivia were estimated to have been in the region of $20 million. The three main national labels—Discolandia, Lauro, and Heriba—accounted for around $2 million of these profits, but the lion's share ($18 million) went to multinationals operating in the country (Ortiz and Herrera 2003). During the 1990s, these included EMI Music, BMG, Warner Music, Universal Music, Sony Music, Leader Music, and Santa Fe Records.

Levels of audio and video piracy were already high in the mid-1990s, but according to Andrés López (formerly of Sony Music), it was Bolivia's economic crisis in 1999 that decisively tipped the balance: piracy levels rose from around 65% in 1998 to 85–89% in 1999 (*Tiempos del Mundo* 2000). During this period, the national and international labels jointly organized a series of campaigns to combat piracy, involving television advertisements, newspaper articles, raids on street vendors using hired police officers, and the mass destruction of pirated media. The industry also lobbied for strengthening of Bolivia's 1992 copyright law (Law 1322), pressured the government to tackle copyright infringement, and criticized the state for treating piracy as a "social" rather than a "legal" issue (*La Rázon* 2000). They brought several cases against pirate producers to the courts, but the defendants, although generally caught red-handed and admitting guilt, went free after receiving judicial pardons.

Throughout, the major labels made few concessions on pricing, and as the market for $15 CDs was undercut by pirated CDs selling for only one or two dollars, they came under severe financial pressure. By 2003, recording industry profits in the

Acronyms and Abbreviations

IP	intellectual property
VCD	video compact discs
ALBA	Alianza Bolivariana para los Pueblos de Nuestra América (Bolivarian Alliance for the Peoples of Our America)
ATPDEA	Andean Trade Promotion and Drug Eradication Act
IIPA	International Intellectual Property Alliance
SENAPI	Servicio Nacional de Propiedad Intelectual (National Service for Intellectual Property)

country were estimated to have shrunk to around $0.6 million (Ortiz and Herrera 2003). All the major international labels had closed their Bolivian offices, Lauro and Heriba had both ceased operating, and Discolandia had cut its staff from 150 to 20. Today, Discolandia—which recently celebrated its fiftieth anniversary—is the only major record label still operating in Bolivia. Rather than competing downmarket, it focuses on the niche market for high-quality recordings of local acts, often incorporating glossy informative booklets.

The New Wave

The declining costs of recording and sound editing also created opportunities for new, low-cost, local labels. Bolivia now has many small digital studios, which are primarily recording local artists for regional markets. Typically, these labels sell their goods at very low prices that make them competitive with pirate products.

A large proportion of these new digital studios might be described as "informal" as they neither pay taxes nor register recordings with performers' rights organizations (or with SENAPI). According to Wilson Ramirez of Banana Records, some of these labels have their origins in pirate production, which provides both an education in the Bolivian music scene and a means of raising the capital necessary to set up recording studios. Some, he observed, continue their pirate practices clandestinely.[5]

Because of this growth at the low end, the larger cultural impact of the collapse of the Bolivian recording industry is hard to gauge. Although a number of internationally recognized neo-folklore groups have stopped recording for the domestic market, our market observations suggest that overall the number of new music releases in Bolivia has increased. Opportunities for producers and musicians from poor and indigenous backgrounds have expanded as the market for local music has grown. Profits, however, are extremely low, and contracts now typically require that the artist pay production costs to the studio and often take responsibility for distribution. The strongest rationale for these arrangements is not CD sales but the promotion of live performances.

Although the new music labels are commonly criticized for inferior—and sometimes negligible— production values, this has not played a decisive role in the local markets where such products are sold. Nonetheless, a degree of differentiation among the new producers is occurring as some seek out the higher-value segments of the market, including roles in promotion. In this respect, pirate production seems likely to follow the path of other countries documented in this report, in which the most successful pirate producers seek ways of going legit. This may be particularly salient in Bolivia, where the move from the informal to the formal economy is an important and widely held aspiration.

---

5   Interview with William Ramirez, CEO of Banana Records, October 2008.

Formats and Reproduction Equipment

Because Bolivia has one of the lowest Internet connectivity rates in South America, file sharing, digital downloads, and other forms of web access to music and film remain relatively insignificant factors in the pirate economy.[6] While the audio cassette remains important, especially in rural regions without electricity mains, the optical disc is now the dominant format. CD sales grew rapidly during the 1990s, but this market was largely restricted to the urban middle-class market. Poorer Bolivians, by contrast, tended to move directly from audio cassettes to VCDs (video compact discs) containing music videos, which became commonplace in the early part of the decade. The rapid development of this market was fueled by a flood of cheap VCD players and reproduction equipment, made in China and Taiwan and dumped on the Latin American market in the early 2000s.

The first low-budget VCD productions by regional indigenous (*originario*) artists were released around 2003 and were followed by a wave of productions by small-scale labels targeted at low-income regional audiences. More recent equipment can usually play multiple formats—CDs, MP3s, VCDs, and now DVDs—allowing the smaller labels to piggyback on the existing consumer infrastructure and increasingly saturate the market. The CD/VCD continues to be the medium of choice in Bolivia, however, because of its lower cost—usually two-thirds of the price of blank DVDs.

A Short History of Bolivian Piracy

As in many other parts of the world, large-scale music piracy emerged in Bolivia with the rise of cassette technology. A market for cassette versions of vinyl records developed in the 1970s with the introduction of cheap radio/cassette players and expanded rapidly in the 1980s as the players became available even in rural peasant communities. The standard distribution method in poorer areas involved vendor ownership of single "master" cassettes, from which copies could be made on demand. By the mid-1990s, much of this artisanal labor had given way to imports, generally smuggled from Colombia and Paraguay, and later Peru as VCDs and DVDs gained popularity. Estimates of the scale of this transborder traffic vary. One vendor consulted in Sucre (in southern Bolivia) in 2007 indicated that some 70% of the pirated music discs sold there were produced in Peru. Our research suggests that, by 2007, the actual percentage was probably much lower, with the difference attributable to increased local Bolivian production.

Regardless, the Peruvian trade remains significant and well known to most vendors. Most of the pirated discs imported from Peru pass into Bolivia via the border town of Desaguadero, situated near Lake Titicaca. According to vendor sources, Peru has several centers of pirate production, including the cities of Juliaca, Arequipa, and Lima, which one vendor in Desaguedero described

---

6   In 2008, only 62,000 Bolivians had broadband Internet access, in a country of over nine million. Only 1.23% owned computers (Arratia 2009).

as "the mega-capital of piracy." In the early 2000s, distribution to local Bolivian vendors was controlled by a relatively small number of dealers who traveled to Peru to collect merchandise or who acted as local agents for shipping, often by long-haul buses. According to a vendor based in Cochabamba, Bolivia, who labels his products "El Super Pirata DJ" (DJ Super Pirate):

> When I began this business, at first there were only something like four majors, four large-scale pirates who delivered their CDs everywhere. They had their sellers who would take, let's say two thousand CDs to one place, [where the vendors] would choose what they wanted and then take them off to the next place. During the day they would dispose of the two thousand CDs wholesale and in the night go to collect the money owed.

Local vendors often have contacts with dealers based in Desaguadero or La Paz, from whom they can order stock. The distribution network can also be used by vendors to dispatch new Bolivian releases to Peru for mass copying. This often includes the creation of a new, color printed cover (lámina) by Peruvian graphic artists, using elements from the original cover or video images captured from the VCD itself. The creativity involved in the design of these alternative covers is notable and contrasts with other parts of the world (and certain areas of the Peruvian market) where pirates favor artwork that is identical to, and ideally indistinguishable from, the original.

On a visit to Desaguadero in April 2008, on one of the town's three weekly market days, there were no border-control police in sight. Bolivian traders and vendors crossed the frontier unimpeded. A mass of stalls and shops located just inside the Peruvian border sold VCD recordings of Bolivian artists, priced in bolivianos (the Bolivian currency) to simplify cross-border sales. Single discs typically sold for 2Bs ($0.27 cents), with discounts for bulk purchases.

One of the Peruvian vendors explained that she had taken up selling VCDs in the past year, having previously run a stall serving hot food. She claimed that the profits from these two economic activities were much the same and very limited: "just enough to feed the family." Another vendor—a man in his late twenties based in a shop opposite the Bolivian frontier—clearly worked on a larger scale. He had trained as a graphic designer but, like so many vendors interviewed, had been unable to find formal-sector employment. Although his business remained worthwhile, he explained that profits had dropped dramatically: Five years ago, "if you invested $1,000, the next month you would have $3,000 or $4,000." Now the wholesale price of each pirated disc had gotten so low that profits depended on massive sales volume. The principle reason for this drop in prices was a reduction in the cost of CD-burning equipment. He claimed that five years earlier a ten-disc burning tower had cost around $8,000, which meant that only a few pirate producers with access to significant capital were able to purchase equipment. These producers could then corner the market and maintain higher prices. When we spoke in 2008, the same equipment was retailing for about $600, making it accessible to many more people, greatly increasing competition and leading to a reduction in the prices for copied discs.

With wider availability of disc-burning equipment, disc copying is increasingly done by the vendors themselves or by local suppliers. Because the actual discs are generally indistinguishable from one another, high-quality printed covers have become a key differentiator of product categories, signaling quality and commanding higher prices. Some former producers and distributors of copied discs have now branched out into printing and selling the covers—a business in which investment in expensive equipment still confers a technological advantage. Certain disc traders in La Paz now also specialize in selling *laminas*, which are largely produced in Peru.

While finished pirated discs pass into Bolivia through Desaguadero, the trade in raw materials tends to flow the other way. Blank discs, plastic presentation "jewel" cases, and the small, clear plastic bags in which discs are sold first arrive in Iquique, Chile, from China or Taiwan. They are then trucked through Bolivia into Peru via Desaguadero. Similarly, Peruvian pirate producers often travel to La Paz to purchase disc-burning towers. The sale of all these production components is presented as perfectly legal, although the IIPA claims that border duties on such goods are often avoided. Purchased in bulk (in 2008, in La Paz), blank CD/VCDs could be had for about $0.10, DVDs for $0.15, and jewel cases for $0.11.

Desaguadero's economic dependence on the black market means that overzealous border officials are not tolerated, and attempts to crack down on smuggling have generally been met with fierce local resistance. Tensions rose in 2008, for example, after the Bolivian government attempted to end the smuggling of natural gas canisters into Peru, where they can be sold for nearly five times the subsidized Bolivian price. In June 2008, a public ceremony to mark the opening of a military garrison in Desaguadero, put there to control the border trade, was met by a violent mob of townspeople, which drove the military detachment out of town before ransacking and burning the customs offices. This incident led to much tighter policing of the roads leading to and from Desaguadero, which in turn has made the journey to purchase Peruvian-produced VCD and DVD discs more hazardous for Bolivian vendors and dealers.

A greater chance of arrest is just one of several factors driving a shift toward local Bolivian production. A DVD vendor in La Paz who for many years had traveled to Desaguadero every Friday to purchase stock listed the advantages of using local Bolivian suppliers. In addition to a lower risk of police trouble, she saved a day per week in transit and the 40Bs ($6) bus fare. Customer satisfaction was also better: faulty DVDs could be returned to the supplier. However, she also stressed that her profits had fallen dramatically. A few years earlier, she had paid a wholesale price of 10Bs ($1.36) for DVDs, reselling them for 20Bs ($2.72) each, realizing a profit of 10Bs ($1.36) per disc. When we spoke in April 2008, the wholesale price was 5Bs and the retail price 8Bs, yielding a profit of only 3Bs ($0.40) per DVD.[7]

---

7 Other accounts suggest that this localization of production is not universal: variations in costs over time or by region appear to change this calculation for individuals or groups of vendors. In 2008, journalist Wilfredo Jordán recounted the story of a vendor based in El Alto, La Paz, who had abandoned the selling of children's shoes for the more lucrative pirated-CD business. Initially, the family burnt their own discs, but finding that the material costs in Desaguadero were identical and the labor costs lower, they

There is considerable variation in the prices of pirated DVDs, VCDs, and CDs. Many factors affect price, including format, genre, quality of the cover, degree of vendor competition, region, and even locality within a particular town or city. In larger towns such as Potosí and Sucre, it is common for both pirated and legal VCDs (by regional artists) to be sold for the same standard price of 10Bs (when presented in a jewel case). Many local *originario* artists in these towns, whose work is typically destined for an indigenous migrant audience, own their own stalls in poorer market areas. Some distribute their discs to vendors personally, which provides them opportunities to monitor the circulation of their work and build relationships of trust with vendors. Such practices also reflect the ways in which patterns of consumption map onto class and ethnic hierarchies in Bolivia. Original VCDs are widely available in the poorer peripheral market areas of Sucre but almost never in the more upmarket city center, with its university, beautiful colonial churches, and tourists. Stalls in the city center tend to stock national and international genres—almost without exception pirated—but rarely local or regional *originario* artists.

From Pirates to Legitimate Distributors?

In many Bolivian towns and cities, street vendors need municipal licenses, permits, or union membership in order to trade. In large cities, such as La Paz, there are several main unions representing vendor interests. Concern for local artists and the desire for wider legitimacy in the vendor community have led to some interesting efforts to address piracy at the union level, often in the form of deals with local artists in which the unions agree to act as distributors. In a few cases—notably involving local rock groups—this model has been a success: vendor unions pay artists 7Bs per CD and retail them at 10Bs. However, attempts to scale up this model and to work with music producers have generally failed. As one producer put it:

> It was an ideal proposal, but it never worked because obviously we are talking about people who have lived all their lives from informal work. In other words, they never pay anyone a penny, never pay taxes, live from the work of others, and are not disposed, nor are going to become disposed, to change this.

Union representatives, for their part, complain that producers wanted to dump out-of-date and second-rate recordings on them, which they would have been unable to sell.

A particularly ambitious 2006 agreement brought together the La Paz Union of Cinema Workers and the National Federation of Small-Scale Audio-Visual and Music Merchants (dubbed a "pirate union" in many press reports). This agreement required vendors to refrain from selling VCDs or DVDs of national and international films until after their exhibition in La Paz cinemas—typically a three-month period following first release. According to union officials, it also stipulated

opted for weekly trips to purchase stock. According to Jordán, the number of traders traveling to the Peruvian border to acquire pirated discs continues to grow.

protection in perpetuity for nationally produced films. Still more unusual was the role played by the city administration: the agreement was facilitated by the mayor's office in La Paz, and police officers were assigned responsibility for enforcement.[8]

Implementation of the agreement, however, broke down almost immediately. Press articles condemned the mayor's office for giving "a green light to piracy." Musicians' rights organizations, such as the Bolivian Society of Authors and Music Composers (Sociedad Boliviana de Autores y Compositores de Música), condemned the lack of respect for the rights of international artists. But the real damage was done by non-unionized vendors and members of other vendor unions, who were not bound by the agreement and undercut its effectiveness.

Indigenous Politics

Given the complexity of current Bolivia-US relations and the recent loss of US bargaining power following Bolivia's expulsion from the ATPDEA, external pressure seems unlikely to achieve much change in Bolivian law or enforcement practices in the near future. Internal efforts to protect, control, and develop indigenous intellectual property may prove more consequential, however. According to the Foreign Trade Information System (Sistema de Informacion al Comercio Exterior) of the Organization of American States, "The current head of SENAPI, appointed by President Evo Morales, has declared a 'revolution' at SENAPI, and currently the office seems to be focused on the registration of traditional knowledge" (USTR 2008:40). Bolivia's new national constitution, enacted in 2009, and the 2007 United Nations Declaration on the Rights of Indigenous People, which Bolivia was the first member country to sign, may also lead to new law protecting local and indigenous cultural production. The UN Declaration states:

> Article 31
>
> 1. Indigenous peoples have the right to maintain, control, protect and develop their cultural heritage, traditional knowledge and traditional cultural expressions, as well as the manifestations of their sciences, technologies and cultures, including human and genetic resources, seeds, medicines, knowledge of the properties of fauna and flora, oral traditions, literatures, designs, sports and traditional games and visual and performing arts. They also have the right to maintain, control, protect and develop their intellectual property over such cultural heritage, traditional knowledge, and traditional cultural expressions. (UN General Assembly 2007:11)

Originario-artist organizations have lobbied the government extensively to extend and protect their rights. The targets of their anger, however, include existing music-rights-collection societies, whom they accuse of having excluded them and sometimes of having plagiarized their work. This

---

8 For background on the movie sector in La Paz and, secondarily, on film piracy and circulation, see Himpele (2008).

is the other side of intellectual property politics in contemporary Bolivia—not simply the state's inability to enforce anti-piracy laws and revise copyright legislation but also the belief among some important stakeholders that existing international IP arrangements reinforce domestic and global inequality. This view is not marginal in the region: notably, it figures in the alternative regional trade agreement sponsored by Venezuela in 2003, The Bolivarian Alliance for the Peoples of Our America (Alianza Bolivariana para los Pueblos de Nuestra América, or ALBA), which counts Bolivia and eight other countries as members:

> Within the general framework of imbalanced relationships between North and South, the advantages of the North are particularly evident in scientific and technological areas. The international regime of intellectual property is strategically positioned to accentuate the imbalance. The system protects the strongest countries while leaving unprotected the areas in which the poorer countries of the South have a real advantage: the genetic biodiversity of their territories and the ancient knowledge of indigenous and farming communities. (ALBA n.d.)

Precisely how the Bolivian government will manage these issues and develop and enforce its policies remains to be seen. But it seems clear that strong enforcement is not likely to be a viable solution—even if deployed in support of indigenous works. As in many other parts of the world, recorded music in Bolivia is acquiring a primarily promotional role with respect to artistic livelihoods, rather than a direct income-generating function. It would be unrealistic to expect Bolivia or other poor countries to reverse this trend. The important point, in Bolivia and in other countries with comparable economic positions, is that for the vast majority of consumers and artists, this is not a change for the worse. The promotional function, in Bolivia, is not the remnant of a once vibrant music business but rather a new source of value for local artists and a sign of the emergence of a much wider music-consuming public than existed under the old model. How this music economy will provide pathways to legitimacy and more-than-marginal profitability is a different and critically important question. But it is not a new question in Bolivia, where most artists, even some of the best-known regional singer-songwriters, have always had to combine music with other forms of economic activity. For all but a few musicians, this was no different in the era before ubiquitous piracy.

About the Study

This account of piracy in Bolivia is based on the author's interviews with musicians, record producers, and vendors in several parts of the country (as well as in neighboring Peru). It is part of a larger ethnographic research project entitled "Digital Indigeneity," conducted in Bolivia between September 2007 and August 2008 and supported by the British Academy and the UK Arts and Humanities Research Council (AHRC).

References

ALBA (Alianza Bolivariana para para los Pueblos de Nuestra Américas). n.d. "What is the ALBA?" http://www.alternativabolivariana.org/modules.php?name=Content&pa=showpage&pid=1981.

Arratia, Orlando. 2009. *Bolivia: Universal Broadband Access; Advances and Challenges*. Melville, South Africa: Association for Progressive Communications. http://www.apc.org/en/system/files/CILACInvestigacionBolivia_EN_20090707.pdf.

BBC News. 2008. "Bolivia Tells US Envoy to Leave." September 11. http://news.bbc.co.uk/go/pr/fr/-/2/hi/americas/7609487.stm.

Himpele, Jeffrey D. 2008. *Circuits of Culture: Media, Politics, and Indigenous Identity in the Andes*. Minneapolis: University of Minnesota Press.

IIPA (International Intellectual Property Alliance). 2006. *Special 301 Report on Copyright Protection and Enforcement*: Bolivia. Washington, DC: IIPA. http://www.iipa.com/rbc/2006/2006SPEC301BOLIVIA.pdf.

———. 2008. "Comment on Andean Trade Preferences Act: Effect on the U.S. Economy and on Andean Drug Crop Eradication." Accessed August 11, 2008. http://www.iipa.com/pdf/IIPAAndeanATPAfilingtoUSITCfinal07292008.pdf.

———. 2010. 2010 *Special 301 Report on Copyright Protection and Enforcement: Peru*. Washington, DC: IIPA.

Jordán, Wilfredo. 2008. "El Negocio de CD 'Piratas' se Yergue Entre la Informalidad y el Contrabando." *Wilfredo Jordán Blog*, December 15. Accessed March 25, 2010. http://wilfredojordan.blogspot.com/2008/12/el-negocio-de-cd-piratas-se-yergue.html..

La Rázon. 2000. "La Pirateria Hiere a la Industria y Priva de $US 15 Milliones al Pais." January 7.

Ortiz, Pablo, and Ricardo Herrera. 2003. "Lauro & Cia Cerró Como Productora Fonográfica." *El Deber*, February 11.

Schneider, Freidrich. 2002. "Size and Measurement of the Informal Economy in 110 Countries Around the World." Paper presented at a workshop of the Australian National Tax Centre, Canberra, July 17. http://rru.worldbank.org/Documents/PapersLinks/informal_economy.pdf.

Tiempos del Mundo. 2000. "Industria Fonográfica en Crisis." April 20.

UN General Assembly. 2007. Resolution 61/295. "United Nations Declaration on the Rights of Indigenous Peoples." September 13. http://www.un.org/esa/socdev/unpfii/documents/DRIPS_ en.pdf.

USTR (Office of the United States Trade Representative). 2008. *2008 National Trade Estimate Report on Foreign Trade Barriers.* Washington, DC: USTR. http://www.ustr.gov/about-us/press-office/ reports-and-publications/archives/2008/2008-national-trade-estimate-report-fo-0.

———. 2010. *Fifth Report to the Congress on the Operation of the Andean Trade Preference Act as Amended June 30, 2010.* Washington, DC: USTR. http://www.ustr.gov/sites/default/files/ USTR%202010%20ATPA%20Report.pdf.

Vuletin, Guillermo. 2008. "Measuring the Informal Economy in Latin America and the Caribbean." IMF Working Paper WP/08/102, International Monetary Fund, Washington, DC. http://www. imf.org/external/pubs/ft/wp/2008/wp08102.pdf.

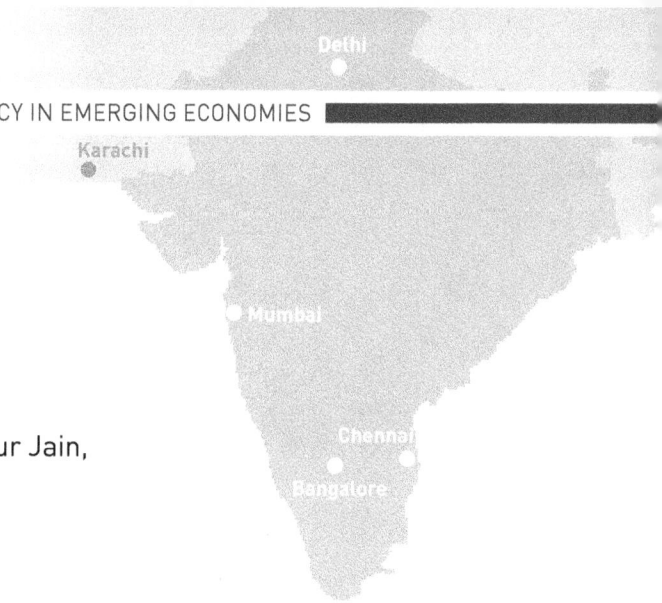

Chapter 8: India

Lawrence Liang and Ravi Sundaram

Contributors: Prashant Iyengar, Siddarth Chaddha, Nupur Jain, Jinying Li, and Akbar Zaidi

Introduction

Piracy entered public consciousness in India in the context of globalization in the 1980s. The rapid spread of video culture, the image of India as an emerging software giant, and the measurement of comparative advantage between nations in terms of the knowledge economy pushed questions about the control of knowledge and creativity—about "intellectual property"—into the foreground of economic policy debates. The consolidation of Indian media industries with global ambitions in film, music, and television gave the protection of copyright, especially, a new perceived urgency. Large-scale piracy—at the time still primarily confined to audio cassettes and books sold on the street—began to be seen as a threat not just to specific businesses but to larger economic models and national ambitions. The "problem" of intellectual property (IP) protection in India—in terms of both the laws on the books and enforcement practices on the streets—took shape through this conversation between lawyers, judges, government ministers, and media lobbyists.

The high-level policy dialogue has produced several important revisions to Indian copyright law, including amendments to the Copyright Act in 1994 and 1999 to address the growth of cassette and optical disc piracy, respectively. A new round of proposals, reflecting a more recent array of battles over the control of cultural goods, began to emerge in 2006 and will probably be voted on in early 2011.

Since the 1990s, piracy in India has been shaped by a now familiar set of global transformations in the production, circulation, and regulation of media and culture. These range from macro-level changes, including new international IP obligations and India's integration into global media markets, to extremely local developments, such as the adoption of cheap DVD players, burners, and computers in poor urban neighborhoods. In this respect, India belongs to the wider story of technological, cultural, and policy change recounted throughout this report. And yet two related factors make the Indian case profoundly different.

First, unlike nearly every other middle- and low-income country, India's film and music markets are dominated by domestic firms, which compete fiercely on price and services. The gap between high-priced international media goods and very low-priced pirated goods is filled, in India, by Indian companies. The resulting turmoil around distribution and pricing resembles the current upheavals in the US and European markets, where the rollout of low-cost digital media services is throwing older business models into crisis—and a difficult process of reinvention.

Chapter Contents

339   Introduction

344   The Circuits of Piracy

347   Piracy on the Streets

350   Piracy and Time

352   The Transnational Trade

356   Internet and P2P

356   Broadband Deployment

356   The Growth of P2P in India

358   Content

360   Internet and Digital Media Policy

362   The (Re)organization of Indian Film Markets

364   Market Segmentation

365   Cinema Halls

366   Windowing

370   Price and Competition in the Optical Disc Market

374   The Market for Piracy Studies

376   The Judicial Enforcement of Copyright

383   Enforcement Networks

385   The Motion Picture Distributors Association

388   The Indian Music Industry

390   Software Piracy

391   The Conditions of New-Media Art in India

393   Conclusion

395   About the Study

395   References

But such innovation remains rare in developing-country media markets dominated by multinational companies, where small markets structured around high prices remain the norm.

The dynamism of Indian media markets, we should be clear, does not provide a model for fighting piracy—indeed India presents unique barriers to enforcement on a number of levels. But it does offer a powerful counterpoint to the dilemma facing most developing countries: the chronic weakness of legal media markets, trapped between high prices and widespread piracy. And it provides a unique answer to one of the central questions that emerges from this report: not whether piracy can be eliminated or even significantly diminished, but whether the accompanying legal markets will be structured around high-cost or low-cost competitors.

The second factor that distinguishes India is its decentralization of cultural production and governance. Although Bollywood is often used as shorthand for Indian film in general, it technically only describes the portion of the industry centered around the Hindi studios based in Mumbai (formerly Bombay). In practice, Bollywood is only one of several important regional and local language cinemas, including distinct Tamil, Kannada, and Malayalam production. Television follows similar patterns of regional production, grounded in media policies that date back to Nehru's modernization campaigns of the 1950s. These regional media are enormously popular. Indian film, television, and music production dominate the domestic market and have confined Hollywood and other transnational media companies to very marginal roles. Hollywood accounts for around 8% of the Indian box office—reversing the percentage found in many other middle- and low-income countries. International music accounts for just 6% of the market (Kohi-Kandekar 2010).

Patterns of corporate ownership reinforce this strongly regional organization of markets. Despite considerable efforts in the past decade to implement modern corporate ownership structures, the Indian entertainment industries retain a high degree of informality and are often organized around family units. Such structures grow out of and can convey advantages in local markets, but to date they have provided a relatively weak interface to the global media economy, and to its global rights-holder organizations in particular.[1]

This decentralization has been a source of perennial frustration for industry lobby groups such as the Motion Picture Association of America (MPAA) and the Business Software Alliance (BSA), which have long sought—and generally failed to find or establish—unified Indian rights-holder groups or authorities with whom to work. When international enforcement organizations want to talk to Indian rights holders, they deal with a plethora of regional producers focused mostly on local markets regulated by state (rather than national) laws. Multinational industry groups in India have as a result seen their role as producing a discourse on piracy and enforcement for the local players as much as for themselves, with the goal of creating a stronger context for their own long-term operations and lobbying. On these terms, the multinationals have been modestly successful in localizing anti-piracy discourse within individual Indian state governments and in enlisting local cinema, music, and software industry associations in those efforts. Many of these local groups have, in turn, adopted the anti-piracy rhetoric and practices of the multinationals, and several have independently sponsored enforcement campaigns in the Indian media.

Policy advocacy and enforcement efforts confront similar challenges. Enforcement, in India, is organized at the state level, not by the national government.[2]

---

1   See Rajadhyaksha's (2009) account of the progressive corporatization and globalization of the Hindi film industry in the 1990s.

2   The constitution of India separates the powers of the national state into three categories or "lists": those belonging to the central government, those belonging to the states, and those belonging to both—the concurrent list.

Acronyms and Abbreviations

ACTA	Anti-Counterfeiting Trade Agreement
BSA	Business Software Alliance
DRM	digital rights management
ESA	Entertainment Software Association
FICCI	Federation of Indian Chambers of Commerce and Industry
FPBAI	Federation of Publishers and Booksellers Association of India
GDP	gross domestic product
IDC	International Data Corporation
IFPI	International Federation of the Phonographic Industry
IIPA	International Intellectual Property Alliance
IMI	Indian Music Industry
IP	intellectual property
IPRS	Indian Performing Right Society
ISP	Internet service provider
IT	information technology
MPAA	Motion Picture Association of America
MPDA	Motion Picture Distributors Association
NASS-COM	National Association of Software and Services Companies
P2P	peer-to-peer
TRIPS	Agreement on Trade-Related Aspects of Intellectual Property Rights
UPDF	United Producers and Distributors Forum
USIBC	U.S.-India Business Council
USTR	Office of the United States Trade Representative
VCD	video compact disc
WCT	WIPO Copyright Treaty
WIPO	World Intellectual Property Organization
WPPT	WIPO Performance and Phonograms Treaty
WTO	World Trade Organization

Although India's legal regime and Western-style court system make it relatively hospitable to imported legal arguments and practices, the state-based organization of police, law, and courts means that enforcement efforts quickly become enmeshed in complex local political contexts, where industry actors have different degrees of leverage in mobilizing the police and pushing cases through the overburdened Indian courts. India's cumbersome criminal judicial procedures, in particular, are a regular theme of International Intellectual Property Alliance (IIPA) reports on India and form the basis of demands for new "fast-track" IP courts that can process more than the current trickle of infringement cases. Industry successes—notably in expanding the use of organized-crime statutes to prosecute film piracy—have all come at the state level, with the support of powerful local industry stakeholders.

Although foreign lobbies have become more active at the state level, they have also tried continually to make the central Indian government a more effective agent for enforcement, with stronger control over domestic media markets and media flows. Major touchstones of this effort have included stronger border surveillance measures and stronger national coordination of India's highly decentralized state police forces. More recent strategies include the push for stronger enforcement provisions in the pending revision of the Copyright Act. The difficulty of this project is another recurrent feature of the IIPA's India reports. In 2009, the IIPA argued that "what is desperately needed in India, and particularly for the Indian copyright industries, is a national anti-piracy strategy at the central government level, with the ability to link in the State governments . . . in a meaningful, enforceable way" (IIPA 2009b). The lack of traction on these issues has contributed to India's more or less permanent place on the USTR watch lists.

Indian enforcement is generally rated a failure by the IIPA and its member groups—although the IIPA's accounts seem to reflect at least as much frustration at the lack of responsiveness of the Indian government as with the prevalence of piracy itself, which is high but fairly typical of developing countries. Reported rates of piracy have remained relatively stable over the past several years—66% in 2008 for software, 55% for recorded music, 89% in the last Entertainment Software Association (ESA) survey of game piracy (released in 2007), and 29% in the last MPAA survey of film piracy (released in 2005). Significantly higher numbers circulate among some of the local industry groups, including estimates of 90% piracy in the DVD market and 99% in the digital music market.[3] As usual, the research underlying these claims is not made public and so cannot be evaluated.

Despite IIPA complaints, the overall pattern of enforcement in India and changes in the organization of piracy show close parallels with other countries. Middle-tier retailers are increasingly exiting the pirate market—pushed by a combination of police harassment in major marketplaces, such as Palika Bazaar in Delhi and Lamington Road in Mumbai, and by the fall in

Law and order is a state subject. Thus, enforcement initiatives that rely on the police are organized at the state level rather than through any central agency.

3   Sanjeev Varma, head of corporate communications for Moser Baer, interview, 2009. The 99% figure is an IFPI figure cited by the IIPA (2010).

pirate disc prices, which has dramatically reduced profit margins. Both factors contribute to the growing "deformalization" of the pirated optical disc trade and to the emergence of more mobile, lower-cost, enforcement-resistant street vending practices.

Parallels in enforcement practices are also striking. Domestic producers have been very successful at mobilizing police in the service of specific enforcement campaigns, most frequently to protect the release windows of anticipated hit films. Targeted, large-scale police actions in Mumbai (for Hindi releases), Chennai (for Tamil releases), and Bangalore (for Kannada releases) have become a relatively common part of film industry release strategies and create a very uneven enforcement terrain for the broader community of movie producers. International films, in contrast, circulate through a wider array of pirate distribution channels and have no such local chokepoints, making them less suitable targets of concentrated police effort.

The balance of this chapter explores these intersections between media markets, piracy, and enforcement in India, with a focus on three broad issues:

- First, the organization of the informal media economy in India, from street vending to transnational networks for cheap hardware and pirate disc production. These networks are part of what has been a predominantly Asian geography of piracy over the past two decades, encompassing China, Malaysia, Pakistan, and the destination countries of the South Asian diaspora. This geography is now losing its coherence as once-dominant practices of transnational cassette and disc smuggling give way to growing local production and Internet distribution.

- Second, the rapid growth and changing structure of the Indian film industry, whose fortunes dominate debates about piracy and enforcement in India. Two factors stand out in this analysis: the ongoing reorganization of theatrical exhibition around the multiplex, marked by rising prices and an increasingly locked-down distribution channel; and rampant competition in the secondary markets for DVDs, soundtracks, video-on-demand releases, and other similar media, marked by rapidly declining prices and increasingly direct competition with the pirate market.

- Third, the evolution of enforcement activities in India, shaped by the regionalism of Indian media markets and the decentralization of political authorities but also by emerging cooperation between international, national, and local organizations. Increasingly, such cooperation extends across the full spectrum of enforcement activities, from street-level raids and investigations, to the training of police and judges, to the cultivation of a wider discourse on intellectual property and piracy in the popular media.

The Circuits of Piracy

In India, as elsewhere, most accounts of media piracy are built around the moral and economic claims of rights holders and are intended as justifications of those claims. These accounts tend to focus on the illegality and criminality of acts of piracy. They project a sense of loss and danger onto such acts, often by implicating piracy in a range of Indian social crises, from unemployment to organized crime. Such efforts to shape perceptions about piracy have been central to the organization of enforcement efforts but, in our view, are largely disconnected from the actual practices of media piracy. By raising the level of drama, they rarely capture the ordinariness and ubiquity of piracy in the contemporary Indian media landscape.

This ordinariness is rooted, first and foremost, in the central role the informal sector plays in cultural innovation in India, especially in the context of the digital revolution of the past decade. In India, as in other middle- and low-income countries, digital culture has become a mass experience primarily through the informal sector. Cheap Chinese hardware imports, grey-market recycled goods, and pirated content have been fundamental, for better and for worse, to this growth and to the resulting expansion of media access. They have also been central to the remarkable democratization of media production as the costs of production and distribution decline. In this context, pirate and grey-market practices have been vectors not only of "consumption" in a narrow sense but also of cultural participation, education, and innovation for a wide range of Indian publics.

Generally, these changes rise to the attention of the Indian media only in the context of the piracy debates, where they are usually stigmatized. There has been very little work on the nature of these new-media networks and what they mean for the social and economic development of India—or for our broader understanding of globalization. This chapter—and the longer trajectory of work of which this report is a part—is an attempt to document these transformations in the experience of media and urban life in India.

Most explorations of this subject in Indian journalism and scholarship circle around two poles of the Indian imagination: the interventions of the state, exemplified by the modernization plans of the 1960s and 1970s, and more recently, the dynamism of the market—a perspective that has become familiar with the emergence of India as a major player in information technology (IT) in the past two decades. In practice, however, India counts few simple transitions from rural society to fast-paced, market-driven modernity. The experience of modernity, for most Indians, comes not through the arrival of the new but through the recycled technologies and cheap copies that follow. It comes secondhand, as technologies circulate through communities, regions, and classes long after their original utility is past.

By locating piracy within histories of non-elite media practices, we have tried to avoid definitions of piracy as theft or crime and focus instead on how pirate practices weave into existing social relations while at the same time transforming them. Consistently, this has meant focusing on groups excluded from the technical education common in the Indian middle and upper classes,

Nehru Place

Nehru Place, in Delhi, hosts Asia's largest secondhand computer market, India's largest garment-export center, and a large pirated software market. This is far from its original purpose. In the 1962 master plan for Delhi, Nehru Place was slated to be the largest of fifteen district centers that would redefine urban life in the city. It was the epitome of the modernizing imagination at work in Delhi in the 1960s, and in India more generally.

H. P. Singh was one of the earliest garment exporters to move to Nehru Place. He now owns one of the largest warehouse-and-retail stores in the complex, and he feels deeply betrayed by the failure of that modernizing vision: "When I saw the plans of Nehru Place, I was promised the boulevards of London and Paris, the district centers of Europe. Little did I know that this is the shape it would take."

Today, Nehru Place is a prime example of the fluid boundary between formality and informality in the Indian economy. It is also a bustling example of commercial rejuvenation associated primarily with the establishment of the "secondhand hardware" market in the area, which sells refurbished computers, software, and a variety of support services. The computer hardware shops first made their presence felt in the early 1980s when local companies and government departments were beginning to adopt computers in large numbers. They continue today as part of the large Indian grey market for computers and software.

The majority of the goods in the secondhand hardware market are discarded computers shipped from Southeast Asia, which are disassembled and then reassembled for local reuse. Sachin, a hardware seller, described the basis of the practice: "In most countries abroad the life of a computer is considered to be anything between a year and a year and a half. Upgrading is the key to technology there. Over here, people do not sell their computers. They look at them like their TVs, passing them on to others when they are through with them."

The market operates as a distributed network of shops that communicate with each other and trade parts and services. Most of the shops are connected through an intercom, referred to as the *chhoti* line (or small line), the existence of which is illegal since it bypasses the official telephone networks. When specific goods are needed, calls are made to the vendors who may have them. When this fails, the request is often passed to others, ensuring that most demands can eventually be met. One implication is that there are no centralized warehouses within the complex. Another is that most transactions are in cash, in order to facilitate the trade in parts and services across the network.

Pirated software is also widely sold at Nehru Place and often comes pre-installed on secondhand machines. Overlooking the software section of the market is Microsoft's Delhi headquarters, which stands in the middle of the main courtyard at Nehru Place. Like all the other businesses at Nehru Place, the Microsoft site was originally intended for other purposes. On the original map, it is identified as a cinema hall.

yet who have managed to climb the value chain in the information economy. Many of these social trajectories begin in call centers, photocopying centers, cyber cafes, or the computer service and repair industry. Many are marked by the aspiration to escape the constraints imposed by a lack of formal education. An understanding of these worlds of social mobility and aspiration has been key to our understanding of piracy and new-media networks (Sundaram 2001, 1996; Liang 2003).

Indian urban experience teaches us that the question of legality is often the least interesting place from which to begin such an inquiry. Strict lines between the legal and the illegal are often irrelevant to the construction of Indian media practices, especially in the context of the large informal sector. More often, it is better to ask how people navigate the urban media environment—how they access or make the media they want in relation to the range of available resources and constraints, including legal constraints.

Our interest is piracy is therefore not primarily about its illegality—indeed the construction of that boundary in the law and on the street has been enormously complicated in India. Instead, we are interested in its ordinariness—a question we have approached by analyzing the social worlds in which piracy emerges, the forms of circulation and consumption in which it is implicated, and the fears and forms of social control that it generates.

It is impossible to understand these practices without understanding the conditions of informality that shape access to goods, services, housing, political authority, and most other features of life in urban India and in other cities of the Global South. Most such cities are built on—and through—informal networks of housing and infrastructure, only partially imagined by municipal planning and only partially integrated into the networks of business, services, and governmental authority that face outward to the global economy.

These conditions of informality have been widely tolerated by the Indian state as an unavoidable aspect of urban growth. But attitudes have shifted in the past decade as the integration of the Indian economy into the global economy has increased pressures on land use in most cities. Assaults on zones of informality have become far more common, with the courts playing a central role in determining the survival of slums and squatter settlements in many Indian cities. Inevitably, this process results more in the displacement of informality than in its elimination. As old zones of informality are destroyed, the poor and marginal Indians who inhabited them move elsewhere. The resulting disruptions of the urban landscape also create new temporary zones, where informal trade and marginal livelihoods reemerge.

We make this digression through urban redevelopment because piracy cannot be divorced from the circuits in which it takes place. A nuanced account of piracy in India begins with the "many piracies" that cut across the daily lives of Indians. Three of these circuits have proved central to our inquiry: the traditional and still ubiquitous world of street vending, the production and distribution networks that organize the optical disc trade, and the more recent emergence of peer-to-peer (P2P) networks and other channels of digital distribution.

Piracy on the Streets

Arguably, the greatest threat to the informal street economy is not the long arm of the law but the unpredictable nature of the weather. It is a common sight in Bangalore to see hundreds of street vendors sent scattering for cover by sudden, heavy monsoon rain. Police raids, in general, produce much the same effect and often have the same outcome, with street vendors resuming their places once the raid is over.

As optical disc piracy became commonplace in the 1990s, markets and street spaces emerged as semi-permanent points of sale. Places like National Market and SP Road have achieved an almost iconic status as the pirate centers of Bangalore, home to wholesalers of an assortment of counterfeit and pirated products: DVDs and DVD players, Chinese-made mobile phones and PDAs, MP3 players and jukeboxes, fake Ray-Bans, and gaming consoles. Even VHS players can still be found, servicing the legacy collections of video cassettes built up in the 1990s and early 2000s.

Vendors know their customer base and vary their goods accordingly. On a cart outside the Ayyapan Temple on Millers Road (a destination spot for Tamil residents of Bangalore), Tamil films make up the bulk of vendor stock, followed by Hindi movies. English-language films sit in a single pile, mostly undisturbed by customers. On MG Road, a major office corridor, pirates cater to young professionals looking for after-work entertainment. Here, stock tends toward an even mix of Hindi, English-language, and regional Indian cinema. When construction work for the Bangalore Metro began in the posh neighborhood around 100 Feet Road, the road was transformed into a downmarket street bazaar. Vendors set up outside the big-brand shops and sold discs to corporate executives returning home from the high-tech corridors around the city. In this part of town, Hollywood rules the informal box office. While the mode of sale remains similar across locations, the sellers are mobile and quick to adapt to changes in the city and in their clientele.

Street pirates also offer different types of goods, reflecting the changing availability of higher- or lower-quality copies of new films and the perceived market for supplementary materials such as liner notes. At the high end of the market are the high-definition releases of new films, generally compressed from Blu-ray masters down to 720p MP4 files or similar formats capable of being burned onto a DVD. These are still a rarity in most street markets but are increasingly common in neighborhoods frequented by HDTV-owning corporate employees, such as those living around 100 Foot Road. At the low end are compilation discs of Hollywood and Bollywood films, usually with three to five films per disc, but sometimes with up to ten or more. The higher the number of films, the lower the quality of reproduction. Single vendors generally stock products across these categories. A copy of *The Untouchables* purchased during our visits, for example, was a duplicate of the official DVD release, while a DVD with three new Hindi films contained barely watchable "camcordered" prints. Over several weeks, successively better copies of the Hindi films will become available as improved camcorder releases are paired with improved audio tracks. Within a couple months, vendors will be selling bit-perfect copies of the DVD release.

As in most of the countries documented in this report, enforcement campaigns against organized retail piracy have intensified over the last five to six years. In states like Tamil Nadu and Karnataka, local governments have extended the Goondas Act—traditionally used to curtail activities like bootlegging and extortion—to cover video piracy. The Goondas Act has been a lightning rod for criticism due to its high penalties, which include prison terms of up to two years, fines of up to US$2000, and the possibility of pre-trial detention without bail for a period of up to a year.

The extension of the Goondas Act to cover video piracy was passed at the behest of local film industry representatives and produced a typically parochial arrangement in which the measure was applied only to local film. In Burma Bazaar in Chennai—arguably the largest pirate market in India—most DVD retail kiosks post notices that they respect the copyrights of the Tamil film industry and do not stock or sell Tamil films. Pirated copies of the latest films from Hollywood and Bollywood, in contrast, are available in large quantities, in plain sight. Such arrangements are common in India and reflect the intense localism of many aspects of cultural identity, trade, and governance. Pirates in the southern state of Karnataka do not stock Kanadda-language films. Enforcement in Andra Pradesh targets only local Telegu film. The local politics of enforcement are often the only politics of enforcement.

Major markets for pirated goods, such as National Market in Bangalore and Palika Bazaar in Delhi, have nonetheless come under growing pressure from police. Although local police presence in such markets is usually mediated by a variety of informal agreements with vendors and market operators designed to minimize the incidence of serious crime, the last five to six years have seen a significant rise in the frequency of anti-piracy raids to a level that has forced many pirate vendors out of the relatively exposed market settings. The most obvious result of this pressure has been the further decentralization and deformalization of pirate sales. In Delhi, for instance, the concentration of pirate vendors in the major markets has given way to a much more local organization of pirate distribution, often intermingled with street vending of vegetables, fruit, and other goods.

In Bangalore, interviews with vendors, wholesalers, and police generally pointed to Chennai, a city on the southeast coast of India, as the main distribution hub for Hollywood, Bollywood, and Tamil films. Chennai has long been a center of smuggling and other informal market activity in India. When imported luxury goods such as perfumes and electronic goods were subject to high tariffs, before the liberalization of the Indian economy in the 1990s, Chennai was the electronics mecca of southern India. Chennai markets like Burma Bazaar were often the only places where popular imported consumer goods could be procured. When tariffs were removed, undercutting the profitability of many types of smuggled goods, vendors in Burma Bazaar shifted to selling pirated VCDs (video compact discs) and later DVDs. Burma Bazaar remains one of the most highly contested spaces in the Indian informal economy, with regular raids disrupting the flow of pirated goods but with little evidence of lasting effects or a diminution in the overall trade.

Blue and Silver Maal

Pirated discs are commonly called maal in India—a colloquial term meaning "goods" but usually used in reference to illegitimatte or pirated goods. There are two kinds of maal in circulation: blue and silver. The average blue maal is a low-quality VCD—generally a locally produced copy of a Bollywood film. These cost anywhere between Rs.40 and Rs.50 ($0.80–$1.00) in Mumbai. Imports are generally higher-quality silver maal— discs copied from DVD masters. Silver maal are available for both Bollywood and international films and can command a premium price of up to Rs.100 ($2), especially when they replicate the cover treatments of licensed discs.

Although the street prices for both categories of discs have dropped, the price used by police in estimating the values of seizures has risen, feeding skepticism about police reporting on the size of pirate markets. Prior to 2006, the cost of each VCD/DVD was pegged by police at Rs.100 ($2)—the high end of street prices. Today, when maal are seized by the police, the cost of each is estimated at Rs.300 ($6)—the high end of retail prices for most local film. This shift from street prices to retail prices in estimating the value of pirated goods is consistent with how rights-holder groups like to calculate damages in court but, as we have argued, no longer consistent with how they calculate broader losses (see chapter 1). In practice, the shift inflates the scale of both piracy and enforcement operations, which, in a system that rewards the public display of enforcement, is almost certainly the point.

Despite industry attempts to link such markets to larger networks of organized crime, examination at the local level reveals the small-scale nature of much of the business. Pirated goods are brought into Bangalore from other cities, such as Mysore and Hosur, in small cartons carried in personal vehicles or on tourist buses. These are delivered to the wholesale markets, which in turn distribute to the retailers. There are many intermediaries, and at every step, the margins in the trade have become very thin. Prices for pirate media have fallen dramatically in the past four to five years, putting significant pressure on all players in the commodity chain.

As recently as 2004, DVDs typically sold for over $2. Today, distributors generally buy films at wholesale for around $0.70 and sell to street vendors at $0.80 per copy. The vendors, in turn, sell the disc for around $1 on the street or in the pirate bazaars. There is still a great deal of variation in price as well, reflecting factors such as the proximity to upscale neighborhoods or the distance from highly competitive wholesale markets like Burma Bazaar. At Burma Bazaar, DVD prices at retail can be as low as $0.40.

With such thin margins, volume sales are critical. According to our interviews, a wholesaler in National Market in Bangalore sells on average a thousand discs per day. Outside National Market, there are other small shops and street-side vendors spread across the city that purchase from wholesalers and sell with a markup of Rs.10 to Rs.15 ($0.20 to $0.30). The average price of a copy of an English-language or Bollywood film at Brigade Road or Indira Nagar, both hot spots for Bangalore's commercial activity, is roughly $1 per disc. Here, the vendors sell between fifty and a hundred discs per day, generally making a profit of $10–$20.

Nearly all informants agreed that the pirated disc market had become much less lucrative in recent years. Most blamed a confluence of factors, from the recession to the changing release practices of the studios, which have narrowed the window between theatrical and DVD release and thereby diminished the period in which the pirates have a monopoly on distribution. All, however, saw the Internet as the primary threat, as improvements in bandwidth undercut the two traditional advantages of the street vendor: faster availability and lower prices.

Piracy and Time

The notion of access to media, developed throughout this report, usually centers on questions of cost and availability. But another crucial factor is the relationship to time. Global licensing regimes for film, especially, attempt to maintain well-ordered flows of commodities across time and space. "Windowing" is the industry term for the control of circulation over time. A modern Hollywood or Bollywood production has many release windows, beginning with the long, anticipatory advertising campaign, which primes the public for the initial theatrical release. The theatrical release window is the critical period for revenues. In India, with local products in intense competition for screens, this period can be very brief.

Recurrently in our work, we find that timing is as important as price in shaping both licit and illicit media markets. The temporal nature of distribution is tied not just to an economic logic but also to what we call an economy of anticipation. The buildup to the latest film; the release of trailers, posters, and soundtracks; the premiere—all are part of this economy of anticipation. Within it, however, the share of waiting is very unevenly distributed. The wait grows longer as you move from the northern hemisphere toward the Global South, and from metropolises to small towns and villages. The trend toward simultaneous global release—now common for many large Hollywood productions—is an effort to minimize the pirates' opportunities to exploit these gaps. But even this trend is operative only in major cities. In the provinces, people wait.

In these contexts, the newness of the films, the quality of their reproduction, and the quality of the cinematic experience come to stand for temporal and cultural differences—differences between the North and the South, between the town and the city, and between global modernity and those who are "not quite modern." In films such as *Main Madhuri Dixit Banna Chahti Hoon!* (2003) and *Haasil* (2003) or Pankaj Kumar's documentary *Kumar Talkies* (1999), we get a glimpse of this waiting-room world of cinema. In a delightful scene in *Main Madhuri Dixit*, the protagonist goes to watch the Hindi film *Devdas*, but after a few reels the film stops and the audience has to wait for the arrival of the other reels from the neighboring village. Members of the audience complain that the last time, they had to wait for over two hours after the delivery bike carrying the other reels broke down.

The big city is the place where these fractures can be repaired, where films are shown in their entirety, and where audiences do not have to confront their geographical and cultural marginality every time they attend the cinema. The social life of piracy occurs at this intersection of anticipation—now often measured in days or weeks—and aspiration to belong to the modern, to inhabit the space of global time represented by and through the movies, where things are not perpetually breaking down or delayed (Vasudevan 2003; M. Prasad 1993; Bagchi 2006).

Waiting for the latest Hollywood or Bollywood release, in this context, becomes an apt metaphor for the experience of those placed differently within the circuit of space and time. Brian Larkin and Ravi Sundaram, both students of the "pirate modern," argue that in contrast to the dizzying, real-time global integration of the information era, the great majority of people in the Global South experience time not through the trope of speed but through the experience of interruptions and breakdowns. Breakdowns create a temporal experience that has less to do with speed and more to do with the process of waiting.

From waiting for e-mail messages to open, machines to be fixed, or electricity to be restored, the experience of technology outside the high-tech centers is subject to a constant cycle of breakdown and repair. In most countries, the promise of technological prosthesis—of enhanced memory, enhanced perception, enhanced communication—is thwarted by the everyday experience of technological failure. Each repair enforces another waiting period. The experience of slowness, moreover, comes as a consequence of speed-producing technologies, so that speed and acceleration, deceleration and stasis are relative, continually shifting states. In most countries, consequently, technological modernity is predicated not on the smooth functioning of new technology but on its imperfect adaptation or indigenization. Digital piracy in developing countries is an example of this wider process, built on the cheap, repurposed infrastructures of the information economy (Sundaram 2001; Larkin 2004).

An interesting instance of this adaptation in film technology is the history of the video compact disc. Sony and Philips introduced the VCD in 1993 as a format for recording video on compact discs. It was cheap, convenient, and initially seemed to signal an emerging standard. At the time of the introduction of the new format, however, development of the technologically superior DVD was already underway. From the beginning, Philips was aware of the impending arrival of the DVD and its threat to the VCD. Anticipating a bleak future for the new format, Philips and Sony abandoned plans to launch the VCD in Western markets and opted instead to launch it in China, where its technological inferiority would not be as rapidly challenged (Wang 2003; Hu 2008). Because Philips and Sony had a tight grip on the production of discs and players, the film industry believed that VCDs would help fight widespread video cassette piracy. Instead, the introduction of the VCD triggered a boom in the Chinese production of cheap disc technologies.

Asian markets enthusiastically adopted the VCD and—shortly thereafter—VCD piracy as a means of bypassing global distribution networks for Hollywood and Bollywood film. Sharp Chinese quotas on the number of Hollywood films that could be released domestically in a given year gave a huge boost to the practice. Pirated VCDs became the only means of watching many of the latest Hollywood titles, few of which ever saw theatrical release.

Within a short period of time, the VCD became the primary movie format in large parts of the developing world. It was also a short-lived format that inaugurated a process of rapid diffusion and turnover of new, cheap, digital consumer goods. By 1998, VCD adoption was already widespread in China, with roughly sixteen VCD players per hundred households. By 2000, the number had more than doubled, significantly outpacing cell-phone adoption. But the shift to VCD-compatible DVD players was already underway. In 1999, VCD player sales in China peaked at twenty-two million. By 2000, annual DVD player sales had jumped from one million to three million, on their way to a 2006 peak of nineteen million (Linden 2004; Digital TV News 2008).

VCD technology spread rapidly from East Asia to other parts of Asia. Within a few years of their introduction, VCDs had replaced video cassettes as the standard video format in the region and had become vastly more prolific than the VHS format ever was. In India, the price of VCRs never fell below $200. VCD players, in contrast, had plummeted to as little as $20 by the middle of the decade. As with other obsolete technologies, the VCD infrastructure remains important outside the major Indian cities; the total number of DVD players surpassed VCD players only in 2008 (Kohi-Khandekar 2010).

The VCD also spread rapidly to other world regions. In Nigeria, home to the second-largest film industry in the world in terms of numbers of films produced (more than 1,200 in 2008), most films are available only on VCD and DVD. The Andean countries were also flooded with cheap players in the early 2000s, and VCDs remain prolific in Andean pirate markets—a topic explored in our Bolivia chapter. But most Western markets never saw the VCD, and the format remains a marker of the technological periphery.

The Transnational Trade

Well-developed networks trafficking in pirated Indian films emerged in the 1980s, in the early days of the video cassette era. In a pattern that would be repeated over the next three decades, illicit networks took advantage of market opportunities created by the major producers. In this case, the Bombay studios decided to ban video releases of new films for fear of cannibalizing theatrical exhibition. This did little to stop video retail and exhibition in India, but it did ensure that the growing sector remained entirely illicit (Sundaram 2009). Additionally, because Bollywood's international distribution networks were poor, the pirate networks provided the primary means of circulating new Bollywood films to international audiences. The United Kingdom, Pakistan, and Dubai—the last of these the offshore hub of much legal and illegal Indian business in the period— became the main nodes in this international distribution network.

Other Indian media and IT businesses grew out of similar transnational networks in the 1980s. Indian grey-market suppliers for computers and electronic components traveled back and forth to Taiwan and Southeast Asia to source components and raw materials for emerging domestic producers. Financing for these trips was often provided through local bazaar networks or through

The T-Series Story

In the late 1970s, Gulshan and Gopal Arora owned a fruit-juice shop in Delhi, but their real interests were in music and electronics. In 1979, the two brothers opened a small studio where they began to record Gharwali, Punjabi, Bhopjpuri, and other Indian regional music. Borrowing money, they visited Japan, Hong Kong, and Korea to learn more about the recording industry and cassette production technologies. On their return to India, they set up a factory to produce magnetic tape and audio cassettes and eventually built a large manufacturing plant where they offered duplication services to smaller regional-cassette producers. By the late 1980s, their company, T-Series, was the market leader in cassette production in India and had begun to diversify into manufacturing videotapes, televisions, washing machines, and detergents, and later VCD and MP3 players.

T-Series was a profoundly disruptive force in the Indian music market, in large part because it was a tremendously successful pirate. The company built its catalog through a variety of quasi-legal and illegal practices, notably by abusing a provision in the fair-use clause of the Indian Copyright Act, which allowed for version recording. On this basis, T-Series released thousands of cover versions of classic film songs. It also engaged in more straightforward copyright infringement in the form of pirate releases of popular hits, and it often illegally obtained film scores before the release of the film to ensure that its recordings were the first to hit the market. Many other accusations were leveled against T-Series, including the wholesaling of inferior magnetic tape to competitors in an effort to discredit their brands.

On the plus side, T-Series changed the rules of distribution in ways that permanently transformed the music industry and music-buying public in India. Breaking with the narrow existing channel of retail outlets, T-Series moved aggressively to distribute cassettes in neighborhood shops, grocery stores, paan waalahs (wrapped betel-nut stands), and tea shops—making the cassette a ubiquitous product in Indian commercial life.

It also expanded the music-consuming public by focusing on genres and languages that had been ignored by the dominant Indian record labels and distributors, notably HMV. HMV had viewed recording in languages other than Hindi as unprofitable due to the small scale of the respective markets. T-Series proved that it was possible to expand these markets with stronger distribution and lower price points. By providing duplication services to smaller labels, it also assisted in the revival of other, small-market music traditions.

These innovations were inseparable from the company's assault on the price structure of recorded music in India. In a market dominated by two government-licensed companies—HMV and EMI—audio cassettes were priced between $3.60 and $4.60. T-Series reduced the price of cassettes to $2.50, fueling the first mass market in recorded Indian music.

Branding T-Series a pirate doesn't quite do justice to the larger revolution in the music business of which they were a part or to its close relationship with the informal market. In an interview with the media scholar Peter Manuel, a T-Series employee commented on the forces that the company both capitalized on and unleashed:

"What the people say about our activities in the early years—it is mostly true. But I tell you that back then, the big Ghazal singers would come to us and ask us to market pirate versions of their own cassettes, for their own publicity, since HMV wasn't really able to keep up with the demand." Even major players like HMV dealt with pirates. When HMV found that it could not meet the demand for one of its biggest hits, *Maine Pyar* *Kiya*, it reportedly entered into an agreement with pirate cassette producers to raise their price on the album from Rs.11 to Rs.13 and pay HMV half a rupee for every unit sold. HMV, in return, promised not to sue them or raid their businesses. Other producers also colluded with pirates in order to minimize their costs, taxes, and royalty payments to artists (Manuel 1993).

---

the diasporic networks of the merchant castes. Major Indian media companies like T-Series had their origins in such trips.

These networks also provided key support for the rapid adoption of subsequent technologies, such as compact discs, especially as low-cost Chinese hardware and Malaysian discs began to flow into India in the late 1990s. By then, economic liberalization and growth in India and China had greatly increased the volume and sophistication of transnational trade in the region. The relatively simple informal sector of the 1980s had become a complex ecology of organizations that ran from local street vendors to factories throughout Asia. Because profit margins depended on efforts to accelerate the production and delivery of goods, these networks grew and innovated very rapidly.

Street-level pirate vendors and wholesale markets were strongly embedded in these wider metropolitan, regional, and transnational networks. In our interviews, large facilities in Pakistan, Malaysia, China, and Hong Kong were still identified by vendors and intermediaries as primary sites of production, with DVDs entering India through a variety of regional supply routes. Malaysian imports, for example, were said to follow two regional distribution circuits on their way to wholesale and retail markets in Mumbai—one passing through the cities of Dhaka (Bangladesh) and Kolkata (India), and the other passing through the city of Chennai (India).

Many of these routes are anchored in long-standing, transnational ethnic and kinship networks. The link between Chennai and Kuala Lumpur, for example, is marked by the presence of a large (Indian) Tamil population living in Malaysia. Pakistani pirates, in turn, build on and service the large South Asian diaspora in the Anglophone world, hungry for Indian music and film. Often, traffic within such networks goes both ways. Diasporic Tamilians in Malaysia eagerly await the latest Tamil films, while Tamilians living in Chennai await the latest Hollywood releases, copied in Malaysian factories.

The entertainment industry is aware of these regional circuits but has tended to view them primarily through the lens of global hits. According to the IIPA, for example, the May 19, 1999, release of *The Phantom Menace* set in motion a sort of regional domino effect in which pirated VCD copies of the film were available on May 24th in Singapore; on the 25th in Hong Kong,

Taiwan, and Macao; on the 26th in Thailand; on the 27th in Indonesia and Australia; on the 28th in Korea; and on the 31st in Pakistan. After Pakistan, the VCD was available on June 2nd in India.

This complex itinerary is emblematic of the highly structured pirate media flows of the late 1990s and early 2000s, in which the diffusion of physical discs from a central source—here, Malaysian factories—shaped the pace and geography of the pirate release. India's appearance at the end of the chain very likely signals a different production path, passing through masters delivered to Pakistani factories and copies smuggled across the border into India. Today, such an account would look very different. The geographical trajectory for any current blockbuster film is now radically compressed, with camcordered or better copies globally available via the Internet on or before the initial release date and street distribution following shortly after. Factory-produced copies do continue to appear in this context, but inevitably later and typically as premium products.

Our interviews in 2009 found evidence that these networks still play a role in the South Asian arena, but it is also clear that the pirate ecosystem is changing rapidly, driven by cheap copying technologies that diminish the advantages of industrial-scale production and further decentralize distribution. The Internet is a crucial factor in this context but not the only one: local factories and cheap consumer burners, storage, and other consumer infrastructure play major roles. When Ernst & Young investigated the origins of pirate discs in India in 2008, it estimated that 40% came from local disc manufacturers, 50% from informal cottage production, and only 10% from transnational networks (USIBC/Ernst & Young 2008). By all accounts, these shifts have vastly expanded the flow of pirated goods within India and Asia more generally, even as they displace the complex organizational networks that, until very recently, structured them.

Inevitably, this displacement is less of a factor in the hard-goods trades—especially electronics— where the Asian geography of the grey market remains highly visible and unchallenged. This geography is signaled to consumers in a variety of ways, most directly in the packaging and other signs marking the origin of the goods. In India, it also announces itself in the names of the street markets. Visitors stepping out of National Market in Bangalore can look across to Bangkok Plaza and, a few meters away, to Bangalore's own Burma Bazaar. Across from Burma Bazaar is New Hong Kong Bazaar. All specialize in non-legal media commodities, from counterfeit phones to DVDs and software. All are part of the grey-market-media world of modern India.

Internet and P2P

Optical disc piracy was an easy fit within the wider informal economy in India and quickly became a ubiquitous presence in the bazaars and street markets. Optical discs were one more variety of the cheap copy, paralleling the trade in other recycled, resold, and counterfeit goods. The spatial organization of pirated disc sales was continuous with these other goods, as were the strategies for policing it.

Until very recently, file sharing and P2P networks were incidental to this structure of piracy. Internet connectivity in India was low, and the limited broadband infrastructure was of poor quality, with low speeds and frequent disruptions. Indian officials, like officials throughout the Global South, took a wait-and-see approach to copyright issues in this environment, watching how piracy and enforcement efforts played out in more technologically advanced countries of the West or in parts of Southeast Asia. In the last two to three years, this has begun to change.

Broadband Deployment

In 2004, the Department of Telecommunications announced a national Broadband Policy for India, with a target of 115% annual growth and 20 million users by the end of 2010. Actual numbers will fall short of the target, hindered by the challenges of rural and semirural buildout of services and, in the past two years, by the global financial crisis. Nonetheless, growth has been prodigious, averaging around 65% per year. Broadband subscriptions jumped from 1.35 million in 2006 to 6.27 million in 2009 (ISPAI 2009).

Overall broadband penetration, in a country of 1.1 billion, is still very low—notably in comparison with China, which reached 103 million subscribers in 2009 (or roughly 8% of the population) (Zhao and Ruan 2009). Personal computer adoption in India is also low at 30 per 1,000 people—roughly a quarter that of China (Anandan 2009).[4] These rates, however, disguise the concentration of connectivity in the major Indian cities, where business adoption has outpaced consumer use and become the norm in most commercial contexts.

The Growth of P2P in India

The broadband push by the government notwithstanding, the continued scarcity and poor quality of consumer broadband connections, even in the major urban centers, mean that P2P use remains a relatively marginal practice from the perspective of the wider Indian copy culture. Slow connections—India defines broadband as anything over 256 kbps—and widespread use of bandwidth caps has hindered P2P use among early broadband adopters. The most common model

---

4   Not all technical infrastructure is so underdeveloped, however. In June 2010, the Indian government reported that there were 635 million cell phones in India, representing an adoption rate of almost 56% in the total population (TRAI 2010).

for home connections is still a $10/month plan capped at 1GB of data. Unlimited bandwidth plans have become available in the past two years, though still at prices prohibitive to everyone but the commercial elite.

Nonetheless, in a country the size of India, that commercial elite still numbers in the tens of millions, and even very low adoption rates can generate large numbers of new users. Indian P2P use is, by most accounts, growing rapidly. Industry groups like the MPDA (Motion Picture Distributors Association—the local Indian affiliate of the MPAA) as well as several major BitTorrent sites that post their traffic sources routinely list India among the top countries for P2P activity. The MPDA recently claimed that India is the fourth-largest contributor to global P2P traffic (Ernesto 2008; Borpujari 2009).[5]

Major international P2P services, for their part, traffic widely in Indian media, especially Bollywood films. The popularity of international sites is complemented by a significant India-focused P2P scene, also primarily using the BitTorrent protocol. The progenitor of these sites is DesiTorrents.com, launched in January 2004 (and currently registered in the United States). Most other Indian BitTorrent sites emerged out of the DesiTorrents community, including the popular DCTorrent and BwTorrents. Unlike the top international sites, most of these Indian sites have registration fees—generally on the order of $10.

Indian torrent sites, like many other sites below the top-tier torrent trackers, tend to specialize in local and non-English-language media. The more successful sites have large communities that actively seed new content. Site communities compete to post the newest releases quickly, and many of the most active groups watermark their copies. Although there are, in principle, norms favoring the exclusivity of watermarked material, these rarely constrain the user communities, and high-quality files move very quickly from one to another. The rapid release of Bollywood films is a top priority in these communities: camcordered versions generally appear within a day or two of theatrical release. These are quickly superseded by higher-quality or remastered versions, especially when digitally reproduced audio tracks become available (Sharky 2009).

The globalization of Bollywood and the large Indian diaspora ensure that Indian tracker sites have substantial international followings. DesiTorrents receives 77.7% of its traffic from India (based on our scrape of the site in April–August 2009). Pakistan and Bangladesh account for 4.5% of visitors, with the balance of traffic coming from countries with large Indian populations, including the United States, the United Kingdom, Canada, Qatar, and Australia. DCTorrent, in comparison, draws 65.4% of its traffic from India; with 15% from Pakistan, Bangladesh, and Sri Lanka; and the balance from the United States, the United Kingdom, and other migration hubs.[6]

---

5   These reports show enough consistency to be taken seriously but are difficult to reconcile with India's objectively small number of broadband connections. We would expect this category to grow rapidly but are unable to account in this study for the apparent current discrepancies.

6   Alexa.com, accessed August 14, 2009.

CONTENT

Although the piracy of Bollywood movies gets the most attention in India, our data crawls of second-tier sites indicate that the most popular category of shared content is local television programming—in some cases by a wide margin. TV content is a particular specialty of DCTorrent, though it dominates even the movie-heavy DesiTorrents rankings. Soap operas, recent cricket matches, stage shows, and news programs figure prominently in these listings (see tables 8.1 and 8.2). Although our data does not permit tracking the downloads of particular content by geographical location, we see two likely factors behind these preferences: the growth of bandwidth-rich Indian communities abroad who seek news and televised programming from home; and the underdevelopment in India of digital video recorders and streaming-video services for high-income consumers, which makes P2P a logical tool for more basic consumer practices like time-shifting and repeat viewing.

The availability of pirated versions before or within the initial cinematic release window is widely assumed to impact the profitability of films. In 2009, the poor showing of the Oscar-winning *Slumdog Millionaire*—a Fox Searchlight film about children in the Indian slums—became a touchstone in this debate. While the film grossed over $100 million internationally and was a major critical success, it netted only $648,500 in the Indian box office. Because five months passed between the US release and the Indian theatrical release, pirates had an exceptionally long monopoly on distribution during which they alone benefitted from the considerable media attention surrounding the film. As a Delhi-based distributor, Joginder Mahajan, noted: "By the time it came to India . . . the majority of people had seen the English movie" (IANS 2009b). In March and April of that year, two to three months after the film appeared on Indian screens, our data crawls showed the film still circulating widely on DCTorrent and DesiTorrents, with multiple DVD and high-definition versions available.

Table 8.1 DCTorrent: Most Popular Content Categories, March 2009

Uploaded Content	Number of Files	Popular Shows/Types
Movies	117	
Music	240	Original soundtracks, Remix/Pop/Asian fusion, Classics/Ghazals, Pakistani/Afghani music, Punjabi/Bhangra music
Sports	4836	Punjabi/Urdu/Pakistani stage shows
Television Shows	4836	News content, daily soaps, weekly shows, music and dance programs

Source: Authors.

Table 8.2 DesiTorrents: Most Popular Content Categories, March 2009

Uploaded Content	Number of Files	Popular Shows/Types
Bollywood Movies	380	
Music	656	Regional music (Kannada, Punjabi, Bengali, Malayalam, etc.), world music
Regional Cinema	370	Gujarati, Punjabi, Malayalam, Kannada, Bengali, etc.
Television Shows	1224	News content, daily soaps, weekly shows, music and dance programs

Source: Authors.

Industry assertions that piracy undercut the domestic release are plausible in this context, but also illustrative of the difficulty of drawing any specific conclusions about losses—either for individual films or the market overall. The basic factors shaping the film market in India are the remarkable 450% growth in revenues between 2000 and 2008 (before tumbling 15% in 2009 in the context of the economic crisis and a protracted conflict between producers and exhibitors) and the intensely hit-driven nature of that growth: roughly 90% of Indian films lose money (S. Prasad 2008). Although it is likely that piracy impacts sales at the margin, and conceivably more so in the case of *Slumdog Millionaire,* where the studio release strategy and press attention guaranteed widespread piracy, the specific impact of piracy is very difficult to isolate and inevitably turns on counterfactuals. *Slumdog Millionaire* performed poorly for an Indian blockbuster but reasonably well in relation to the independent films it arguably most closely resembles. International media attention guaranteed it a very high profile in India during the runup to its release, but much of the local media attention was negative, concerned with the role of Indian stereotypes and the derogatory title.

Our data crawls suggest that in India, as elsewhere, the most downloaded films are nearly always the biggest hits—but any causal attribution here is almost certainly imprecise. The *Slumdog* case suggests, following Balazs and Lakatos (2010), that the wider culture of anticipation surrounding a film, rather than the post facto box office performance, may be the better indicator of the takeup of films in pirate networks—and a better explanation for the cases in which studio and pirate release strategies fall dramatically out of synch.[7] Neither, however, provides much leverage on understanding monetary losses.

---

7   Balazs and Lakatos use the number of screens on which a movie appears as a (admittedly imperfect) proxy for this culture of anticipation.

An arguably more typical case is *Ghajini*, a Bollywood remake of a 2005 Tamil film that was itself loosely inspired by the 2000 American film *Memento*. *Ghajini* was the biggest Bollywood hit in 2008 and, according to our interviews with pirates in Bangalore, also the most sought-after film in the informal market, where it was available almost immediately after its release. *Ghajini* nonetheless earned some Rs.2 billion ($42 million) in its first two weeks in theatres and went on to become the third-highest-grossing Bollywood film of all time. When we conducted our data crawl in March and April 2009, it was still being heavily downloaded on DesiTorrents. The publicity surrounding the complicated origins of the film also led to a boomlet in pirate sales of the original *Memento*, which had never been released in India. One of our pirate vendor sources reported sales of over a hundred copies per day of *Memento* during the first month of *Ghajini's* release.[8]

INTERNET AND DIGITAL MEDIA POLICY

The lack of a strong empirical case for specific damages has not prevented aggressive action against online infringement. Legal action against P2P sites in the United States and Europe has become a major front in the anti-piracy wars, and courts in most countries have established precedents for liability (see chapter 1 for a broader discussion). These legal templates have, in turn, been exported to other copyright battlegrounds, where suits against locally hosted sites are becoming common.

India is well behind this curve. No suits against Indian P2P sites have been filed. DesiTorrents, DCTorrent, and several other sites serving primarily Indian content are hosted outside India, conferring some protection from the relatively disorganized international enforcement efforts of Indian rights holders.[9]

In part, this inaction relates to uncertainty regarding the liability of intermediaries for copyright infringement. Domestically, the Indian IT Act confers immunity on ISPs (Internet service providers) and other online services if they are able to prove that they have followed relatively common—if also notoriously underspecified—standards of due diligence to prevent infringement. This relatively broad safe harbor is complicated, however, by the priority accorded the Copyright Act, which does allow for intermediary liability in cases when the party has "reasonable ground for believing" (Section 51) that infringement is occurring or "knowingly infringes or abets the infringement" (Section 63). At present, there is considerable disagreement on the interpretation of these provisions, and the issue will have to await resolution in the courts.

Unlike in the United States, intermediary liability has not been developed further into a doctrine of "contributory infringement," leaving file sharing sites in the same category as other search and service providers who may host or link to infringing content. Among the major rights

---

8   March 26 interview with pirates on M.G. Road.

9   As this report was going to print, press reports described the first alleged BitTorrent-related arrests in India, involving four men in Hyderabad accused of uploading Bollywood films to BitTorrent networks (Ernesto 2010). The men also sold pirated DVDs.

holders, T-Series has been the most aggressive in challenging these limits on liability in court, initially through a 2007 injunction against YouTube for infringing its music copyrights, and more recently with requests for injunctions against MySpace. Because both services host user content, these cases represent an effort to expand liability from the current ex post system, in which a service like YouTube must comply with rights-holder requests to take down infringing files, to an ex ante system of liability for any infringing content posted to the site. If the latter scenario prevails, "due diligence" will increasingly require services to make use of filters to pre-screen infringing content—however imperfectly. Also distinct from US and much international law, Indian law does not provide for "counter-notification" in the event of a takedown, leaving no remedy if the request is unwarranted or frivolous.

Unlike YouTube and MySpace, P2P services do not host content. They are simply indexes or search engines for files hosted on—and shared directly between—users' machines. T-Series' recent suit against Guruji.com, a popular Indian search engine with a dedicated music-search feature, will test the scope of search engine immunity when simply linking to infringing files. A win against Guruji.com would significantly diminish the practical meaning of immunity under the IT Act and likely open the door to additional suits against BitTorrent and other P2P sites.

In the sixteen years since the passage of the TRIPS agreement, India has been wary of international efforts to extend IP protection beyond its obligations under TRIPS. Much of this caution arises from India's role as a manufacturer and exporter of generic pharmaceuticals—a position that has kept it in the cross hairs of the major pharmaceutical companies and, consequently, of the USTR. But this wariness also reflects a broad-based, long-term, public sector commitment to increasing access to knowledge goods, going back to debates over educational provisions in the Berne Convention in the 1960s. Although Indian law contains strict civil and criminal provisions for copyright infringement, it also contains (and successive governments have acted to preserve) what are among the most expansive public-interest exceptions and limitations to copyright in the world.[10]

Among post-TRIPS copyright initiatives, India has joined neither the WIPO Copyright Treaty (WCT) nor the WIPO Performances and Phonograms Treaty (WPPT). Notably, it has preserved rights to reverse-engineer or circumvent technological protection measures (such as digital rights management, or DRM) on copyrighted goods—an important condition for the exercise of fair use (in India, "fair dealing") in the digital era that is sharply restricted by the WCT. This became a point of contention in recent plans to reform the Indian Copyright Act. The music industry, in particular, advocated for "anti-circumvention" provisions along the lines of the US Digital Millennium Copyright Act, which goes beyond the WIPO treaty in important respects.[11]

---

10   Consumers International (2010) rates the Indian Copyright Act as one of the most balanced in the world, with broad scope for private, educational, and critical use and broad online rights. For a summary, see the India country report at http://a2knetwork.org/reports/india.

11   Notably in expanding protection to technical protection measures that deny access to works, not merely

The anti-circumvention debate set the copyright industries in opposition to groups who routinely reverse-engineer or modify technological protections, including the free-software community and organizations for the visually disabled. The resulting compromise language in the bill makes the intention to infringe the necessary threshold for liability. This is a particularly important point in a digital economy largely built on practices of recycling and reuse, where the criminalization of circumvention could apply to a very wide array of activities. The copyright industries are opposed to this weaker standard, and the IIPA has characterized the provision somewhat extravagantly as "almost completely eviscerating any protection" (IIPA 2010).

Stronger anti-circumvention and narrower safe-harbor provisions remain controversial among Indian lawmakers. Similarly charged debates have taken place around the exceptions available for educational use, where public interest groups are seeking to expand and formalize rights to access educational materials (Liang 2010). Despite growing international pressure on India to meet the stricter WIPO standards (and, beyond those, US/IIPA standards), the Indian government appears prepared to fight for national discretion on these issues, informed by development needs. Notably, the Indian government has been among the most vocal in expressing concerns about the emerging Anti-Counterfeiting Trade Agreement (ACTA)—a US-led effort to strengthen international enforcement standards, to which India is not a negotiating party. As of late 2010, India has signaled a clear anti-ACTA position and a willingness to take its concerns to the WTO (World Trade Organization), where questions about the eventual jurisdiction and implementation of ACTA are sure to be raised.

The (Re)organization of Indian Film Markets

Movies are basic to public life in India. Over three billion tickets were sold in India in 2009, representing roughly half the global total. A city like Bangalore sees a minimum of six to seven new releases every Friday, including the latest Hollywood, Bollywood, Kannada, Tamil, and Malayalam films. Cinema halls draw huge crowds, with most films selling out on weekends. Films turn over very quickly: flops are typically identified within the first week and pulled from theatres within two weeks of their release. The most popular films rarely stay in theatres for more than two months.

This frenzied pace is in large part a function of the extreme competition in the Indian film market, which during most of the past decade pushed over 1,000 domestically produced feature films per year, plus the major Hollywood hits, into rivalry for some 10,000 screens. (US studios, in contrast, produce 500–600 films per year for 40,000 screens.) Despite massive growth in revenues in the past decade (see figure 8.1), only 10% of films turn a profit under such conditions. This has

---

the ability to copy them. Unlike the major international music labels, which have collectively abandoned DRM for digital music sales, Indian music companies like T-Series continue to use it to restrict the copying and playback of files in the nascent digital download market.

Figure 8.1 Indian Box Office Revenue (in billions of US dollars)

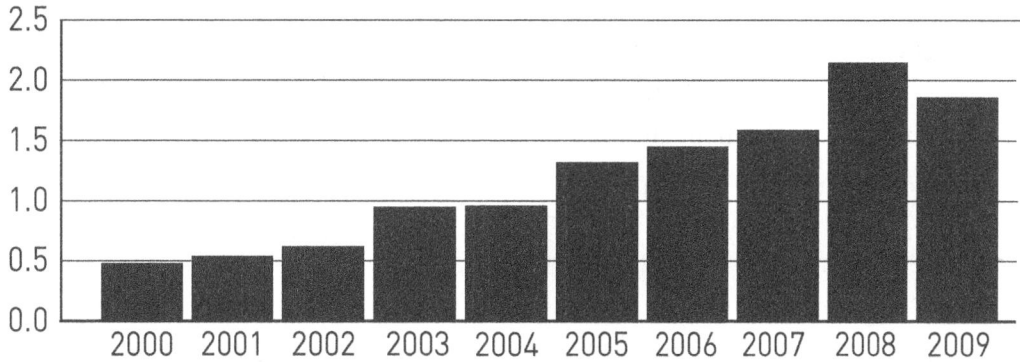

Source: Authors based on data from Kohi-Khandekar (2010) and FICCI/KPMG (2010).

Figure 8.2 Indian Domestic Film Production (number of films)

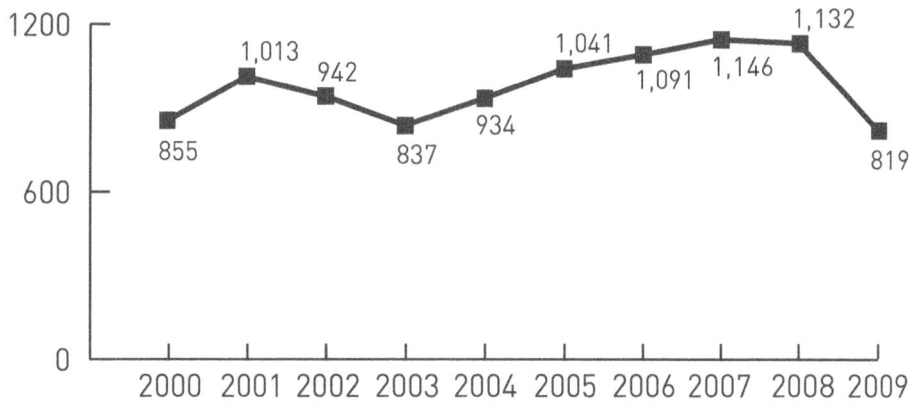

Source: Authors based on European Audiovisual Observatory (2001–10) data.

Figure 8.3 Per Capita Film Admissions in India

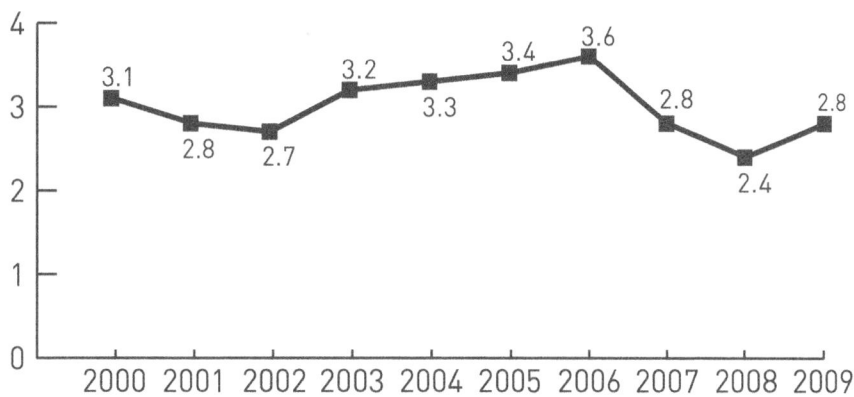

Source: Authors based on European Audiovisual Observatory (2001–10) data.

not deterred production, which set records every year through 2008 (see figure 8.2). But for the many who lose money it has fueled the belief that piracy, not competition, is the main obstacle to profitability in the Indian film business.[12] Industry resentment of the large, uncontrolled pirate DVD market is a predictable and highly visible outcome.

The domestic Indian film industry was built by catering to a mass market—first the urban middle class, and then wider populations reached through cheaper circuits of exhibition of the kind described earlier in this chapter. Per capita film admissions have hovered around three per year since the late 1990s (see figure 8.3)—nearly triple the rate of Mexico and Russia and six times that of South Africa and Brazil in a country with a fraction of those countries' per capita GDP (gross domestic product). Prior to the introduction of multiplexes and the progressive integration of the Indian cinema into global distribution networks in the late 1990s, the average cost of a movie ticket was around $0.20 (European Audiovisual Observatory 2010).[13] As core urban areas grew richer, urban redevelopment brought the conversion of many older properties into shopping arcades—often anchored by a cinema on the top two floors. The new multiplexes mixed Hollywood and domestic films and introduced a significant rise in ticket prices. As Hollywood globalized, so too did Indian film markets.

Market Segmentation

Among developing countries, India is nearly unique in having domestic industries in control of all aspects of the major media markets, from music and film production to distribution and exhibition. Domestic cinema accounts for some 92% of the market (Kohi-Khandekar 2010)—roughly inverting the ratio of domestic to foreign market share found in Brazil, Mexico, South Africa, and many other countries. Because of this commitment to domestic markets, Indian companies have a long and often fierce history of competition for domestic consumers. Because of the wide variation in incomes and infrastructure across India, they also practice extensive price discrimination in order to reach those audiences.

Geography plays an important role in this distribution of prices and access, as the developed theatrical exhibition markets in Delhi, Mumbai, and Bangalore give way to the less sophisticated theatres and halls (and the still-less-sophisticated home-video infrastructure and pirate distribution) of India's innumerable towns and villages. Where DVDs and occasional high-definition formats dominate the pirate markets of the major cities, residents of smaller towns have

---

12 2009 was a comparatively bad year for Indian cinema, with a drop-off in revenues as the economic crisis continued and in the number of films released as a result of a two-month-long conflict between producers and multiplex owners. Most media analysts (PricewaterhouseCoopers 2010; FICCI/KPMG 2010) anticipate a return to rapid growth in the next years.

13 Given the elaborate price discrimination practiced in the Indian market (detailed later in this chapter) and the lack of information about how the Observatory arrived at this number, we cite it with reservations. We are aware of no other estimates.

a strong continuing investment in VCDs. This complexity allows overlapping markets in which the same goods sell at different prices. The pirate market is not separate from this segmentation but, in effect, represents its largest and lowest tier.

In the cinema market, two broad developments have reshaped this landscape in recent years: (1) the shift from single-screen theatres to multiplexes, with a corresponding rise in ticket prices and change in the composition of the movie audience; and (2) rampant competition in the rest of the distribution chain, leading to the collapse of windowing practices around cable and home video, a move toward much lower-cost models for video rentals and sales, and a variety of other price and service innovations. This second trajectory has many parallels with the emergence of lower-cost distribution models for film, TV, and recorded music in the United States and Europe and in our view has much the same root cause: the presence in the market of strong companies that have to compete for domestic audiences.

Cinema Halls

The shift from single-screen theatres to multiplexes has important material and social dimensions. Like many other countries, India has seen dramatic changes in film exhibition in the last decade, as multiplexes begin to supplant the traditional single-screen theatres in the major cities. In the late 1990s, there was only one multiplex in India; by 2008 there were over 100, with some 850 screens (European Audiovisual Observatory 2010). Although multiplexes still represent a very small portion of the total screens in India, they have had a disproportionate effect on the cinema market, affecting the distribution of revenues, the composition of the audience, the spatial organization of cinema, assumptions about the formulas for viable films, and—ultimately—the market for pirated goods, which has become more firmly entrenched as the primary form of access among the poor.

Multiplexes in India are concentrated in the major cities, where their growth has largely mapped the rise in urban real estate prices. They are generally built as the anchor institutions for new shopping malls. Unlike the old movie theatres, which were usually run by families who owned the property on which they were built, the new theatres are much more commonly the products of corporate real estate development, requiring heavy initial investment and a correspondingly high rate of return. These pressures have dramatically pushed up the prices of tickets. Today, the cheapest tickets in the multiplexes run Rs.120 ($2.60), with premium tickets for the best seats in the best theatres on the weekends reaching Rs.500 ($11) (see table 8.4).

This price differential is dividing the Indian market, separating the poorer audiences who frequent the older single-screen theatres from those who can afford multiplex prices. As the middle class grows more affluent and real estate prices rise further, pressure on the single-screen theatres has increased. In most major Indian cities, single-screen theatres are in slow decline—either shutting down or converting to multi-screen formats. Prices in all cinemas have risen, driving a gentrification of the movie audience and pushing poorer Indians into greater reliance on the pirate market for the viewing of new releases. The social diversity characteristic of the older theatre-going audiences is one casualty of this shift. The big single-screen theatres were home to the unusual Indian spectacle of elites and poor brought together in the same venue. The expensive, highly policed spaces of the new shopping malls are much less amenable to this type of social interaction.

Today, low-income groups throng the remaining single-screen cinemas, where ticket prices typically remain under $1. Most of these theatres screen Bollywood productions or other regional films. The multiplexes, in contrast, cater to high-income audiences interested primarily in the most globalized of the Bollywood films and in international releases—including the recent trend of re-releasing Academy Award winners after the Oscars. The practice of releasing Hollywood films in single-screen theatres has largely ceased. Always a very small player in India, Hollywood has successfully repositioned itself at the high end of the movie market.

Figure 8.4 Number of Screens in India

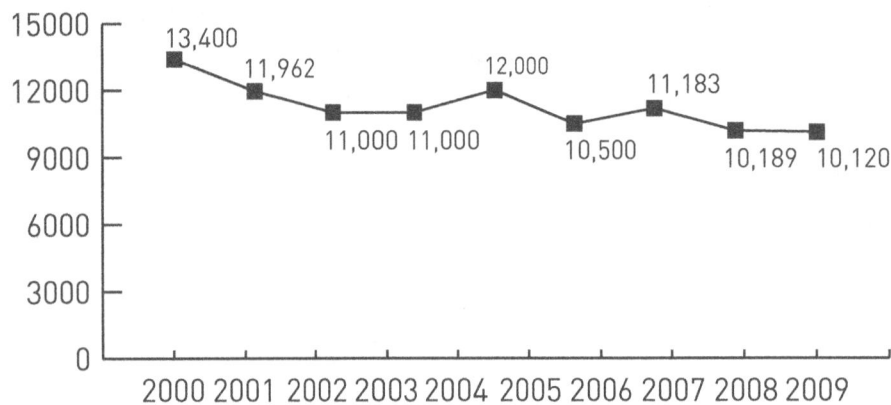

Source: *Authors based on Kohi-Khandekar (2010) and European Audiovisual Observatory (2001–10) data.*

High multiplex prices have generated many anxious commentaries about the transformation of movies into a more elite art form,[14] but they have also proved attractive to producers who see opportunities for films to become profitable with smaller audiences. The model has, in any event, produced a remarkable boom in revenues. The Indian market grew from $470 million in 2000 to $2.15 billion in 2008.[15] Nearly all this growth is due to higher prices, especially at the multiplexes, which control less than 10% of the screens but take in 25%–40% of the revenue. Neither the cinema-going public nor the overall size of the exhibition market, in contrast, has grown in the

---

14   Among others, by director Mukesh Bhatt (IANS 2009c).

15   Before dropping to $1.86 billion in 2009, as a result of a two-month shutdown of the multiplexes during a dispute with producers.

period: the total number of screens has actually declined over the past decade (see figure 8.4), and the number of tickets sold has remained roughly stable, hovering between 3 and 3.5 billion (until the recent downturn). Total exhibition capacity has consequently shrunk. Most of the older single-screen theatres could accommodate 800 to 1000 people; the new multiplexes are generally designed for audiences of 150 to 300.

Table 8.3 Price Discrimination at the Movies, 2009

Cinema	Type	Screens	Language	Weekday Morning Price	Normal Weekday Price	Weekend Price
PVR Classic	Multiplex Standard	18	English, Hindi, Kannada, Tamil, Telegu	Rs.60	Rs.120	Rs.200
PVR Europa	Multiplex Premium	6	English, Hindi, Kannada	—	Rs.170	Rs.300
PVR Gold	Multiplex Exclusive	6	English, Hindi	—	—	Rs.500
REX CINEMA	Single-screen	1	Hindi	Balcony – Rs.60 Rear Stall – Rs.50	Balcony – Rs.100 Rear Stall – Rs.80	Balcony – Rs.170 Rear Stall – Rs.130
FAME LIDO	Multiplex	12	English, Hindi, Kannada, Tamil, Telegu	Rs.100	Rs.140	Rs.200
FUN CINEMAS	Multiplex	10	English, Hindi, Kannada, Tamil, Telegu	—	Gold – Rs.190, Premium – Rs.170	Gold – Rs.250 Premium – Rs.200

Source: Authors.

Windowing

In India, the pressure of the initial release window is intense. Films must produce big first-week sales or face rapid closure—often after only two weeks. The biggest Bollywood hit of 2008–9, *Ghajini*, collected nearly half its $43 million global total in its first two weeks of domestic release, launching it on a long domestic (and later international) run. The competition for screens and the speed of turnover mean that pre-release advertising plays an inordinately important role in this process. In practice, there is almost no opportunity for a sleeper or word-of-mouth hit.

Once a film enters theatres, the producers and distributors scramble to manage the subsequent exhibition windows. The first opportunities are the less valuable theatrical markets in the provinces,

Direct Download to Theatre

The infrastructure for real-time, encrypted satellite downloads of movies to theatres began to be deployed in India five to six years ago. Currently, there are about three thousand digital screens in India, out of a total of roughly ten thousand (PricewaterhouseCoopers 2010).

The new distribution technologies were intended to put an end to the creation of high-quality pirate copies within theatres, either copied directly from the reel or by filming the screen during off-hours. Real-time, single-use distribution also means that films can be watermarked to enable relatively precise tracking of pirated copies back to their source. In 2008, a pirated DVD of the film *Tashan* (2008) was traced back to an April 25th showing at a single-screen theatre in Bilimora, a small town in Gujarat (UFO Moviez 2008). The theatre was raided and a camcorder-recording racket was broken up.

As they become more widespread, such technologies will raise new barriers around the theatrical exhibition window—though if the past is any indicator, relatively low ones that will not impact the longer-term pirate availability of the film. For most of the Bollywood studios, however, creating such short-term obstacles is the only realistic goal of enforcement.

The Melting Block of Ice

The control of release windows is widely viewed as critical to the business model for commercial film. Chander Lall, the MPAA's lead lawyer in India, describes films as a "block of ice" melting in the producers' hands.[1] Unless producers and distributors can make their money quickly, through a planned release program, they run the risk that it will turn to water without covering their costs. The initial theatrical release plays the central role in these strategies. But the time gap between each subsequent window is also important. At each stage, the producers reach out to new audiences with one hand, and with the other try to prevent the film from being distributed through unauthorized channels or media. High rates of piracy undermine this control and, consequently, are often blamed for the high rates of failure of feature films.

There are, however, dissident accounts of why so many films fail. Because the film market is highly competitive and because the reception of films is highly subjective, predictors of success are notoriously weak. The prominent Mumbai filmmaker Anurag Kashyap[2] notes that the basis for the initial investment in a film is almost always the assembled star power, which is viewed as the chief predictor of success. Distributors and exhibitors, in turn, accept films based on perceptions of the buzz surrounding the film, including the initial advertising campaigns. The role of the audience, in contrast, is largely confined to its initial reaction to these campaigns, measured in first-week ticket sales. Success can, in principle, be built by audiences over

i   Interviewed in 2009.
ii   Interviewed in 2009.

time through word of mouth, but only if the film overcomes the opening box office hurdle and is granted an extended exhibition window. By making initial visibility the main commodity, the market reinforces the reliance on star vehicles, sequels, copycat strategies, and massive advertising to push films past the hundreds of other films competing for the same small range of outlets. Production costs have risen accordingly, creating a higher investment floor for success but no greater probability of it.

Oye Lucky! Lucky Oye! (2008) was one of the first films to break with the traditional windowing schedule. Like many Bollywood films, the initial advertising campaign was built around the release of the film soundtrack and promotional music video. Rights to the soundtrack were owned by T-Series, which made it available for download for $2.52 (or $0.33 per song).

The film's theatrical release suffered from very bad timing, however. *Oye Lucky! Lucky Oye!* appeared in theatres on November 27, a day after a wave of terrorist attacks on downtown Mumbai hotels. Widespread fear of crowded places following the attacks kept many Indians away from the theatres. Though the film was a critical success, it lost money in this environment. *Oye Lucky! Lucky Oye!* stayed in the cinemas for the next four weeks and made about $1 million. It earned an additional $400,000 in subsequent worldwide release.[3]

In an effort to recoup its investment, UTV Motion Pictures released the film on two major direct-to-home platforms on December 18, less than a month after its theatrical release. The home-video distribution rights were sold to Moser Baer Entertainment Ltd., a subsidiary of the world's second-largest optical disc manufacturing company, which released VCDs and DVDs in the first week of January 2009 at a price of $2.17 per DVD and $0.88 per VCD—directly competing with the pirate market. This pattern has since been repeated for many loss-making films.

---

i   Box Office Mojo, http://www.boxofficemojo. com/movies/?id=oyeluckyluckyoye.htm, accessed March 4, 2009.

---

which must generally wait for the arrival of prints closed out of the major markets. Subsequent distribution on DVD and on cable and satellite TV networks enables further segmentation and expansion of the audience. Control of these secondary delivery and distribution channels allows producers to spread the considerable risks of production (Liang 2008). Although hit status in the theatres ensures a much more advantageous deal in these secondary channels, even flops have the opportunity to recoup costs.

These segmentation strategies begin in the theatres themselves, which practice elaborate forms of price discrimination. Ticket prices for the same movie can range from $1.30 to $11, depending on the choice of the theatre, the timing of the show, the location of the seat, and other peripheral services the cinema may offer, such as dinner. The date of release of the eventual VCD and DVD is determined by the popularity of the film in theatres and resulting calculations about the profitability of longer theatrical runs and the value of residual advertising effects. Cable, satellite TV, and pay-per-view windows are also based on these calculations.[16]

Price and Competition in the Optical Disc Market

The growth of the Indian movie market is mostly the story of rising prices—and of the successful creation of a premium product in the multiplexes that can justify those price increases. The corresponding home-media sector, in contrast, has evolved very differently and become a site of tremendous competition on price and services. In the past three to four years, major Indian media companies such as Reliance Big Entertainment, T-Series, and Moser Baer Home Video have launched price wars that have transformed large parts of the Indian media ecosystem, from price points, to distribution, to licensing practices for movies and music.

Figure 8.6 India's Home-Video Market (in millions of US dollars)

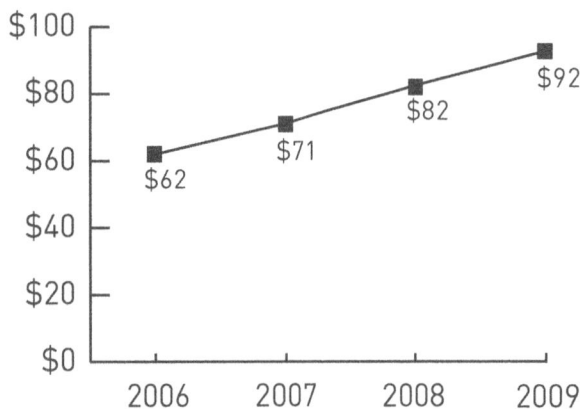

Source: FICCI/KPMG 2010.

Unlike the $2 billion theatrical market in India, the market for licit DVD and VCDs remains miniscule—some $92 million in 2009 (figure 8.6). Traditionally, the Indian home-video market was of low interest to film producers. The window between theatrical release and home-video release was typically over three months. DVD prices for domestic film were high, ranging between $5 and $7. This was a recipe for pirate control of the market—up to 90% of it according to estimates by Moser Baer (interview, 2009).

Until recently, the retail distribution channel for discs was very weak. Discs were sold through local video and audio shops, specialty chain stores such as Planet M, bookstores such as the Crossword chain, and shopping mall chains such as Reliance Time Out. The location of stores generally determined the content on sale. Upmarket retail stores sold English-language, Hindi,

---

16   The smooth functioning of this sequence depends in part on the administrative machinery for issuing the various licenses for each type of use and for collecting royalties (Wang 2003).

and regional Indian content, whereas small shops typically had only the most popular Hindi and local content. The rise of Moser Baer and Reliance Big Entertainment in the home-video market since 2007, however, has triggered a small revolution in price and accessibility, including expansion to a much wider range of retail outlets.

The Moser Baer business model will be a familiar one to readers of this report: release the film to DVD as cheaply, quickly, and widely as possible. Moser Baer's emulation of the pirate market was deliberate, as was the intention of competing with it to establish a viable Indian home-video market. Over the past three years, Moser Baer estimates its share of the total DVD market at 10%, built on competitive pricing, aggressive marketing, and pleas to customers to "Kill Piracy."[17]

Moser Baer's price war began with a major play for distribution rights. In 2005 and 2006, Moser Baer acquired home-video distribution rights to over 10,000 films with low potential for re-release in the high-priced DVD market. Nearly all were Indian; most came from small distributors. Over three years, the company has released 60% of these films on VCD and DVD at a starting price of Rs.40 ($0.85) for VCDs and Rs.99 ($2.12) for DVDs. After favorable responses from consumers, the company began collaborating with producers on new releases. In 2009, in a deal worth Rs.250 million ($5.5 million), Moser Baer acquired home-video release rights to the catalog of film production company UTV, including films in production through mid-2009 (Mitra 2008).

The other disruptive firm in the Indian home-video market is Reliance Big Entertainment, part of the $80 billion Reliance ADA group, the largest conglomerate in India. Reliance Big Entertainment has been a major player on several sides of cinema market growth in India, including film production, the development of multiplexes, and now the low-priced home-video market. In 2008, Reliance shook up home-video rental with a new service called BigFlix. At a base price of $6.50/month, BigFlix offered an unlimited number of movies from a collection of over 15,000 films. The BigFlix online portal recently extended this service to the Internet, allowing direct download to computers and a wide variety of free content.

Like Moser Baer's pricing in the DVD sales channel, BigFlix inaugurated a price war in the home-video rental sector and is progressively driving local video parlors out of business. Reliance's Big TV direct-to-home service has had a similar impact on that video market and currently offers over two hundred channels at an entry price of $32.50/month. Tata Sky and Dish TV, the two other major direct-to-home providers, have responded by lowering their entry-level prices in turn—in some circumstances down to zero.

These price wars occur almost entirely within the domestic media sector. Neither the "big four" record labels (EMI, Sony Music Entertainment, the Universal Music Group, and Warner Music Group) nor the Hollywood studios have opted to play this game. The cheapest Hollywood films in our DVD-price survey cost Rs.399 ($8.50)—normally for an older title at a discounted price. New releases and popular films are sold at Rs.500 ($11) and upward.[18] Major Hollywood

---

17   Interview with Sanjeev Varma, head of corporate communications, Moser Baer, June 20, 2009.

18   There is also some legacy demand for VCDs, sold at marginally cheaper prices, and for Blu-ray discs in

hits make up nearly the entirety of the selection of foreign films, when these are available. Even now, small-market films in English and other world cinema offerings are seldom released in the legal market. Pirate distribution, via either P2P sites or some of the specialized vendors working in the larger markets, often represents the only way to view such films in India.

The practical impact of these price differences is illustrated by the comparative purchasing power method used throughout this report, which translates the price of legal goods and their pirated equivalents into relative prices that reflect how expensive the item would be for Americans, if priced at an equivalent percentage of US per capita GDP. The results are predictably stark. GDP per capita in India is about 1/46 that of the United States.

Unlike the theatrical market, the legal DVD market follows relatively uniform pricing practices, segmented into categories of goods (for example, between Hollywood, Bollywood, and low-cost Moser Baer discs) but with little price discrimination among the same goods, at the same time. Pirate disc prices are a different story altogether and show wide variations in observed prices. These reflect a variety of factors, including geography, perceived demand, and the degree of bundling of films or albums on single discs. The spectrum of DVD pricing, for example, runs from wholesale markets, like Burma Bazaar in Chennai, where high-quality DVDs can be had for $0.40, to prices upward of $2 in the more organized, tourist-friendly markets of Delhi and Bangalore. It includes a large subcategory of film compilations, especially of Bollywood productions, which enjoy considerable popularity among consumers and are available at a much lower per-film price. Our price data for licit DVDs is based on spot checks of Minimum Retail Price (MRP) stickers on licit DVDs available in Bangalore stores. Prices for pirated films are listed at around $1—a typical price for single-feature, high-quality DVDs.

The comparison of licit and illicit DVD prices (table 8.5) suggests the pricing dynamic at work in Indian home video. The 2008 Hollywood blockbuster *The Dark Knight* is sold at the more-or-less uniform international price of $14–$15—uniform because of the control that the Hollywood studios exercise over the global licensing and distribution of their products. *A Beautiful Mind* is an older Hollywood hit available, like other older films, at a discounted price. In both cases, the price represents a hugely disproportionate share of local income compared to price/income ratios in the United States or Europe. At an equivalent share of US GPD per capita, a *Dark Knight* DVD would cost $663; *A Beautiful Mind*, $421.

high-end specialty stores, which command a sharp price premium of Rs.1,299 ($28) or more per disc.

Table 8.4 Licit and Illicit DVD Prices in India, 2009

	Legal Price ($)	CPP Price	Pirate Price	Pirate CPP Price
The Dark Knight (2008)	14.25	663	1	46.5
A Beautiful Mind (2001)	9.10	421	1	46.5
Ghajini (2008)	8.50	395	1	46.5
Flashbacks of a Fool (2008)	6.42	298	1	46.5
Oye Lucky! Lucky Oye! (2008)	2	93	1	46.5
Jaane Tu . . . Ya Jaane Na (2008)	3.8	176	1	46.5

Source: Authors.

In contrast, distributors of Bollywood films have sharply reduced prices of their products over the years. New players like Moser Baer have negotiated rights to popular Indian films on terms that permit much lower pricing—as low as Rs.40 ($0.85 cents) for VCDs and Rs.99 ($2.12) for DVDs. Some blockbuster films have been kept out of the price war, such as *Ghajini*, which costs Rs.199 ($4.24) on VCD and Rs.399 ($8.50) on DVD. But most traditional home-video distribution companies, such as Shemaroo and Eagle, have been forced to reduce their prices to stay competitive. In 2008, T-Series dropped the average price of its VCD releases of new films to Rs.38 in a bid to compete.

As a result, the difference in cost between a pirated and an original copy of a Bollywood film is far less than for a Hollywood title—often a factor of two rather than ten or more. This difference has proved small enough to produce dramatic increases in legal DVD sales. Sales of over a million discs for major releases have become relatively common. The Moser Baer DVD of the hit *Jab We Met* sold over six million discs when it was released on home video, five weeks after it hit theatres in 2008.

The Moser Baer impact on the home-video market has been dramatic and illustrates a basic, recurring dynamic in this report: high-priced media leads to widespread piracy, and widespread piracy—when in the presence of competitive legal businesses—catalyzes lower-priced, mass-market, legal alternatives. In India, these cheap legal alternatives are the new incumbents. Since the beginning of the price war in 2007, home-video revenues have grown rapidly—albeit from a very low baseline.

Distributors have also moved into the market segments pioneered by pirate producers and vendors. The pirate practice of offering three to four films bundled on one DVD has been widely copied in the formal market, as have the genre and theme-based collections (for example, World War II films) that proved their popularity in pirate markets. The slow emergence of a formal market for foreign non-English-language films is another example. For years, cinephiles turned to the pirate markets for their supply of international films. Prominent non-English-language directors such as Wong Kar-Wai and Pedro Almodovar were rarely screened in India and never released on DVD. Here again, Moser Baer (in collaboration with Palador Pictures, which holds the Indian rights to a wide range of classic international films) has moved in to fill the obvious gap and now sells world cinema to the "premium DVD market" at Rs.399 ($8.50). Somewhat amusingly, Palador head Gautam Shiknis describes the deal as "the beginning of a World Cinema movement in India" (Indiantelevision.com 2007).

For obvious reasons, many industry incumbents are ambivalent about being forced to play (or having entered, forced to stay) in a much more competitive, lower-margin market. In 2007, industry groups proposed an "Optical Disc Law" designed to regulate the manufacture and sale of discs. The Ministry of Information and Broadcasting rejected the idea, with Ministry Secretary Asha Swarup describing piracy as a sign of the work still to be done in democratizing access in the legal markets: "The best way to tackle piracy is to manage the supply side, by releasing the films on digital platforms: simultaneous releases on Theatres, Discs and Internet is a solution. If there is enough availability of DVDs in national and international markets on time, then people would not be going to these sites" (Pahwa 2008).

The Market for Piracy Studies

Prior to 2004, research on piracy came mostly from US industry groups working in India. Their efforts contributed to the annual IIPA reports on India, which informed the USTR's Special 301 reports, which in turn routinely targeted India for special criticism regarding its pharmaceutical policies, copyright protection for film and books, and a spectrum of real and alleged deficiencies in the Indian legal regime.

As efforts to build local corporate and government support for enforcement ramped up early in the decade, the market for piracy studies also began to grow and diversify. An assortment of consulting firms, including PricewaterhouseCoopers, KPMG, Ernst & Young, and the International Data Corporation (IDC), entered the picture to explain piracy to Indian and multinational stakeholders and to provide more compelling evidentiary support for industry lobbying and enforcement campaigns. Media consulting, in the booming Indian media market, also frequently meant IP consulting, leading to an echo chamber in which piracy findings were repeated and reinforced from one report to the next. All the major Indian industry groups now commission work, including the IMI (Indian Music Industry) and the IPRS (Indian Performing Right Society) for music, NASSCOM (National Association of Software and Services Companies) for software, and the FPBAI (Federation of Publishers and Booksellers Association of India) for books.

In 2007, Ernst & Young won a contract to produce piracy studies for the U.S.-India Business Council (USIBC), a liaison group based at the intensely pro-enforcement US Chamber of Commerce in Washington, DC. Ernst & Young's first study, *The Effects of Counterfeiting and Piracy on India's Entertainment Industry* (2008), was released during USIBC lobbying for the Optical Disc Law and was widely cited as evidence of why that legislation was needed. Despite this close connection to lobbying, the study provides one of the more comprehensive examinations of the pirate marketplace for film, TV, music, and entertainment software (but not business software) available in recent years. To its credit, it also offers the most extensive methodological appendix of any industry study we have examined in our overall project, and it avoids some of the gimmicks used to inflate industry-loss numbers into wider claims of economic losses (such as the application of economic multipliers).

Less to its credit, the report relies on other, not so transparent industry reports for some of its estimates (notably MPAA work on pirate markets in 2004–5) and repeatedly uses the retail value of goods (or its equivalent) as the basis for calculating losses. The latter assumption, in particular, ignores the price/income imbalances that create pirate markets in the first place and puts the report out of step with the industry's belated acknowledgement that substitution rates—that is, the probability that a pirated disc substitutes for a legal purchase—are generally far less than one. Such effects are now a relatively common part of the US and European discourse on piracy, but they are largely absent from the conversation in developing countries, where price/income ratios dictate very low rates of substitution and consequently much lower-than-retail estimates of actual losses.

Using these methods, the report found that the Indian media industries lost $4 billion and 820,000 jobs to piracy in 2007—over half of that in alleged losses to broadcasters from piracy by cable networks. Losses to the film industry were said to approach $1 billion. These numbers were extensively quoted in the media, and it is relatively easy to see the report as part of the longer-term effort to foster a national, rather than regional, rationale for enforcement efforts.

Other recent studies generally follow this line, making promises of increased employment, foreign investment, and tax revenues if piracy can be reduced. The basis for these claims can be somewhat fluid from year to year. In 2003, the Business Software Alliance argued that a 10% reduction in piracy would produce 50,000 new jobs and added investment of $2.1 billion in the Indian economy. By 2005, the BSA's claimed piracy losses had climbed 50%, and the estimated benefits had more than doubled: now a 10% reduction in piracy would add no fewer than 115,847 new jobs, $5.9 billion to the economy, and $386 million in tax revenues. By 2008, claimed losses had further quadrupled from their 2005 level, but the economic benefits had shrunk: now, a 10% reduction in piracy would now add only 44,000 jobs, lead to $3.1 billion in added investment, and increase tax revenues by $208 million (BSA/IDC 2008). No explanation for this variation is given, and the underlying method is only cursorily described. Competition between industrial powers in the region—chiefly India, China, and Russia—provides a still newer angle for the studies. Accordingly, in 2008, "A 10 point reduction in piracy could make China's IT workforce the largest in the world, surpassing the United States, and make Russia a bigger IT market than India" (BSA/IDC 2008)

Chapter 1 of this report casts ample doubt on the methodologies underlying such claims, and we will not revisit these points in detail here. But the USIBC/Ernst & Young and BSA/IDC reports do repeat the framing errors common to such analyses. Sums "lost" to domestic vendors are almost never lost to the larger economy but are simply spent on other things. Losses to foreign companies are, prima facie, gains to the domestic economy, putting multibillion-dollar loss claims in a very different light. Such countervailing factors may or may not outweigh the corresponding domestic losses to specific industries—such analysis becomes very complex. But it is disingenuous to ignore them in order to bolster a case for stronger domestic enforcement.

The Judicial Enforcement of Copyright

The courts have been a constant battleground in industry efforts to strengthen Indian copyright enforcement. The most immediate problem is the dysfunction of the Indian court system. Massive backlogs of both civil and criminal cases mean that new infringement suits or criminal prosecutions can take years to reach their conclusions.[19] Such cases often require investments far in excess of any eventual fine, compensation for the injured party, or possible dissuasive effect on pirates, making the prosecution and defense of suits financially burdensome for all involved. The MPAA reported having over 1,900 pending cases in 2009; the IMI over 8,000 in 2008 (IIPA 2010, 2009b).

But the problems for enforcement go beyond slow judicial process. Judges, for their part, have been very reluctant to rule for plaintiffs in infringement cases. The IIPA reports that, between 1992 and 2007, there were only sixteen convictions under the piracy provision of the Indian Copyright Act, and only six since 2000—all for film piracy. Recent reports suggest slightly more success in pushing cases through to conviction: the IMI reported sixty convictions for piracy in 2008 and roughly double that number under milder statutes governing the uses of certificates of authenticity. These conviction numbers come in a context of some three thousand raids attributed to the IMI and the MPAA alone in 2008 (IIPA 2009a).

From the industry perspective, judicial attitudes and the prevailing interpretation of copyright law have been major contributors to this broader enforcement failure. Industry representatives routinely talk about the need to "sensitize" judges to the claims of rights holders. Industry groups, accordingly, have made pedagogical intervention with the judiciary a high priority, typically in the form of training workshops led by IP lawyers from major Delhi firms. These promote the

---

19   There is an estimated backlog of 31 million cases pending in Indian courts. A high- court judge recently estimated that it will take 320 years to clear the backlog (PTI 2010).

"culture of intellectual property," to use WIPO language, anchored in belief in the social harms caused by copyright infringement and—for judges and prosecutors—in the promotion of rights-holder-friendly legal strategies for expediting cases and ramping up penalties. Our interviews with members of the judiciary found relatively widespread resentment of this pedagogical model.[20] Certainly it has not produced significant change in the rate of convictions.

The difficulties of obtaining convictions mostly relate to the complexity of due process, which introduces a variety of possible delays and points of failure. Criminal charges normally involve the police, begin at the lowest levels of the judiciary, and impose a significant burden of proof on the prosecutor. The complainant, for his or her part, must be present on each hearing date—a requirement that raises the overall cost and inconvenience of pursuing a case. Defendants—also sensitive to these costs—often do not appear.[21]

In any given proceeding, several agencies must coordinate to bring a case to trial. The rules are cumbersome and slow, but they exist, in part, because the police are widely considered to be one of the most corrupt institutions in India, often working in tandem with the "criminal elements" they are supposed to police. In our interviews in street markets, such low-level corruption was taken for granted. Informants described a variety of forms of complicity between pirate vendors and local police, including police tip-offs to raids and corresponding gratuities or payoffs by vendors. In pirate markets, the sight of police officers buying pirated CDs and DVDs is not uncommon.

Raids against vendors and retail operators are controversial among the judiciary and the police. All see their effects in clogging the courts, and few view them as efficient strategies for reducing piracy. A partial exception to this rule are the efforts to protect release windows for high-profile films. Major releases now routinely involve the mobilization of courts and police in short, intense anti-piracy campaigns targeting illegal street markets and high-value points of distribution, such as cable-network operators that show pirated films. Where pirate distribution is still meaningfully organized around particular localities, such as Mumbai for Bollywood productions or the regional centers of Tamil and Kannada film, street enforcement campaigns have shown some limited capacity to suppress the availability of pirated copies in the key theatrical markets. These are, invariably, short-term efforts, with little impact on subsequent availability. But in the Indian movie market, the short term matters.

Given these difficulties, enforcement in India has begun to rely more on civil remedies. Civil complaints are generally simpler, faster, and in practice less focused on playing out the dispute in court. On the basis of such complaints, courts routinely issue injunctions against infringing

---

20   Based on conversations with high court judges at judicial trainings on copyright enforcement conducted between 2006 and 2008, including at the National Judicial Academy in Bhopal in 2008.

21   Microsoft was recently fined by the Delhi High Court for bringing cases in Delhi for alleged damages in other states—effectively requiring the defendant to travel. The court called this "harassment" of the defendants and an abuse of Microsoft's "money power" (IANS 2009a).

Fan Club Enforcers

Movie star fan clubs are important units of political organization in India, particularly in the south, where film culture and politics routinely mix. For stars entering politics, the clubs play important roles in traversing caste and class lines and often provide the infrastructure for political campaigns. Such clubs are typically supported directly by the stars themselves and engage in various visible "social work" efforts, such as blood drives and the adoption of local orphans.

In recent years, the clubs have also become involved in policing video piracy. In 2007, the main fan club of Rajnikant, a major Tamil star, instructed its branches to set up anti-piracy squads to surveil audiences and theatre staff during the release window of Rajnikant's film *Sivaji* (2007). There have also been instances where members of fan clubs attacked video pirates and forced them to close up shop. Commenting on such vigilantism, Rajnikant himself noted: "It is not correct to ask the fans to bash these people. This will create a law and order problem" (SouthDreamz 2010).

Although fan club surveillance is not a very effective form of enforcement, it is highly visible and provides the only grass-roots manifestation of the wider corporate culture of enforcement. None of the fan club efforts, however, have moved beyond the protection of the work of particular stars.

---

parties and can also appoint judicial officers ex parte to conduct raids and seize goods—including private actors drawn from the ranks of the enforcement organizations themselves.

Such injunctions are of limited use against street piracy, where the informal organization of the trade makes raids and seizures relatively ineffective as a general strategy. But they have been widely used against more organized pirate retail and, in particular, against operators of local cable networks, who routinely show pirated versions of films during the initial release period.[22] In such cases, after obtaining an injunction order, the complainant typically accompanies the police to the premises and seizes equipment. Raids of this type are geared less toward the collection of evidence for lengthy court proceedings than toward what we describe in chapter 1 as the confiscation regime, in which the primary goal is to destroy or impound pirate stock and infrastructure in an effort to disrupt the business.[23]

Because of the greater traction of such civil enforcement measures, industry lawyers have pushed to dramatically expand injunction powers in the past several years. A wide array of legal strategies developed in other areas of law have been imported into the copyright infringement arena, including ex parte injunctions granted without hearing the other party, John Doe orders

---

22   Ernst & Young estimates that the percentage of pirated content (that is, content for which no rights were cleared) on legal Indian cable networks is 60% (USIBC/Ernst & Young 2008).

23   Some of the prominent uses of such injunctions include the 2003 case filed against cable operators by Mira Nair for her film *Kama Sutra* (Mirabai Films Pvt. Ltd. v. Siti Cable Network And Ors). See also the 2007 case Time Warner Entertainment Company, L.P. v. RPG Netcom And Ors.

issued against anonymous offenders, and Anton Pillar orders that expand search and seizure authority. Collectively, these measures have made injunctions much more powerful tools to bring to bear against the informal economy and are now routinely invoked by plaintiffs. They remain controversial in the judiciary, however, and are not consistently granted.

Many of the copyright lawyers we interviewed expressed ambivalence about the practice of enforcement. The IIPA and USTR focus on the criminal enforcement of copyright in India, in particular, was greeted with considerable skepticism. As one put it, it is "pointless to go around killing bees with a hammer."

Such views do not reflect indifference to infringement. Many of the same copyright lawyers also complained that the police do not take copyright cases seriously. We see ample evidence for this view. Indian police have comparatively strong legal powers to combat piracy, including suo motu (ex officio) authority, which allows them to make arrests for copyright infringement without a prior complaint. But such arrests rarely happen outside the context of specific enforcement campaigns, such as the recent large-scale police sweeps in Tamil Nadu.

Police attitudes toward IP are thus another front in the effort to strengthen enforcement practices in India. Like interventions with the judiciary, these efforts have a strong pedagogical dimension and in some respects have become simply one more facet of the wider campaign waged by civil society groups to shape police culture. One of the lawyers we interviewed serves as a lecturer at the Police Training College in Delhi, where he conducts sessions on software and film piracy for police inspectors. He described the importance of "sensitizing" police officers to the harms of media piracy—a word elsewhere used mostly by human rights and gender rights groups to describe their concerns with law enforcement.

Despite concerns about the general disinterest of the police in these matters, relationships between industry and police remain critical to the practice of raid-based enforcement. A thriving private marketplace for expertise in this area has emerged in recent years, yielding a complex web of professional and social links between public and private enforcement efforts. The agencies hired to investigate infringement or assist in raids are usually headed by former police officers or retired police officials. Expertise with pirate networks is part of the job description. The ability to call in favors with former colleagues is another—and a critical one in the context of the thousands of raids per year initiated by industry groups.

There are other, practical difficulties associated with police action on copyright, including continued poor police understanding of copyright law, especially regarding the different standards of proof of ownership that apply in infringement complaints. Consistent with international law, copyright in India is automatic and does not require registration. Yet police often follow the norms of tangible property crime in requiring evidence of ownership, for fear of becoming involved in false complaints lodged for the purposes of harassment.

By most accounts, police and judicial education efforts have had, at best, modest success. Overall rates of conviction remain vanishingly low. Most courts remain reluctant to expand the use of ex parte procedures, in which the accused need not be present at a hearing or—at the limit—

Organized Crime and Piracy

Anti-piracy campaigns around the world now routinely feature claims about the links between piracy, organized crime, and terrorism. In India, the role of the criminal gang D-Company is often used as proof of both assertions. The most recent examination of D-Company's role comes from an MPAA-funded RAND Corporation report on criminal and terrorist linkages to movie piracy, published in 2009. According to RAND, D-Company controls all aspects of the Indian film industry and pirate marketplace: "Since the 1980s, [D-Company's founders] have been able to vertically integrate D-Company throughout the Indian film and pirate industry, forging a clear pirate monopoly over competitors and launching a racket to control the master copies of pirated Bollywood and Hollywood films" (Treverton et al. 2009). Upon its release, the RAND report was cited extensively in the Indian media, and its claims about D-Company were quickly integrated into MPAA and Indian film industry anti-piracy campaigns.

Much of the attention focuses on Dawood Ibrahim, D-Company's founder and Mumbai's leading mafia chief during the 1980s. Ibrahim allegedly sponsored the 1993 bombing attacks in Mumbai that left over 250 people dead (as a reprisal for Hindu-led violence against Muslims in the previous year). Under police pressure, he departed India for Dubai and later for Pakistan. Ibrahim now operates from Karachi, where he reportedly manages narcotics-smuggling operations across large parts of South Asia, Africa, and Southeast Asia. There are widespread allegations of connections to Pakistani intelligence and to Al Qaeda, the latter of which has earned him a place on the US terrorist watch list.

Ibrahim's role in film piracy is usually attributed to his alleged acquisition of Sadaf Video, a major Karachi-based DVD producer and distributor that grew to prominence by circumventing the Pakistani ban on Indian films. Until 2008, when the ban was finally rescinded,[1] pirated copies manufactured by Sadaf were the primary form of access in Pakistan to wildly popular Bollywood movies. Large VCD and DVD factories in and around Karachi met most of this demand. Sadaf's merchandise was then often smuggled back into India and shipped to other regional markets for sale (such as South Africa—see chapter 3).

Reliable information on the activities of these criminal networks is scarce, but our work in Mumbai and Karachi suggests that the RAND account is an exaggeration: D-Company controls long-standing regional smuggling routes between Dubai, Karachi, and Mumbai. But we see no evidence that this supply chain plays a major role in contemporary Indian piracy, or that it extends to other parts of India. We see no evidence of a D-Company monopoly on the pirate market, even in Mumbai. Quite the contrary, the Mumbai market for pirated DVDs appears highly contested among a range of local suppliers—and increasingly so as falling production costs and prices have opened the door to smaller-scale production. Ernst & Young, in its study of Indian piracy, estimated the presence of foreign-produced DVDs from all

---

[i] The ban was lifted in stages. The release of selected Indian films was allowed in 2006. The ban was completely rescinded in 2008, mostly as a measure to stimulate Pakistani multiplex development.

sources at 10% of the Indian market in 2008, with the remainder split between local factories and cottage production (USIBC/Ernst & Young 2008).[2] D-Company's film piracy activities also suffered from a US-instigated crackdown in Pakistan in 2005, which allegedly broke up the major Karachi factory networks.

Furthermore, we see no evidence of the wider control of the Mumbai film industry by D-Company or other mafia actors described by the RAND report. Mafia financing undoubtedly played a role in film production in the 1980s and early 1990s—the Ibrahim era—though by no means the dominant role asserted in the RAND report. These connections were highlighted by the arrest (and later acquittal) of film financier Bharat Shah in 2001, who was charged with running an extortion racket on behalf of Mumbai gangsters and with using slush money to finance his own productions. Rumors of mafia financing of particular films have circulated in the past two decades, but none have been confirmed. Nor have mafia links been demonstrated in regard to any of the large regional film industries, such as the Tamil and Telegu industries. The film financing problems described by RAND as a point of entry for mafia money have dissipated in the past decade as the Indian film industry has grown and diversified. Today, financing comes from a continuum of Indian and foreign sources, including Hollywood studios. Although connections to the Mumbai underworld almost certainly exist, we see no evidence that they are systematic, much less dominant.

ii   This fact did not stop Ernst & Young from describing the "strong organized crime nexus backing the piracy industry," without further comment, as a major challenge to law enforcement.

at their criminal trial. And police have generally resisted adopting copyright enforcement as part of their core mission, despite revisions to Indian law (including *suo moto* authority) that expressly encourage them to do so. Few of the police we spoke with viewed media piracy as a high priority in a country where law enforcement and the court systems are hugely overburdened by more serious crimes.

However, in a system of police and courts as decentralized as India's, some institutions have proved easier to sensitize to the enforcement agenda than others. The most visible example is the Delhi High Court, which sees a high percentage of IP cases due to the concentration of large IP law firms in the city.

Since the early 2000s, the Delhi High Court has been a reliably activist court on behalf of rights-holder claims and enforcement powers. It has consistently expanded the scope of injunctions and established, for the first time, punitive damages as a regular outcome of trademark and copyright infringement suits. It has played an important role in defining evidentiary standards used in infringement cases and notably legitimized industry claims about the damages associated with infringement. And it has set precedent in making decisions ex parte—without the presence of the accused.

One of the more important cases in this respect was the *Microsoft Corporation v. Yogesh Popat and Another* suit, initiated in 2003, in which a computer store owner in Delhi was accused

Copyright Act Reform

The Copyright Act is currently being amended in India and, predictably, has become a football for the diverse, contending interests in the growing Indian media sector. Stronger enforcement measures and penalties for infringement have figured repeatedly in this conversation—initially in the context of a push for stronger criminal penalties. But since 2006, the enforcement wish list has been progressively sidelined, with a probable compromise position on anti-circumvention measures representing the only clear step beyond existing provisions. Other likely measures are much more specific to the Indian context. The music industry has lobbied strongly for the dilution of the version recording provision in the current Copyright Act, which was widely used by music companies to issue unauthorized covers of popular songs. These "pirates" have gone on to become the incumbent labels and are now seeking to close the loophole.

Struggles within the film industry also feature prominently in the current draft legislation. The government, for instance, has sought to introduce provisions that would allow writers greater control over the rights to their works. The film producers have rallied strongly against this provision, reluctant to see any changes that strengthen the bargaining power of writers and lyricists. A final bill will likely be voted on in Parliament in early 2011.

---

of loading two hundred computers with pirated copies of Office 2000 and twenty computers with Visual Studio 6.0. The Popat case combined a number of legal innovations. It was conducted entirely ex parte—Popat never appeared in court—and it set precedent for a particularly difficult issue in Indian (and other national) law: how plaintiffs establish the losses associated with infringement.

Traditionally, the assessment of damages is made with reference to the accounts and sales figures of the defendant. The defendant typically has to "render accounts" to the court to determine the profits derived from the infringement. Such measures are often problematic when dealing with informal businesses, however, because these may not keep accounts (or, in this case, appear in court to respond to a request for such accounts).

In the Popat case, consequently, the court accepted Microsoft's representations of its "estimated loss of business," which it described as equivalent to the retail value of the pirated goods (some $140,000). The court then calculated Microsoft's forgone net profit on the pirated sales. It deducted a "dealer's profit" of $5,200 and then applied Microsoft's corporate profit rate for that year—32%. The court arrived at a sum of $46,500 in damages—the highest damages in an IP case through 2005.

Because India is a common law country, such decisions by a high court set powerful precedents. The Popat decision created a basis for valuing software losses at retail price, rather than at the much lower price at which the goods would have been sold in the pirate market or—for that matter—through the volume licensing deals that make up 80% of Microsoft's profits. As we have discussed at length in chapter 1, retail price offers a convenient but erroneous account of damages and is no longer used by any of the major industry groups in their research on piracy (the last, the BSA, abandoned the retail price calculation of losses in 2010). In practice, the major software companies price-discriminate throughout the market, making the pirated street price a much more accurate reflection of foregone sales.

The Popat decision also set precedent for relying only on plaintiff affidavits—a step that allowed the court to adjudicate the case with what Microsoft and other industry observers praised as unusual speed. Both precedents have been subsequently invoked in other piracy cases, such as *Microsoft v. Kamal Wahi* (2004), which broke Popat's damages record, and *Indian Performing Right Society Ltd. v. Debashis Patnaik And Ors* (2007), a performance-rights case in which losses were also determined and awarded ex parte, solely on the basis of plaintiff claims.[24]

Enforcement Networks

Coordinated enforcement in the Indian film and music sector has proved difficult over the years, with regional markets, intense competition, and local politics taking precedence over periodic efforts to build broad-based industry coalitions. Even narrower groups of actors, such as the Bollywood studios based around Mumbai, have found it hard to work together.

One of the obstacles was—and is—the market structure of the audiovisual sector. When the Bollywood studios rejected video distribution in the 1980s, they did so to protect the theatrical market. They viewed the VCR not as a potential new revenue stream but as an uncontrollable distribution channel that would lead to the proliferation of small-scale, informal exhibition (which, in fact, it did despite the ban). Such defensive actions commanded general agreement among the major players because they cost nothing. A corresponding commitment to enforcement, in contrast, was harder to reach. The cost of policing the vast Indian informal economy was too high to undertake as a private venture and too marginal a problem to attract serious public support— even in the south, where film culture plays a large role in local politics. The decentralization of Indian policing and the array of other development challenges vying for public attention raised additional barriers to any concerted public effort, local or national.

In a movie business without a strong investment in home video, consequently, the perceived benefits of comprehensive enforcement rarely outweighed the real private costs. Through the

---

24   There is some evidence that the Indian judiciary has begun to push back against the use of ex parte injunctions in infringement cases. Because of the basic justice concerns involved in ex parte action, such injunctions traditionally require a high burden of proof, including evidence of irreparable loss and damages caused to the plaintiff. Indian courts have rarely allowed such claims without affording a hearing to the other party, even in cases of harassment or threats of violence. In the early 2000s, the Delhi High Court routinely went beyond this tradition in granting ex parte injunctions in infringement cases. Other high courts, including the Chennai High Court, have since passed orders narrowing their use—in particular specifying that ex parte orders in infringement suits be granted only after effective judicial scrutiny of oral and documentary evidence (*FDC Limited v. Sanjeev Khandelwal*).

early 2000s, enforcement remained a relatively small-scale, private practice, carried out by the major studios, through the courts, on behalf of individual films during their release windows. Enforcement in the much smaller music business was similarly undersubscribed, in part due to the struggle between domestic music companies in the 1980s and 1990s, which left T-Series in possession of 60% of the market and a reputation as an unscrupulous piratical firm in its own right.

The rise of modern Indian media corporations with global ambitions, such as Yash Raj Films, T-Series, and Moser Baer, began to alter this landscape in the late 1990s. Pressure for greater public investment in enforcement began to grow, as did pressure for stronger coordination among industry actors. Bollywood studios were generally at the center of these efforts because their domestic and international reach was greater than that of other regional cinema. A national framework for enforcement would logically begin with the Bollywood studios, and those studios would be its primary beneficiaries.

Much of the actual impetus for coordination, however, has come instead from international rights-holder groups. Two organizations have taken the lead in pursuing a cohesive national enforcement agenda in India: the MPDA (the local branch of the MPAA) in film and the IMI (the Indian affiliate of the IFPI—International Federation of the Phonographic Industry) in music. These two groups have emerged as the organizational centers of wider networks of Indian and international rights holders in the country and have led several successive efforts to ramp up enforcement activity, with the most recent beginning in 2009.

Because of the prominence of the Bollywood studios among rights holders, the key industry groups are headquartered in Mumbai. IMI and studio interests generally align because the music market is dominated by film soundtracks and accordingly, at the business level, by recording and distribution deals between studios and labels. The software industry is a significant outlier from this organizational network, both geographically and organizationally. The major domestic software organization, NASSCOM, is headquartered in Delhi to facilitate lobbying of the national government and global trade bodies. The Business Software Alliance maintains a very low profile in India, preferring to work through NASSCOM or through bilateral forums like the US Chamber of Commerce–based U.S.-India Business Council.

Another important player in this landscape is the FICCI (the Federation of Indian Chambers of Commerce and Industry), the main body of Indian industry. The FICCI has aggressively promoted media industries in the past decade in an effort to replicate the success of India's IT sector. It is primarily a convening and lobbying organization and has been a strong advocate of moving the predominantly informal organization of Indian media companies toward corporate models and global norms. Its biggest event is the annual media-business conference FICCI FRAMES, attended by Indian and global media companies. Piracy and enforcement discussions have played a growing role at FICCI FRAMES, and the event has become an important locus for promoting new enforcement coalitions. The FICCI also organized a major anti-piracy conclave in January 2009, which brought together Bollywood studios, record labels, and industry activist groups like the MPAA.

The Motion Picture Distributors Association

In most countries, the local branch or equivalent of the MPAA is, for all intents and purposes, the film lobby. The MPAA represents the globalized Hollywood studios, and those studios generally dominate the local box office and distribution channel. This is not the case in India, however, where US-based studios account for only 8% of the roughly $2 billion annual box office take (Kohi-Khandekar 2010).

The local branch of the MPAA in India is called the Motion Picture Distributors Association. The MPDA is a very small player in terms of the market power it represents but an increasingly active player in the enforcement business. Although the MPDA officially opened in Mumbai only in 2009, MPAA presence in India dates back to 1994, when it engaged the high-profile Delhi law firm Chander Lall & Sethi to represent its Asia-Pacific wing. Chander Lall & Sethi still does most of the public lobbying for new enforcement legislation and continues to work with enforcement teams in Mumbai.

Rajiv Dalal, the MPDA's managing director, described the studio enforcement strategy in India in this way:

> The MPA [Motion Picture Association—the international arm of the MPAA] is taking a multipronged approach, since one of the main problems of previous attempts made by producers in India was that they were fragmented and their efforts inconsistent, lasting only for the first ten days of the theatrical release of the film. What the MPA has been doing since February [2009] is to have consistent raids and not just wait for a theatrical release. This was a collaborative effort between the Indian studios and the MPA. What the MPA was also doing separately was working with the exhibitors, trying to get at the source of the generation of pirated material. There is a need to create awareness amongst the exhibitors about camcorder recording and to lobby for anti-camcorder-recording legislation with the government, along with optical disc regulation legislation. The MPA is also trying to work with ISPs to stop Internet piracy while waiting for new amendments to the copyright and IT laws in the country.[25]

Coordination among these groups has been a challenge due to not only the strong regionalism of the Indian market but also the sharply divided corporate interests within the sector (and the history of sometimes intense competition between them). This is particularly true of pricing in different distribution channels. Indian companies have been unable to maintain the de facto cartel behavior that shapes media prices in markets controlled by the multinationals. There is no enforcement organization that unifies even the Mumbai-based groups, much less the range of regional producers and state and local political authorities. Although there have been several prior

---

25   Interview with Rajiv Dalal, May 2009.

efforts by the MPAA and other groups to create such a coordinating body, the last major effort broke down "as it was too expensive for the Film and TV Producers Guild of India to collaborate with the MPA."[26] The guild was reluctant, in particular, to commit money to a US-style enforcement effort, preferring the usual tactics of high public rhetoric against piracy and a film-to-film approach in the courts, funded by the individual producers.

Because India is one of the fastest growing theatrical markets, the MPAA has treated India as a project for the long term. Notably, it has continued to militate for stronger coalitions and coordination among Indian stakeholders, in both the public and private sectors. The current effort, dating to 2009, was occasioned by what the media called "the producer-multiplex standoff," which pitted the United Producers and Distributors Forum (UPDF), representing a range of mostly Mumbai-based film producers and distributors, against the Multiplex Association of India in a dispute over revenue sharing. The dispute shut down film exhibition in the multiplexes for over two months between April and June of that year—an unprecedented event in India that cost the industry an estimated $70 million. In the end, the multiplex owners were more exposed to these costs than the producers and were the first to crack. The eventual negotiations were mediated by Reliance Big Entertainment, due to its role in both production and exhibition.

Although piracy was not a subject of dispute during the conflict, it did present a welcome common enemy in what most stakeholders perceived as fragile times for the industry. At MPDA urging, the final agreement between the producers and multiplex owners involved not just revised revenue-sharing but also a substantial new commitment to collective anti-piracy efforts.

The new anti-piracy group was hailed as the "first ever Hollywood-Bollywood" coalition against film piracy in India.[27] All the biggest film, music, and home-video players were represented, including Moser Baer, Studio 18, Eros International, UTV, Reliance Big Entertainment, Yash Raj Films, the UPDF, and the IMI (the notable exception was T-Series). The social services wing of the Mumbai police, which is responsible for local anti-piracy efforts, was also involved from the start.

In the short term, the formation of the new coalition prompted new enforcement action on several levels. Most visible was a new round of raids against DVD/VCD shops in Mumbai in mid-July 2009, which put many of the more established DVD pirates under pressure. Our work was unable to determine whether these efforts had any lasting effects: the less formal networks of street vendors never disappeared from their locations near train stations and bus stops. In our interviews in June and July 2009, vendors showed considerable relief at the conclusion of the dispute, since the halt in new Bollywood releases also cut off their supply of new material.

As of mid-2010, the Hollywood-Bollywood coalition had scored one major legislative success: the addition of audio and video piracy to the list of offenses prosecutable under the Maharashtra

---

26   Interview with Supran Sen, the secretary general of the Film and Television Producers Guild of India, January 23, 2009.

27   Interview with Girish Wankhede, corporate communications manager at Cinemax, Mumbai, August 3, 2009.

Yash Raj Films: The Global Enforcers

Although enforcement within India gets the lion's share of attention from the industry and government authorities, Bollywood also has growing global markets and distribution channels and, consequently, a growing interest in global enforcement. Yash Raj Films, the largest and most influential Bollywood production house, is by most accounts also the leader in this relatively new phase in the globalization of Indian media. Yash Raj's international efforts are directed primarily at piracy in high-income markets—especially among the large Indian expatriate communities in the United States and the United Kingdom. Yash Raj has initiated a number of US civil suits and worked with UK enforcement agencies. The high cost of fighting piracy in the US courts limits these efforts, however: the cost of a civil suit for infringement starts at around $75,000, including the costs of registering the complaint, hiring lawyers and investigators, and building evidence.[1] For complex cases, the bills scale quickly upward.

Outside the United States and the United Kingdom, Bollywood continues to have little or no effective local representation—and certainly no coordinated representation of the kind US studios have cultivated over the years through the MPAA. Because copyright enforcement around the world is structured around partnerships between law enforcement and industry groups—both for legal reasons, such as the need for a complainant in most civil and criminal actions, and for less obvious issues of influence and cost sharing—the lack of global Bollywood anti-piracy networks means that its rights go largely unenforced, even where enforcement efforts are otherwise extensive (see the South Africa chapter in this report).

---

i   Interview with Aswin Punathambekar, assistant professor, Department of Communication Studies at the University of Michigan, Ann Arbor.

---

state organized-crime statute, the Maharashtra Prevention of Dangerous Activities (confusingly, also known as the MPDA) Act. Similar to the Goondas Act in Tamil Nadu, the MPDA Act allows detention without bail for anyone with a prior arrest for video piracy. Not surprisingly, MPDA director Rajiv Dalal welcomed the act, drawing the connection between media piracy and organized crime:

> We applaud the passage of this deterrent legislation that places piracy offenses under the Maharashtra state organized crime statute. Over the past several months, the release of the RAND report on "Film Piracy, Organized Crime, and Terrorism" has established strong links between film piracy and organized crime/terrorist funding in India. This legislation is indeed timely, and will significantly curb piracy and funding to organized criminal and terrorist syndicates in one of the most important global film markets." (Business of Cinema 2009)

Although the national anti-piracy strategy envisioned by the IIPA and the MPAA has not met with much success, the MPAA has shown that state-level co-optation of the law and the police, built around regional stakeholders, is a viable option. A state-by-state expansion of the enforcement trends seen in Maharashtra, Tamil Nadu, and Karnataka is clearly possible.

The Indian Music Industry

With a reported $606 million in revenues in 2007, the Indian music market is significantly smaller than the $2 billion film market (Kohi-Khandekar 2010).[28] But the two are closely linked: roughly 70% of album sales are Indian movie soundtracks. International repertoire, in contrast, makes up only 6% of the market,[29] with the result that only a handful of international albums see domestic release.

Like the film industry, the music industry has had difficulty speaking with a united voice on piracy issues and coordinating enforcement efforts beyond the local level. The major industry group is the IMI, the Indian Music Industry. The IMI represents the four major international labels, but also, and more importantly, a uniquely powerful collection of locally owned labels that control the domestic music market (and, in some cases, the wider regional marketplace). Tips Music, Saregama (formerly HMV), Yash Raj, and Venus are among the most prominent among them. Collectively, the domestic labels control approximately 82% of the marketplace—reversing the usual pattern of domination by the international majors. The IMI is also becoming the preferred enforcement organization for video-game piracy and counts Sony's video-game division among its members.

The major gap in the IMI roster is T-Series—the biggest music company in India, in control of over 60% of the market. Relations between T-Series and the other companies are poor because of the company's history as a pirate producer in the 1980s—a period in which it effectively broke the monopoly pricing of then-dominant record companies like HMV. T-Series is still a very aggressive player in the business and, independently of the other labels, also an aggressive enforcer of its own copyrights. It maintains independent lobbying and anti-piracy operations and has been a pioneer, especially, in suing online companies like YouTube and MySpace for infringement. Reportedly, it maintains an anti-piracy unit that employs some two hundred people in "raid teams" (Bailay 2009).

Until the recent efforts by the MPAA, the IMI was the most active sponsor of raids against street pirates and unlicensed performances. Led by former police officer Julio Ribero, the organization claims to have filed 3,500 suits in 2008 (IIPA 2009b).

In the past decade, the media market Palika Bazaar in Delhi has been a frequent destination for IMI raid teams. This "shopping mall" is a massive market for unbranded goods and in the

---

28   The IFPI puts the wholesale, or "trade value," of recorded music in the Indian market in 2008 at $140 million (IFPI 2009).

29   IMI, "Size of the Music Industry in India," http://www.indianmi.org/national.htm.

last ten years has become famous for film, music, and software piracy. The Delhi office of the IMI has made repeated efforts to raid the bazaar but has faced organized opposition from shop operators: raiding parties have been sometimes literally beaten out of the marketplace, even when accompanied by police. Raids in force, in turn, are hard to keep secret. Vendors have become adept at identifying IMI teams entering the market and have developed contingency plans to ensure that illegal goods are gone by the time raiding teams arrive.

As in other countries, the frequency of raids is often used as a metric for enforcement efforts in India, but the overall impact of such enforcement is unclear. It seems likely to us that the raids have simply punctuated or briefly accelerated the longer-term deformalization of optical disc piracy, without significantly affecting the overall supply. Shifts in the technologies of music consumption, for example, have made the raid increasingly irrelevant. The adoption of MP3 players is part of this story, but even more so the explosion in cell-phone use—the most ubiquitous digital consumer good in India, with over six hundred million users. Increasingly, pirated music is sourced not only from CDs and the Internet but also from the array of cell-phone vendors specializing in off brands and custom "mobile chips," pre-stocked with MP3s. Legal mobile-music sales, including ringtones, have grown dramatically and now represent almost 50% of all music sales (Kohi-Khandekar 2010:173–76). This shift is on IMI's research and policy agenda and figures in recent IIPA reports. What it isn't, clearly, is a recipe for effective enforcement. As the IIPA plausibly proposed in its 2010 India report: "The piracy rate for music in the online space is estimated at 99%."

This shift is readily visible on the streets of Delhi. Jack's is a popular hangout for Western-music lovers, where customers can ask for any international album and have a pirated copy available in a week. In 2009, Jack's stopped stocking CDs, moving entirely to digital distribution on cell phones, thumb drives, and other digital devices. The pirated CD is now a relative rarity on the streets, giving way to products with higher markups. Consumers buying MP3s in National Market in Bangalore generally buy in bulk. Discs that sell for Rs.50 ($1.06) often contain ten to twelve albums of the latest film releases or other popular compilations. For those with broadband connections, music is, of course, easily available through file sharing and file locker sites, facilitated by specialized search engines like Gujuri.com.

Consistent with the wider shift in digital music practices, we also note the strong trend toward the acquisition of single tracks rather than albums—a shift that to date has greatly advantaged the pirate market, with its more sophisticated digital distribution and bulk prices. Distributors have responded by lowering the prices of Bollywood albums, especially new releases, in an effort to pull customers back to the CD format. Online sales portals established by T-Series and others have also emerged in the past two years to capitalize on the shift to singles, with songs running Rs.12–Rs.15 ($0.30–$0.35). These are still hobbled by digital rights management technologies, however, and have not yet made a dent in the marketplace.

Software Piracy

As in other countries, the retail prices of business software such as Microsoft Office and Windows Vista are pegged to wider international prices, leading to predictably high levels of software piracy by consumers and businesses; widespread "pre-installation" piracy in the large computer grey market, where vendors sell machines assembled from low-cost components; and volume licensing programs wherever software companies can identify—and effectively manage—an institutional market. Given the difficulty of measuring the grey market, estimates of software piracy in India should be viewed with more caution than usual. According to the BSA, software piracy is slowly declining in India, falling to 65% in 2009 from 74% in 2004 (BSA/IDC 2010). Game piracy—including console-game piracy facilitated by the modding of machines in the grey market—is typically estimated at around 90% of the market (IIPA 2009b; USIBC/Ernst & Young 2008), reflecting prices pegged to international levels and the complete absence of price discrimination.

In our view, the Indian experience is consistent with the market development strategy outlined in chapter 1 of this report, in which the major software vendors (1) tolerate high levels of piracy in order to capture market share and lock out open-source competitors and then (2) progressively enforce licenses against the largest public institutions and organizations. Recent licensing deals with state governments in Karnataka and Maharashtra exemplify this second phase of operations, as do volume licensing deals with Hewlett-Packard and other locally-active equipment vendors, which ensure that new machines come pre-loaded with copies of Windows to discourage both pirate and open-source alternatives.

As elsewhere, the licensing deals are a gamble: they push public institutions into the legal software market but also increase the risk of large-scale adoption of open-source software as institutions think about their long-term software strategies. School-based open-source adoption programs, in particular, are widespread in India, with a large-scale pilot program in the state of Kerala providing the template for more recent adoption efforts in Karnataka, Gujarat, Assam, and West Bengal.

And as elsewhere, these dynamics operate entirely outside the retail market. Consumers and small businesses rarely benefit from institutional licensing, and few can pay Western prices. From their perspective, the market remains bifurcated between high-cost legal options and the very low-cost pirate market. Open-source platforms such as Linux remain very marginal competitors in consumer and small business markets especially.[30] As one respondent noted: free software in India means Microsoft Windows.

---

30   A 2010 survey by Springboard Research put the share of Windows Server in the small-to-medium business market in India at 91.8% of the installed base—compared to 94.7% in Asia overall (Withers 2010). Estimates of Linux desktop penetration are notoriously poor but were pegged by the World Wide Web Consortium at around 1.3% of the global installed base as recently as 2007 (Paul 2007).

Piracy Free

In 2004, Microsoft sponsored an anti-piracy workshop at the prestigious National Law School in Bangalore to "sensitize" members of the judiciary to issues of software piracy. At the end of the workshop, company representatives offered to declare the law school a "piracy-free zone." A subsequent software audit, however, revealed that only a handful of faculty members used licensed software. The company responded by offering a bulk license to the university at a discounted rate. The discounted price, however, still nearly equaled the annual budget of the library and was rejected by the university. After the university signaled that it would switch to open-source software, the company issued it a free blanket license—and "piracy-free" status.

Microsoft has played a very public role in pushing on both the volume licensing and enforcement fronts, moving aggressively to offer lower-cost solutions at the school and government level and expanding raids on small businesses and grey-market vendors operating on the edge of the formal economy. *Microsoft Corporation v. Yogesh Popat* was a minor example of this ramp up in activity in the early 2000s. Domestic stakeholders, in contrast, have become significantly less active on enforcement in recent years—notably India's powerful software association, NASSCOM. In its early days under the late Dawang Mehta (who served as chairman from 1990 to 2001), NASSCOM adopted an aggressive anti-piracy stance, featuring highly publicized raids against vendors. In recent years, NASSCOM has shifted its focus to trade and IT industry business lobbying, including perennial issues such as export quotas and H-1B visas to the United States. Anti-piracy has all but disappeared from the association's discourse and activities. This has made software enforcement an almost entirely foreign-led enterprise in India.

The Conditions of New Media Art in India

This chapter has privileged three broad perspectives on the media economy in India: (1) the consumer experience of media access in India through the multitude of legal, grey, and illegal distribution channels that mark daily life; (2) the market and business practices that shape India's uniquely domestic, broad-based, and competitive media sector; and (3) the practices of enforcement and enforcement advocacy that maintain the shifting boundary between the two. But as we described at the beginning of this chapter, the digital media revolution in India is not only a story about expanded opportunities for consumption, or corporate strategies for securing the distribution channel. It is also a story about the vast democratization of media production. Filmmaking, in particular, is no longer an elite industry practice, but an increasingly popular art form that circulates outside the traditional industry channels. Pirate and grey-market practices have been essential to participation and education in these contexts, both for artists and audiences.

The studios routinely mobilize star actors and directors on behalf of anti-piracy campaigns, continuing the long-standing and, perhaps most important, cheap strategy of moral exhortation against piracy. Major stars like Rajnikant have taken prominent anti-piracy stands and provide the most visible face of enforcement efforts. But our interviews reveal considerably more complexity in artists' views. Emerging artists, especially, express ambivalence about copyright—balancing the desire to control the commercialization of their work against the necessity of pirate access to the tools of modern media production. Distinct from their personal choices, most artists are also aware that the music and film industries indirectly depend on infringement within the feeder system for new talent. Low-cost access to tools and widespread copying and appropriation are, in many respects, the conditions of artistic renewal in a high-tech media culture—and especially in one marked by the economic disparities, cultural diversity, and energy of India.

These are not marginal experiences. Most independent media artists in India begin their careers in some close relationship to copyright infringement. The basic facts of low incomes and high prices make it impossible to avoid this pattern. Anurag Kashyap, one of the leading screenwriters and directors in Bollywood, points out that although many other Bollywood filmmakers consider piracy to be a threat to the industry, his filmic education was inseparable from piracy—from watching movies in makeshift theatres, on battered VCRs playing pirated cassettes. This was cinematic culture in the small town of Tanda in Faizabad, available at two rupees per ticket. His literary education, he notes further, was based on cheap pirated reproductions of Dostoevsky, Tolstoy, Chekhov, and other Russian classics. This experience led him to condone piracy in the context of a broader condemnation of the culture of "cheap remakes" that dominates Bollywood (Kashyap 2005).

By necessity, and often in the absence of wider norms, artists chart their own ethical paths through these media landscapes. Kashyap Murali, a Bangalore-based DJ and video artist, is typical in drawing a range of complicated distinctions between contexts and types of piracy. In our interviews, he acknowledged pirating film but never music, which he insists on buying in the original to support the artists and because he values the accompanying materials. Although he professes a strong preference for legal software, the professional software packages required for his work cost $2,000 and upward—a prohibitive investment, even for many successful Indian artists. Software piracy, in Murali's world, is simply a condition of artistic production. Music piracy, in contrast, is an ethical matter.

Archana Prasad is a director of music videos. When she began her career as a music-video jockey, her work was composed almost entirely of infringing clips of other artists' work. The appropriation and reuse of copyrighted work was commonplace among her peers and embedded within a wider set of norms governing the sharing of work within the VJ community. Nonetheless, she is sensitive to the issues of originality and derivative work and sets a personal limit of thirty seconds on the use of other artists' clips. Over time, she has been able to create a substantial body of her own footage and no longer relies as heavily on other sources. She knows that this personally identifiable work will, in turn, contribute to the broader culture of reuse in the Delhi VJ scene and that this, too, is a form of professionalization and marketing. As Prasad observes, the best way to sell concert tickets is to make sure one's music is widely available: "People will come and pay, if they can afford it. It makes sense for musicians to push out their work so that people can hear it. It makes sense for starting bands to promote their own music online."

Conclusion

The price wars between Indian media companies illustrate a familiar dynamic among incumbents and new entrants in media markets. The new entrant is often a pirate, or characterized as such, until the market incorporates the new business innovation and the upstart becomes an incumbent in its own right—and often a ferocious defender of its own IP interests. Piracy, in this context, is as much a matter of political and market power as it is of legality. Similar dynamics hold sway at the individual level, as artists make their way from largely informal economies of cultural production into the formal, increasingly corporatized environments of Bollywood and the other regional industries.

The real significance of India to the larger account developed in this report is that these pathways exist at all. In other middle- and low-income countries, media markets are far more bifurcated between tiny, high-priced legal markets, dominated by multinational companies, and vast, low-priced pirate markets. India, in contrast, has a hugely differentiated marketplace in which legal competitors exist at all levels and can capitalize on and integrate aspects of the informal sector. Perhaps most important, the Indian model predates emergence of the ubiquitous Internet culture that is driving the cost and access revolution in high-income countries.

It would, in all likelihood, be naive to view this as an exportable model—or even a stable one. The roots of Indian media autonomy go back many decades and are grounded in the intense regionalism of Indian culture, which erected barriers to foreign entry. Of the other major film-producing countries, only Nigeria has made a similar transition, built on similar conditions of cheap production, informality, and long-term, state-supported cultural independence. The scale of this achievement is modest from a social welfare perspective but remarkable from a cultural one. Average cinema attendance in the US hovers around 4–5 movies per year. In India, possessing 1/46 the US GDP per capita, the average is around 3. Mexico—the largest film market in Latin America by a wide margin—averages 1.5, with eight times the per capita GDP of India. The majority of low- and middle-income countries hover between .5 and 1.

Figure 8.7 Movie Admissions Per Capita Per Year

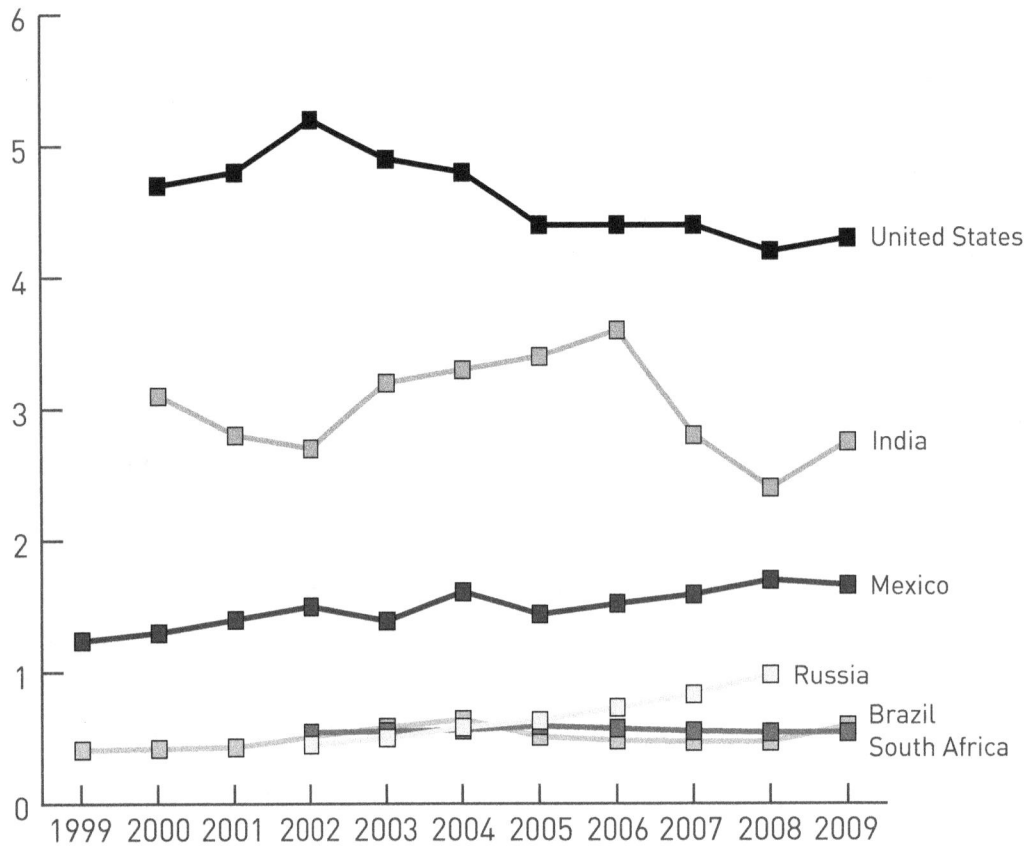

Source: Authors based on European Audiovisual Observatory (2001–10) data.

The other side of this story is the changing enforcement environment, as Indian companies adopt modern corporate models of organization, develop global-market ambitions, and—above all—adopt the enforcement rhetoric and practices of the multinational groups. To date, these groups have been frustrated by the dizzying complexity of India's regional markets, state laws, and squabbling media sectors and by a copyright debate focused on national autonomy and local issues rather than alignment with the global enforcement agenda. We do not see much prospect that this will change in the short term. But industry lobbying is persistent, and domestic corporate interests are likely to further align with international ones. The recent Bollywood-Hollywood partnership exemplifies this alignment, and though it will do little to diminish piracy, it does represent another step in the creation of a wider enforcement culture in India, capable of pressing state and, eventually, national government for stronger measures and more public investment.

About the Study

The India study was conducted by a research team based at Sarai, a media research program at the Centre for the Study of Developing Societies in Delhi, and at the Alternative Law Forum, a legal research and practice center in Bangalore. Project and research leadership was provided, respectively, by Ravi Sundaram at Sarai and Lawrence Liang at the ALF. Several other researchers contributed to the study, including in particular Siddharth Chaddha on the pirate markets in Bangalore, Prashant Iyengar on media and research discourse on piracy, and Nupur Jain on the film industry and enforcement in Mumbai. Most of the primary information came from field interviews conducted in late 2008 and 2009 with media pirates, members of the film industry, police, lawyers, and members of anti-piracy organizations. More broadly, the study draws on nearly a decade of Sarai and ALF investigations of piracy, enforcement, and emerging media cultures in India.

The project also benefited from other valuable assistance, including that of Bodó Balázs and Dmitri Pigorev, who collected data on the Indian torrent trackers, Tripta Chandola at the CSDS in Delhi, and Joe Karaganis, who was a relentless source of feedback, improvements, and editorial support. The chapter was improved through generous feedback from readers, including Shamnad Basheer.

References

Anandan, Rajan. 2009. "Harnassing IT for India's Growth." *Economic Times*, July 13. http://economictimes.indiatimes.com/Comments-Analysis/Harnessing-IT-for-Indias-growth/articleshow/4770640.cms.

Bagchi, Jeebesh. 2006. "Acceleration and Conflicts: Comments on the Cinematic Object in the 1990s and After." *Journal of the Moving Image, no. 5* (December).

Bailay, Rasul. 2009. "Action Groups Devise Ways to Check Piracy." *Livemint.com*, January 26. http://www.livemint.com/2009/01/26221735/Action-groups-devise-ways-to-c.html.

Borpujari, Utpal. 2009. "India Major Online Film Piracy Hub." *Deccan Herald*, December 16. http://www.deccanherald.com/content/41541/india-major-online-film-piracy.html.

BSA/IDC (Business Software Alliance and International Data Corporation). 2008. *Piracy Reduction Impact Study*. Washington, DC: BSA. http://www.bsa.org/country/Research%20and%20Statistics/~/media/E5EAABBAC7814D6CB3A486E47982DA92.ashx.

———. 2010. 2009 Global Software Piracy Study. Washington, DC: BSA.

Business of Cinema. 2009. "Audio-Video Piracy Included under MPDA Act in Maharashtra." *Business of Cinema*, July 17. http://www.businessofcinema.com/news.php?newsid=13741&page=1.

Consumers International. 2010. *IP Watchlist Report 2010*. London: Consumers International. http://a2knetwork.org/watchlist.

Digital TV News. 2008. "CCID Consulting: Review and Forecast of China's DVD Market in 2008." December 26. http://www.digitaltvnews.net/content/?p=6125.

Ernesto. 2008. "India Huge Growth Market for BitTorrent Sites." *Torrent Freak* (blog), November 19. http://torrentfreak.com/india-huge-growth-market-for-bittorrent-sites-081119/.

_____. 2010. "Indian Police Arrest Four Member BitTorrent Gang." *Torrent Freak* (blog), November 22. http://torrentfreak.com/indian-police-arrest-4-member-bittorrent-gang-101122/

European Audiovisual Observatory. 2001–10. *Focus: World Film Market Trends*. Annual reports. Paris: Marché du Film.

FICCI/KPMG (Federation of Indian Chambers of Commerce and Industry and KPMG). 2010. *Back in the Spotlight: FICCI-KPMG Indian Media & Entertainment Industry Report*. New Delhi: FICCI.

Hu, Kelly. 2008. "Made in China: The Cultural Logic of OEMs and the Manufacture of Low-Cost Technology." *Inter-Asia Cultural Studies* 9:27–46.

IANS (Indo-Asian News Service). 2009a. "Microsoft Fined for Using 'Money Power.'" *Economic Times,* December 15. http://economictimes.indiatimes.com/MS-fined-for-using-money-power/articleshow/5336125.cms.

_____. 2009b. "Piracy, Poor Publicity Spoil 'Slumdog's' India Collections." *Economic Times*, February 22. http://economictimes.indiatimes.com/News/News-By-Industry/Media--Entertainment-/Entertainment/Piracy-poor-publicity-spoil-Slumdogs-India-collections/articleshow/4168686.cms.

_____. 2009c. "Mukesh Bhatt Meets PM to Discuss Producers-Multiplex Tiff." April 13.

IFPI (International Federation of the Phonographic Industry). 2009. *The Record Industry in Numbers*. London: IFPI.

IIPA (International Intellectual Property Alliance). 2009a. *IIPA Special 301 Report on Copyright Protection and Enforcement*. Washington, DC: IIPA.

_____. 2009b. *India: IIPA Special 301 Report on Copyright Protection and Enforcement,* 2009. Washington, DC: IIPA.

_____. 2010. *2010 Special 301 Report on Copyright Protection and Enforcement: India*. Washington, DC: IIPA.

Indiantelevision.com. 2007. "Moser Baer in DVD Deal with Palador for 50 World Cinema Titles." November 23. http://www.indiantelevision.com/aac/y2k7/aac123.php.

ISPAI (Internet Service Providers Association of India). 2009. "Growth of Broadband Subscribers: 256 Kbps and More." http://ispai.in/Stat2-BroadbandSubscribers256Kbps.php.

Kashyap, Anurag. 2005. "Pirates of the Arabian Sea." *Tehelka*, March 15. http://www.tehelka.com/story_main11.asp?filename=hub031905Pirates_of.asp.

Kohi-Khandekar, Vanita. 2010. *The Indian Media Business*. Delhi: Sage Publications.

Larkin, Brian. 2004. "Degraded Images, Distorted Sounds: Nigerian Video and the Infrastructure of Piracy." *Public Culture 16*:289–314.

Liang, Lawrence. 2003. "Porous Legalities and Avenues of Participation." *In Sarai Reader 05*: Bare Acts, edited by Monica Narula, Shuddhabrata Sengupta, Jeebesh Bagchi, and Geert Lovink. Delhi: Centre for the Study of Developing Societies.

———. 2008. "Meet John Doe's Order: Piracy, Temporality and the Question of Asia." *Journal of the Moving Image*.

———. 2010. "Exceptions and Limitations in Indian Copyright Law for Education: An Assessment." *Law and Development Review 3*.

Linden, Greg. 2004. "China Standard Time: A Study in Strategic Industrial Policy." *Business and Politics 6*.

Manuel, Peter Lamarche. 1993. *Cassette Culture: Popular Music and Technology in North India.* Chicago, IL: University of Chicago Press.

Mitra, Ashish. 2008. "Moser Baer, UTV Ink Strategic Alliance." *Screen*, December 26.http://www.screenindia.com/news/moser-baer-utv-ink-strategic-alliance/403204/.

Pahwa, Nikhil. 2008. "@FICCI Frames: Asha Swarup, Secretary Ministry Of I&B On Mobile TV, Copyright And Optical Disc Law." contentSutra.com, March 25. http://contentsutra.com/article/419-ficci-frames-asha-swarup-secretary-ministry-of-ib-on-mobile-tv-copyrigh/.

Paul, Ryan. 2007. "Linux Market Share Set to Surpass Win 98, OS X Still Ahead of Vista." *Ars Technica*, September 3. http://arstechnica.com/apple/news/2007/09/linux-marketshare-set-to-surpass-windows-98.ars.

Prasad, Madhav. 1993. "Cinema and the Desire for Modernity." *Journal of Arts and Ideas*, nos. 25–26 (December).

Prasad, Sudha. 2008. "Flops Mar Film Industry in South; Only 10% Films Recover Money." *Financial Express*, December 28. http://www.financialexpress.com/news/flops-mar-film-industry-in-south-only-10-films-recover-money/403788/.

PTI (Press Trust of India). 2010. "Courts Will Take 320 Years to Clear Backlog Cases: Justice Rao." *Times of India*, March 6. http://timesofindia.indiatimes.com/india/Courts-will-take-320-years-to-clear-backlog-cases-Justice-Rao/articleshow/5651782.cms.

PriceWaterhouseCoopers. 2010. *Indian Entertainment and Media Outlook 2010.* Mumbai: PwC, India. http://www.pwc.com/gx/en/entertainment-media/pdf/PwC_India_EM_Outlook_2010.pdf.

Rajadhyaksha, Ashish. 2009. *Indian Cinema in the Time of Celluloid: From Bollywood to the Emergency.* Bloomington: Indiana University Press.

Sharky. 2009. "The Desi/Bollywood P2P Scene: BitTorrent's Other Side." FileShareFreak. February 9. http://filesharefreak.com/2009/02/01/the-desibollywood-p2p-scene-bittorrents-other-side/.

SouthDreamz. 2010. "Fumes and Flutters in Kollywood." January 7. http://www.southdreamz.

com/2010/01/fumes-and-flutters-in-kollywood.html.

Sundaram, Ravi. 1996. "Beyond the Nationalist Panopticon: The Experience of Cyberpublics in India." Paper presented at the Fifth International Conference on Cyberspace, Telefonica, Madrid, June 6–9. http://www.sarai.net/research/media-city/resouces/film-city-essays/beyod_the_nationalist_panopoticon.pdf.

_____. 2001. "Recycling Modernity: Pirate Electronic Cultures in India." *In Sarai Reader 01: The Public Domain*, edited by Raqs Media Collective and Geert Lovink. Delhi: Centre for the Study of Developing Societies.

_____. 2009. "The Pirate Kingdom." *Third Text* 23:335–45.

TRAI (Telecom Regulatory Authority of India). 2010. "Information Note to the Press: Telecom Subscription Data as on 30th June 2010." Press Release No. 34/2010, July 23. http://www.trai.gov.in/WriteReadData/trai/upload/PressReleases/746/PressRelease23july.pdf.

Treverton, Gregory F. et al. 2009. *Film Piracy, Organized Crime, and Terrorism*. Santa Monica, CA: RAND Corporation.

UFO Moviez. 2008. "Stop Piracy with UFO Movies." http://www.ufomoviez.com/piracy_news.html.

USIBC (U.S.-India Business Council)/Ernst & Young. 2008. *The Effects of Counterfeiting and Piracy on India's Entertainment Industry*. Washington, DC: USIBC.

Vasudevan, Ravi. 2003. "Cinema in Urban Space." *Seminar 525* (May). http://www.india-seminar.com/2003/525.htm.

Wang, Shujen. 2003. *Framing Piracy: Globalization and Film Distribution in Greater China*. Lanham, MD: Rowman and Littlefield.

Withers, Stephen. 2010. "Linux Spreading, but Windows Server Still Rules in India." *iTWire*, July 12. http://www.itwire.com/business-it-news/open-source/40340-linux-spreading-but-windows-server-still-rules-in-india.

Zhao, Judy, and Levi Ruan. 2009. *Broadband in China: Accelerate Development to Serve the Public*. Milan: Value Partners. http://www.valuepartners.com/VP_pubbl_pdf/PDF_Comunicati/Media%20e%20Eventi/2010/value-partners-PR_100301_BroadbandInChinaZhaoRuan.pdf.

Coda: A Short History of Book Piracy

Bodó Balázs

Introduction

The history of media piracy explored in this report is predominantly a history of the digital era. Digital technologies have brought a sharp drop in the cost of reproduction of many cultural goods and, consequently, in the degree of control that producers exercise over how and where those goods circulate. The breakdown of this control has been so rapid that it is no surprise that many see it as a revolution—and indeed, from the perspective of many industry incumbents, as an unprecedented disaster.

But a longer historical lens suggests that the current crisis of copyright, piracy, and enforcement has much in common with earlier periods of change and conflict among cultural producers. From the early days of the book trade in the fifteenth century, cultural markets were shaped by deals within the publishing trade and with political authorities over who could reproduce works and on what terms. While printers and publishers sought protection from competition, state and church authorities wanted to control the circulation of texts. Regulations designed to serve these goals led to a highly centralized printing trade in most European countries, in which state-favored publishers monopolized local markets.

Such monopolies inevitably attracted competitors from the ranks of the less privileged printers, as well as from those outside local markets. Repeatedly, over the next centuries, state-protected book cartels were challenged by entrepreneurs who disregarded state censorship, crown printing privileges, and guild-enforced copyrights. Already in the early seventeenth century, incumbent publishers labeled such printers pirates, evoking maritime theft and plunder.

Such conflicts were not limited to local markets. Pirate printers tended to flourish at the geographical peripheries of markets—often across borders, where the enforcement power of the state stopped. Scottish and Irish publishers competed with London publishers for English audiences; Dutch and Swiss publishers printed for the French market under the ancien régime. To a considerable extent, the European sphere of letters emerged through this transnational explosion of print.

Pirate publishers played two key roles in this context: they printed censored texts, and they introduced cheap reprints that reached new reading publics. Both actions fueled the development of a deliberative public sphere in Europe and the transfer of knowledge between more and less privileged social groups and regions.

New pirate entrants always responded to the market inefficiencies created by the cartels. In the short run these distortions could be upheld by state power. But in the long run, pirate practices were almost always incorporated into the legitimate ways of doing business. Over time, regulatory frameworks changed to accommodate the new publishing landscape.

Similar stories could be told in many modern industrial contexts, including computer hardware, chemical engineering, pharmaceuticals, and software. The piggybacking of local industry on the intellectual products of more developed, geographically remote competitors is not an aberrant form of economic development—it is one of its basic features (Johns 2010; Chang 2003; Ben-Atar 2004). This development narrative threads its way through many of the preceding pages as pirate challengers catalyze change in local markets. As a conclusion to this report, we return to the early history of book publishing and piracy to as a way of emphasizing this continuity and clarifying how these dynamics play out in cultural markets. Consistently, we see five loose "laws" of piracy at work in cultural markets:

1. Persistent gaps between supply and demand due to artificial constraints on price or supply will be filled by pirate producers.

2. When faced with piracy, industry incumbents almost always turn to the state to defend their market positions and usually adapt their business models only when other recourse has failed.

3. Conversely, pirate producers tend to operate at the edges of the sphere of influence of incumbents, where differences in law and difficulties of enforcement create spaces of ambiguous or conflicted legality.

4. Piracy, at these economic and political peripheries, has a well-established role as a development strategy that facilitates the circulation of knowledge goods.

5. In many of these contexts, piracy also plays a clear political role as a counterweight to the centralized control of information—whether by states or private interests. The censorship of texts in pre-modern England and France was continually undermined by pirate networks. As this report has described, piracy played much the same role in Russia and South Africa in the 1980s.[1]

---

[1] This last point is not a major focus of our country reports but deserves attention in the context of the current enforcement push. Recurrently, the enforcement of copyright has become mixed up with political or commercial motives for restraining speech. This is hardly surprising: copyright enforcement is, by definition, a form of control over expression, and on that basis has been the subject of innumerable disputes about the proper limits of that control. In practice, in both liberal and authoritarian societies, the last guarantor of freedom of speech has been not formal rights and protections but simply the inefficiency of enforcement. We have little language in our political systems for valuing this inefficiency and the leaky, hard-to-control cultural economy that results from it. In our view, this leakiness is no less important today in an era of growing technological capacity for enforcement and commercial demands to use it.

Synthetic and Crown Rights

In the latter half of the fifteenth century, several decades after Gutenberg's invention of the printing press, the publishing trade was still poorly developed. The knowledge and technical infrastructure needed to support publishing were slow to spread. Demand soon outpaced supply: many cities had an abundance of manuscripts and codices awaiting publication or republication but lacked printers. Governments granted monopolies and other exclusive rights to encourage the local establishment of printing businesses, often by enticing skilled printers and tradesmen to emigrate from other cities.

By the end of the fifteenth century, this scarcity had begun to give way to a more developed culture of print. A wider European book trade was emerging, reflecting not only growth in the number of printers but also higher demand for contemporary works. Expanded trade, in turn, created a market for cheap reprints. A printer in Lyon could turn a high profit reprinting a book first published in Venice or Basel. Reprinting emerged very quickly in the book trade and led printers to seek state support for proprietary publishing claims.

The first such protection was issued in Milan in 1481 to Andrea de Bosiis, granting him exclusive rights to print and sell Jean Simoneta's *Sforziade* (Feather 1987). Such privileges were only as good as the geographical reach of the political authority that issued them. Exclusive rights in Milan did not extend to Venice or Rome. Large unified empires, like France and Spain, had some success limiting internal competition but were ineffective against competition from abroad.

The export-driven mercantilist economics of the age further complicated issues of geography and legality. A publisher might be a respected member of society in his home country—if his activities were legal under local law and profited the community—but widely regarded as a pirate outside it if he disregarded the printing rights and privileges of other territories. As more publishers began

Chapter Contents

399 Introduction

401 Synthetic and Crown Rights

402 The Elizabethan Book Pirates

403 Hills the Pirate

405 The War for the Public Domain

407 Market Research in Continental Europe

408 The American Pirate Century

412 References

to operate within the pan-European sphere of letters, the potential grew for profit-destroying conflicts between them.

In the absence of an international copyright regime, publishers established informal agreements regarding rights of republication and sale. Often, these "synthetic copyrights" provided more security than locally issued regulations because of the interdependence of the publishing trade (Bettig 1996:17). Internationally active publishers relied on foreign publishers to carry their books and consequently were embedded in a network of relationships that required trust and reciprocity. These thick social and business ties meant that transgressions could be, and often were, punished by the publishing community itself, regardless of local regulations.

Even without state backing, publishers in the international trade had a strong collective interest in establishing rights of exclusivity. Such agreements often encompassed large numbers of publishers within and across state borders. By the late eighteenth century, a system of synthetic rights was in place among Dutch and Swiss publishers printing books for the French market (Birn 1970; Darnton 1982, 2003). Irish publishers had a similar system until the Union with Great Britain in 1800 (Johns 2004). US publishers devised a system of synthetic copyrights to manage competition for foreign works, which were denied copyright protection under US law throughout the nineteenth century (Clark 1960; Khan and Sokoloff 2001). Eighteenth-century German publishers specified the circumstances under which members of the trade could produce and circulate pirated editions: "[if] the original publisher's prices increased . . . [if] codes of conduct were broken, [if] colleagues as well as the public were damaged, or if pirate editions were only distributed in regions where the original itself was not available" (Wittmann 2004). The exclusivity rights of individual publishers were thus secured within and through the publishing community.

The relationship between rules imposed from above and agreements and norms initiated from below was always complex. Emerging regulatory frameworks of the time sometimes deliberately built upon and exploited community norms, while in other cases they were intended to rewrite existing rules of the trade. Most conflicts between legal publishers and pirates occurred when state rules diverged from community norms, violating community notions of fair competition. Such divergence typically occurred when some players were able to "capture" state favor or regulation in new ways; when new entrants capitalized on weaknesses in regulation or in the capacity to enforce it; or when key stakeholders (such as authors) were left out of the bargaining in ways that destabilized the system in the long run.

The Elizabethan Book Pirates

In sixteenth century England, Elizabeth I granted monopoly privileges to select publishers over such basic texts as the Bible, alphabet books, almanacs, books of grammar, and law books. These steady-selling, high-volume texts were exceptionally valuable to publishers. Many of the smaller publishers were locked out of these lucrative markets, making it very difficult to earn a living, raise capital, buy manuscripts, or secure copyrights. With rights to the best texts doled out as political favors, this period saw the emergence of a class of impoverished printers, struggling to stay in business with more obscure texts.

Tensions between wealthy and poor printers increased over time and eventually degenerated into a publishing war. Poor publishers began to pirate protected books in large numbers and militate for a more egalitarian distribution of privileges. Because the prices of authorized copies were kept high, the black-market book trade was very profitable. Even in a context of high risk in which homes of suspected pirates were routinely searched, illegally printed copies confiscated, and printing machines destroyed, illegal publishing proved impossible to suppress.

Roger Ward's case illustrates the scale of the conflict. In 1581/2, Ward confessed to illegally printing 10,000 alphabet books—a massive number in an era in which 1,500 copies was considered a large print run (Judge 1934:48–49). Other records cite similar figures: 4,000 psalm books printed in a ten-month period; 10,000 more alphabet books printed in eight months. Another record of the work of eleven printers lists 10,000 alphabet books and 2,000 psalm books printed and sold in less than a year. Such sales were significant enough to seriously disrupt the legal market.

After many fruitless years of conflict, privilege holders began to change course. Gradually, they adopted a strategy of appeasement and co-optation of the opposition as a means of regaining at least some control of the book market. Some of the pirates were simply bought off. John Wolfe, one of the most notorious pirates, was given part of Richard Day's profitable monopoly on *The A.B.C. with Little Catechism* and admitted to the printers' guild (the Stationers' Company). He soon became one of its most reliable policemen. For others, the Stationers' Company made important concessions: in 1583/84 it authorized non-Company printers to print a wide variety of works, including certain law books; Scottish, French, Dutch, and Italian versions of the Psalms; a list of eighty-two other protected titles; and all out-of-print works.

This strategy of accommodation proved successful and maintained a loose equilibrium in the British book market that lasted for most of the seventeenth century. At the end of the century, however, Parliament upset the status quo.

Hills the Pirate

By the 1690s, the Licensing Act—the legislation governing publishing privileges—was overdue for revision. The act was a deal between the Stationers' Company and the Crown, involving Crown support for copyright and guild privileges in return for guild support for Crown censorship. Among these privileges, the act capped the number of master printers in England at twenty; regulated the numbers of presses, journeymen, and apprentices; restricted printing to London, Oxford, Cambridge, and the city of York; and limited the importation of books through the port of London (Astbury 1978). For the Crown, the act served as the legal foundation of censorship in England, as well as its mechanism through the control it afforded over publishers.

The prospect of renewal of the Licensing Act generated significant controversy. John Locke made impassioned arguments against its renewal, most famously on the grounds of freedom of the press, which the act clearly constrained. Daniel Defoe connected these arguments to the claims of the emerging class of intellectuals who wanted to earn a living by their pen, rather than through patronage. Other commentators argued against the Stationers' monopoly on the basis of its market consequences: high book prices and restricted access to classical texts.

Parliament let the act lapse in 1695, marking a major victory for freedom of the press in English law. The publishers' trade was also transformed, though in ways that were not immediately apparent. The privileges and copyrights secured in prior years were maintained but only through common law: the legislation that acknowledged these rights and provided the institutional and legal framework for their enforcement had been abolished. The number of printers and publishers was uncapped, and restrictions on imported books went unenforced. These changes set the stage for a brief but turbulent period in which old publishing privileges and copyrights were unenforced and insurgent publishers could experiment with radical new models for selling books.

As the eighteenth century began, printers still treated books as luxury goods, catering to wealthy customers willing to pay for expensive editions. Several categories of books, however, enjoyed wider circulation, including psalm books, alphabet books, and almanacs. These had begun to create not only broader literacy but a nascent mass market for a wider range of literature.

Smaller publishers began to reprint copyrighted works in large quantities, challenging the market structure and pricing of the incumbent publishers. Henry Hills the Pirate, as he became known, was the most famous of these. Beginning in 1707, Hills republished popular poems, pamphlets, and sermons, selling them for between a halfpenny and twopence—a fraction of the typical sixpence price. He published an unauthorized compilation of the first one hundred issues of *The Tatler*, one of the most popular magazines of the time, years before an official compilation was released. The motto on each of Hills's one-penny prints testified to the popular ambition of his publishing model: "For the benefit of the poor." Estimates of the total number of copies printed by Hill reached 250,000 (Solly 1885).

Three factors made Hills's radically lower pricing possible: (1) he ignored rights-holder claims, (2) he used the cheapest possible materials, and (3) he kept his per-copy profit to a minimum. The resulting business model was extremely powerful. Hills was arguably the first businessman of the era to cultivate a mass-market model for books, based on large volume and low profit margins.

The War for the Public Domain

Established publishers used the radicalism of Hill and others like him to mobilize political support for the renewal of English publishing laws. A long and tumultuous debate ensued that began with claims of harm from piracy but quickly expanded to include the freedom of the press, the dangers of print monopolies, the benefits of copyright, and the political and financial independence of the intelligentsia.

The Parliament finally passed a law in 1710, the Statute of Anne, which is usually described as the first modern copyright law. Because the debate had moved well beyond piracy, the new law brought a number of profound changes to how print was regulated. The best known of these was the establishment of the author as the source and original holder of the copyright. This change diminished the monopoly power of publishers and clarified the transactions of rights involved in the production of a book. It was not, however, a clear assertion of authors' rights:

> Emphasis on the author in the Statute of Anne implying that the statutory copyright was an author's copyright was more a matter of form than of substance. The monopolies at which the statute was aimed were too long established to be attacked without some basis for change. The most logical and natural basis for the changes was the author. Although the author had never held copyright, his interest was always promoted by the stationers as a means to their end. Their arguments had been, essentially, that without order in the trade provided by copyright, publishers would not publish books, and therefore would not pay authors for their manuscripts. The draftsmen of the Statute of Anne put these arguments to use, and the author was used primarily as a weapon against monopoly. (Patterson 1968:147)

The second and—it later turned out—very consequential change for the larger book market was the establishment of a short, fixed term for copyright. Under the previous system, registration in the Company's Registry guaranteed perpetual ownership of the text. Under the act of 1710, however, newly created works were protected for a period of fourteen years (with the possibility of renewal for an additional fourteen). Already-published works retained copyright protection for twenty-one years. This dramatic restriction of copyright reflected the changing intentions of lawmakers. Where earlier laws were intended primarily to ensure the Crown's control of information, the Statute of Anne was intended to regulate trade—acting in the interest of society by preventing monopoly and in the interest of the publisher by protecting works from piracy (Patterson 1968:144).

The protection for already-printed works had been a compromise with London printers, who feared the loss of property invested in the copyrights registered by the Stationers' Company. When the twenty-one-year grace period came to an end, publishers renewed their efforts to secure the perpetual copyright established under earlier common law. These efforts were triggered by the rush of Scottish publishers into the market for works emerging from copyright.

Although the Scottish publishers were treated as pirates in London, their situation reflected an underlying problem of legal pluralism within the English system. The key issue was whether the limited copyright period established by the act of 1710 took precedence over the perpetual copyright consolidated under English common law, which did not apply in Scotland. Thus began a new period of public controversy and political conflict around copyright—this time between

London-based and Edinburgh-based publishers. As copyrights on popular works expired in the 1730s, Scottish printing houses flooded northern English markets with cheap reprints. London publishers objected to this piracy of their back catalogs on the basis of the perpetual copyright established under English common law. Distance favored the Edinburgh publishers, and Edinburgh soon became an important publishing center. The conflict was finally resolved in 1774, when the House of Lords ruled against the common-law precedent in the case of Donaldson v. Beckett.

Donaldson v. Beckett ended the concept of perpetual copyright in English law and affirmed what we now know as the public domain—the body of work that can be used and republished without permission. According to the publishers' estimates, the ruling erased property vested in copyright totaling approximately 200,000 pounds. Nonetheless, the dissolution of the book market foretold by London publishers failed to occur. On the contrary:

> The decision of 1774 transferred, through lower prices, a huge quantum of purchasing power from book producers to book buyers. With more firms entering the business, increasing price competition, and the prices of pretender copyrights plummeting towards zero, the British book industry as a whole moved to a faster growth rate. Bankruptcies tripled, a sign of boom, and the industry as a whole prospered as never before.
>
> After 1780, the minimum price for high-demand out-of-copyright texts fell to half, and then to a quarter, of previous levels. Print runs for major editions grew by a factor of three or four, and there were many more editions, often on sale at the same time. . . . Within a generation, the book-binding industry doubled in size—a more reliable indicator of the growth of book production than printing capacity or titles published. . . . The period also saw a rise in the annual growth rate of book titles published nationally, much of it accounted for by reprints of older titles, as well as a rise in the rate of growth of provincial book publishing, provincial bookshops, and provincial circulating libraries. There was a boom in anthologies, abridgements, adaptations, simplified and censored versions, as well as books sold in parts. We see the rapid growth of a new children's book industry, which also drew on anthologies and abridged out-of-copyright authors, and which drove out or absorbed the long-frozen ballad and chapbook canon within a few years.
>
> The quantified estimates I have assembled match the more impressionistic judgement made by the remainder bookseller Lackington writing in 1791: "According to the best estimation I have been able to make, I suppose that more than four times the number of books are sold now than were sold twenty years since . . . In short all ranks and degrees now read." (St. Clair 2004:115–18)

Market Research in Continental Europe

Beyond England, pirate publishers also played an important role in contesting the extensive censorship of texts practiced in seventeenth- and eighteenth-century France. Authors such as Voltaire, Rousseau, Mercier, and Restif de la Bretonne were banned in France but widely available in editions printed abroad. Foreign editions smuggled back into France often became the standard editions for such works (Darnton 1982). In practice, much of the Enlightenment in pre-revolutionary France passed through Dutch and Swiss publishers (Birn 1970:134).

The business environment for such extraterritorial publishers was extremely complex, as they were competing not only with the legitimate publishers in France and elsewhere but also with each other. As pirates, they could not rely on formal protection mechanisms, such as royal privileges, to mitigate some of the risks associated with publishing. Business practices adapted to this highly competitive environment:

> What really set the pirate publishers apart was their way of doing business. They
> practiced a peculiarly aggressive kind of capitalism. Instead of exploiting privileges
> from the protected position of guilds, they tried to satisfy demand, whatever,
> wherever it was. (Darnton 2003:28)

The key component of this business strategy was market research on both local demand and potential competitors' plans (Darnton 2003:28). Fréderic Samuel Ostervald, an alderman in the Swiss town of Neuchâtel and one of a small network of pirate publishers serving the French market, left an extensive record of how such networks operated (Darnton 2003:4).

Over two decades, Ostervald received close to 25,000 letters from a network of French booksellers, Dutch and Swiss pirate publishers, traveling agents, and authors writing in French across Europe. Such letters were first and foremost a way of gauging audience and potential competition. But they also provide evidence of informal agreements among pirate publishers about who would publish which works for different markets. These were quintessential gentlemen's agreements, operating on easily violated trust, but they proved strong enough to create a stable market that minimized cannibalization among publishers and counterbalanced the effects of fragmented and often restrictive local regulations.

Collectively, these pirate networks invented an international regulatory regime for copyrights more than a century before the Berne Convention codified copyright relations on the international level. Pirate correspondence and gentlemen's agreements limited unfair competition in local markets among the members of the network in an era in which state-sponsored mercantilism still favored the enforcement of local claims and the raiding of foreign copyrights.

The American Pirate Century

In the second half of the nineteenth century, there were several efforts to curb state-sponsored cross-border piracy through bilateral agreements, but a truly international copyright standard came together only in 1886, when Germany, Belgium, Spain, France, the United Kingdom, Italy, Switzerland, and Tunisia signed the Berne Convention for the Protection of Literary and Artistic Works. From Berne forward, local and cross-border piracy became a more explicit subject of national attention, if not always of new regulations or sanctions. For many countries, copyright and enforcement remained exercises in triangulation between the desirability of cheap access to foreign works, the interests of local publishers, and the demands of international trading partners. One of the chief pirate nations, in this context, was the United States of America.

For roughly a century, American copyright law was a clear-cut case of situational piracy—of behavior legalized under US law but widely condemned abroad. The US federal copyright statute implemented in 1790 was based closely on the Statute of Anne and replicated its limited fourteen-year renewable term. But—possibly due to a misinterpretation of the English statute (Patterson 1968:200)—the US law granted copyrights only and exclusively to US citizens. As a major importer of British titles, this clause created a massive subsidy for US publishers and helped establish a de facto cultural policy of cheap books, which in turn became an essential component of mass public education. This situation persisted until the 1891 Chace Act granted limited copyright to foreign authors. Another century would pass before the United States joined the Berne Convention, in 1989.

The US rejection of British claims, in particular, persisted for a century because it served the interests of a developing nation and its nascent publishing industry. This rejection was itself often construed as both a sovereign right and an explicit policy of national improvement. As one publisher put it during one of the many Senate debates on the subject:

> All the riches of English literature are ours. English authorship comes to us as free
> as the vital air, untaxed, unhindered, even by the necessity of translation; and the
> question is, Shall we tax it, and thus impose a barrier to the circulation of intellectual
> and moral light? Shall we build up a dam, to obstruct the flow of the rivers of
> knowledge? (Solberg 1886:251)

By the second half of the nineteenth century, the combination of high literacy, plummeting printing costs, and the most advanced postal and transportation system in the world had produced rapid growth in the US book and magazine markets (Beniger 1986). Cheap pirated literature helped strengthen the book publishing industry and educate the rapidly expanding American reading public:

Piracy had created audiences and large-scale publishing operations, including the elaboration of editorial, production, and critical functions. Meanwhile, the availability of pirated British literature may have stimulated the development of the profession of authorship in the long run, as well as the invention of distinctive themes and new literary forms and techniques. (Bender and Sampliner 1996/97:268)

Even though American authors actively lobbied for greater respect for international copyrights, the country's pirate century only ended when the biggest stakeholders, the East Coast publishers, threw their weight behind such efforts. This conversion was, as usual, more a product of competitive concerns than moral ones: by the late nineteenth century, eastern publishers faced competition from new West Coast firms. Notably, these new entrants operated outside the system of gentlemen's agreements that governed competition among the East Coast publishers (and mollify British publishing houses through informal royalty payments) (Clark 1960). When efforts to restrict competition through the courts failed, the East Coast firms decided that international copyright would provide them an advantage in securing and defending publishing rights against their less-capitalized and -connected West Coast competitors. The shift in attitudes toward foreign copyrights was quick, and the move to join the international community got underway.

The 1891 Chace Act extended copyright protection to foreigners but was clearly written to serve the competitive interests of domestic publishers. It had enough loopholes, one scholar has noted, to "make the extension of copyright protection to foreigners illusory" (Ringer 1967:1057). This situation persisted long after the United States officially complied with international norms and was a constant source of tension with European publishers. By the mid-1930s, some Dutch publishers had given up on legal remedies and adopted a policy of retaliation:

> Two outstanding incidents involved *The Yearling* by Marjorie Kinnan Rawlings and *Gone With The Wind* by Margaret Mitchell. At the trial in the Netherlands involving the latter book, it is interesting to note, the Dutch publishers, in the words of the presiding judge, "stated that they would have been quite prepared to pay for the right of translation if it were not for the fact that works copyrighted in the Netherlands are published in the United States time and again without any compensation. The only way to compel the United States to accede to the Berne Convention is to disregard, in the countries which have acceded to that Convention, the copyrights of the citizens of that country." (Kampelman 1947:421)

In the long run, the decisive factor in shifting US policy on international copyrights was the growth of American export industries based on intellectual property (IP). By the 1930s, the United States was an exporter of a wide array of knowledge goods and services. The rise of Hollywood, in particular, consolidated this role in the cultural sphere, but it was a small piece, overall, of a much larger shift toward an IP- and services-driven US economy. Eventually, this shift produced an international policy agenda. By the 1980s, "American exporters heavily reliant upon intellectual property—such as the computer, entertainment and pharmaceutical industries—were growing ever more frustrated with both legitimate competition and proliferating piracy, while the White House found itself casting about for a politically painless way to address the growing trade deficit" (Alford 1992:99).

A uniform IP protection regime that would support the global trade of assets and services became more attractive in this context. But the weakness of international conventions based on voluntary adherence had also become clear. A number of countries stayed away from the international copyright conventions. Others joined one or the other but failed to fulfill their obligations. Developing countries had their own cheap-books policies, and confrontations with Western rights-holders were common in the 1960s and 1970s.

The parallels were obvious. After the breakup of the colonial empires, developing countries faced challenges similar to those faced by the United States a century before. Nearly all were IP-importing countries; nearly all saw the path to development passing through mass education and literacy. What were developing countries to do in this situation? By the 1980s, the new trade agreements promoted by the United States and other developed countries offered an answer: higher standards of protection and enforcement.

With the architecture of liberalized, global markets taking shape, it became important to articulate how and why poor countries would benefit from stronger IP protection when the United States and many other countries had clearly taken a different path. One simple strategy was to suggest that the loose positive correlation between IP protection (or, inversely, piracy rates) and wider indicators of socioeconomic development is in fact a causal relationship[2] — that stronger IP protection spurs development. But of course correlation is not causation, and a wide range of commentators have observed that "the causation might very well run the other way, with richer countries both more able and more willing to protect intellectual property since they have a larger share of their economy devoted to such pursuits" (Thallam 2008).

We see growing evidence for the latter view. Unqualified claims that strong IP protection is necessary for foreign direct investment (FDI) are rarely heard anymore, not least because they have been contradicted by rapid rates of FDI growth in many high-piracy industrializing countries—notably China, which has climbed the industrial value chain through massive copying of foreign goods and technologies.[3] The claim that strong IP protection is necessary for the growth of local

---

2   A wide range of socio-economic indicators play a role in these contexts, variously rendered in terms of GDP (gross domestic product) (Varian 2004), institutional development (Thallam 2008), foreign direct investment (Mansfield 1994), and business leaders' perceptions of national "competitiveness" (World Economic Forum 2010). For a typical example of the causal argument, see the International Chamber of Commerce's 2005 white paper "Intellectual Property: Source of Innovation, Creativity, Growth, and Progress."

3   When the Organisation for Economic Co-operation and Development examined the literature on this issue in 2008, it reached the conclusion that "other factors outweigh the negative effect of counterfeiting and piracy on foreign direct investment." See also Chang (2003) and, on China in particular, Yu (2007).

industries, for its part, is hard to sustain in such sectors as film, music, and software, where US and multinational firms completely dominate most local markets.

As we have argued throughout this report, our study suggests that the main differentiator between widely served, relatively affordable media markets (for example, in India or the United States) and anemic, high-priced media markets, like those of most of the rest of the developing world, is not income but competition, and that such competition is likely to be strongest where domestic firms control large shares of production and distribution. Local firms, broadly speaking, are much more likely to aggressively compete for local audiences and to innovate on pricing and services. Multinationals operating in low-value markets, in contrast, seek primarily to protect their high-value markets and to maintain their positions as they wait out the slow process of economic growth. Fostering local ownership, control, and competition within national media markets is, in our view, a key challenge for developing-country governments.

At the same time, we see little reason to think that changes in IP protection or enforcement will significantly affect this playing field. Such changes do little to alter the balance of power in local media markets, and as we have shown, the ease with which enforcement resources are captured tends to reinforce those inequalities. In our view, the key question looks much the same in both low-income countries and high-income countries: how to serve the new, larger publics catalyzed by the pirate economy? To return to Robert Bauer's formulation of the problem for the MPAA: "Our job is to isolate the forms of piracy that compete with legitimate sales, treat those as a proxy for unmet consumer demand, and then find a way to meet that demand."[4]

Not surprisingly, the economic arguments for stronger enforcement tend to ignore how IP regimes are actually made. The history of book publishers and pirates, on the other hand, tells us something of this story—one in which the distribution and enforcement of IP rights marks less a state of development than a set of power relations among firms within cultural markets. In periods when no major political, economic, cultural, or technological transformation challenged the status quo, copyright laws tracked conventions between dominant producers and served to reinforce and refine the prevailing order.

But while some of these arrangements were long-lived, they were also fragile and easily disrupted by competition outside the jurisdiction of the dealmakers, by technological change, and—above all—by the combination of the two. In such cases, incumbents ultimately had to assimilate the pirates, together with their marketing strategies, their novel approaches to production and distribution, their expanded audiences, and above all their lower prices. Now three hundred years after the passage of the Statute of Anne, we find ourselves in a moment of similar necessity.

---

4 Interview with Robert Bauer, director of strategic planning for the Motion Picture Association, 2009.

About the Coda

This chapter draws on portions of the book *Necessity Knows No Laws: the Role of Copyright Pirates in the Cultural Ecosystem from the Printing Press to the File-Sharing Networks*, to be published in early 2011 in Hungarian. This study revisits copyright history from the perspective of copyright pirates in order to understand the functions they fulfilled in the production and circulation of knowledge.

References

Alford, W. P. 1992. "Intellectual Property, Trade and Taiwan: A Gatt-Fly's View." Columbia Business Law Review 97.

Astbury, R. 1978. "The Renewal of the Licensing Act in 1693 and its Lapse in 1695." Library s5-XXXIII (4): 296–322.

Ben-Atar, Doron S. 2004. *Trade Secrets*. New Haven, CT: Yale University Press.

Bender, T., and Sampliner, D. 1996/97. "Poets, Pirates and the Creation of American Literature." New York University Journal of International Law and Politics 29.

Beniger, J. R. 1986. *The Control Revolution: Technological and Economic Origins of the Information Society*. Cambridge, MA: Harvard University Press.

Bettig, R. V. 1996. *Copyrighting Culture: The Political Economy of Intellectual Property*. Boulder, CO: Westview Press.

Birn, R. 1970. "The Profits of Ideas: Privileges en Librairie in Eighteenth-Century France." Eighteenth-Century Studies 4 (2): 131–68.

Chang, Ha-Joon. 2003. *Kicking Away the Ladder: Development Strategy in Historical Perspective*. London: Anthem Press.

Clark, A. J. 1960. *The Movement for International Copyright in Nineteenth Century America*. Washington, DC: Catholic University of America Press.

Darnton, R. 1982. *The Literary Underground of the Old Regime*. Cambridge, MA: Harvard University Press.

_____. 2003. "The Science of Piracy: A Crucial Ingredient in Eighteenth-Century Publishing." Studies on Voltaire and the Eighteenth Century 12:3–29.

Feather, J. 1987. "The Publishers and the Pirates. British Copyright Law in Theory and Practice, 1710–1775." Publishing History 22.

International Chamber of Commerce. 2005. "Intellectual Property: Source of Innovation, Creativity, Growth and Progress." White paper, ICC, Paris.

Johns, Adrian. 2004. "Irish Piracy and the English Market." Paper presented at the History of Books and Intellectual History conference, Princeton University, Princeton, NJ, December 3–5.

_____. 2010. Piracy: *The Intellectual Property Wars from Gutenberg to Gates*. Chicago, IL: University Of Chicago Press.

Judge, C. B. 1934. *Elizabethan Book-Pirates*. Cambridge, MA: Harvard University Press.

Kampelman, M. M. 1947. "The United States and International Copyright." American Journal of International Law 41 (2): 406–29.

Khan, B. Z., and Sokoloff, K. L. 2001. "The Early Development of Intellectual Property Institutions in the United States." Journal of Economic Perspectives 15 (3): 233–46.

Mansfield, Edwin. 1994. *Intellectual Property Protection, Foreign Direct Investment, and Technology Transfer*. Washington, DC: World Bank.

Organisation for Economic Co-operation and Development. 2008. *The Economic Impact of Counterfeiting and Piracy*. Paris: OECD.

Patterson, L. R. 1968. *Copyright in Historical Perspective*. Nashville, TN: Vanderbilt University Press.

Ringer, B. A. 1967. "The Role of the United States in International Copyright—Past, Present and Future." Georgetown Law Journal 56.

Solberg, Thorvald. 1886. "International Copyright in Congress." Library Journal 11.

Solly, E. 1885. "Henry Hills, the Pirate Printer." Antiquary, xi, 151–54.

St. Clair, W. 2004. *The Reading Nation in the Romantic Period*. Cambridge, UK: Cambridge University Press.

Thallam, Satya. 2008. *The 2008 International Property Rights Index*. Washington, DC: Property Rights Alliance.

Varian, H. 2004. *Copying and Copyright*. Berkeley: University of California Press.

Wittmann, R. 2004. "Viennese and South German Pirates and the German Market." Paper presented at the History of Books and Intellectual History conference, Princeton University, Princeton, NJ, December 3–5.

World Economic Forum. 2010. *The Global Competitiveness Report 2009–2010*. http://www.weforum.org/en/initiatives/gcp/Global%20Competitiveness%20Report/index.htm.

Yu, Peter. 2007. "Intellectual Property, Economic Development, and the China Puzzle." In Intellectual Property, Trade and Development: Strategies to Optimize Economic Development in a TRIPS Plus Era, edited by Daniel Gervais. Oxford, UK: Oxford University Press.

Credits

Chapter 1: Rethinking Piracy

Joe Karaganis, Program Director, Social Science Research Council: For the past ten years, Joe Karaganis has directed SSRC projects on media, technology, and culture. His research focuses on the relationship between digital convergence and cultural production and has recently included work on broadband adoption and data policy. He is the editor of Structures of Participation in Digital Culture (2007) and of The Politics of Open Source Adoption (2005). Since mid-2010, he is also vice president at the American Assembly.

WITH THANKS TO

Jaewon Chung, Program Assistant, Social Science Research Council

Jinying Li, PhD Candidate, Department of Cinema Studies, Tisch School of the Arts, New York University

Emmanuel Neisa, Graduate Student, Sciences Politiques

Sam Howard-Spink, Assistant Professor, Steinhardt School of Music and Performing Arts, New York University

Nathaniel Poor, independent scholar

Pedro N. Mizukami, Center for Technology and Society, Getulio Vargas Foundation

Chapter 2: Networked Governance and the USTR

Joe Karaganis, Program Director, Social Science Research Council

Sean Flynn, Associate Director, Program on Information Justice and Intellectual Property, Washington College of Law, American University: Sean Flynn teaches courses on the intersection of intellectual property, trade law, and human rights and is the associate director of PIJIP. At PIJIP, Professor Flynn designs and manages a wide variety of research and advocacy projects that promote public interests in intellectual property and information law and coordinates the Washington College of Law's intellectual property and information law academic program.

CONTRIBUTORS

Parva Fattahi, Fellow, Program on Information Justice and Intellectual Property, Washington College of Law, American University

Mike Palmedo, Assistant Director, Program on Information Justice and Intellectual Property, Washington College of Law, American University

Susan Sell, Professor, Political Science and International Relations, Elliot School of Foreign Affairs, George Washington University

Chapter 3: South Africa

Natasha Primo, National ICT Policy Advocacy (NIPA) Coordinator, Association for Progressive Communications: Natasha Primo is a gender and development practitioner whose interest in broader ICT public policy issues led her to be a part of the APC policy team between 2007 and September 2010. She served as chair of APC from 2005 to 2007. She was previously executive director of the Women'sNet, launched in 1998 as a joint initiative of the South African Non-Governmental Organisation Network (SANGONeT) and the Commission on Gender Equality (CGE). She recently left APC and is now a consultant working on projects focused on gender and communications as well as monitoring and evaluation of programs and projects from a gender perspective. She is based in Johannesburg, South Africa.

Libby Lloyd, Media Consultant: Libby Lloyd started out as a journalist, working mostly in radio, including Capital Radio in South Africa, National Public Radio in the United States, and BBC Ireland. She was also active in the Speak media project and was subsequently head of radio training at the Institute for the Advancement of Journalism in Johannesburg. A councilor with the Independent Broadcasting Authority (the predecessor of Icasa) and founding CEO of the Media Development and Diversity Agency, she is now a consultant, working on projects involving media and development issues and policy, focusing on gender and communications. She was chosen as Vodacom Media Woman of the Year in 2005.

CONTRIBUTORS

Tanja Bosch, Senior Lecturer, University of Cape Town
Natalie Brown, Research Assistant and Recent MA Graduate in Communication and Culture at York University, Toronto
Adam Haupt, Senior Lecturer, University of Cape Town
Julian Jonker, Lecturer, University of Cape Town
Nixon Kariithi, Associate Professor, University of Witwatersrand

Chapter 4: Russia

Olga Sezneva, Assistant Professor of Sociology, University of Amsterdam: Olga Sezneva joined the Department of Sociology and Anthropology at the University of Amsterdam in December 2009, where she is a fellow at the Institute for Migration and Ethnic Studies. She received her PhD (2005) from New York University and was a Harper Fellow and a Collegiate Assistant Professor at the University of Chicago (2005–9). In 2008, she was invited to join the New Generation group at the Center for Transcultural Studies at the University of Pennsylvania. Her main interests lie at the intersection of migration and urban studies, with particular focus on social memory and space. She is currently working on the book manuscript *My Place, Your Memory*, which follows transformations of German Königsberg into Russian Kaliningrad after the Second World War.

It examines different strategies of forging a collective past, the goals they accomplish, and the implications that they have for urban space and urban design.

Joe Karaganis, Program Director, Social Science Research Council

INVESTIGATORS

Oleg Pachenkov, Deputy Director, Centre for Independent Social Research: Oleg Pachenkov became deputy director for international relations and strategic development at CISR in 2001, after joining the Centre as a researcher in 1997. He holds a PhD (2009) in sociology from St. Petersburg State University. He is a board member and co-editor of the Internet resource the Open-air Market Network (www.openair.org) and on the editorial board of the new social science journal Laboratorium: Russian Review of Social Research. His current research interests include urban milieus and spaces, urban planning, informal and "street" economies, ethnic minorities, and ethnicity and migration studies.

Irina Olimpieva, Research Fellow, Centre for Independent Social Research: Irina Olimpieva is a researcher and head of the Social Studies of Economy Research Department at CISR. She holds a PhD (1990) in economic sociology and sociology of labor from St. Petersburg State University. Previously, she was a visiting scholar in the Junior Faculty Development Program at the University of Kansas (1999–2000) and a researcher in the labor research lab at Kujbyshev Economic Institute (1986–87). Her research interests include industrial relations, informal economy, organizational studies, and post-socialist transformation.

Anatoly Kozyrev, Head of Sub-Department, Information Technology, Moscow Institute of Physics and Technology: Anatoly Kozyrev is a Russian economist and mathematician and the author of two books on valuation of intellectual property and intangible assets. He is head of the Intellectual Capital Center at the Central Economic-Mathematical Institute of the Russian Academy of Sciences as well as head of the Information Technology sub-department at the Moscow Institute of Physics and Technology, where his students served as research assistants for this project.

WITH THANKS TO

Dmitry Pigorev, Lecturer, Moscow Institute of Physics and Technology

Chapter 5: Brazil

Pedro N. Mizukami, Center for Technology and Society, Getulio Vargas Foundation: Pedro N. Mizukami is a researcher at the Center for Technology and Society at the Getulio Vargas Foundation (Fundação Getulio Vargas—FGV) School of Law in Rio de Janeiro, where he has done work related to copyright law and licensing, Internet regulation, peer-to-peer file sharing, and open-access publishing. He holds a master's degree in constitutional law from Pontifical Catholic University of São Paulo.

Oona Castro, Director, Overmundo Institute: Oona Castro is executive director at the Overmundo Institute (Instituto Sociocultural Overmundo), a non-governmental institution that develops projects in digital communications, such as the website Overmundo (www.overmundo.com.br) and studies on economy of culture. The institute was created to support production of free culture, access to information and culture, and projects characterized by innovative intellectual property models.

Luiz Fernando Moncau, Center for Technology and Society, Getulio Vargas Foundation: Luiz Fernando Moncau obtained his LLB from the School of Law of Pontifical Catholic University of São Paulo. Beginning in 2006, he worked as a lawyer/policy analyst in the Advocacy Department of the Brazilian Institute for Consumer Defense (Instituto Brasileiro de Defesa do Consumidor—Idec), with an emphasis on policy research and analysis of consumer protection, and served as Idec's liaison to Congress and for press/media and seminars on telecommunications and access-to-knowledge issues. Prior to this appointment, he worked at Idec as a law clerk, with an emphasis on consumer protection, class actions, and judicial procedures. He is currently part of the Open Business project team at the Center for Technology and Society at GVF.

Ronaldo Lemos, Center for Technology and Society, Getulio Vargas Foundation: Ronaldo Lemos is head professor of intellectual property law at FGV Law School and director of the Center for Technology and Society. He is also the director of Creative Commons Brazil and a member of the board of iCommons. He earned his LLB and LLD from the University of São Paulo and his LLM from Harvard Law School. He is the author of several books, including Direito, Tecnologia e Cultura (2005), and Tecnobrega (2008). He coordinates various projects, such as the Cultura Livre project and the Open Business Project, an international initiative taking place in Brazil, Nigeria, Chile, Mexico, South Africa, and the United Kingdom. He is one of the founders of the Overmundo Institute, winner of the Digital Communities Golden Nica, granted by the Prix Ars Electronica 2007. He writes weekly for *Folha de São Paulo*, a major newspaper in Brazil.

Contributors

Susana Abrantes, FGV Opinion
Olívia Bandeira, Overmundo Institute
Thiago Camelo, FGV Opinion
Alex Dent, Department of Anthropology, George Washington University
Joe Karaganis, Program Director, Social Science Research Council
Eduardo Magrani, Center for Technology and Society, FGV
Sabrina Pato, FGV Opinion
Elizete Ignácio dos Santos, FGV Opinion
Marcelo Simas, FGV Opinion
Pedro Souza, FGV Opinion

Chapter 6: Mexico

John C. Cross: John C. Cross is a sociologist who has published widely on the political struggles of street vendors and other members of the urban poor in Mexico City and elsewhere. His research in Mexico City began with studies of vendor organizations, including several in Tepito, for his PhD dissertation. He became interested in the phenomena of piracy when he noticed the Tepito neighborhood shifting dramatically from the sale of electronic goods to piracy in the late 1990s.

Chapter 7: Bolivia

Henry Stobart, Senior Lecturer, Royal Holloway University of London: In addition to his position at Royal Holloway University, Henry Stobart is the founder and coordinator of the UK Latin American Music Seminar, associate fellow of the Institute for the Study of the Americas, and former committee member of the British Forum for Ethnomusicology. He studied tuba and recorder at Birmingham Conservatoire, performed with a number of baroque ensembles, and taught music in several schools before completing a PhD (1996) at St John's College, Cambridge, focused on the music of a Quechua-speaking herding and agricultural community in Northern Potosí, Bolivia. He is active as a professional performer with the early/world music ensemble SIRINU, who have given hundreds of concerts and recorded on many European radio networks since their first Early Music Network tour in 1992. His current research focuses on indigenous music VCD (DVD) production, music "piracy," and cultural politics in the Bolivian Andes.

Chapter 8: India

Lawrence Liang, Attorney, Alternative Law Forum: Lawrence Liang, a graduate of National Law School, subsequently pursued his master's degree in Warwick, England, on a Chevening Scholarship. His key areas of interest are law, technology and culture, and the politics of copyright. He has been working closely with the Sarai program at the Centre for the Study of Developing Societies in New Delhi on a joint research project on intellectual property and the knowledge/culture commons. A keen follower of the open-source movement in software, he has been working on ways of translating open-source ideas into the cultural domain.

Ravi Sundaram, Co-director, Sarai, Centre for the Study of Developing Societies: Ravi Sundaram was one of the founders of the Sarai program, which he co-directs with his colleague Ravi Vasudevan. Ravi Sundaram's work rests at the intersection of the postcolonial city and contemporary media experiences and has looked at the phenomenon that he calls "pirate modernity," an illicit form of urbanism that draws from media and technological infrastructures of the postcolonial city. He is visiting faculty at the Department of Urban Design at the School of Planning and Architecture in Delhi. He recently published Pirate Modernity: Media Urbanism in Delhi (2009) and is finishing two edited volumes, No Limits: Media Studies from India and Delhi's Twentieth Century, both forthcoming from Oxford University Press.

CONTRIBUTORS

Siddharth Chadha, Alternative Law Forum, Bangalore
Prashant Iyengar, Alternative Law Forum, Bangalore
Nupur Jain, PhD Candidate, Cinema Studies, School of Arts and Aesthetics, Jawaharlal Nehru University, Delhi
Jinying Li, PhD Candidate, Department of Cinema Studies, Tisch School of the Arts, New York University
Abkar Zaidi, Economist and Independent Scholar

WITH THANKS TO:

Tripta Chandola, Center for the Study of Developing Societies, Delhi

Coda: Book Piracy

Bodó Balázs, PhD, Researcher, Budapest University of Technology and Economics: Economist Balázs Bodó has been an assistant lecturer and researcher at the Center for Media Research and Education in the Department of Sociology and Communications at Budapest University of Technology and Economics since 2001. He was a Fulbright Visiting Researcher at Stanford Law School in 2006/7. He is a fellow at the Center for Internet and Society at Stanford University and is the project lead for Creative Commons Hungary.

Other Acknowledgments

Many people provided generous help as sources or reviewers for this project. We would particularly like to acknowledge:

Robert Bauer, Director of Strategic Planning, MPAA
Shamnad Basheer, National University of Juridical Sciences, Kolkata, India
Pria Chetty, Principal Attorney, Chetty Law, South Africa
David Cross, Department of Film and Media Studies, National University of Mexico
Willie Currie, Research Director, Association for Progressive Communications
Maria Haigh, Department of Communications, University of Wisconsin
Kathryn Hendly, Department of Political Science, University of Wisconsin
Alex Kochis, FiveBy Solutions, United States
Ramon Lobato, Swindburne University, Australia
Boris Mamlyuk, Ohio Northern University
William Pomeranz, Deputy Director, Kennan Institute, Washington, DC
Igor Pozhitkov, IFPI Russia
Andrew Rens, SJD Candidate, Duke University, Attorney (High Court of South Africa)
Tobias Schonwetter, Intellectual Property Research Unit, University of Cape Town

Production and Project Management

Alyson Metzger, heroic editing and copyediting.
Rosten Woo, print and web design.
Zachary Zinn, web support.
Mark Swindle, illustrations and layout.
Jaewon Chung, project management.

Major Institutional Partners

The Social Science Research Council (SSRC), based in New York City, is an independent, nonprofit organization devoted to promoting innovative work across the social sciences. Founded in 1923, the Council seeks, through a diverse range of projects, to build interdisciplinary and international networks, to mobilize new knowledge on important public issues, and to educate and train the next generation of researchers. The SSRC awards fellowships and grants, convenes workshops and conferences, sponsors scholarly and public exchanges, organizes summer training institutes, and produces a range of publications, both in print and online.

The Overmundo Institute (Instituto Sociocultural Overmundo) is a nonprofit organization dedicated to promoting access to knowledge and cultural diversity in Brazil. Created in 2006 and headquartered in Rio de Janeiro, the Institute is concerned with the establishment of new channels and opportunities for the dissemination of cultural production throughout Brazil; the development of studies and strategies of new possibilities for creation, sharing, and circulation of culture and knowledge generated by the Internet and digital technologies; and the encouragement of innovative models for the management of intellectual property and business in the areas of culture and communication that will provide legal and economic bases for the Institute's two other lines of action.

The Center for Technology and Society (CTS) is part of the Getulio Vargas Foundation (Fundação Getulio Vargas—FGV) Law School in Rio de Janeiro and is the only institution in Brazil specifically aimed at dealing with the interplay of law, technology, and society, with focus on intellectual property rights. The CTS is engaged in several research and education projects, always with an interdisciplinary approach. Among its projects, the CTS is responsible for launching and managing the Creative Commons project in Brazil (http://creativecommons.org), in association with the University of Stanford Law School, and leads projects such as Open Business Models, Digital Inclusion, Internet Legal Framework, Access to Knowledge, and Free Culture Production.

The Program on Information Justice and Intellectual Property (PIJIP) of the Washington College of Law at American University promotes public-interest approaches to domestic and international intellectual property law through research, education, events, and the provision of legal and consulting services. PIJIP's curriculum and activities promote a balanced approach to intellectual property and information law that rewards creators while ensuring broad public access to information and its products.

The Association for Progressive Communications (APC). For those of us who have access to it, the Internet has become an essential part of our daily information and communication needs. However, millions of people still do not have affordable, reliable, or sufficient access. APC believes that the Internet is a global public good. Founded in 1990, we are an international network and nonprofit organization that wants everyone to have access to a free and open Internet in order to improve our lives and create a more just world. Eighty percent of APC's members are from developing countries.

The Centre for Independent Social Research (CISR) was established in 1991 and has since been one of the few non-state institutes in Russia engaged in both academic research and professional training for young researchers. CISR's researchers are primarily guided by qualitative sociological methodology and the Centre's research reflects a broad spectrum of sociological interests, with a focus on studies of civil society and social structure. Up to fifty projects are conducted annually, most in cooperation with specialists from all over Russia and abroad. Since 1995, the Centre has published its own research periodicals in Russian and English, along with ten issues of a working papers series, which are available in the CISR library and research archive. CISR supports professional training for young Russian sociologists, is a member of several international research networks, and participates in the creation of new research centers throughout the Russian Federation.

The Moscow Institute of Physics and Technology (MIPT) was created in 1946 by leading Soviet scientists and the government as an advanced educational and research institution with a primary concentration in physics. MIPT quickly assumed a leading position in this field and became known internationally. It is difficult to overestimate the significance of MIPT for Soviet physics and science in general. The Institute's graduates have become leading specialists in nuclear research and rocket science, biophysics, radiophysics, and numerous other branches of science. MIPT's faculty is the highest authority in Russia in the area of physical sciences education at the university and advanced high school levels.

Sarai is a program of the Centre for the Study of Developing Societies (CSDS), one of India's leading research institutes, with a commitment to critical and dissenting thought and a focus on critically expanding the horizons of the discourse on development, particularly with reference to South Asia. We are a coalition of researchers and practitioners with a commitment to develop a model of research practice that is public and creative, in which multiple voices express and render themselves in a variety of forms. Over the last ten years, Sarai has matured into what could arguably be South Asia's most prominent and productive platform for research and reflection on the transformation of urban space and contemporary realities, especially with regard to the interface between cities, information, society, technology, and culture.

The Alternative Law Forum is a collective of lawyers and researchers working on various socio-legal issues. ALF perceives itself simultaneously as a space that provides qualitative legal services to marginalized groups and as an autonomous research institution with a strong interdisciplinary approach. It has been working on public-interest aspects of intellectual property law and policy for the past ten years and has played a role in several legal campaigns, including access to medicine and fair-use issues in copyright law. ALF has been collaborating on a joint research program with Sarai on the social life of media piracy

Funders

The International Development Research Center (IDRC) is a Crown corporation created by the Parliament of Canada in 1970 to help developing countries use science and technology to find practical, long-term solutions to the social, economic, and environmental problems they face. Our support is directed toward creating a local research community whose work will build healthier, more equitable, and more prosperous societies. The Centre supports research under four broad themes: Environment and Natural Resource Management; Information and Communication Technologies for Development; Innovation, Policy, and Science; and Social and Economic Policy. This project owes special thanks to IDRC program staff Phet Sayo, Khaled Fourati, and Alicia Richero.

The Ford Foundation, established in 1936, is an independent, global organization with a legacy of commitment to innovative leaders on the frontlines of social change. This project owes special thanks to program staff Alan Divack, Ana Toni, and Jenny Toomey.

www.ingramcontent.com/pod-product-compliance
Lightning Source LLC
Chambersburg PA
CBHW080810280326
41926CB00091B/4140